PLANT REPRODUCTIVE ECOLOGY

Plant Reproductive Ecology

PATTERNS AND STRATEGIES

Edited by
JON LOVETT DOUST
and
LESLEY LOVETT DOUST

New York Oxford
OXFORD UNIVERSITY PRESS

Oxford University Press

Oxford New York Toronto
Delhi Bombay Calcutta Madras Karachi
Petaling Jaya Singapore Hong Kong Tokyo
Nairobi Dar es Salaam Cape Town
Melbourne Auckland

and associated companies in
Berlin Ibadan

First published in 1988 by Oxford University Press, Inc.,
200 Madison Avenue, New York, New York 10016

First issued as an Oxford University Press paperback, 1990

Library of Congress Cataloging-in-Publication Data
Plant reproductive ecology : pattern and strategies / edited by Jon Lovett Doust
and Lesley Lovett Doust.
p. cm.
Includes index.
ISBN 0-19-505175-0
ISBN 0-19-506394-5

2 4 6 8 10 9 7 5 3

Printed in the United States of America

This book is dedicated to David Lloyd
in appreciation of his theoretical insights,
empirical investigations, and generosity of spirit.

Preface

Our aim in editing this volume has been to produce a cohesive series of synthetic reviews of the field of plant reproductive ecology. These up-to-date accounts appraise past work and seek to highlight new and exciting research fronts. The book is intended for researchers in the discipline of plant reproductive ecology and those just entering the field. The first seven chapters present a critical discussion of important conceptual issues. The next five chapters cover particular biotic interactions shaping the evolution of plant reproductive strategies. The final three, taxonomically based chapters, review the reproductive ecology of the major non-angiosperm plant groups.

We are very grateful to the colleagues who kindly assisted by reviewing particular chapters. Peter Alpert and Robert Nakamura were especially helpful in this regard; in addition we are grateful to Robert Edyvean, Henry Ford, Tom Lee, David Mulcahy, and Jennifer Ramstetter. Authors of individual chapters make their own acknowledgments; we would simply add our appreciation of the spirit of cooperation and enthusiasm that has grown from this shared venture.

South Hadley, Mass. J. L. D.
June 1987 L. L. D.

Foreword

George C. Williams recently wrote that "Historians may one day marvel at the tardiness of the realization that life history attributes are subject to natural selection, and evolve no less than teeth and chromosomes. . . . I attribute this tardiness to a persistent and widespread failure to make full use of the Mendelian formulation of natural selection. What is sometimes called modern Darwinism is a field in its infancy, at best." Jon and Lesley Lovett Doust's *Plant Reproductive Ecology: Patterns and Strategies,* with a dozen and a half first-rank authors, aims to place the study of the evolution of plant reproduction at least in toddlerhood. In the same way that two to three-year-old children challenge their parents (and themselves) with their blend of dependency and zeal for exploration, so does this field mix the old with what "might be the new."

A fuller use of the "Mendelian formulation of natural selection" has made us keenly aware of new possibilities in plant reproduction. The *fact* that pollen and ovules contribute equal numbers of chromosomal genes to seeds leads immediately to the question of how male (pollen) reproductive success influences floral characters and many other aspects of plant reproduction. Historians may one day wonder why this broad question was not posed until the late 1970s. Although male fertility is hard to measure (compared to counting seeds) this does not explain the virtual absence of *speculation* on the importance of the male role in most plant biology texts, which often seem to imply that plants attract pollinators in order to set seed.

Further use of Williams's "natural selection on life histories" question has led to even more subtle possibilities. Whereas many plant reproductive features (e.g., dioecy) have been viewed classically in a Panglossian fashion as regulators of inbreeding, now we are not so sure, and we ask how the features might have been molded by natural selection operating on differential opportunities for male and female reproduction. Whereas seed provisioning has been assumed to represent a strategy employed by the maternal plant to balance offspring size and number, we now realize that mother, father, and offspring have somewhat different interests in the reproductive allocations. The various reproductive tissues present in the ovary can be viewed as close relatives, able to influence each other's reproduction and therefore subject to kin selection. As surprising as some of these suggestions might seem at first, the greatest surprise is how simply Darwinian they are; they follow almost automatically from the use of "the Mendelian formulation of natural selection."

Of course *ecology* is also a key word. As they grow, plants compete for light, nutrients, space, time, and water. Fruits and seeds face the often conflicting demands of dispersal and predator escape. The authors of this book never forget that the plant's life history takes place on an ecological stage and that that history's tradeoffs and compromises are balanced in the face of particular ecological challenges. In addition, many *kinds* of plant reproduction are brought together in this book, crossing divisions of algae, bryophytes, ferns, and flowering plants. Discussion of such phyletic diversity is both unusual and fruitful.

The fifteen chapters are state-of-the-art statements that blend old and new ideas in a pot called skepticism, with a dash of the spice called enthusiasm. For many of the ideas, the data needed to test them are not yet in. But this is clearly an exciting time to be studying plant reproduction.

The authors of this book wish to dedicate it to Professor David G. Lloyd. David's work combines the best of the old and the new and spans the full range between muddy boots and the equations of population genetics. His work inspires us all.

Summit Park, Utah
May 1987

Eric L. Charnov

Contents

III Reproductive Strategies of Non-Angiosperms

Contributors

Spencer C. H. Barrett
Department of Botany
University of Toronto
Toronto, Ontario, Canada M5S 1A1

Robert I. Bertin
Department of Biology
College of the Holy Cross
Worcester, MA 01610

Michael I. Cousens
Department of Biology
University of West Florida
Pensacola, FL 32514

Paul Alan Cox
Department of Botany and Range Science
Brigham Young University
Provo, UT 84602

Robert E. De Wreede
The University of British Columbia
Department of Botany
#3529-6270 University Boulevard
Vancouver, British Columbia, Canada V6T 2B1

David Haig
School of Biological Sciences
Macquarie University
North Ryde 2109 Australia

Stephen D. Hendrix
Department of Botany
University of Iowa
Iowa City, IA 52242

Terry Klinger
The University of California at San Diego
Scripps Institution of Oceanography, A-008
La Jolla, CA 92093

Thomas D. Lee
Department of Botany and Plant Pathology
University of New Hampshire
Durham, NH 03824

Jon Lovett Doust
Department of Biological Sciences
University of Windsor
Windsor, Ontario, Canada N9B 3P4

Lesley Lovett Doust
Department of Biological Sciences
University of Windsor
Windsor, Ontario, Canada N9B 3P4

Thomas R. Meagher
Department of Biological Sciences
Rutgers University
Piscataway, NJ 08855

Brent D. Mishler
Department of Botany
Duke University
Durham, NC 27706

Mark A. Schlessman
Department of Biology
Vassar College
Poughkeepsie, NY 12601

Donald M. Waller
Department of Botany
University of Wisconsin
430 Lincoln Drive
Madison, WI 53706

Jacob Weiner
Department of Biology
Swarthmore College
Swarthmore, PA 19081

Mark Westoby
School of Biological Sciences
Macquarie University
North Ryde 2109
Australia

Michael Zimmerman
Department of Biology
Oberlin College
Oberlin, OH 44074

PLANT REPRODUCTIVE ECOLOGY

I

Conceptual Issues in Plant Breeding Systems

1

Sociobiology of Plants: An Emerging Synthesis

JON LOVETT DOUST and LESLEY LOVETT DOUST

An exciting new synthesis is taking place within the literature of evolutionary biology. Botany, especially plant ecology, is being reexamined in light of theory derived from animal sociobiology. Greater communication between ecologists who study plants and those who study animals has led to some sharing and trading of concepts and contexts. Recently, for example, a number of workers have used the theory of sexual selection to explain intersexual differences in particular plant traits. These include floral traits that may attract pollinators in animal-pollinated plants and involve particular patterns of floral resource allocation.[7,8,14,27,111,124] Others have shown how double fertilization in plants may have arisen through a process of kin selection.[27]

With this interchange has come what Russell Baker has referred to as "young fogeyism," a new pedantry that (in this case) resists the transfer of terms or ideas between the study of plants and animals. Some authors have reservations about applying concepts like "mate choice" and "parental care" to plants. Others argue that the identification of the *possibility* of mate choice in plants has in itself stimulated exciting research. However, although theories of animal behavior are valuable stimuli for good research in plant ecology, a direct one-to-one transfer is not always possible.[91,125]

Plants show greater biological diversity and peculiarities than the best-studied animal systems. Theory and research appropriate to plants still need to be developed, and such development may better fit colonial animals than existing animal theory does. One reason for the difference between plants and animals is the nature of plant form. Plant sexuality is a diffuse and fundamentally quantitative phenomenon: Reproduction in plants may involve many parts of the individual through the formation of many separate flowers and fruits. An analogy can be drawn between the behavior of an animal and the form, or morphology, of a plant: Animals gain evolutionary advantage in large part by virtue of flexible behavioral responses to their environment; they are able to move about and escape circumstances that may be stressful or otherwise unfavorable. Plants, being sessile, enhance their evolutionary fitness mainly through modular architecture and a capacity for reiterative growth.[120] This characteristic allows for much acclimatization by way of variable or plastic patterns of growth and reproduction.

Research over the next few years should clarify the extent to which animal models, based on theories such as kin selection, parent–offspring conflict, and sexual selection, retain their force in the translation to plant systems. In this overview, we draw together ideas and evidence concerning sex allocation, sex habits and dimorphism, sexual selection, paternity, female choice, and aspects of sexual incompatibility in plants; that is, what has been called "sociobotany."[125]

SEX ALLOCATION

Organisms may acquire evolutionary fitness as either paternal or maternal parents. Some organisms (in particular plants, modular animals, and many invertebrates) may be maternal and/or paternal to varying degrees at the same time or at different times in their life cycles. An organism's *gender* (i.e., its relative maleness or femaleness) represents the proportion of its evolutionary fitness transmitted through sperm and eggs, respectively (e.g., Refs. 65, 78). Sex allocation theory addresses opportunities for gains in fitness that may be made via maternal routes and/or paternal routes.

Historically, academic interest in plant sex has been sporadic, but recently it has greatly increased. Charnov, in particular, has been instrumental in pointing out the value of viewing breeding systems as the result of natural selection acting upon separate male and female strategies.[27,29] Charnov is concerned with sexual selection and the evolution of sex allocation, that is, the allocation of reproductive resources to male versus female function. He has lucidly reviewed the argument that patterns of parental resource allocation to different sex functions reflect the evolutionary fitness that may be gained by that sex function. (It should be noted that a number of confusions and resultant problems can arise due to imprecise uses of the concepts of evolutionary fitness, biological success or reproductive success, lifetime reproductive success, inclusive fitness, and various combinations of these. Some of these are described in Clutton-Brock.[35])

In most cases, it is the evolutionarily stable strategy (ESS sensu Maynard Smith[84]) that Charnov considers. Selection "should favor a mutant gene which alters various life history parameters, if the percent gain in fitness through one sex function exceeds the percent loss through the other sex function" (Ref. 29, p. 17). Charnov points out that the ESS pattern of resource allocation to male and female function is often that which maximizes the product of the fitness that can be gained through male function, and the gain through female function. In organisms where there is complete separation between the sexes (dioecy), selection should favor those males and females that control their clutch size (number of eggs per reproductive session), the sex ratio of offspring, and the allocation of resources among offspring, such that this product is maximized.

Charnov presents the central theme of sex allocation as a series of questions. For species with separate sexes, what is the equilibrium sex ratio? For a sequential hermaphrodite, which sex should come first, and when should any sex change take place? For a simultaneous hermaphrodite what is the equilibrium pattern of allocation of resources to male and female function? What sorts of situations favor one sex habit over another? And under what kinds of selective regimes may sexual lability be favored? Answers to all of these questions for any species must, of course, take into account both biotic and abiotic selection pressures. Yet, as Charnov makes clear, the answers to all of these questions may assume the same general form.

Ronald Fisher[47] first pointed out explicitly that in a population the fitness to be gained by males must be, on average, precisely equal to the fitness to be gained by females. The same identity must also hold for male and female function in hermaphrodites.[34] As regards the equilibrium sex ratio, Fisher concluded that, in the absence of any inbreeding, the sex ratio at conception should be adjusted by selection so that the total parental expenditure in raising male offspring should be the same as in raising female offspring. He indicated that where the cost of any one male equals the cost of one female, then natural selection favors production of a 1 : 1 sex ratio.

Fisher's concept of "parental expenditure," however, is difficult to assess biologically. He suggested energy and time as measures of expenditure, and later theoreticians have used either energy allocation or Fisher's undefined term. Energy investment *may* be an appropriate estimate of parental investment, but the evolutionary importance of such investment can be seen only if it is measured in terms of the limiting resource, which is likely to differ across species and environments and vary with season.

Unfortunately, Charnov does not develop the problem of the currency to be used in empirical studies of sex allocation. The adequacy of a carbon-based economy has recently been questioned in whole plant studies, where dry-matter distribution studies have become common. This is an important area needing further work. Goldman and Willson review many of the theoretical and methodological problems which hamper studies of sex allocation; in particular they review the problems of hermaphroditic plants.[54]

Using a resource allocation model of sex function and outcrossing in hermaphrodites, Ross and Gregorius[102] showed that sexual polymorphism could be maintained in a population by frequency-dependent selection if there is variation in the population in pollen and seed fertilities and/or in the extent of selfing. These allocation models (as well as that of Charnov[29]) assume a trade-off takes place between seed and pollen production. However, recent work by Devlin,[43] for example, studying sex allocation in *Lobelia cardinalis,* found no negative correlation between seed and pollen production per flower, which the above model requires.

Models for sex allocation versus the selfing rate predict that allocation to male function should decrease with increasing rates of selfing.[26,27,29,30] Charnov recently described sex allocation as a function of selfing rate in 31 strains of wild rice, *Oryza perennis.*[30] He showed that the male/female allocation ratio was linearly related to the selfing rate (just as was shown by Schoen[106] for *Gilia achilleifolia*). According to Charnov, such linearity indicates that the intermediate selfing rates would have to be maintained by frequency-dependent selection. Using the selfing-sex allocation model proposed by the Charlesworths,[26] Charnov suggested that selfed offspring were half as fit as outcrossed, and that this may be a consequence of frequency-dependent selection, which itself stabilizes the intermediate rates of selfing. Subsequently, Charnov[31] concluded that the Charlesworths' model is flawed by the confounding of sex allocation with selection for/against selfing. He pointed out that their model considers the proportion of seeds which are selfed as being fixed, and allows selection to adjust the proportion of resources given to pollen versus seeds. Charnov argued that the Charlesworths' "fixed selfing rate" is, in effect, *not* fixed. Since in their model selection is able to alter the proportion of reproductive resources given to selfed progeny (by allocating more or less to seeds), the "fixed selfing rate" can be adjusted, as Charnov says, "through the back door by altering sex allocation." He suggests that in a model that corrects this defect a plot of sex allocation versus the proportion of seeds which are selfed "will be linear, independent of the level of inbreeding depression for selfed offspring."

Plants, in numerous ways, can regulate self-pollinations. One certain method is the production of distinct gender morphs, where pollen and ovules are produced on separate individuals (though this will not limit sib-mating, another form of close inbreeding). Other polymorphisms, such as heterostyly, may have a similar outcome because, in general, "pins" and "thrums" can only cross-fertilize.[42,117] The usual interpretation has been that these arrangements serve to promote outcrossing, but Charnov and his colleagues (e.g., Ref. 23) have argued for an explanation for the *maintenance* of heterostyly based on sexual selection and sexual resource allocation. As another example, Lloyd and Webb[72] indicate that, in outcrossing species, there is an inevitable clash between natural selection acting to place pollen and stigmas together for effective pollination, and selection acting to keep the male and female structures apart to avoid interference between them. Lloyd and Webb suggest that the *temporal* separation of pollen and stigma maturity (dichogamy) acts to reduce this self-interference, and it often also reduces self-fertilization. They state that mechanisms which prevent self-fertilization primarily increase maternal fitness, whereas mechanisms which avoid self-interference primarily promote paternal fitness. Webb and Lloyd[121] show that the *spatial* separation of pollen presentation and pollen receipt (herkogamy) can also function to avoid interference between pollen receipt by stigmas and pollen dispatch from anthers in cosexual plants.

Models used to describe the evolution of sex allocation in hermaphrodites are formally identical to models describing the sex ratio of organisms in which the sexes are completely separated.[29] In both cases, male and female gametes contribute genes equally (apart from sex-linked or cytoplasmic genes); thus, total male and female reproduction must be the same, and the fitnesses that a pair of mating individuals derive from their offspring are identical. Charnov[27] first suggested that the sex allocation hypothesis provides an effective evolutionary model for understanding a number of earlier observations in the literature concerning pollen/ovule ratios. For example, Cruden and co-workers[38,39,41] have shown that pollen/ovule ratios of flowering plants are negatively correlated with self-fertilization, clumped pollen transfer, pollen size, and the ratio of stigma area to the pollen-bearing area on the pollinator. Queller[98] has demonstrated that all four negative correlations are consistent with evolutionary models of optimal resource allocation to male and female functions, and has concluded that "selection on sex allocation in hermaphrodites may be governed by the same principles as selection on the sex ratio."

Problems of sex allocation in outcrossing seed plants have been well reviewed by Lloyd[68,69] and Goldman and Willson,[54] who summarized the sparse data on the subject. The same problems arise here as for considerations of reproductive costs in general, since androecial and gynoecial costs may not always be discrete. For example, the production of petals and sepals and nectar-secreting parts of the stamens or carpels are neither maternal costs nor paternal costs alone, but can be seen as investment in features which enhance both male and female fitness. Charnov and Bull[33] show, theoretically, how one might partition attractants to males and females. Problems of resource "currency" remain to be addressed, particularly since floral organs containing chloroplasts may be less limited by carbohydrate supply than by the availability and mobility of imported nutrient elements.

Charnov[27] and Willson[124] have conjectured that, in hermaphrodite flowers, allocation to the secondary organs (of attraction) may be almost exclusively male in function; flower production can be regarded as a male function, and fruit production as a female function in outcrossed plants. Indeed, Bateman's Principle (see Charnov[27]) leads immediately to the conclusion that advertisements and rewards primarily serve the male function (unless maternal choice somehow makes seed quality increase with increased levels of pollen reception; see Ref. 33). This theory effectively elaborates an earlier argument of Willson and Rathcke[126] who suggested that "extra" flowers, in *Asclepias,* served to increase pollen export; the effect (numerically) on female reproductive success was usually less. (However, Willson and Rathcke also postulated that "extra" flowers gave the opportunity for increased pollen reception, and a consequent opportunity for female choice.)

Bell[11] estimates a plant's effective gender simply as the ratio of flower production (male) to fruit production (female) and factors in complications such as fruit abortion and self-fertilization (see also Sutherland[113,114]). Bell describes the results of a series of experiments consistent with the conclusion that "the flower is primarily a male organ, in the sense that the bulk of allocation to secondary floral structures is designed to produce the export of pollen rather than the fertilization of ovules." The implication of this work is that female function (fertilization of ovules) is almost fully satisfied by one or a few insect visits, while effective pollen dispersal (male function) requires repeated visits that can be secured only at the cost of significant investment in organs of attraction. Bell provides evidence of several kinds.

1. Insects visited larger flowers more frequently (in *Fragaria virginiana*). Removal of petals reduced the frequency of insect visits in proportion to the biomass removed (in *Impatiens capensis*).

2. Removal of organs of attraction may lower the probability of the formation of a fruit, but should not affect the number of seeds set per fruit. Bell manipulated plants of *Impatiens capensis* (having essentially solitary flowers producing several-seeded capsules), *Asclepias syriaca* (which forms several multiseeded fruits per inflorescence), and *Viburnum alnifolium* (which produces a cyme whose central flowers are small and fertile and form drupes containing one seed, while the outermost flowers are larger and sterile). The *Viburnum* results are clear: removal of the sterile flowers significantly reduced fruit production and seed set per inflorescence. Similarly, removal of flowers from *Asclepias* inflorescences did not reduce seed set per fruit, but did reduce fruit set per inflorescence, while mutilation of the *Impatiens* flowers had no effect on seed set per fruit. These results appear to be consistent with the hypothesis that a single insect visit suffices to fertilize all or almost all the ovules per ovary in these plants, while further visits produce little additional effect on seed formation in that ovary.

Bell's[11] observations are in agreement with the few extant data describing patterns of allocation of floral biomass (e.g., See Refs. 54,75,76). Bell showed that the average ratio of male to total floral allocation was about 50% if calculated for primary structures alone, or if secondary floral structures are divided equally between male and female function. The ratio was over 80% if secondary structures were assumed to be purely male in function. However, Bell's recalculations were averaged across the array of breeding systems represented by a mixture of species.

Floral sex allocation patterns may be significantly affected by the breeding system of a species. Lovett Doust and Cavers[75] concluded that cross-fertilizing species had greater relative investment in male structures and self-fertilizing species allocated more to female structures; thus, it may be misleading to ignore the breeding system as Bell has done. Similarly, as mentioned above, Schoen showed in *Gilia* and Charnov showed in *Oryza* that there were decreases in male allocation with increases in self-fertilization.[30,106] Cruden and Lyon recently studied some 39 species with different mating systems, determining patterns of dry-matter distribution in male and prezygotic female structures.[40] Evidence from the xenogamous species supports models predicting unequal resource allocation to male and female functions in outcrossed hermaphroditic angiosperms (e.g., Charnov[27,29] and Charlesworth and Charlesworth,[26] but see also Charnov[31]). Cruden and Lyon found in xenogamous species a pattern of biomass allocation that seemed independent of sex habit (whether the plant was hermaphroditic, monoecious, or dioecious) and pollen vector. Prezygotic costs were male-biased. If parental investment was included in female function (postzygotic costs of fruit and seed) the bias was reversed. These workers observed an overall decrease in both the absolute and proportionate biomass allocated to male function when they contrasted xenogamous and facultatively autogamous mating systems.

Sex allocation cannot be fully described by single harvests of flowering plant material: it is necessary to consider the breeding system and the phenology of male and female reproductive function, both pre- and postzygotic, and to summarize sex allocation for the individual. It is difficult to obtain useful measures of fitness, particularly male fitness, though the technical problems of this are less daunting than they used to be (see section below titled Paternity, and Bertin,[14] this volume). It remains to be seen whether there is indeed a reasonable correspondence between expenditure on male and female function, and the fitness payoff of that pattern of expenditure. Patterns of resource allocation within and between flowers as well as measures of the degree and distribution of variability within populations need to be characterized. Sex allocation, even in hermaphrodites, may be sensitive to the sexuality and numbers of neighbors of the same species, though no satisfactory mechanism has yet been proposed for such a response. Queller's view[98] that the same principles may govern the evolution of sex allocation patterns within hermaphrodite individuals as govern population sex ratios, calls for attention and testing. Obviously a general model that was able to incorporate both of these levels of selection would be of great interest in evolutionary biology.

SEX HABITS

Even species having the same sex allocation pattern may exhibit differences in the arrangement of male and female sporangia. Floral arrangements in higher plants are distinguished by the degree of separation of male and female sporangia between the flowers of an individual and/or between individuals. Angiosperm flowers either produce both ovule and anther in a single ("monoclinous") flower, or have only ovules, while other flowers bear only anthers (they are "diclinous" or unisexual). A population may be sexually either monomorphic or dimorphic. In monomorphic populations, all individuals are similar in their sex habit. The predominant sex habit of most higher plants is the monomorphic one known as hermaphroditism.[129] Even in hermaphrodites there can be functional separation. For example, in the hummingbird-pollinated *Lobelia cardinalis,* Devlin[43] showed that, in relative terms, hermaphrodite flowers on the basal portion of the inflorescence were functionally more female and flowers on the terminal part of the inflorescence were functionally more male. (Devlin determined that basal flowers gave the greatest numbers of seed while terminal flowers gave least; however, pollen number per flower did not vary with position on the inflorescence.)

In monomorphic diclinous species, the two types of sporangia are borne in different flowers on the same plant; there are several variants of this system. A major example is monoecism: some flowers produce only female sporangia, others only male sporangia. Andromonoecism and gynomonoecism are the remaining important classes of monomorphic sex habit. Here again individual plants are similar and bear flowers of two kinds: in andromonoecism, some flowers are hermaphroditic and others are male; in some flowers on each plant, femaleness is suppressed. In gynomonoecism, maleness in some flowers has been suppressed. Lloyd[67] has indicated that most plants are sexually monomorphic; in the majority of populations of seed plants there is a single "cosexual" morph combining maternal and paternal functions in one individual. Lloyd considers that the direction of evolution of the sex habit has principally been from cosexuality (typically hermaphroditism or monoecism) toward separate sexes.

Sexually dimorphic populations consist of two classes of plant with respect to gamete production. Individuals may bear different types of flowers. In typical dimorphic dicliny the genders are completely separated, so that some plants bear staminate and others pistillate flowers. In this "dioecious" habit, the well-defined male and female sporophyte plants in a population rarely vary the sex of gametes they produce. The habit is uncommon among annuals and short-lived perennials but rather is associated with a long-lived, woody habit.[7] Dioecy has been found to be associated frequently with pollination by wind or by unspecialized insect pollinators. It has also been associated with the production of fruits that are fleshy and bird-dispersed[7,10,53,67] (but see Fox[48]).

Lloyd[67] has critically reviewed the potential advantages of being either a specialist female or male, and has showed that a division of sexual labor increases reproductive efficiency; this gives advantage to both females and males over cosexes under several kinds of conditions. For example, conflicts between maternal and paternal functions may arise in a cosex if presentation of pollen and stigmas in one flower causes mutual interference between pollen removal and pollen deposition (see Lloyd and Yates[73]). Females may gain benefits from increased outcrossing, due probably to advantages of heterosis or increased genetic variability. In contrast, males may benefit from more efficient pollen pick-up by insects. Other workers[7,27,124] have suggested that under certain conditions, individuals with partial or total female sterility (and resource reallocation to greater pollen production), as compared to cosex individuals, may benefit from increased success in intrasexual competition; however, we know of no clear evidence to support this conclusion.

In a review of the longevity of individual flowers, Primack[95] noted that in dioecious and monoecious species female flowers generally last significantly longer than male flowers. Primack's study revealed major differences among environments in flower longevity. For example, long-lived flowers tended to be associated with cooler and possibly more humid places. This could be related to the predictability of pollinator visitation. Romero and Nelson studied sexual dimorphism in unisexual *Catasetum* orchids and concluded that the dimorphism promotes pollination. They suggest that competition among male flowers may explain the origin of the pollination system in both *Catasetum* and *Cynoches* species.[100]

In contrast, intersexual floral mimicry is seen as playing an important role in pollination in a variety of species.[1,2,6,52] In dioecious *Rubus chamaemorus*, female flowers are similar in general appearance to males, but are somewhat smaller and produce no pollen and only minute amounts of nectar. Ågren at al.[1] showed that pollinators strongly prefer male flowers (with their greater reward). These workers conclude that female flowers of *Rubus chamaemorus* attract pollinators by "deceit" based on mimicry. The authors went on to suggest that pollen availability limited seed production in female-dominated habitats, but not in areas having an equal floral sex ratio. In general the relative importance of factors limiting female reproductive success is not constant but is a function of the floral sex ratio of the population.[1]

Looking at whole-plant patterns of biomass allocation in dioecious *Silene alba*, however, Lovett Doust et al.[79] note a danger in interpreting limited data as illustrating male–female dimorphism in terms of secondary sex characteristics. Although this study documented significant contrasts between males and females in a natural population, such differences seem "soft" because, for the most part, they disappeared under certain experimental conditions.[79]

We can view the diversity of sexual systems as reflecting different patterns of relative resource allocation to maternal and paternal functions, which separately optimize maternal and paternal reproductive success.[27] Sexual selection pressures may indeed be significant in shaping these alternate allocation patterns; however, as Bawa and Beach[9] showed clearly, the efficacy of (prepollination) sexual selection in zoophilous species will be determined mainly through interactions with pollinators. The dynamics of the pollination system seem fundamental to the evolution of a particular sexual system.[33]

Biased population sex ratios may be, in part, a consequence of the differential costs associated with reproducing as a male or as a female. For example, Cavigelli et al.[25] concluded that in dioecious *Ilex montana* females flower less frequently than do males. If female costs are greater, then survivorship could be reduced[71] and in long-lived dioecious species, males may become predominant.[77] Putwain and Harper[96] showed that in perennial *Rumex acetosella* females allocated between three and four times more of their biomass to sexual reproduction than did males. In annual *Rumex hastatulus*, females also allocated a greater proportion of biomass to reproduction than did males.[37] In *Arisaema dracontium* and *Arisaema triphyllum* the biomass costs of reproducing as a fruiting female are two to three times those of reproducing as a flowering male.[78]

Female reproductive effort in several other species is not necessarily greater than in males, but is a function of fruit and seed set. For example, Wallace and Rundel[119] found that at 100% seed set in the shrub *Simmondsia chinensis*, females allocated 30–40% of their resources to reproduction, as compared to 10–15% for males; female effort equaled male effort at about 30% seed set. Gross and Soule[55] found that, in herbaceous *Silene alba*, only females having greater than 20% fruit set had a higher reproductive cost than males. Linhart and Mitton[64] reported a tradeoff between growth rate and female but not male cone production in monoecious ponderosa pine. The tradeoff between female investment in growth and reproduction seemed to be genetically based, since the variability in growth rate and variability in relative female cone production were found to be linked with patterns of protein heterozygosity.

The diversity of plant sex habits has always been an intriguing aspect of their reproductive ecology. Complete separation of sexual function, though a minority state, seems to have evolved many times. Resource allocation patterns can be used to characterize both floral and somatic differences between gender morphs. Vegetative characteristics may therefore sometimes be of use in predicting the sex expression of an individual, and may indeed be symptoms of differences in the physiology or phenology/ontogeny of the sexes. It is still an open question, however, whether the degree of divergence in sexual function and appearance is driven by pollinator behavior, considerations of reproductive efficiency, resource allocation, or a mixture of these influences.

SEX LABILITY

Since a plant continuously produces new organs throughout its life, it has a capacity for ongoing developmental responses to its environment. Bradshaw critically discussed the remarkable phenotypic plasticity of plants as one of the more important evolutionary consequences of being a plant.[19,20] However, he did not deal with plasticity of sex expression, and we know that the gender expressed by the sporophytes of flowering plants can often be substantially modified in the face of environmental constraints.[49,70] Evidence suggests that sporophytic sex expression can often be altered by such agencies as levels of soil nutrients, light regime, and the amount of storage tissue, or leaf area.

Lloyd and Bawa[70] developed the idea that there are two measures of plant sex expression: "phenotypic gender" and "functional gender" (see Lloyd[65,66]). Phenotypic gender describes male and female functions in the initial investment of parental resources, whereas functional gender reflects the relative success of a plant as a paternal parent and a maternal parent. Modification of both phenotypic and functional gender may take place as a result of an individual's circumstances, which may be a function of both the external environment and a plant's internal resource status, attributable to a combination of its size and vigor.

Lloyd and Bawa[70] distinguish two patterns of gender modification, based on the extent of departure from the mean gender of a class of plants. (These two general patterns of gender modification are also discussed in Charnov and Bull.[32])

1. *Gender adjustments,* in which gender varies continuously about one modal value, in response to environmental or status signals. Most theoretical studies have assumed close correspondence between phenotypic and functional gender (e.g., Refs. 29,101). Clearly, if there was little linkage between the two then selection on sexual characteristics would tend to be inhibited, and this would prevent the evolution of sexual specialization in hermaphrodite plants. Thomson and Barrett[115] suggested that the two measures of gender would be unrelated due to environmental variation and this would inhibit the action of natural selection on sexual traits; thus environmental variation could be responsible for maintaining sexual polymorphisms. More recently, Devlin and Stephenson[44] showed that reproductive success in *Lobelia cardinalis* is affected by a variety of factors, including the phenotypic gender of the individual and that of neighboring individuals, as well as temporal patterns of seed set and individual variation in seed and pollen production. Devlin and Stephenson conclude that the phenotypic and functional genders of individuals remain related in spite of the actions of numerous external influences that partially obscure the relationship.

2. *Phase choices,* in which the distribution of gender is bimodal and individuals "choose" between discrete phases—alternative male and female forms, according to conditions. Such so-called "sex choosing" (gender diphasy) has been described in several species, including members of the genus *Arisaema* (e.g., Refs. 15,74,94), where change seems to be regulated by the supply of resources available for reproduction. Resource availability is a function of environmental variables and the sex expressed in preceding years.

Charnov's review of sex regulation (the adjustment of son or daughter production in response to particular environmental conditions) is especially interesting.[29] Earlier, Trivers and Willard had postulated that, for animals, the physiological condition of the mother may be a critical variable affecting the reproductive capacity of offspring.[116] These authors argued that mothers in "good" condition should produce more sons, those in poorer condition more daughters. They suggest that sons should gain relatively more than daughters by being in good condition because a male's reproductive success may depend to a greater extent on its physiologic condition compared to other males.[116] This argument assumes that variance in male fitness is greater than variance in female fitness, an observation that has been made for *Drosophila,* but that has not so far been shown in many other organisms.*

Michael Ghiselin's "size advantage" model[51] was developed around the idea that, within a species, if a small organism (or a young one) can reproduce more effectively as one gender, while a large one (or an older one) can reproduce more easily as a member of the other gender, then it becomes advantageous for an individual to switch its gender as it grows and ages. Whether male or female reproduction comes first should be a function of which sex is first to give a greater return in terms of fitness. In fact, the regulation of son and daughter production and sex change are both essentially examples of environmental sex determination. Charnov reviews in detail evidence in support of this theme, covering examples of nematodes and wasps and plants, among others.[29]

There is in plants, then, a certain "casualness" to sex expression. Schlessman[105] (this volume) concludes: "The notion that male, female, and ambisexual states always represent distinct phases of gender expression must be replaced with the concept that gender may vary within and among individuals, even when the individuals are genetically predisposed toward maleness and femaleness."

PATERNITY AND MALE COMPETITION

Assessment of maternal fitness has been much more tractable than measuring paternal success. However, since all sexually produced organisms have both a mother and a father, each responsible for half of its genes, it is important to probe the pattern of fitness among male as well as female parents. An early conclusion of Bateman,[5] from studies of evolutionary fitness (or rather the less direct metric, reproductive success) among male as opposed to female fruit flies, was that variance in reproductive success was greater for males than for females. This meant that some males (presumably supe-

*But see, for example, Clutton-Brock et al.[36] and Woolfenden and Fitzpatrick.[127]

rior in some respect) were highly successful as fathers while other males made little or no contribution to the next generation. In contrast, most females were likely to reproduce successfully, whatever their relative quality. The majority of plants, being hermaphroditic, have two possible pathways for transmitting their genes to the next generation: pollen and/or ovules. It is of considerable interest to test the generality of Bateman's findings, particularly as it pertains to both the process of sexual selection and the relationships between sexual selection, resource allocation, and the evolution of plant breeding systems.

It is difficult to determine paternity in plants (as in most other organisms) with certainty (see Bertin, this volume, for a detailed review of issues in plant paternity[14]). It *is* possible, however, to set up experimental situations that limit the number of possible fathers, and use fathers whose pollen carries distinct genetic markers. Stanton et al. showed that in experimental populations of wild radish composed of two homozygous petal-color morphs, color discrimination by pollinators had no apparent effect on relative maternal function (i.e., fruit and seed production) in the two color morphs. However, yellow-flowered individuals proved to be far more successful paternally (as pollen donors) when pollen receivers were also yellow, rather than the white morphs, which were visited less frequently by pollinators.[110] These results support the earlier conclusion that the evolution of floral signals may be driven primarily by selection on male function.[27,110,124]

The measurement of paternity in natural populations is even more difficult than in controlled studies involving experimental manipulations. The two best data sets come from the investigations of Ellstrand and Marshall[45,46] and Meagher.[85] Ellstrand has demonstrated that several pollen parents may sire the seeds within an individual fruit of wild radish, *Raphanus sativus.*[45] Ellstrand and Marshall measured the extent of multiple paternity in the multiseeded fruits of wild radish from three natural populations.[46] All of the parents sampled produced one or more multiply sired fruits; in most plants, more than half the fruits that were examined proved to be a product of multiple paternity.

Meagher identified the most-likely male parents in a study of the forest herb *Chamaelirium luteum* and showed that males had a higher variance in the number of mates, suggesting sex-specific selection.[85] Meagher also showed that the variance in the distance to mates was greater among females than among males, again suggesting that sex-specific selection was a possiblity.

Any attempt to address sexual selection by measuring male competition and the possibility of female choice calls for the determination of the relative input of potential fathers in terms of both their contribution to the load of pollen on local stigmatic surfaces and their contribution to the siring of new zygotes. It should be possible to elucidate the latter using electrophoretic markers and the application of paternity exclusion analysis[45] for small populations, or a maximum likelihood procedure[86] for larger populations. Unfortunately, this latter technique could assign paternity with total certainty for fewer than 10% of seeds. Methods of analysis must be refined before conclusive statements can be made about paternal fitness; the relative success of seeds of different male parentage must be measured. In summary, measurement of pollen production is several steps away from measurement of paternal fitness, and problems of technique and quantification remain unsolved. It may be that "fingerprinting" through immunological techniques or analysis of DNA restriction fragment polymorphisms will become routine and will be turned to the task of determining genetic identity in pollen, zygotes, and adults in natural populations.

Mulcahy has argued that the closed carpel of the angiosperm gynoecium, in conjunction with abundant pollination by insects, results in high levels of male gametophyte competition.[89] The Mulcahys and their co-workers have shown that pollen competition takes place, and have suggested that it has played an important part in the rapid evolution of angiosperms (see Ref. 89). It has been difficult, however, to separate the part played by maternal tissues from pollen competition in determining the outcome of the race run down the style from stigma to ovule. Working with *Silene dioica* and *Mimulus guttatus,* Searcy and Mulcahy found that, in female plants tolerant of and raised in toxic metals, the rate of pollen tube growth was unaffected.[107] The toxic stylar environment was not directly affecting the growth rate of either tolerant or nontolerant pollen. However, the number of successful fertilizations and rate of formation of viable seeds were affected, and the nontolerant pollen suffered more than the metal-tolerant pollen in those respects. Searcy and Mulcahy suggested that early abortion of fertilized ovules fathered by nontolerant pollen may have caused the under-representation of that type, but they concluded that the stylar environment did not affect pollen competition and was not responsible for prezygotic selection for metal tolerance in pollen.[107]

Snow has reviewed levels of pollen tube competition in natural systems.[109] She showed that, in *Epilobium canum,* the likelihood of pollen tube competition was highly variable, and depended upon year, plant, fruits within plants, and even ovules within fruits. At present there is little evidence concerning the prevalence of pollen tube competition in nature because few researchers have actually measured *natural* levels of pollen arrival on stigmas in any species.

INCOMPATIBILITY

The study of mechanisms of incompatibility is another situation in which it is helpful to examine the stages of pollination and pollen–pistil interactions. Barrett[3] (this volume) urges careful study, from pollination to seed set, in order to distinguish incompatibility phenomena from postzygotic inbreeding effects (and see Seavey and Bawa[108]). In the first instance, the interaction is a function of maternal genotype; in the second, the genotype of the zygote is central. It remains to be seen whether other phenomena involving pollen–pistil interactions, such as the superior performance of cross- as opposed to self-pollen in self-compatible plants, optimal outcrossing, and the apparent advantage of pollen from outside populations, are related to mechanisms of self-incompatibility. The Mulcahys' "heterosis" model of gametophytic self-incompatibility suggests overall synergy between contrasting gametophytic and maternal sporophytic genomes that may stimulate pollen tube growth, resulting in differential growth rates of self- and other-pollen. Other workers (e.g., Lawrence et al.[60]) do not feel there is enough evidence to reject the "oppositional" model [which invokes some (unknown) chemical interaction between self-pollen and sporophytic tissue]. It will be necessary to separate these alternative explanations of self-pollen failure, and while it may be true that strong self-incompatibility may be explained by the oppositional model, more subtle "leaky" or cryptic self-incompatibility may indeed be more a function of the "ecology" of the pollen tube growing within the environment of the pistil.

Evidence from many plant species suggests that self-incompatibility barriers *in the ovary,* so-called "late-acting self-incompatibility systems," may be in fact much more common than has been previously assumed. Seavey and Bawa show that the distinctions between postzygotic incompatibility and inbreeding depression are not easily made and depend upon assumptions underlying the genetic models of self-incompatibility.[108] They outline four possible approaches that could distinguish between the two, and suggest a possible mechanism for the operation of postzygotic self-incompatibility (that is essentially the same as that proposed by Westoby and Rice[122]). The work of Seavey and Bawa opens new possibilities for late-acting incompatibility, but does not prove it. In the model of Westoby and Rice for maternal resource allocation, there is a variable threshold below which allocation to progeny would be terminated in early embryogeny. The availability of resources to the female would determine the level of the threshold in this model.[122] Seavey and Bawa incorporate a small modification to this, suggesting that some plant species have genetically fixed threshold responses that operate to reject self-embryos by cutting off resources.[108] The implication is that the less heterotic (selfed) genotypes would regularly fall below the maternal threshold of resource allocation required for further embryo development.

FEMALE CHOICE?

Sexual selection involves two kinds of processes: competition between members of one sex (usually males) for access to members of the other sex and, second, choice on the part of the second sex (usually females) of a mate from the potential mates available. Interestingly, while the concept of competition among males is widely accepted, female choice has been largely ignored as a mechanism of sexual selection in animals as well as plants. (Indeed, several authors have suggested that there is inherent male chauvinism in the widespread assumption that female choice is a passive phenomenon, that females simply accept the hierarchy of quality established through male competition.[17]) Most experiments, to date, have not been designed to discriminate between male competition and female choice,* and the issue has only recently been raised with respect to reproduction in plants.

It is probable that some male plants are more likely to father zygotes than others, depending on a number of factors that are, in part, beyond the control of either parent. For example, the male plant's proximity to the nearest receptive female is important. The distribution of the pollen "rain" is crucial in wind-pollinated species, whereas a plant's ability to attract specialized pollinators is most important for species pollinated by insects. Plant size is not only correlated with the size of the pollen crop, but may also affect the probability and frequency of insect visits for that genetic individual. Flowering phenology plays a part, and there is an advantage to synchrony with the receptivity of members of the other sex. It seems reasonable to postulate a role for competition among females as well as among males; after all, the plant whose floral display attracts more pollinators enhances its female fitness as well as the dispersal of its own pollen. Charnov[27] clearly argued for the possibility of female choice in plant reproduction.

Willson and Burley[125] point out that sociobiological arguments regarding mate

---

*But see Clutton-Brock et al.,[36] Ryan[103] and Clutton-Brock.[35]

competition and mate preferences depend upon a constraint on the reproductive success of one sex (usually female) by "food" resources, and limitation of reproductive success of the other sex (usually male), by the number of fertilizations effected. They review the available evidence and show that female plant reproductive success is indeed generally limited by resource availability; however, in a small number of cases it has been shown that seed production by a female plant may also be limited by the availability of pollen, rather than "food" resources (Bierzychudek,[16] but see Lovett Doust et al.[81] and Willson and Burley[125]). Estimation of pollen limitation may be methodologically difficult. It should be noted that both the movement of pollen and seed production have been shown to be a function of the spatial distribution of individuals. For example, Wyatt and Hellwig[128] showed this for fruit set in distylous *Houstonia caerulea,* as did Kay et al.[58] for *Silene alba;* however, no correlation was found to link fruit set and local floral sex ratio in dioecious *Aralia nudicaulis*[4] or distance to nearest male in *Jacaratia dolichaula.*[21] Likewise, Ågren et al.[1] found no association between fruit set and floral sex ratio in dioecious *Rubus chamaemorus.*

How might female choice be made in plants? Willson and Burley show that embryos may be forced to compete with each other while still maturing on the parent plant; thus superior genotypes prevail. Polyembryony involves the production of more than one embryo per seed, and Willson and Burley review three kinds. Simple polyembryony (SPE) is the presence in each female gametophyte of more than one egg and the potential for each egg, if fertilized, to produce an embryo. SPE apparently is common in many ferns and is widespread in gymnosperms, where typically only one embryo survives in the mature seed. In SPE, all eggs arise from the same gametophyte and possess identical maternal haploid genomes, but the resulting zygotes may or may not have different fathers, depending on the source of the pollen grains and the number of sperm per pollen tube. Willson and Burley describe cases where some gymnosperms may have 16 to 20 sperm per pollen grain. Therefore, selection between embryos of different genotype could occur in cases of SPE.

Cleavage polyembryony (CPE) involves the production of multiple, usually genetically identical zygotes from a single fertilized egg. CPE is extraordinarily well developed in the gymnosperms and is present in some angiosperms (and in armadillos!). In CPE, the nuclear genomes of all embryos within a seed are identical, so competition among these embryos is of little evolutionary consequence.

The embryos in adventitious polyembryony (APE), especially in such genera as *Citrus* and *Opuntia,* may develop directly from (diploid) maternal sporophyte tissues; several such embryos often mature. In APE, embryos in general have only the maternal genome (or sometimes an aneuploid variant thereof). Selection between these embryos is also of little evolutionary consequence if they are genetically identical. Now an interesting issue arises; competition between identical zygotes should be *more* intense than competition between zygotes that differ genetically, since identicals' resource demands and developmental timetable are likely to be very similar. However, by arguments based on kin selection, identical zygotes ought to compete less with each other.[90] A small genetic change—the inclusion of a new extrachromosomal fragment, mutation of a nuclear gene, or a change in chromosome number—will be subject to intense and discriminating competition. Polyembryony might provide an arena where advantageous genetic changes could be swiftly selected.

Why have multiple zygotes developing within one seed? Willson and Burley suggest that simple polyembryony in gymnosperms may be a means of female choice to select among zygotes from different fathers.[125] They also suggest that automatic abortion of a fixed proportion of ovules in angiospersm (observed in *Cryptantha flava* by Casper and Wiens[24]) may serve a similar function. This would only be so if ovules of some genotypes were more likely to be aborted than others. Using a form of source-sink argument, that several rapidly growing embryos may draw more resources to the seed than a single zygote, Willson and Burley speculate that CPE ultimately could increase zygote growth rates. This could be viewed as a female tactic allowing a certain amount of "assessment" of relative quality in different seeds. (It may also be interpreted as a tactic of competition between males that will reduce the risk of abortion of their offspring, since a large sink of identical sibs would draw a larger supply of resources.)

The function of APE in the flowering plants remains unclear. It requires explanation of both parthenogenetic reproduction and the development of multiple seedlings from a single seed, and we lack a good understanding of both topics. It is possible that subtle differences may exist among these embryos, if not in terms of changes in the nuclear genome, then perhaps in terms of extrachromosomal genetic factors or viral infection.

Willson and Burley also considered some mechanisms whereby females might control the growth of zygotes. They suggest, for example, that females may be able to control the amount of resources provided to enhance the rate of pollen tube growth. It is difficult, however, to see how this might be differentially donated to particular pollen tubes.

In their analysis, Willson and Burley consider the possible consequences of both prezygotic and postzygotic female choice. They point out a rather striking difference in the potential capacity for mate choice by gymnosperms and angiosperms: gymnosperms appear to have only very weakly developed abilities to identify potential mates before fertilization, whereas among angiosperms early detection abilities, such as rejection of pollen in the stigma or style, appear to be well developed. As a result, any element of female choice in gymnosperms should be dependent upon perfection of abortion techniques. This process could be accompanied by selection acting on male function to prevent abortion; in angiosperms the situation must be more complex, since both pre- and postzygotic mate choice may be practiced to a considerable extent.

Willson and Burley conclude that overall there is little support for the idea that significant female choice occurs in plants. There is no strong evidence of a female ability to discriminate on the basis of genetic quality, etc. Yet, broadly defined, plant incompatibility systems are a type of prezygotic female choice which are widespread. "On the other hand," these authors state, "we have not been able to find evidence that invalidates the model, and by and large, biologists have not asked questions or gathered data in ways that permit evaluation of our ideas." Willson and Burley provide a stimulating set of more than 50 hypotheses, some original, others drawn from the literature, that focus on testable aspects of the mate choice model.

For female choice to be a possibility, it is necessary that seed set for these females be resource-limited, rather than pollen-limited. Polyembryony may force competition on zygotes within a single seed. This will only be important in an evolutionary sense if there are genetic differences between the zygotes—there is no "mate choice" involved if the male and female gametophytes were the same for all the zygotes. Where mate choice might come into play is in the situation where the growth of multiple (identical) zygotes could amplify differences between different clones developing in neighboring seeds. Then the overall effect could be to exaggerate the strength of the sink for resources of that particular mating. Alternatively, polyembryony may allow rapid selection of new traits; mutations or genetic inclusions that are "tested" against an otherwise identical background.

SEXUAL SELECTION

Several other workers have recently assessed sexual selection and the ecology of embryo abortion in plants. Stephenson and Bertin[111] emphasize the importance of determining the variance in the number of offspring resulting from males, as opposed to females, as an indication of the intensity of sexual selection. Stephenson and Bertin use the definition of "variance" developed by Wade and Arnold.[118] Wade and Arnold demonstrated mathematically that variation in the reproductive success of males can be partitioned into two useful components: one due to variance in the fertility of females and one due to variance in the number of mates per male. According to Wade and Arnold, the latter component is the major proximate cause of a difference between the sexes in the intensity of selection on reproduction.[118] Male competition should be intense if the variance in male function exceeds that of female function. Stephenson and Bertin suggest that, in the case of obligate self-fertilization, variances in the success of male and female functions have to be equal.[111]

Unfortunately, very few data exist to evaluate these ideas on variance in male and female reproductive success, and the results are sometimes complex and conflicting. Bertin has carried out valuable studies of the trumpet creeper, *Campsis radicans.*[111] He hand-pollinated all possible pairs of nine *Campsis* plants, and determined that variance in the reproductive success of pollen donors (measured as the number of fruits fathered) was similar to the variance in reproductive success of pollen recipients. Bookman studied milkweed *(Asclepias speciosa)* and, after similar hand pollinations, found that variances in male function were greater than female variances.[18] In this species, it may be that surplus fertilized ovules and initiated pods (? more severe competition among zygotes) caused the differential maturation of pods, based on paternity and seed number. However, as Stephenson and Bertin point out, both these sets of results have a shortcoming; any variance in the ability of donor individuals to get their pollen deposited on stigmas is not considered, since the flowers are hand-pollinated rather than pollinated naturally. More recently, Bertin showed that *early* in the flowering season in *Campsis radicans,* plants were more selective about which experimental donor they accepted pollen from. This seemed to be related to the number of prior pollinations and developing fruits in an inflorescence, and so possibly to resource availability.[12] In the same species, the number of pollen donors had no significant effect on percentages of fruit production, seed number, or seed weight.[13]

Ellstrand and Marshall reported that in wild radish the total number of fruits set per plant increased significantly with multiple paternity, and that singly sired fruits were more likely to be aborted than multiply sired fruits (for some, but not all, females).[46] In discussing possible fitness consequences of multiple paternity, they rule out plant size and allometric effects, since there was no correlation between total number of fruits and number of flowers per plant. Also, pollen limitation for fruits with single paternity did not seem the cause of their lower fruit set, since multiple paternity was not a function of the number of pollinator visits (plants were hand-pollinated and the pollen loads were equalized).[82]

Marshall and Ellstrand describe in detail the effects of pollen-donor identity on the number of ovules fertilized, the position of ovules fertilized, the fruit set, seeds per fruit, and seed weight.[83] They were unable to separate male–male competition from female "choice" in bringing about the effect of pollen-donor identity. However, it was also found that multiply-sired wild radish fruits had a greater total weight of seeds per fruit than did fruits sired singly by any of the pollen donors used.[83] Studies of intraclonal competition have shown that the more similar competitiors are, the more intense is competition among them (as long as they are not physiologically integrated) (L. Lovett Doust;[80] S. Bliss and L. Lovett Doust, unpublished observations). Similarly, embryos of similar genotype may compete more intensely on the parent plant (their nutrient demands and the schedule of these demands will be more similar). In contrast, diverse offspring may make asynchronous demands on the resources of the maternal parent. An extended flowering and fruiting season would enhance this effect.

Stephenson and Bertin,[111] like Willson and Burley, conclude that "uncritical enthusiasm for the role of sexual selection in shaping breeding systems and reproductive strategies is not warranted." They suggest that sexual selection is unlikely to be an important force in the evolution of self-pollinating species or in those species in which seed production is consistently limited by pollination rather than by food resources. In these latter species, there would be little opportunity for competition between males. However, it is important to confirm pollinator limitation according to the criteria of Bierzychudek[16] (see Zimmerman,[130] this volume). Also, it should be recalled that male–male competition through attraction of pollinators is one major facet of sexual selection.[33]

Nakamura examines the sociobiological aspects of wild and cultivated beans by focusing upon parental investment and theories pertaining to brood reduction, and exploring hypotheses that allow him to test genetic predictions pertaining to offspring quality.[91,92] For example, he performed various crossing experiments to determine if the degree of genetic relatedness between maternal parent and offspring affected embryo quality and if it changed the fates of fruits and seeds. For each cross, Nakamura measured resource investment and survivorship between pods, and also measured at the finer level, ovule and seed survivorship within pods.

Nakamura's data for *Phaseolus vulgaris,* like those reviewed by Willson and Burley and Stephenson and Bertin, lend little support to the idea that maternal control may be exercised over embryo survival on the basis of embryo quality. Any selection pressures arising from inbreeding depression, outbreeding depression, and kin selection seem to be of minor importance in controlling fruit and seed abortion and allocation of resources to embryos. However, the crossing experiment used domesticated beans. Nakamura concluded that genetic factors may contribute to the well-documented effect that ovule position in the fruit is highly correlated with the probability of embryo abortion. This is obviously a complex interaction.[92,93] Lee and Bazzaz described a pattern of nonrandom ovule abortion within fruits of *Cassia fasciculata,* an annual legume.[63] They noted a "position effect" where ovules toward the fruit base (pedicellar end) had greater frequencies of abortion than did those at the distal end. Lee[62] reviews possible mechanisms that give position effects in seed abortion. Casper, studying variability in the pattern of ovule maturation in borage, *Cryptantha flava,* found that the probability that a flower would mature more than one seed depends upon its location within the inflorescence.[22] She noted that flowers closer to the main axis were significantly more likely to produce two seeds than those farther away; however, the probability that a flower would fail to mature seeds was the same for flowers in all positions. Casper concluded that the factors determining whether a flower yields a seed at all are different from those determining the number of seeds produced.[22] Her study suggests that there are both environmental and genetic components to the abortion of fertilized ovules within developing fruits of *Cryptantha flava.*

Lee argues persuasively that, when the number of pollen grains deposited on the stigma is greater than the number of ovules in the ovary, competition for ovules can result, and the fastest growing pollen tubes may be the ones resulting in fertilization.[61] Lee's "gametophyte competition" hypothesis predicts that if these embryos have higher fitnesses, then a plant that selectively matures fruit from flowers having high pollen/ovule ratios should have greater fitness. The hypothesis predicts that in those species having more than one seed per fruit, fruits having maximal numbers of seeds should be matured preferentially over fruits containing only a few seeds.[61] Stephenson and Winsor conclude that *Lotus corniculatus* can influence offspring quality through nonrandom fruit abortion. *Lotus corniculatus* selectively aborts those fruits having the fewest seeds and, in so doing, increases the average quality of its offspring.[112] In contrast, Holtsford showed that in the lily, *Calochortus leichtlinii,* flowers were not matured selectively on the basis of pollen quantity; rather, the first flower to open was in all cases matured.[57]

Galen et al.[50] concluded that, in clonal *Clintonia borealis,* male competition and female choice may occasionally be coupled. Schedules of stigma receptivity have the potential to affect both pollen tube density and donor diversity by structuring the rate of pollen tube recruitment and the length of time in which tubes may be recruited prior to the fertilization of all the ovules.

Taking another approach, Wiens has reviewed the "natural history" of ovule survivorship by comparing the relationships among seed/ovule ratios (S/O ratios, i.e., the percentage of ovules maturing into seeds), breeding systems, life history, and life form.[123] He found that S/O ratios are about 85% in annuals, but only about 50% in perennials. Many annuals are normally self-pollinating, whereas perennial plants more typically are cross-pollinating. Wiens showed that annual plants have significantly higher numbers of seeds maturing within individual fruits (bigger "brood size") than perennials. Among perennials, woody plants have lower S/O ratios and smaller brood sizes than do herbaceous perennials. Wiens concludes that S/O ratios seem to be largely genetically determined, whereas resource limitations are perhaps more important in controlling flower production. However, Wiens did not report fruit abortion, and it may be that constraints of resource limitation act *after* flower production. For example, in beans, changes in pot size have been found to change the proportion of ovules forming mature seeds.[91,93]

If sexual selection operates in plants, one would expect to see certain symptoms of differential selection on male and female traits. There might be greater variance in the number of offspring of males as opposed to females. A good study should involve comparison of the variance in the male and female success of the *same* hermaphrodite plants. Such variance should be measured under natural conditions where the effects of differential pollinator visits will be incorporated in the experimental design. There is conflicting evidence on whether or not offspring of lower quality are more likely to be aborted. In perennials, which are more likely to cross-pollinate, S/O ratios are lower than in annuals (which normally self). This may be a function of contrasting patterns of resource allocation in the two types of plants or it may signal greater scope for discrimination among offspring of mixed paternity, in the outcrossing perennials.

To make headway on the issue of sexual selection in plants it will be necessary to design experiments that separate effects attributable to each of the two components of sexual selection. There is a pressing need to agree on a definition of female choice; for example, is the differential nurturing of offspring of different fathers, which may be an active or a passive correlate of source-sink relationships, an example of mate choice? A useful definition of mate choice must incorporate a position on prefertilization phenomena such as incompatibility systems, events in the style, and stages of egg and endosperm fertilization. It should also consider the importance of barriers to nutrient transport within the seed (see Sociobiology of the Seed, below, and Haig and Westoby,[56] this volume).

SOCIOBIOLOGY OF THE SEED

"Evolutionary conflict" between mother and offspring may confound the female reproductive strategy.[27,28] Selection on the mother for exploiting resources and maximizing the number of her descendants will tend to be at odds with selective pressures acting on an offspring's exploitation of *its* environment. In fact, in seed plants the seed itself is a potentially competitive amalgam of up to four genetically different tissues (see e.g., Refs. 56,97,122). There is a "new generation" in the form of the developing embryo; in addition, there may be (1) protective tissue and dispersal organs contributed solely by the mother, (2) the remains of the haploid female gametophyte, and (3) a food supply, the endosperm, to which calories have been contributed solely by the mother, but for which genes *controlling* this supply typically are only two-thirds maternal and one-third paternal.

Willson and Burley review aspects of double fertilization in the angiosperms.[125] Charnov[27] first showed how double fertilization may have arisen through a process of kin selection. Subsequently, several authors have considered the degree of relatedness between the endosperm and the developing zygote. They suggest that the endosperm may be the evolutionary product of a genetic conflict between mother and father. Each parent has made a genetic contribution to the endosperm, gaining a proportionate amount of control over resources going to the developing zygote.[27,122] Willson and Burley also entertain the possibility that endosperm evolved as a "buffer" against parent–offspring conflict; however, they conclude that this is unlikely, in part because the selection pressures acting to produce double fertilization must have been concentrated disproportionately upon males (inasmuch as females end up retaining more genetic influence).[125] Queller[97,99] showed that degrees of relatedness, by themselves, don't enable prediction of the direction of evolution. His detailed model revealed that allele frequency and dosage were more fundamentally important in determining whether or not selection will favor an endosperm allele for taking more investment.[99]

Haig and Westoby[56] (this volume) offer a new interpretation of the various tissues within the seed, an interpretation that invokes conflict of interest between parent and offspring, and persuasively argues for regulatory rather than nutritive or protective roles for the hypostase, endothelium, and integuments. Haig and Westoby assemble evidence to show that the inner surface of the endothelium, which is in direct contact with the embryo sac, often becomes cutinized; this phenomenon has been interpreted in the past as the exchange of a nutritive for a protective function. If the seed composite is examined in the light of the separate identities of parent and offspring, however, there is a striking *lack* of direct connection between mother and offspring, and the deposition of blocking materials in the endothelium suggests the maternal plant may thereby achieve greater control over resource movement. Interestingly, in *Petunia* the callose barrier disappears from ovules that have been successfully fertilized, while in *Pisum* callose is found in the hypostase of aborting but not developing ovules.[56] As Haig and Westoby point out, their interpretation makes particular sense in the light of the fact that the so-called "protective" layer forms from the inside rather than from the outside of the integuments!

They also note that embryo abortion is associated with the proliferation of endothelium, integuments, and other maternal tissues. Earlier interpretations indicated that this proliferation was a consequence of a decrease in nutrient demands by the endosperm. Haig and Westoby argue that maternal proliferation may be a *cause* of slow endosperm growth—there is therefore a plausible mechanism for mother plants to choose among offspring, although mechanisms for discriminating among offspring have yet to be pinpointed. We should note, however, that the dynamics of ovule and seed abortion are not well understood. Haig and Westoby present correlations, but causes remain uncertain.

It seems likely that the next significant wave in plant reproductive ecology will emerge from quantitative genetics. The questions addressed in this paper are about the evolution of traits that are often quantitative characters.[87,88,104] We wish to know now, for example, how heritable are differences in traits? Is there additive genetic variance for sex allocation, sex lability, seed size, etc.? An understanding of the genetic bases of these traits will continue the synthesis between animal sociobiology and plant ecology. It is also important to continue to investigate the relationship between the contrasting properties of the sexes, and the forces of sexual selection. It would be rewarding to pursue the characterization of physiological and metabolic differences between males and females, and to determine their roles in male and female function.

A major challenge is to discriminate between male competition and female choice. The differential success of pollen grains from different fathers is a consequence of a series of interactions between the female sporophyte and the male gametophyte and, later, the new zygote. Interactions will also occur between the female gametophyte and both the pollen and the zygote, as well as between sibs and half-sibs. The fact that all these interactions occur within the tissues of the maternal sporophyte makes it particularly difficult, and perhaps impossible, to separate effects of male competition and female choice. The future development of sociobotany lies in evaluating the validity of sociobotanic interpretations through carefully designed experiments.

ACKNOWLEDGMENTS

We are grateful to E. L. Charnov, J. Laylander, R. Nakamura, K. Searcy, and M. Willson for helpful comments on an earlier version of this paper.

REFERENCES

1. Ågren, J., Elmqvist, T., and Tunlid, A., Pollination by deceit, floral sex ratios and seed set in dioecious *Rubus chamaemorus* L., *Oecologia* **70,** 332–338 (1986).
2. Baker, H. G., "Mistake" pollination as a reproductive system with special reference to the Caricaceae, in *Tropical Trees: Variation, Breeding and Conservation* (J. Burley and B. T. Styles, eds.), pp. 161–169. Academic Press, London, 1976.
3. Barrett, S. C. H., The evolution, maintenance and loss of self-incompatibility systems, in *Plant Reproductive Ecology: Patterns and Strategies* (J. Lovett Doust and L. Lovett Doust, eds.), Chapter 5. Oxford Univ. Press, New York, 1988.
4. Barrett, S. C. H., and Thomson, J. D., Spatial pattern, floral sex ratios, and fecundity in dioecious *Aralia nudicaulis* (Araliaceae), *Can. J. Bot.* **60,** 1662–1670 (1982).
5. Bateman, A. J., Intra-sexual selection in *Drosophila, Heredity* **2,** 349–368 (1948).
6. Bawa, K. S., The reproductive biology of *Cupania guatemalensis* Radlk. (Sapindaceae), *Evolution* **31,** 52–63 (1977).
7. Bawa, K. S., Evolution of dioecy in flowering plants, *Annu. Rev. Ecol. Syst.* **11,** 15–39 (1980).
8. Bawa, K. S., Mimicry of male by female flowers and intrasexual competition for pollinators in *Jacaratia dolichaula* (D. Smith) Woodson (Caricaceae), *Evolution* **34,** 467–474 (1980).
9. Bawa, K. S., and Beach, J. H., Evolution of sexual systems in flowering plants, *Ann. Missouri Bot. Gard.* **68,** 254–274 (1981).
10. Bawa, K. S., and Opler, P. A., Dioecism in tropical forest trees, *Evolution* **29,** 167–179 (1975).
11. Bell, G., On the function of flowers, *Proc. R. Soc. London B* **224,** 223–265 (1985).
12. Bertin, R. I., Nonrandom fruit production in *Campsis radicans:* Between year consistency and effects of prior pollination, *Am. Nat.* **126,** 750–759 (1985).
13. Bertin, R. I., Consequences of mixed pollinations in *Campsis radicans,* Oecologia **70,** 1–5 (1986).
14. Bertin, R. I., Paternity in plants, in *Plant Reproductive Ecology: Patterns and Strategies* (J. Lovett Doust and L. Lovett Doust, eds.), Chapter 2. Oxford Univ. Press, New York, 1988.
15. Bierzychudek, P., Assessing "optimal" life histories in a fluctuating environment: The evolution of sex-changing by Jack-in-the-pulpit, *Am. Nat.* **123,** 829–840 (1984).
16. Bierzychudek, P., Pollinator limitation of plant reproductive effort, *Am. Nat.* **117,** 838–840 (1981).
17. Blaffer Hrdy, S., and Williams, G. C., Behavioral biology and the double standard, in *Social Behavior of Female Vertebrates* (S. K. Wasser, ed.), pp. 3–17. Academic Press, New York, 1983.
18. Bookman, S. S., Evidence for selective fruit production in *Asclepias, Evolution* **38,** 72–86 (1984).
19. Bradshaw, A. D., Evolutionary significance of phenotypic plasticity in plants, *Adv. Genet.* **13,** 115–155 (1965).
20. Bradshaw, A. D., Some of the evolutionary consequences of being a plant, *Evol. Biol.* **5,** 25–47 (1972).
21. Bullock, S. H., and Bawa, K. S., Sexual dimorphism and annual flowering pattern in *Jacaratia dolichaula* (D Smith) Woodson (Caricaceae) in a Costa Rican rain forest, *Ecology* **62,** 1494–1504 (1981).
22. Casper, B. B., On the evolution of embryo abortion in the herbaceous perennial *Cryptantha flava, Evolution* **38,** 1332–1349 (1984).
23. Casper, B. B., and Charnov, E. L., Sex allocation in heterostylous plants, *J. Theor. Biol.* **96,** 143–149 (1982).
24. Casper, B. B., and Wiens, D., Fixed rates of random ovule abortion in *Cryptantha flava* (Boraginaceae) and its possible relation to seed dispersal, *Ecology* **62,** 866–869 (1981).
25. Cavigelli, M., Poulos, M., Lacey, E. P., and Mellon, G., Sexual dimorphism in a temperate dioecious tree, *Ilex montana* (Aquifoliaceae), *Am. Midl. Nat.* **115,** 397–406 (1986).
26. Charlesworth, D., and Charlesworth, B., Allocation of resources to male and female functions in hermaphrodites, *Biol. J. Linn. Soc.* **15,** 57–74 (1981).
27. Charnov, E. L., Simultaneous hermaphroditism and sexual selection, *Proc. Natl. Acad. Sci. U.S.A.* **76,** 2480–2484 (1979).
28. Charnov, E. L., Parent-offspring conflict over reproductive effort, *Am. Nat.* **119,** 736–737 (1982).
29. Charnov, E. L., *The Theory of Sex Allocation.* Princeton University Press, Princeton, New Jersey, 1982.
30. Charnov, E. L., On sex allocation and selfing in higher plants, *Evol. Ecol.* **1,** 30–36 (1987).
31. Charnov, E. L., Some comments on "Sex allocation and selfing in higher plants,' *Evol. Ecol.* **1,** in press.
32. Charnov, E. L., and Bull, J. J., When is sex environmentally determined? *Nature (London)* **266,** 828–830 (1977).

33. Charnov, E. L., and Bull, J. J., Sex allocation, pollinator attraction and fruit dispersal in cosexual plants, *J. Theor. Biol.* **118,** 321–325 (1986).
34. Charnov, E. L., Maynard Smith, J., and Bull, J. J., Why be an hermaphrodite? *Nature (London)* **263,** 125–126 (1976).
35. Clutton-Brock, T. H., ed., *Reproductive Success,* Univ. of Chicago Press, Chicago, in press.
36. Clutton-Brock, T. H., Guinness, F. E., and Albon, S. D., *Red Deer: Behavior and Ecology of Two Sexes.* Univ. of Chicago Press, Chicago, 1982.
37. Conn, J. S., and Blum, U., Differentiation between the sexes of *Rumex hastatulus* in net energy allocation, flowering and height, *Bull. Torrey Bot. Club* **108,** 446–455, 1981.
38. Cruden, R. W., Pollen-ovule ratios: A conservative indicator of breeding systems in flowering plants, *Evolution,* **31,** 32–46 (1977).
39. Cruden, R. W., and Jensen, K. G., Viscin threads, pollination efficiency and low pollen-ovule ratios, *Am. J. Bot.* **66,** 875–879 (1979).
40. Cruden, R. W., and Lyon, D. L., Patterns of biomass allocation to male and female functions in plants with different mating systems, *Oecologia* **66,** 299–306 (1985).
41. Cruden, R. W., and Miller-Ward, S., Pollen-ovule ratio, pollen size and the ratio of stigmatic area to the pollen bearing area of the pollinator: An hypothesis, *Evolution* **35,** 964–974 (1981).
42. Darwin, C., *The Different Forms of Flowers on Plants of the Same Species.* Appleton, New York, 1877.
43. Devlin, B., The relationship between seed and pollen production from *Lobelia cardinalis* flowers, *Evolution,* in press.
44. Devlin, B., and Stephenson, A. G., Sexual variations among plants of a perfect-flowered species, *Am. Nat.,* in press.
45. Ellstrand, N. C., Multiple paternity within the fruits of the wild radish, *Raphanus sativus, Am. Nat.* **123,** 819–828 (1984).
46. Ellstrand, N. C., and Marshall, D. L., Patterns of multiple paternity in populations of *Raphanus sativus, Evolution* **40,** 837–842 (1986).
47. Fisher, R. A., *The Genetical Theory of Natural Selection.* Oxford Univ. Press, Oxford, 1930.
48. Fox, J. F., Incidence of dioecy in relation to growth form, pollination and dispersal, *Oecologia* **67,** 244–249 (1985).
49. Freeman, D. C., Harper, K. T., and Charnov, E. L., Sex change in plants: Old and new observations and new hypotheses, *Oecologia* **47,** 222–232 (1980).
50. Galen, C., Shykoff, J. A., and Plowright, R. C., Consequences of stigma receptivity schedules for sexual selection in flowering plants, *Am. Nat.* **127,** 462–476 (1986).
51. Ghiselin, M. T., *The Economy of Nature and the Evolution of Sex.* Univ. of California Press, Berkeley, 1974.
52. Gilbert, L. E., Ecological consequences of a coevolved mutualism between butterflies and plants, in *Coevolution of Animals and Plants,* (L. E. Gilbert, and P. H. Raven, eds.), pp. 210–240. Univ. of Texas Press, Austin, 1975.
53. Givnish, T. J., Ecological constraints on the evolution of breeding systems in seed plants: Dioecy and dispersal in gymnosperms, *Evolution* **34,** 959–972 (1980).
54. Goldman, D. A., and Willson, M. F., Sex allocation in functionally hermaphroditic plants: A review and critique, *Bot. Rev.* **52,** 157–194 (1986).
55. Gross, K. L., and Soule, J. D., Differences in biomass allocation to reproductive and vegetative structures of male and female plants of a dioecious, perennial herb, *Silene alba* (Miller) Krause, *Am. J. Bot.* **68,** 801–807 (1981).
56. Haig, D., and Westoby, M., Inclusive fitness, seed resources and maternal care, in *Plant Reproductive Ecology: Patterns and Strategies* (J. Lovett Doust and L. Lovett Doust, eds.), Chapter 3. Oxford Univ. Press, New York, 1988.
57. Holtsford, T. P., Nonfruiting hermaphroditic flowers of *Calochortus leichtlinii* (Liliaceae): Potential reproductive functions, *Am. J. Bot.* **72,** 1687–1694 (1985).
58. Kay, Q. O. N., Lack, A. J., Bamber, F. C., and Davies, C. R., Differences between sexes in floral morphology, nectar production and insect visits in a dioecious species, *Silene dioica, New Phytol.* **98,** 515–529 (1984).
59. Kress, W. J., Sibling competition and evolution of pollen unit, ovule number, and pollen vector in Angiosperms, *Syst. Bot.* **6,** 101–112 (1981).
60. Lawrence, M. J., Marshall, D. F., Curtis, V. E., and Fearon, C. H., Gametophytic self-incompatibility re-examined: A reply, *Heredity* **54,** 131–138 (1985).

61. Lee, T. D., Patterns of fruit maturation: a gametophyte competition hypothesis, *Am. Nat.* **123,** 427–432 (1984).
62. Lee, T. D., Patterns of fruit and seed production, in *Plant Reproductive Ecology: Patterns and Strategies* (J. Lovett Doust and L. Lovett Doust, eds.), Chapter 9. Oxford Univ. Press, New York, 1988.
63. Lee, T. D., and Bazzaz, F. A., Maternal regulation of fecundity: Nonrandom ovule abortion in *Cassia fasciculata* Michx., *Oecologia* **68,** 459–465 (1986).
64. Linhart, Y. B., and Mitton, J. B., Relationships among reproduction, growth rates, and protein heterozygosity in ponderosa pine, *Am. J. Bot.* **72,** 181–184 (1985).
65. Lloyd, D. G., Sexual strategies in plants. 3. A quantitative method for describing the gender of plants, *N. Z. J. Bot.,* **18,** 103–108 (1980).
66. Lloyd, D. G., The distributions of gender in four angiosperm species illustrating two evolutionary pathways to dioecy, *Evolution,* **34,** 123–134 (1980).
67. Lloyd, D. G., Selection of combined versus separate sexes in seed plants, *Am. Nat.* **120,** 571–585 (1982).
68. Lloyd, D. G., Gender allocations in outcrossing cosexual plants, in *Perspectives on Plant Population Ecology* (R. Dirzo and J. Sarukhán, eds.), pp 277–303. Sinauer Assoc., Sunderland, Massachusetts, 1984.
69. Lloyd, D. G., Benefits and costs of biparental and uniparental reproduction in plants, in *The Evolution of Sex* (R. E. Michod and B. Levin, eds.), pp. 233–252, Sinauer Assoc., Sunderland, Massachusetts, 1987.
70. Lloyd, D. G., and Bawa, K. S. Modification of the gender of seed plants in varying conditions, *Evol. Biol.* **17,** 255–338 (1984).
71. Lloyd, D. G., and Webb, C. J., Secondary sex characters in seed plants, *Bot. Rev.* **43,** 177–216 (1977).
72. Lloyd, D. G., and Webb, C. J., The avoidance of interference between the presentation of pollen and stigmas in angiosperms. I. Dichogamy, *N. Z. J. Bot.* **24,** 135–162 (1986).
73. Lloyd, D. G., and Yates, J. M. A., Intrasexual selection and the segregation of pollen and stigmas in hermaphrodite plants, exemplified by *Wahlenbergia albomarginata* (Campanulaceae), *Evolution* **36,** 903–913 (1982).
74. Lovett Doust, J., and Cavers, P. B., Sex and gender dynamics in Jack-in-the-pulpit, *Arisaema triphyllum* (Araceae), *Ecology* **63,** 797–808 (1982).
75. Lovett Doust, J., and Cavers, P. B., Biomass allocation in hermaphroditic flowers, *Can. J. Bot.* **60,** 2530–2534 (1982).
76. Lovett Doust, J., and Harper, J. L., The resource costs of gender and maternal support in an andromonoecious umbellifer *Smyrnium olusatrum* L., *New Phytol.* **85,** 251–264 (1980).
77. Lovett Doust, J., and Lovett Doust, L., Modules of production and reproduction in a dioecious clonal shrub, *Rhus typhina, Ecology,* in press.
78. Lovett Doust, J., and Lovett Doust, L., Parental strategy: Gender and maternity in higher plants, *BioScience* **33,** 180–186 (1983).
79. Lovett Doust, J., O'Brien, G. A., and Lovett Doust, L., Effect of density on secondary sex characteristics and sex ratio in *Silene alba* (Caryophyllaceae), *Am. J. Bot.* **74,** 40–46 (1987).
80. Lovett Doust, L., Intraclonal variation and competition in *Ranunculus repens, New Phytol* **89,** 495–501, 1981.
81. Lovett Doust, L., Lovett Doust, J., and Turi, K., Fecundity and size relationships in Jack-in-the-pulpit, *Arisaema triphyllum* (Araceae), *Am. J. Bot.* **73,** 489–494 (1985).
82. Marshall, D. L., and Ellstrand, N. C., Proximal causes of multiple paternity in the wild radish, *Raphanus sativus, Am. Nat.* **126,** 596–603 (1986).
83. Marshall, D. L., and Ellstrand, N. C., Sexual selection in *Raphanus sativus:* Experimental data on nonrandom fertilization, maternal choice, and consequences of multiple paternity, *Am. Nat.* **127,** 446–461 (1986).
84. Maynard Smith, J., *The Evolution of Sex.* Cambridge Univ. Press, Cambridge, 1978.
85. Meagher, T. R., Analysis of paternity within a natural population of *Chamaelirium luteum.* I. Identification of most-likely male parents, *Am. Nat.* **128,** 199–215 (1986).
86. Meagher, T., Sex determination in plants, in *Plant Reproductive Ecology: Patterns and Strategies* (J. Lovett Doust and L. Lovett Doust, eds.), Chapter 6. Oxford Univ. Press, New York, 1988.
87. Mitchell-Olds, T., Quantitative genetics of survival and growth in *Impatiens capensis, Evolution* **40,** 107–116 (1986).
88. Mitchell-Olds, T., and Rutledge, J. J., Quantitative genetics in natural plant populations: A review of the theory, *Am. Nat.* **127,** 379–402 (1986).

89. Mulcahy, D., The rise of the angiosperms: A genecological factor, *Science* **206,** 20–23 (1979).
90. Nakamura, R. R., Plant kin selection, *Evol. Theor.* **5,** 113–117 (1980).
91. Nakamura, R. R., Reproductive capacity and kinship in *Phaseolus vulgaris.* Ph.D. thesis, Yale University, New Haven, 1983.
92. Nakamura, R. R., Maternal investment and fruit abortion in *Phaseolus vulgaris, Am. J. Bot.* **73,** 1049–1057 (1986).
93. Nakamura, R. R., The ecology of bean embryos, submitted to *Oecologia.*
94. Policansky, D., Sex choice and the size advantage model in Jack-in-the-pulpit *(Arisaema triphyllum), Proc. Natl. Acad. Sci. U.S.A.* **78,** 1306–1308 (1981).
95. Primack, R. B., Longevity of individual flowers, *Annu. Rev. Ecol. Syst.* **16,** 15–37 (1985).
96. Putwain, P. D., and Harper, J. L., Studies in the dynamics of plant populations. V. Mechanisms governing the sex ratio in *Rumex acetosa* and *R. acetosella, J. Ecol.* **60,** 113–129 (1972).
97. Queller, D. C., Kin selection and conflict in seed maturation, *J. Theor. Biol.* **100,** 153–172, 1983.
98. Queller, D. C., Pollen-ovule ratios and hermaphrodite sexual allocation strategies, *Evolution* **38,** 1148–1151 (1984).
99. Queller, D. C., Models of kin selection on seed provisioning, *Heredity* **53,** 151–165 (1984).
100. Romero, G. A., and Nelson, C. E., Sexual dimorphism in *Catasetum* orchids: Forcible pollen emplacement and male flower competition, *Science* **232,** 1538–1540 (1986).
101. Ross, M. D., Five evolutionary pathways to subdioecy, *Am. Nat.* **119,** 297–318 (1982).
102. Ross, M. D., and Gregorius, H. R., Outcrossing and sex function in hermaphrodites: A resource allocation model, *Am. Nat.* **121,** 204–222 (1983).
103. Ryan, M. J., *The Túngara Frog.* Univ. Of Chicago Press, Chicago, 1985.
104. Samson, D. A., and Werk, K. S., Size-dependent effects in the analysis of reproductive effort in plants, *Am. Nat.* **127,** 667–680 (1986).
105. Schlessman, M., Gender diphasy ("sex choice"), in *Plant Reproductive Ecology: Patterns and Strategies* (J. Lovett Doust and L. Lovett Doust, eds.), Chapter 7. Oxford Univ. Press, New York, 1988.
106. Schoen, D., Male reproductive effort and breeding system in an hermaphrodite plant, *Oecologia* **53,** 255–257 (1982).
107. Searcy, K. B., and Mulcahy, D. L., Pollen tube competition and selection for metal tolerance in *Silene dioica* (Caryophyllaceae) and *Mimulus guttatus* (Scrophulariaceae), *Am. J. Bot.* **72,** 1695–1699 (1985).
108. Seavey, S. R., and Bawa, K. S., Late-acting self-incompatibility in Angiosperms, *Bot. Rev.* **52,** 195–214 (1986).
109. Snow, A. A., Pollination dynamics in *Epilobium canum* (Onagraceae): Consequences for gametophytic selection, *Am. J. Bot.* **73,** 139–151 (1986).
110. Stanton, M. L., Snow, A. A., and Handel, S. N., Floral evolution: Attractiveness to pollinators increases male fitness, *Science* **232,** 1625–1627 (1986).
111. Stephenson, A. G., and Bertin, R. I., Male competition, female choice, and sexual selection in plants, in *Pollination Biology* (L. Real, ed.), pp. 109–149. Academic Press, New York. 1983.
112. Stephenson, A. G., and Winsor, J. A., *Lotus corniculatus* regulates offspring quality through selective fruit abortion, *Evolution* **40,** 453–458 (1986).
113. Sutherland, S., Floral sex ratios, fruit-set, and resource allocation in plants, *Ecology,* **67,** 991–1001 (1986).
114. Sutherland, S., Patterns of fruit-set: What controls fruit-flower ratios in plants? *Evolution* **40,** 117–128 (1986).
115. Thomson, J. D., and Barrett, S. C. H., Temporal variation of gender in *Aralia hispida* Vent. (Araliaceae), *Evolution* **35,** 1094–1107 (1981).
116. Trivers, R. L., and Willard, D. E., Natural selection of parental ability to vary the sex ratio of offspring, *Science,* **179,** 90–91 (1973).
117. Vuilleumier, B., The origin and development of heterostyly in the Angiosperms, *Evolution,* **21,** 210–226 (1967).
118. Wade, M. J., and Arnold, S. J., The intensity of sexual selection in relation to male sexual behaviour, female choice, and sperm precedence. *Anim. Behav.* **28,** 446–461 (1980).
119. Wallace, C. S., and Rundel, P. W., Sexual dimorphism and resource allocation in male and female shrubs of *Simmondsia chinensis, Oecologia* **44,** 34–39 (1979).
120. Watkinson, A. R., and White, J., Some life-history consequences of modular construction in plants, *Phil. Trans. R. Soc. London B.* **313,** 31–51 (1985).

121. Webb, C. J., and Lloyd, D. G., The avoidance of interference between the presentation of pollen and stigmas in angiosperms. II. Herkogamy, *N.Z. J. Bot.* **24,** 163–178 (1986).
122. Westoby, M., and Rice, B., Evolution of the seed plants and inclusive fitness of plant tissues, *Evolution* **36,** 713–724 (1982).
123. Wiens, D., Ovule survivorship, brood size, life history, breeding systems, and reproductive success in plants, *Oecologia* **64,** 47–53 (1984).
124. Willson, M. F., Sexual selection in plants, *Am. Nat.* **113,** 777–790 (1979).
125. Willson, M. F., and Burley, N., Mate Choice in Plants: Tactics, Mechanisms, and Consequences. Princeton Univ. Press, Princeton, New Jersey, 1983.
126. Willson, M. F., and Rathcke, B. J., Adaptive design of the floral display in *Asclepias syriaca* L., *Am. Midl. Nat.* **92,** 47–57 (1974).
127. Woolfenden, G. E., and Fitzpatrick, J. W., The Florida Scrub Jay: Demography of a Cooperative-Breeding Bird. Princeton Univ. Press, Princeton, New Jersey, 1984.
128. Wyatt, R., and Hellwig, R. L., Factors determining fruit set in heterostylous bluets, *Houstonia caerulea* (Rubiaceae), *Syst. Bot.* **4,** 103–114 (1979).
129. Yampolsky, C., and Yampolsky, H., Distribution of sex forms in the phanerogamic flora, *Bibl. Genet.* **3,** 1–62 (1922).
130. Zimmerman, M., Nectar production, flowering phenology and strategies for pollination, in *Plant Reproductive Ecology: Patterns and Strategies* (J. Lovett Doust and L. Lovett Doust, eds.), Chapter 8. Oxford Univ. Press, New York, 1988.

2

Paternity in Plants

ROBERT I. BERTIN

Evolutionary studies of plant reproduction prior to the 1970s largely or completely ignored the paternal contribution to fitness. Within the last few years this topic has received increased theoretical interest and some empirical study. This attention reflects the realization that, on average, an hermaphroditic plant obtains half its fitness via its male function. A knowledge of male performance has become essential for evaluating the burgeoning body of theory dealing with plant reproduction, particularly in the areas of sexual selection, resource allocation, and the evolution of breeding systems. The paucity of data on paternal performance is the major stumbling block to the testing of hypotheses and revision of theories in this discipline.[224] The importance of data on male performance is emphasized by the fact that selection on male performance is considered to be the most important force in the evolution of floral display in some species,[17,36,158,201,227] and even in the evolutionary preeminence of angiosperms.[135]

This chapter cannot resolve any issues involving the role of paternity in the evolution of plant reproductive characteristics; the lack of adequate data makes an attempt at such a resolution premature. My goals are simply to review and synthesize existing information, and to point out areas in most need of future study.

Information on paternity is fragmentary, scattered in the literature of several disciplines (e.g., agriculture, ecology, forestry, genetics), and is rarely gathered from studies designed to answer evolutionary questions. Our interpretations must bear these limitations in mind. The lack of good information on paternity results especially from the difficulty of determining the male parentage of seeds, particularly in natural populations. A second reason is the difficulty in attributing specific events between pollination and fruit maturation exclusively to paternal (as opposed to maternal) influences.

I will begin with a brief consideration of maternal influences on paternal performance. The bulk of this chapter will review (1) the circumstances favoring particular patterns of allocation to male function, and (2) the evidence that particular patterns of allocation or male traits do in fact affect male fitness. Because of the current state of the field the former is largely theoretical and the latter more empirical. I will then review the importance of male function in the evolution of floral display, discuss the assessment of male performance, and make some suggestions for future study.

THE MATERNAL PERSPECTIVE

Paternal traits are evolutionarily relevant only to the extent that they affect the number of offspring surviving to reproductive age. Such traits influence the number of surviving offspring only after they have acted on and interacted with the maternal sporophyte and gametophyte in the events between pollination and seed dispersal. A complete understanding of the potential importance of particular male traits therefore requires an accurate knowledge of the functioning of the pistil. Unfortunately, our knowledge of relevant gynoecial capabilities is poor, although several thought-provoking discussions of this topic are available (see Haig and Westoby,[64] this volume; and Ref. 225).

Several maternal characteristics are potentially relevant to male performance: genetic makeup, physiological status, and maternal environment. Effects of the genotype of the maternal sporophyte are well illustrated by incompatibility systems, wherein the performance of a given pollen grain is determined by its interaction with the genotype of the stigma or style (see Barrett,[12] this volume; and Ref. 143). A probable effect of the physiological status of an inflorescence on pollen performance is illustrated in *Campsis radicans.*[22] The likelihood of success of a hand-pollination using pollen from a particular donor was influenced by the number of prior pollinations (and presumably the resource levels) in the recipient.[22] The magnitude of the effect differed for different pollen donors. Environmental factors markedly influence fruit production,[95,196] which in turn affects the male success of pollen donors.

The physiological mechanisms by which the above maternal characteristics exert their effects are poorly known. For example, disagreement still exists as to whether gametophytic self-incompatibility systems reflect the operation of complementary or oppositional factors.[92,137] Postulated mechanisms include recognition of identical S-proteins (coded for by self-incompatibility alleles) in pollen and pistil.[137] These in turn could affect enzyme activity, hormone levels, and availability of substances such as calcium and boron in the pistil. Many biochemical differences have been demonstrated between self- and cross-pollinated pistils, or between pollinated and unpollinated pistils. These differences include peroxidase activity,[27] starch and lipid synthesis or mobilization,[74] carbohydrate availability,[46] and hormone levels.[119] It is well known that pollen tube growth is affected by the presence of stylar or stigmatic extracts or exudates,[5,55,91,125] and by inorganic substances such as calcium and boron.[209] The presence of gynoecial exudates can vary with the age and condition of the stigma.[125] Numerous mechanisms exist, therefore, whereby a gynoecium can affect the performance of pollen it receives, even though the precise sequence of events is obscure.

Paternal characteristics to which maternal tissues might respond include genotype (gametophytic or sporophytic), physiological conditions under which the pollen grain was produced, and environmental conditions under which the grain was produced, transported, or deposited (see sections titled Allocation to Male Function and Pollen Attributes below). Maternal responses can occur at two levels: the level of the single pollen grain, ovule, or seed, and the level of the entire pistil or fruit.[64,95] The former permits discrimination among individual grains in a pollen load, or among the seeds containing genomes from different pollen grains.[95] It is well illustrated by the differential success of different pollens in mixtures.[10,20,42] At the level of the entire gynoecium, either a fruit will be produced, or it will be aborted. Regulation at this level will reflect in part the characteristics of the entire stigmatic pollen load. Such regulation is illustrated by the different proportions of fruits matured as a function of pollen load size and pollen source (references in Lee,[95] this volume).

From the above examples it is evident that male performance can never be considered in isolation, but rather that the genotype, physiological condition, and environment of maternal tissues play an important role in determining the success of particular pollens. Much less clear are the physiological mechanisms determining these patterns and the extent to which we can tease apart a female effect, a male effect, and an interaction effect.

EVOLUTION OF PATERNAL STRATEGY

The evolution of paternal characteristics is likely to be influenced both by environmental circumstances (those external to the population) and life history attributes. The former, of course, are the selection pressures; the latter might be considered historical constraints, and serve as a backdrop that is likely to influence the relative merits of particular traits in a given environment. For example, selection pressures on paternal traits in two plant species differing greatly in seed or pollen dispersal ability might differ greatly, even though the species shared the same habitat.

To date, theoretical developments in three areas are particularly relevant to understanding the paternal strategy: resource allocation, pollen/ovule ratio (P/O), and sexual selection, Below, I provide a brief general introduction to each of these topics and discuss some ways in which each may be relevant.

A complete review of resource allocation theory, even that part of it relating to male function, is beyond the scope of this chapter. Recent reviews by Charnov,[35] Lloyd,[112] and Goldman and Willson[62] provide a good introduction. Many arguments pertaining to male allocation are presented in graphical form. Figure 2.1 (the specifics of which will be discussed later) is such an example, with male and female fitness measured on the y-axis, and allocation to male function or a similar variable on the x-axis. Fitness gain curves describe the effect of the x variable on male and female fitness. Several curves are usually presented, corresponding to different constraints. The optimum allocation of resources (x-axis) will be that for which any small change in allocation will cause the fitness loss via one gender to exceed the fitness gain via the other, i.e., any change would cause a net loss of fitness. Thus the relative slopes of the fitness gain curves and points at which they level off are critical in determining evolutionary optima.

Empirical studies of resource allocation face two major difficulties. First, it is generally not known what resource should be measured. Units of energy are frequently used, although specific mineral nutrients or even water may be the relevant metric in some populations.[62,118,130] A second problem is that predictions of expected patterns of resource allocation could be sensitive to small changes in the shapes and heights of fitness curves, yet the technical difficulties in obtaining a single fitness curve, particularly for the male function, are immense. Consequently, the curves used may reflect only certain fitness components (e.g., number of seeds sired). Such curves may or may not closely resemble those based on a complete measure of fitness, and conclusions drawn from them are equally uncertain.

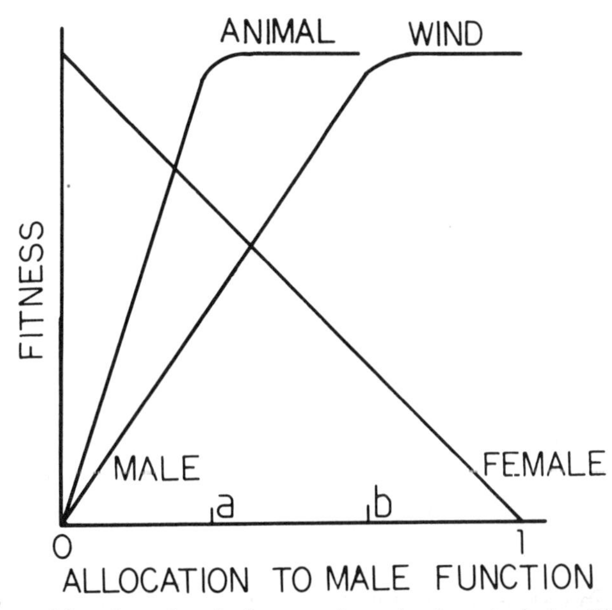

Fig. 2.1. Graphs of female and male fitnesses for animal- and wind-pollinated plants as a function of allocation to male function. These graphs predict a greater allocation to male function in a wind-pollinated species (b) than in an animal-pollinated species (a). This relationship would hold if animal pollinators behave in such a way that they place an upper limit on allocation to male function, as by eating or removing pollen, consuming nonrenewable rewards, etc.[112]

The P/O of an individual or a population bears an intuitive relationship to sex allocation in that both involve quantitative comparison of male and female attributes. Several workers have cautioned that P/Os need not be highly correlated with sex allocation because pollen and seed size and the nature of other sex-specific reproductive structures are not taken into account.[35,41,112] Nevertheless, as seen below, some similar conclusions emerge from the two approaches.

Sexual selection theory is relevant to an examination of paternal strategy in that the process of sexual selection is considered by some to be a major force shaping a plant's sexual strategy.[15,33,111,223,225] A detailed review of the subject is not possible here, but is available in Stephenson and Bertin[197] and Willson and Burley.[225] Sexual selection has been defined as the differential production or siring of offspring by individuals of one sex as a result of mate selection.[24] Of particular importance to the male strategy is intrasexual selection (among males), one of the means by which mating becomes nonrandom. In plants this is manifested particularly in competition among pollen grains on or in a single pistil.[134,135,147,199] Given the above definition, sexual selection is likely to be occurring in every sexual population wherein mating is not entirely random. Of more interest than whether sexual selection exists in specific plant populations are its intensity and its consequences. The intensity of sexual selection is measured by the variance in number of offspring of members of one sex,[24] or variance in number of mates of members of one sex.[216] Such measures are difficult to obtain for the male function and in fact the variance in male success has been estimated in only one natural plant population.[131]

With regard to hermaphroditic plants, theoretical developments in resource allocation and sexual selection overlap. Ross[165] concludes, for example, that quantitative models for the evolution of dioecy based on a resource allocation approach are essentially equivalent to those based on sexual selection. This overlap occurs because sexual selection in hermaphrodites is thought to act on allocation to each sex. However, this overlap is not complete. Selection pressures unrelated to mate selection presumably also act on allocation patterns. Conversely, sexual selection may act on reproductive traits other than patterns of allocation, such as protandry versus protogyny.[115]

In the following sections, I review several variables that have been suggested to influence P/Os, patterns of resource allocation, or intensity of male competition. Each variable will be considered in terms of its effect on fitness gain curves.

Efficiency of Pollen Transport

Several workers have suggested that inefficient pollen transport (much pollen loss between anther and stigma) favors increased P/Os or increased male allocation.[38,39,41,61] Cruden and co-workers[38,39,41] use several lines of evidence to support this contention, including greater male allocation or P/Os (1) in outcrossing species than in selfing species, (2) in wind-pollinated species than in animal-pollinated species, (3) in species with loose pollen than in species with pollen held together by viscin threads, and (4) in species with a low ratio of stigmatic area to the pollen-bearing area of the pollinator as opposed to those with a high ratio. Cruden has been criticized for viewing "pollen production as if pollen existed simply to allow plants to produce fertilized seeds."[33] Charnov[33] suggests that at least the outcrossing/selfing pattern is better explained in terms of local mate competition (see below). Lloyd[112] also concludes that overall efficiency of pollen transfer has little effect on male/female allocation. Instead the allocation to each sexual function should depend on the marginal efficiency, i.e., the way that efficiency changes with changing investment. If the theoretical arguments about the unimportance of pollinator efficiency in determining allocation patterns are correct, this does not necessarily mean that the trends described above are not real. Explanations may lie in some of the factors reviewed below (e.g., limits to paternal fitness determining the difference between wind- and animal-pollinated species).

Efficiency of Male and Female Functions

Trivers and Willard[206] suggest that, in dioecious species, great efficiency at producing offspring of one sex will lead to heavier investment in that sex. This argument has been extended to hermaphroditic plants,[35] and Queller[158] states that high efficiency in one sexual function should lead to greater investment in that function. If efficiency is measured in terms of fitness increment per unit of cost this is certainly true. The more efficient sexual function would have the steeper fitness gain curve. Generally, however, efficiency is not measured directly in terms of fitness. In that case it would be very difficult to relate male and female efficiency in a meaningful way. For example, suppose male efficiency is measured by pollen removal[158] and female efficiency by seeds produced; suppose further that a given increase in investment would double seed production or increase pollen removal threefold. We might initially conclude that the resources would be allocated to (the more efficient) male function; however, the important measure for either function is the number of offspring surviving to reproductive age. Tripling pollen removal might only increase the number of surviving sired offspring by 50%, due to much of the removed pollen being deposited on the plant of origin (which may be incompatible) or on nearby plants which soon become saturated with pollen. Overall, then, increased allocation to the female function may be favored. Measuring male and female efficiency in units other than fitness can produce misleading results.

Level of Outcrossing

As the degree of outcrossing in a population increases, so does the P/O, allocation to male function and, presumably, the intensity of male competition. This result is predicted on theoretical grounds[32,33,109,112,166,223] and supported by numerous empirical studies.[38,40,60,98,110,117,130,176,178]

Charnov[34,35] explains this pattern in terms of the increased intensity of local mate competition (LMC) in inbred populations. The expectation is that in increasingly inbred populations, the availability of stigmas and hence the return in fitness for additional male investment, declines, because a plant's pollen grains compete more strongly with one another. The paternal fitness curve decelerates and a lower allocation to male function is favored. The origins of the LMC idea are in Hamilton's[65] treatment of the effects of LMC on sex ratios in dioecious species, an explanation that has recently been challenged.[144] Lloyd (personal communication) believes that the effects of selfing on gender allocation do not reflect LMC (which he refers to as a shape factor, influencing the shape of the fitness gain curve), but rather reflect limits on the maximum fitness obtainable via the male function (which he refers to as a size factor).[113] Whatever the theoretical basis for this trend, it is well supported by empirical studies.

Other Factors

Lloyd[112] suggests that local competition between offspring of the same maternal parent would favor increased allocation to male function. His reasoning is that competition will flatten the fitness gain curve for female function, making it more advantageous for plants to invest in male than in female function. This prediction might be tested by comparing species with different seed dispersal abilities, but possible confounding effects of local pollen competition would have to be taken into account.[112]

Lloyd's[112] evolutionarily stable strategy (ESS) model predicts that facilitation and interference between sexual functions should influence relative allocation to male and female functions. Allocation to the function that causes interference should decrease in proportion to the degree of interference. A function that enhances the other [e.g., pollen (male function) serving as an attractant to visitors that deposit pollen on the stigma (female function)] should receive increased allocation.

In addition to its effects on allocation, interference may also favor the separation of sexual functions in time (dichogamy) or space (herkogamy),[15,115] or the evolution of unisexuality.[16] While herkogamy and dichogamy also enhance outcrossing, Lloyd and Yates[115] contend that selection to avoid interference of male and female functions is a more frequent explanation of these traits than avoidance of selfing, because such traits often occur in plants already having mechanisms that prevent selfing, such as physiological self-incompatibility. (See Wyatt[229] for further discussion).

Lloyd's[112] ESS model also predicts that the sexual function with higher recurrent costs (those repeated more than once, but less than once for each offspring, such as a pericarp) will receive greater allocation. Male recurrent costs, such as filaments and anther walls, are usually small, so the important variable is the extent of female recurrent costs. Species with, for example, large expenditures of pericarp per seed would be expected to have decreased allocation to male function. A possible mitigating factor is the ability of female parts to photosynthesize and therefore recoup some of their costs in terms of carbon.

Upper limits on allocation to one sexual function are likely to occur if factors exist that impose an upper limit on the fitness benefits accruing through that function.[112] An upper limit to male fitness might be imposed by a pollinator that picks up a limited amount of pollen and invariably dislodges the rest, making it unavailable to other pollinators. Clearly this would prevent selection for additional pollen per flower (although it certainly would not preclude selection for additional pollen-bearing flowers). Conversely, if ability to produce fruits were limited by availability of water or some other nutrient, selection should favor increased allocation of photosynthate to the male function (see Freeman *et al.*[58] for a related argument).

Lloyd[112] suggests that factors imposing upper limits to male function are more common in animal-pollinated than in wind-pollinated species, and that this might explain the lower allocation to male function (and P/O) in the former group.[38,112] Examples of limiting factors could include a tendency for animal pollinators to eat a fraction of the pollen they pick up, or to exhaust nonrenewable rewards (e.g., nectar),[112] or a tendency to groom off excess pollen. If this is true, the male fitness gain curve for animal-pollinated species would level off before that for wind-pollinated species, causing a lower relative allocation to male function in the former (Fig. 2.1).

FACTORS INFLUENCING MALE PERFORMANCE

Male fitness is generally expressed in terms of the number of sired offspring surviving to reproductive age. Such a complete assessment of fitness is usually impractical or impossible. Currently, only incomplete measures or presumed correlates of fitness are available. Such metrics include performance of sired seedlings, germination, weight and number of sired seeds, pollen germination ability, pollen tube growth rate, and ability of pollen to effect fertilization. The degree of correlation of these measures with fitness is unknown, and may well vary among species. Several studies have shown that differences in offspring quality that are unapparent or scarcely apparent in seeds or young plants may become more pronounced with age.[52,56,149,198] Therefore studies concluding, on the basis of seed weight, seedling germination, or short-term seedling growth experiments, that certain male traits or pollination treatments have no effect on offpsring quality must be viewed with caution. Furthermore, differences between treatments sometimes appear when seedlings are stressed by competition, even if no differences occur when the seedlings are grown alone.[97,129]

While the paternal influence on offspring performance is often thought to result exclusively from nuclear DNA, this may not be entirely true.[225] Functional male cytoplasmic organelles seem to enter the zygote in at least some species.[167,168] Their significance, if any, for offspring performance is unknown.

We can evaluate the effects of paternal traits and other factors on male fitness in two ways. An indirect approach is to test predictions by examining trends in samples of species. A direct approach is to make an appropriate experimental manipulation, preferably on a field population. Suppose, for example, that we wish to evaluate the idea that early flowering during a blooming period enhances male fitness.[114,197] Using the indirect approach we might note the relative timing of male and female flowering in monoecious and dioecious species. Earlier opening of male than female flowers would be consistent with our thesis. The difficulty with such evidence is that the patterns may occur for reasons other than the one proposed. A direct approach would be to induce plants to flower at different times in a season and to compare their levels of male success.

Ideally this section of the paper would review experiments of the latter sort. Because of their technical difficulty, however, such experiments have not been performed in natural populations. My review therefore includes mostly partial, or circumstantial evidence, including evidence that the variability in important paternal traits has a genetic basis, and evidence that such variability affects measures of performance that might be related to male success. Much of this information is from studies of cultivated species, and the possibility that typical responses have been altered by artificial selection must be kept in mind.

My discussion focuses on three areas. In the first section below, I review evidence for a relationship between male success and allocation to male function, a relationship assumed by arguments in the previous section. Second, I review evidence linking particular pollen attributes (other than pollen grain size and composition, which are treated in the first section) to male reproductive success. In the third section I examine how characteristics of the stigmatic pollen load influence performance of particular grains in it. Paternal characteristics other than those discussed below undoubtedly influence male performance,[114,115,197,223] but space considerations preclude their discussion here.

Allocation to Male Function

A fundamental plant attribute is the level of resource allocation to male function. This allocation can initially be broken down into number of pollen grains, size (or more generally, cost) of pollen grains, and allocation to male accessory structures and substances.

Number of Pollen Grains

The number of pollen grains produced by a plant reflects the number of staminate flowers produced, the average number of stamens per staminate flower, and the average number of pollen grains per stamen. Selection on any of these traits could affect a plant's pollen production.

Genetic control of some of these components has been demonstrated. In loblolly pine *(Pinus taeda),* for example, about 40% of the variation in number of male strobili was attributed to genotype.[174] Strong genetic control of male output was also found in a second loblolly planting[221] and in slash pine *(Pinus elliottii).*[11,180] Environmental factors can also have a strong effect on male flowering.[1,130] In *Pinus taeda,* for example, there were strong year × clone interactions, indicating that male performance of particular trees varies greatly from year to year.[174] This emphasizes the necessity of multiyear studies.

Despite the theoretical claims in the previous section, there are few data that shed light on the relationship between pollen production and male fitness. Differences in male success of varieties or clones or *Euoenothera,*[76] *Pinus sylvestris,*[140] and *Gossypium*[195] have been attributed to differences in pollen production, but the evidence is weak. Studies of pollinia-bearing species have shown increases in numbers of pollinia removed per inflorescence as inflorescence size increases.[171,226,227,228] In *Amianthium muscaetoxicum,* a species with loose pollen, larger inflorescences exported more fluorescent dust placed on pollen than did smaller inflorescences.[162] These studies of inflorescence size and pollen export do not isolate pollen amount as the important variable influencing pollen export, since the size of the visual target and total nectar production also increase with inflorescence size. These studies also do not examine the important link between pollen export and male success.

Schoen and Stewart have recently attempted to construct a male fitness gain curve for *Picea glauca* in an experimental planting.[179] Proportion of seeds sired was positively related to the number of male cones (and hence pollen) produced by a clone, although the exact shape of the function was not clear. Exclusion of one of the donors changed the curve from one that appeared to reach an upper limit to one that increased throughout the observed range of male cone production. More attempts to generate and refine male fitness curves are sorely needed.

Accessory Structures and Substances

Allocation to male function includes structures such as petals, sepals and bracts, and substances such as nectar. In species with unisexual flowers, the costs and sizes of such structures can be examined directly for each sex.[114,197] In species with bisexual flowers these attractants and rewards can benefit both sexes, making it more difficult to separate male and female allocation. In dichogamous flowers, allocation to nectar can be separated at least partly into male and female components by measuring production during a flower's male and female phases.[45,115]

Evidence that allocation to accessory substances and structures influences male success can also be obtained by examining male performance as a function of natural or experimentally induced variability in these floral traits. In *Raphanus raphanistrum,* for example, a white-petalled morph was less attractive to pollinators than a yellow-petalled morph. The two morphs did not differ in fruit and seed production, but the white-petalled plants were much less successful as pollen donors.[194] Likewise, the presence and size of the corolla of *Impatiens capensis* had no effect on seed production but did seem to influence pollen removal.[17] Both studies suggest that allocation to attractive structures is particularly important for reasons of pollen donation. One uncertainty in these studies is that female success was measured in terms of seed number, without regard to seed quality. If seed quality differed between treatments, perhaps as a result of intense pollen tube competition accompanying receipt of more pollen,[94,138] there would be a real but cryptic female component to selection on attractive structures.

Allocation to nectar production can also affect male success. In *Asclepias quadrifolia* nectar production per flower was correlated with pollinia removal,[154] and bees visiting inflorescences of *Delphinium* with abundant nectar picked up more pollen than those visiting inflorescences with little nectar.[215] In the latter study, however, bees flew shorter distances after visiting richer inflorescneces, which might limit potential gains in male fitness. This demonstrates the importance of monitoring the success of removed pollen as well as quantifying its removal.

Pollen Grain Size and Composition

The size of a pollen grain and the nature of its contents may influence male performance through effects on pollen germination and pollen tube growth. Some evidence suggests that growth of binucleate pollen *in vivo* consists of two phases, an early phase largely dependent on pollen resources, and a later phase supported by stylar substances.[28,139,210] Thus the early phase is likely to be most dependent on pollen grain size and contents.

Strong interspecific correlations exist between pollen grain diameter and style length.[6,7] Similarly, in distylous species, thrum pollen grains, which must grow through the long pin styles, are larger than those borne in flowers of the pin morph.[59,214] In at least one species, pollen grains in cleistogamous flowers are smaller than those in chasmogamous flowers.[116] Thus pollen grain size may determine maximum pollen tube length.

In nature, however, the rates of germination and pollen tube growth could be more important determinants of success than the maximum attainable length. The rate of pollen germination might be unaffected by grain size or may even be greater in small than large grains. The latter could reflect the greater surface area to volume ratio in small grains, causing them to hydrate and become metabolically active more rapidly than large grains. In studying grain size/growth rate relationships in *Collomia grandiflora,* Lord and Eckard[116] found that the smaller grains of cleistogamous flowers had a slower pollen tube growth rate than that of the larger pollen grains of chasmogamous flowers. In cleistogamous flowers selection for male performance may well be relaxed, however, limiting the applicability of this size/growth rate comparison to species with chasmogamous flowers only.

How variable in size are pollen grains from flowers in natural populations that exhibit neither heterostyly nor cleistogamy? *Raphanus sativus* has been studied systematically in this regard.[193] Mean pollen grain size varied among individuals grown under controlled conditions, and also among flowers from a single individual. Field studies demonstrated significant variation in pollen grain size among individuals and over the growing season. In *Zea mays,* inbreeding reduces the size of pollen grains.[84] In analyzing the literature, Willson and Burley[225] and Stanton[193] concluded that variability in pollen size is widespread within and among individuals in natural populations. An unknown part of this variation is caused by environmental factors such as season, temperature, and nutrient availability.[193,225]

Data on the relationship between pollen grain size and male performance are equivocal. Pollen grains of *Zea mays* bearing the *sp* allele are smaller than normal grains.[184] When equal numbers of normal and sp grains are mixed, the latter sire less than 50% of the seeds because they have a slower rate of germination or pollen tube growth. In *Zea mays* no relationship existed between pollen grain diameter and competitive ability of pollens from different donors.[148] Barnes and Cleveland[9] found that, in *Medicago sativa,* the pollen grains of three plants producing longer tubes were on average 11% larger in volume than those of three plants producing shorter tubes, but the effect of this differential on fitness was not evaluated.

Part of the cost of producing a pollen grain is determined by its composition. Considerable interspecific variation exists in the percentages of starch, sugar, lipid, protein, and other constituents of pollen.[6,192] Intraspecific variation in the starch content of pollen also occurs, including variation among grains from a single anther.[6,19] Developmental stage as well as various environmental factors can greatly affect the composition of pollen grains.[6,192] Apparently no studies specifically address the metabolic costs of producing pollens of different compositions. Presumably lipid-rich pollen is more expensive to produce than carbohydrate-rich pollen.[6]

There is no good evidence that more-costly pollen provides benefits to a plant. Pollen rich in nitrogen or lipid may be preferred by some pollinators,[6,192] but it might also be preferred by pollen thieves. Low levels of starch in pollen grains are sometimes associated with improved performance. For example, pollen collected from *Lycopersicon peruvianum* in September contained less starch than pollen collected in June, and exhibited better germination on artificial media (80 versus 10–40%) and greater pollen tube growth.[18] Pollen of the apple cultivar Starkrimson contained less starch than pollen of cultivar Golden Delicious, and it also had a higher germination percentage.[19] In *Lycopersicon* at least, the differences in starch content were developmental, and it is not clear that we can generalize about relationships between starch content and performance of a pollen grain.

Discussion

Despite the intuitive link between allocation to male function and male fitness, such a link has not been adequately demonstrated. The best approach to this problem is undoubtedly the use of genetic markers. Experimental plots could be set up, perhaps involving potted plants so that their numbers and arrangements can be easily manipulated.[177] Experimental modifications of flower number, anther number per flower, etc. could be made on a few individuals, and the effects observed in the progeny of other individuals. Additionally, the male success of two donor lines differing in specific attributes (e.g., pollen grain size or pollen composition) could be compared with a similar setup, as in the *Raphanus* petal color study cited above.[194] Such experiments provide general information on the relationship between male characteristics and reproductive performance, and allow refinement of hypotheses. Because the theoretical predictions of the section Evolution of Paternal Strategy are sensitive to small changes in the heights and shapes of male and female fitness gain curves, collection of data that are complete and precise enough to support or refute particular arguments will be a major challenge.

Pollen Attributes

Self-incompatibility

The incompatibility phenotype of a pollen grain, imposed either by its own genetic constitution (gametophytic incompatibility) or the genotype of the plant that produced it (sporophytic incompatibility), greatly influences the ability of a pollen nucleus to effect fertilization. The most widespread view of incompatibility is that crosses are unsuccessful when one or more *S* alleles are held in common between the maternal parent and either the pollen nuclei or paternal sporophyte.[12,143] Mulcahy and Mulcahy[137] suggest that gametophytic self-incompatibility may reflect complementarity of male and female genotypes (e.g., heterosis) rather than opposition, and that this reaction can involve many genes. In either case the success of a pollen grain depends on the genotype of the stigma on which it lands. Because of this dependence, selection on pollen grains acts in a frequency-dependent manner, favoring males with uncommon compatibility genotypes, rather than consistently favoring certain paternal genotypes. Such selection would tend to equalize the frequencies of different incompatibility genotypes in the population. If many genes are involved in the incompatibility reaction, or if *S* alleles are linked to alleles not involved in the incompatibility reaction, incompatibility reactions would tend to maintain genetic diversity in populations. However, in at least some populations *S* alleles occur at different frequencies,[93] indicating that frequency-dependent mating success has not equalized their frequencies.

Pollen Longevity

A plant's male performance could be affected by the spans of time after anther dehiscence or pollen removal during which pollination can lead to successful fertilization. Interspecific variation in pollen longevity is great, ranging from hours to over 1 year.[57] Intraspecific variability also occurs, and a genetic component to this variability has been demonstrated in *Zea mays.*[151,153] Environmental conditions strongly affect pollen longevity.[69,85,108,188] General conclusions are that low temperatures and low humidity extend viability of binucleate pollen, while high humidity is better for trinucleate pollen.

The time following anther dehiscence that is required for pollen removal varies substantially among species (e.g., 2 hours in some populations of *Campsis radicans,*[21] much longer in some milkweeds and orchids, as indicated by the incomplete removal of pollinia following a flowering period of several days to 2 weeks).[2,25,120,166] It seems reasonable to expect interspecific correlations between average time until removal and pollen longevity. In most plant species the vast majority of pollen deposition seems likely to occur within minutes of pollen removal. Manipulations by pollinators of their pollen loads might extend this time somewhat, but the extant data on pollen carryover provide no indication that long-delayed pollination is common.[101,156,205] Milkweeds may be an exception.[133] Thus there is no reason to expect strong selection for great pollen longevity following removal. Interestingly, pollen of *Clarkia unguiculata* remains viable longer if removed from the plant than if left in place.[188]

Several mechanisms other than great pollen longevity could ensure availability of fresh pollen to infrequent visitors. These include continual release of pollen,[45] sequential dehiscence of small anthers, and the production of many flowers with little pollen in each.[115]

Pollen Germination

Male success can be affected both by the percentage of grains germinating and the rapidity with which germination occurs on the stigma. Pollen germination or viability is sometimes assessed by indirect methods. These include stainability, tests for enzyme activity, and fluorochromatic procedures. Of these, Heslop-Harrison *et al.*[75] concluded that fluorochromatic tests are usually reliable but that most staining procedures and tests for enzyme activity are nearly worthless. This conclusion is supported by studies of *Allium* and *Gossypium.*[13,146] Hence the many studies using pollen stainability as indicators of germination ability are suspect. Even data from *in vitro* studies of germination must be treated with caution, as the agreement with *in vivo* measures of performance may be poor.[91]

Percent pollen germination often differs among pollen from different donor individuals. Taxa demonstrating such an effect include *Costus guanaiensis,*[172] *Rubus* sp.[159] *Mussaenda* spp.,[48] and *Prunus* spp.[222] A large sporophytic genetic component is likely, although physiological and developmental factors are difficult to rule out. Pollen germination also can be highly variable among pollen grains from a single individual. In *Cichorium intybus,* for example, pollen germination ability decreased markedly in the latter part of the flowering period.[49]

In *Allium cepa* the germinability of pollen varied greatly among flowers within a head and even among anthers within a single flower.[146] A previous study revealed great variability in germination of pollen in different samples from a single anther, perhaps reflecting the tendency of inviable grains to occur in clumps.[146] Finally, the effects of environmental conditions on germinability can be very great, reflecting both conditions under which the pollen donor is grown, and conditions under which pollen germination occurs.[1,8,49,86,105,160,163,173,187] Furthermore, the effect of environment may differ among cultivars[47,121,230] and at different times of year.[72]

The time required for germination once a pollen grain reaches a stigma is potentially as important as the percentage of grains germinating in determining male success. This is true especially of species with limited subsequent maternal screening of pollen. Genetically based differences in pollen germination time occur in *Zea mays*[191] and in *Mussaenda* spp.[48] Levin[99] showed that the greater success of cross-pollen than self-pollen in *Phlox drummondii* reflects the quicker germination of the former on receptive stigmas. This study also illustrates the influence of stigma genotype on rate of pollen germination.

Pollen Tube Growth

Genetically based variability in pollen tube growth rates has been demonstrated in *Zea mays,*[83,152,169] *Medicago sativa*[9] (both cultivated), and in *Asclepias speciosa.*[26] Environmental conditions during pollen tube growth[130,236] and during pollen production[73,132] can also exert sizeable effects on pollen tube growth rate. Temperature is particularly important in this regard.[88,121,222] Many studies of pollen tube growth rate are done *in vitro,* typically in simple media containing sucrose, boric acid, and calcium nitrate. Pollen grains from different individuals or lines often perform differently in different media. The relative performance of different pollens in a particular medium is sometimes, but by no means always, correlated with their relative performance *in vivo.*[9,10,26,43,153,169] The lack of correlation between *in vivo* and *in vitro* performance in some studies[43,153] indicates that *in vitro* studies of pollen tube growth rates should be supported by *in vivo* studies whenever possible. One difficulty with the latter is that male performance may differ markedly on different stigmas,[219] making an overall assessment of male quality difficult.

Discussion

Overall, considerable variability exists in performance of male gametophytes, although relatively few species have been examined in this regard, and almost all have been cultivated species. Examination of wild plants is particularly important because continuous intense selection for high male performance in nature might reduce additive genetic variance for these traits.[197]

The few studies of noncultivated species suggest, however, that such variability has not been eliminated from natural populations.[26,172] Several interpretations of this variability are possible. One interpretation reflects the substantial influence that environmental conditions have on male performance. In environments that are variable in space and time, selection would favor different genotypes at different times or in different places. For example, different male gametophytes could be favored depending on whether the maternal sporophyte grew in the shade or in the sun, or whether the blooming season was cool or warm. This variation, combined with the genetic variability of maternal sporophytes, could allow the maintenance of considerable variability of pollen traits in natural populations.

Results reported in this section also illustrate the great difficulty in obtaining a meaningful general measure of any aspect of male performance. *In vitro* studies can always be criticized for inapplicability to the real world. *In vivo* results are likely to be strongly affected by the particular gynoecia in which male performance is measured. These deficiencies can be met partly by making *in vitro* conditions as similar as possible to those *in vivo,* and by varying them to determine the sensitivity of measured male performance to such changes. *In vivo* studies should use gynoecia from several different plants, and interpretations of male performance should accommodate any variations observed.

Nature of the Pollen Load

A pollen grain's innate abilities do not alone determine its fate after arrival on a appropriate stigma. Also important are the various environmental factors mentioned in the previous section, as well as the size and composition of the remainder of the stigma's pollen load. These variables are largely beyond the control of an individual pollen grain or pollen-producing sporophyte, representing instead environmental conditions selecting for particular paternal traits. To the extent that these factors vary unpredictably they impose limits on the fine-tuning of selection acting on paternal traits.

A paternal sporophyte could, however, exert some influence on the environment of its pollen. By varying the nature of pollen dispersal units (PDUs) it influences the genetic relatedness of pollen grains on the stigma, with large PDUs ensuring the presence of many sibling grains. Little is known of the effects of genetic relatedness of pollen grains on paternal performance.[87,223] Another potentially important sporophytic trait is the timing of pollen release from a male sporophyte, either seasonally or diurnally. This would influence the timing of arrival of its pollen on stigmas relative to the arrival of other pollen, and therefore also the number of competing pollen grains.

Presence of Incompatible Pollen

Stigmas of most plant species with self-incompatibility systems receive a certain number of incompatible pollen grains.[16,189] This might have a negative effect on the success of compatible pollen as a result of blocked access to the stigmatic surface, blockage of the style by incompatible pollen tubes, preemption of certain stigmatic or stylar substances required by pollen or, if self-incompatibility is expressed after fertilization, preemption of ovules.[23,90,182] Some evidence exists for stigma clogging by self- or other incompatible pollen,[145,189] although this evidence is not great.[182] In *Campsis radicans* fruit set is much lower when cross-pollinations are preceded or accompanied by self-pollination of the same stigma than when they are not (unpublished data). The possible negative effects of self-pollination on the performance of compatible pollen thus merit further study.

Beneficial effects of deposition of self-incompatible pollen have also been recorded. In *Malus* and *Pyrus* the likelihood of fruit set is increased if a compatible cross-pollination is preceded by a self-pollination.[213] The cause of this pattern is unknown.

Presence of Compatible Pollen

The stigmas of most plant species probably receive pollen from more than one donor plant. Using electrophoretic markers, Ellstrand[53] demonstrated that at least 85% of 59 fruits from several individuals of *Raphanus sativus* showed multiple paternity, and the minimum number of paternal donors ranged from 1 to 4 for different fruits. In a population of *Chamaelirium luteum* the average female had pollen from 5.5 males represented in her seed crop.[131] More data on male parentage in natural populations are clearly needed.

The success of a particular pollen grain can be influenced by the presence of other compatible pollen through the latter's influence on fruit production and on whether the particular grain is represented in the seeds of that fruit.

Results to date indicate no effect of diversity of pollen donors on fruit set. Marshall and Ellstrand[124] found no significant change in fruit set as the number of pollen donors represented in a pollen load increased form one to three. No consistent effect of donor diversity on fruit production was found in *Cassia fasciculata,*[96] *Vaccinium* sp.,[208] or *Campsis radicans.*[23] In *Encyclia cordigera,* fruit production averaged 97% when each flower in an inflorescence received pollen from a different pollen donor, and 92% when all pollen was from the same donor.[81] The difference was not significant.

Numerous studies do show, however, that the competitive ability of pollen grains is influenced by genotype and differs among lines or individuals, and even among pollen grains from a single individual.[26,42,122,151,161] Therefore the success of a particular pollen grain will depend in part on the genotypes of other pollen grains deposited on the stigma. In a few studies of agricultural species the genetic basis of differences in pollen performance has been identified.[20,82,181] Differences in success of cross-pollinations in natural populations as a function of interparental distance have also been attributed to genetic factors.[70,103,155,218]

Size of Pollen Load

The number of pollen grains in a stigmatic pollen load influences the chances of success of an individual grain. Being part of a very small pollen load is disadvantageous because such loads are less likely to stimulate fruit production than larger loads.[21,95,96,127,128] Being part of a very large pollen load can also be disadvantageous, however, because the chance of an individual grain being represented in the seed crop of a particular fruit declines with larger pollen loads. This is because a single pollen grain represents a smaller part of a large than a small load. Additionally, large pollen loads sometimes yield fewer pollen tubes per pollen grain than do small loads.[4,190] The changing probability of success of an individual pollen grain with changing size of pollen load is illustrated by a study of *Oenothera fruticosa.* The ratio of pollen grains to seeds was about 1 : 1 up to one-third of maximal seed set, but approximately 400 pollen grains were necessary to achieve the maximum set of about 150 seeds.[183] This agrees with Cruden's[38] statement that maximum seed set typically requires pollen loads of 2–7 grains per ovule.

The size of a pollen load can affect the percentage of grains germinating and the rate of pollen tube growth. In several species, increasing the number of pollen grains in a constant amount of medium increases percent germination and pollen tube length.[30,89] In *Costus guanaiensis, in vitro* pollen germination averaged 41% for single grains and 84% in clumps of 64 grains.[172] This effect was at least partly responsible for patterns of fruit set: 4 grain treatments yielded 3% fruit set, 64 grain treatments yielded 68% fruit set. Similar effects were demonstrated for *Lycopersicon* pollen *in vivo,*[77] *Brassica oleracea in vitro,*[163] and for other species.[108] No such pollen population effect was found in *Phlox drummondii.*[99] The pollen population effect has been attributed to release of calcium from the pollen grains, because, in some species, the addition of cell-free pollen extracts or calcium overcomes the poor performance of small pollen loads *in vitro.*[29,89] Calcium may not, however, be the critical factor in all species.[164] Large pollen loads may enhance pollen tube growth rates[203] as well as germination.

All controlled studies of pollen amount are relevant only insofar as they reflect pollen amounts likely to be found on stigmas in nature. For many of the above species this information is unavailable, making uncertain the ecological and evolutionary implications of the studies. The size of the pollen load is also likely to influence the degree of competition among pollen grains and therefore the extent and nature of interactions among pollen grains.[94] This fact is illustrated by studies showing that increasing the pollen load can increase the quality and reduce the degree of variability of progeny. Seedlings of *Petunia hybrida*[138] and *Cucurbita pepo*[199] that were derived from pollinations involving large pollen loads performed better than those from pollinations involving smaller loads. In *Petunia* this advantage was detectable even in the F_2 generation.[135] The effect of pollen load on variability of progeny is illustrated in Ter-Avanesian's work.[202,203] Smaller pollen loads yielded more variable progeny, presumably due to relaxed competiton among male gametophytes. It is highly desirable to examine both of the above effects in natural plant populations using ranges of pollen loads typical of those deposited by the local pollinator fauna.

Timing of Pollen Receipt

Early arrival of pollen on a stigma generally increases its likelihood of success, an advantage proportional to the delay in deposition of the second type of pollen.[50,136,185,186] This reflects the time needed for pollen germination and pollen tube growth.[136] In *Raphanus sativus,* for example, simultaneous deposition of pollen from two donors resulted in 57% of fruits containing some seeds sired by each donor. If the deposition of pollen from one of the two donors was delayed by 2 hours, only 14% of fruits contained seeds sired by both donors.[123] In *Primula sinensis,* the order of pollination had a sizeable effect on the relative success of the two pollens even when they were applied within a few minutes of each other.[207] The large effect of such a small delay suggests that the sequence of pollen deposition itself is important, perhaps reflecting access to the stigmatic surface.

In two genera (*Malus* and *Pyrus*), pollen applied to a stigma in a second pollination is more successful than that applied in a first pollination.[211,212] The stimulatory effect of the first (pioneer) pollen seems to be maximal if the second pollination is made when pioneer tubes are one-third of the way down the style.[211] The cause of this pattern and its applicability to natural populations of other species are unknown.

The timing of arrival of pollen grains on a stigma clearly affects the potential for interaction (e.g., competition) among grains. Simultaneous arrival of the entire load maximizes the potential interaction. Sequential arrival of grains may eliminate completely the potential for interaction if adequate time for germination of a pollen grain and growth of its pollen tube elapses before the next grain is deposited.[136] The amount of pollen received in a stigma's lifetime thus provides an incomplete picture of the potential for pollen interactions. Studies of pollen arrival in two species, *Geranium maculatum*[136] and *Epilobium canum,*[190] reveal that pollen arrives with sufficient frequency to result in some pollen competition. The extent of competition in *Epilobium* did vary, however, among years, plants, and fruits.

PATERNITY AND FLORAL DISPLAY

Willson and Rathcke[227] first suggested that selection on the male function results in more flowers per inflorescence than required for maximum female reproductive output. This postulate resulted from their work showing that the common inflorescence sizes in *Asclepias syriaca* are far larger than those needed for maximum fruit production, that fruit production per flower declines as the number of flowers in an inflorescence increases beyond 30, and that pollinia removal per flower increases.[226,227] Bell likewise concluded that inflorescence size in *Asclepias syriaca* had a much greater effect on pollen donation than fruit production.[17] A similar pattern occurs in *Asclepias exaltata,*[158] but, in *Asclepias tuberosa,* large inflorescences have lower fruit production per flower but higher total fruit production than small inflorescences.[228] Wyatt[228] argues that "only the total relative contribution to the next generation is important," and therefore that the male success argument is unnecessary to explain inflorescence size in *Asclepias tuberosa.* The total number of flowers on a plant is clearly likely to be a major determinant of fitness, and it will be strongly influenced by available resources.[44] Nevertheless, with access to a certain amount of resources, the most efficient allocation of these resources among inflorescences to maximize fruit production would be in inflorescences smaller than those typically found in all of the above milkweed species.

Fruit production is not, however, a complete measure of maternal reproductive output, as it ignores the number and quality of seeds in the fruit. Other work with *Asclepias* has shown that the latter two variables sometimes differ significantly between plants having similar levels of fruit production.[25] Larger inflorescences might, therefore, be favored by selection on the female function because they enhance fruit quality by allowing selective abortion of fruits.[26,198] Hence, female as well as male function could benefit from large inflorescences on a per-flower basis. Additionally, the commonness of species in which a high proportion of pistillate flowers do not yield fruit suggests an advantage to the retention of pistils, which may reflect the opportunities provided for selective fruit and seed abortion,[95,111] or perhaps the structural importance of the gynoecium in the pollination system.[15] Thus we cannot yet consider selection acting on the female function to be unimportant in the evolution of inflorescence size in milkweeds.

Queller[157] argued that male competition is the major determinant of the seasonal distribution of flowers in *Asclepias exaltata.* He observed that the flower/fruit ratio was constant throughout the season. This is consistent with strong selection on male function, which would "lead to flowers being shifted from times when small fractions of flowers mature fruits to periods of high maturation probabilities . . . until the flower/fruit ratio is constant throughout the flowering season." In contrast, selection on the maternal function (to maximize maternal choice among fruits) should favor heaviest flower production when the largest fraction is pollinated. Pollinia removal and fruit initiation were highest at the end of the season at two sites, but flower production was constant through the year. This argument makes several assumptions, including constancy in quality of pollen received. Perhaps a high intensity of flowering would not be beneficial to the female function because it would lead to the insertion of many self or closely related pollinia.[228] In such a case, selection on the female function might favor a pattern of flower production similar to that favored by male function.

Sutherland and Delph[201] make a more general argument that male function drives selection for flower number in angiosperms. Their logic is as follows. If the optimal number of androecia is higher than the optimal number of gynoecia in a plant with hermaphroditic flowers, then some perfect flowers are produced that function only as pollen donors. No fruit will be produced from such flowers. If similar selective pressures act on a monoecious species this should increase the number of male flowers relative to female flowers. Because the number of pistillate flowers does not increase on a monoecious species, the fruit/flower ratio remains high. Consequently, we would expect to see a higher fruit/flower ratio in monoecious species than in species with hermaphroditic flowers if flower number in hermaphrodites is selected for primarily because of their contribution to male fitness. This relationship does in fact occur among self-incompatible species but not among self-compatible species.[200,201] Thus Sutherland concludes that selection on male function has been important in the evolution of floral display in the former group. An intraspecific comparison in gynodioecious *Thymus vulgaris* shows that female plants have lower percent seed set than do hermaphrodites, also suggesting that selection on male function has been important in the evolution of flower number.[36] The only potential qualification is whether there are alternate hypotheses that might yield the same expectations.

Overall then, there is a widespread expectation and some data to the effect that male function is important in the evolution of floral display. A major uncertainty is the extent to which pollen quality varies and therefore influences the quality of fruit. If variation in pollen and fruit quality is unimportant, present data do suggest that selection on male function has been important in the evolution of floral display. In any case, data on species other than milkweeds are desirable.

ESTIMATING PATERNAL SUCCESS IN THE FIELD

Techniques

Most attempts to estimate paternal success (see review by Handel[70]) fall into one of two categories. Some studies occur under natural field conditions but produce only a very crude approximation of male success. Other studies obtain good data on male success (e.g., number of seeds sired), but work with artificially constructed or otherwise simplified populations.

Studies of the first type have used several presumed correlates of male success. The simplest correlates are pollen production (estimated by number of staminate flowers)[111,204] and pollen export.[157,171,227] However, no relationship has yet been shown between these quantities and male success in natural populations, despite the intuitive link between the two. In *Zea mays* there was no relationship between weight of pollen shed and male fecundity,[63] indicating that these variables are not always linked.

Other indirect approaches to the estimation of male success involve quantification of pollinator behavior (especially flight distances between flowers and extent of pollen carryover), and the study of movement of dyed and labeled pollen. These methods have two potential weaknesses: (1) they may not represent the actual patterns of pollen movement, and (2) they do not take into consideration possible nonrandom fertilization or ovule and ovary abortion that can occur after pollen deposition.[101,102] Only in a few cases have these assumptions been examined. Using an experimental setup, Waser and Price[227] found similar patterns of carryover of pollen and dye particles. The correlation between number of pollen grains deposited and number of dye particles deposited per stigma was significant, but weak ($r = 0.44$). Because this was not a natural arrangement of plants, because the flowers were emasculated, and because male reproductive success was not measured, these results do not provide convincing support for the validity of indirect methods of assessing male performance.[217] In an artificial planting of *Brassica campestris,* Handel[67] found poor agreement between movement of dye particles and siring of seeds as detected by a genetic marker. Likewise, Handel and Mishkin [68] found no agreement between level of pollinator activity and male success in *Cucumis sativus,* and Schoen and Clegg[177] concluded that use of pollinator movement patterns would have produced a misleading assessment of male performance of two color morphs of *Ipomoea purpurea.* In *Lupinus texensis* actual gene flow via pollen, determined electrophoretically, exceeded that inferred from observations of pollen movement, with a twofold difference in means.[170] Overall, then, the validity of male performance estimates based on indirect methods is doubtful.

Better data on male parentage can be obtained by examining progeny following pollinations involving genetically marked pollen. No doubt exists as to paternity, but the populations used are typically artificial, and they may differ from natural populations in plant density and dispersion, genetic structure, the nature of intervening vegetation, and the ambient pollinator fauna. The potential effect of some of these variables on pollinator behavior is illustrated by the work of Schmitt[175] and Campbell.[31] The bulk of the direct studies have used one or a few genetic markers and have taken place in agricultural situations,[3,79] experimental forest plots,[141] or experimental gardens.[66,76,78]

The most promising approach to assessing paternity in natural populations is to use sufficient electrophoretic markers to allow correct assignment of paternity to most or all seeds. Paternity is assigned based on either paternity exclusion analysis[53] or a maximum likelihood procedure.[131] Neither approach is without problems. The former is likely to be useful only in very small populations, while the latter leaves some degree of uncertainty as to the male parantage of many seeds. In Meagher's study of *Chamaelirium,* for example, the male parentage of fewer than 10% of seeds was assigned with total certainty, and for most seeds there were over 10 possible male parents.[131] Finally, it must be kept in mind that, even with accurate data on male parentage, we still lack information on the success of seeds. Such data are required for a complete assessment of paternal fitness.

Results

Several generalities emerge from studies of paternity in the field. The male success of individuals or clones often varies greatly.[71,76,78,140] The male performance of particular individuals or clones can also vary greatly among years,[141] indicating the importance of multiyear studies for meaningful conclusions. It is usually unknown whether the differences arise due to differences in pollen production, pollen removal and transport, pollination success, fertilization success, or patterns of ovule or ovary abortion. Most studies show a leptokurtic pattern of gene flow via pollen, with a short median dispersal distance, and limited but evident dispersal to distances much beyond the median.[14,37,80,104,150,220] Gene flow may be markedly asymmetric,[66] reflecting pollinator behavior and local environmental conditions. The short distances of median gene flow have contributed to the idea that local genetic differentiation is the norm in plant populations.[51,100] A few recent studies have challenged this view, by showing relatively long-distance gene flow. In *Pseudotsuga menziesii* estimates of seed sired by parents over 100 m away were 22–44%.[142] In *Raphanus sativus* 8–18% of progeny were sired by individuals at least 100–1000 m distant.[54] For some individuals this percentage was as high as 44%. At least for some species, then, gene flow via pollen is not as restricted as previously thought.

While it seems intuitive that pollinator behavior should affect patterns of gene flow, few studies have addressed this point. Webb and Bawa[220] investigated pollen movement with dyes in the hummingbird-pollinated shrub, *Malvaviscus arboreus,* and the butterfly-pollinated herb, *Cnidoscolus urens.* Dye dispersal from the former was greater in terms of maximum distance and number of receiving plants than from the latter and the authors concluded that this was due largely to the difference in pollinators. It was unclear from this study whether the interspecific difference in dispersal could partly reflect differences in numbers of flowers dyed or amount of dye deposited on each flower. Territorial pollinators can lead to more restricted pollen movement than nonterritorial visitors.[106,107] Because pollinator behavior can be modified by factors such as plant density and dispersion[175] and the diversity of floral resources available,[31] these variables must be considered in any attempt to quantify the effects of pollinators on gene flow via pollen.

FUTURE DIRECTIONS

In the last decade or so, the potential ecological and evolutionary importance of male fitness has been increasingly recognized in studies of plant reproductive biology. This has led to the development of a considerable theoretical literature dealing with topics such as allocation to male versus female function, male competition and its predicted consequences, and the role of the male function in the evolution of floral display. Empirical developments have not kept pace with theory, hampered especially by the difficulty of making valid assessments of male reproductive success. Indeed, of all the predictions reviewed in this chapter, the only one that I feel is well supported is the tendency for male allocation to decrease as selfing becomes more prevalent. Most other questions either have received no empirical attention or are not clearly supported by existing data.

While almost every point raised in this chapter requires additional empirical study, I feel that particular attention is needed in several areas. One is evaluation of the relationship between allocation to male function and male fitness. Untested assumptions as to the nature of this relatiosnhip underlie most of the hypotheses in the section Evolution of Paternal Strategy.

Also needing attention is the extent to which pollen competition is an important ecological and evolutionary force. Mulcahy[135] maintains that it may be extremely important, to the extent that it is a major reason for the current dominance of angiosperms. Indeed, much evidence indicates that pollen competition can occur, and does occur in cultivated species. However, very few studies have examined pollen competition and its effects in noncultivated species, under natural pollination conditions, and in a manner that clearly distinguishes maternal and paternal influences. Thus the general importance of pollen competition in nature is not yet clear.

A third area needing attention is the number and spatial distribution of seed whose embryos are sired by particular sporophytes. Knowledge of this paternal zygote shadow is critical in evaluating patterns of gene flow and variances among sporophytes in male reproductive success. The former is essential to an understanding of genetic structure and evolutionary processes in plant populations. The latter is needed to determine the intensity of sexual selection in males, for comparison among populations and between males and females. Without this information, arguments about the strength of male competition in different groups and how male competition is affected by various traits will remain speculative.

Thus, many opportunities await those willing to tackle the tedious task of evaluating male reproductive success. The method of choice for many questions is to use genetic markers in natural populations, and indeed some important questions can only be answered using this kind of procedure (e.g., measuring the variance in male success). Not all questions, however, require this approach. To determine how specific pollen attributes or characteristics of the pollen load affect male success, one could perform hand-pollinations and examine the performance of progeny. For example, except for two studies of cultivated species,[138,199] we have no evidence that offpsring quality is affected by the size of a stigmatic pollen load. Hand-pollinations with varied pollen loads within the range of those deposited naturally, followed by long-term comparisons of progeny, would be the way to evaluate this question.

Note also that answering a question pertaining to paternity in a single species is only a start. Differences are to be expected among species inhabiting different environments and having different life histories. Only when a variety of species has been examined will we be able to evaluate adequately the consequences of selection on male function in plants.

ACKNOWLEDGMENTS

For providing copies of unpublished manuscripts I am grateful to S. C. H. Barrett, D. R. campbell, D, Haig, A. P. Hartgerink, T. D. Lee, D. L. Marshall, M. J. McKone, T. R. Meagher, C. D. Schlichting, A. A. Snow, M. L. Stanton, A. G. Stephenson, N. M. Waser, and J. A. Winsor. I am also grateful to the following individuals for reading and critiquing earlier versions of this chapter: S. C. H. Barrett, N. C. Ellstrand, D. Haig, T. D. Lee, D. G. Lloyd, J. Lovett Doust, L. Lovett Doust, D. L. Marshall, T. R. Meagher, M. Melampy, R. R. Nakamura, T. Richardson, D. W. Schemske, C. D. Schlichting, A. A. Snow, M. L. Stanton, A. G. Stephenson, S. Sutherland, N. M. Waser, and M. F. Willson. The manuscript was improved greatly as a result of their efforts. I was supported in part by NSF grant BSR 8516362 during the preparation of this chapter.

REFERENCES

1. Abdalla, A. A., and Verkerk, K., Growth, flowering and fruit-set of the tomato at high temperature, *Neth. J. Agric. Sci.* **16**, 71–76 (1968).
2. Ackerman, J. D., Reproductive biology of *Goodyera oblongifolia* (Orchidaceae), *Madroño* **23,** 191–198 (1975).
3. Afzal, M., and Khan, A. H., Natural crossing in cotton in western Punjab. II. Natural crossing under field conditions, *Agron. J.* **42,** 89–93 (1959).
4. Akamine, E. K., and Girolami, G., Pollination and fruit set in the yellow passion fruit, *Tech. Bull. Hawaii Agric. Exp. Stn.,* No. 39 (1959).
5. Ascher, P. D., and Drewlow, L. W., Effect of stigmatic exudate injected into the stylar canal on compatible and incompatible pollen tube growth in *Lilium longiflorum* Thunb., in *Pollen: Development and Physiology* (J. Heslop-Harrison, ed.), pp. 267–272. Butterworths, London, 1971.
6. Baker, H. G., and Baker, I., Starch in angiosperm pollen grains and its evolutionary significance, *Am. J. Bot.* **66,** 591–600 (1979).
7. Baker, H. G., and Baker, I., Starchy and starchless pollen in the Onagraceae, *Ann. Missouri Bot. Gard.* **69,** 748–754 (1982).
8. Barnabas, B., Effect of water loss on germination ability of maize (*Zea mays* L.) pollen, *Ann. Bot. (London)* **55,** 201–204 (1985).
9. Barnes, D. K., and Cleveland, R. W., Pollen tube growth of diploid alfalfa *in vitro, Crop. Sci.* **3,** 291–295 (1963).
10. Barnes, D. K., and Cleveland, R. W., Genetic evidence for nonrandom fertilization in alfalfa as influenced by differential pollen tube growth, *Crop Sci.* **3,** 295–297 (1963).
11. Barnes, R. L., and Bengston, G. W., Effects of fertilization, irrigation, and cover cropping on flowering and on nitrogen and soluble sugar composition of slash pine, *For. Sci.* **14,** 172–180 (1968).
12. Barrett, S. C. H., The evolution, maintenance and loss of self-incompatibility systems, in *Plant Reproductive Ecology: Patterns and Strategies* (J. Lovett Doust and L. Lovett Doust, eds.), Chapter 5. Oxford Univ. Press, 1988.
13. Barrow, J. R., Comparisons among pollen viability measurement methods in cotton, *Crop Sci.* **23,** 734–736 (1983).
14. Bateman, A. J., Contamination in seed crops. III. Relation with isolation distance, *Heredity* **1,** 303–336 (1947).
15. Bawa, K. S., and Beach, J. H., Evolution of sexual systems in flowering plants, *Ann. Missouri Bot. Gard.* **68,** 254–274, 1981.
16. Bawa, K. S., and Opler, P. A., Dioecism in tropical forest trees, *Evolution* **29,** 167–179 (1975).
17. Bell, G., On the function of flowers, *Proc. R. Soc. London, B.* **224,** 223–265 (1985).
18. Bellani, L. M., Pacini, E., and Franchi, G. G., In vitro pollen grain germination and starch content in species with different reproductive cycle I. *Lycopersicum peruvianum* Mill., *Acta Bot. Neerl.* **34,** 59–64 (1985).
19. Bellani, L. M., Pacini, E., and Franchi, G. G., In vitro pollen grain germination and starch content in species with different reproductive cycle II. *Malus domestica* Borkh. cultivars Starkrimson and Golden Delicious, *Acta Bot. Neerl.* **34,** 65–71, (1985).
20. Bemis, W. P., Selective fertilization in lima beans, *Genetics* **44,** 555–562 (1959).

21. Bertin, R. I., Floral biology, hummingbird pollination and fruit production of trumpet creeper (*Campsis radicans,* Bignoniaceae), *Am. J. Bot.* **69,** 122–134 (1982).
22. Bertin, R. I., Nonrandom fruit production in *Campsis radicans:* Between year consistency and effects of prior pollination, *Am. Nat.* **126,** 750–759 (1985).
23. Bertin, R. I., Consequences of mixed pollinations in *Campsis radicans, Oecologia* **70,** 1–5 (1986).
24. Bertin, R. I., and Stephenson, A. G., Towards a definition of sexual selection, *Evol. Theor.* **6,** 293–295 (1983).
25. Bertin, R. I., and Willson, M. F., Effectiveness of diurnal and nocturnal pollination of two milkweeds, *Can. J. Bot.* **58,** 1744–1746 (1980).
26. Bookman, S. S., Evidence for selective fruit production in *Asclepias, Evolution* **38,** 72–86 (1984).
27. Bredemeijer, G. M. M., Peroxidase activity and peroxidase isoenzyme composition in self-pollinated, cross-pollinated and unpollinated styles of *Nicotiana alata, Acta Bot. Neerl.* **23,** 149–157 (1974).
28. Brewbaker, J. L., Pollen cytology and self-incompatibility systems in plants, *J. Hered.* **48,** 271–277 (1957).
29. Brewbaker, J. L., and Kwack, B. H., The essential role of calcium ion in pollen germination and pollen tube growth, *Am. J. Bot.* **50,** 859–865 (1963).
30. Brewbaker, J. L. and Majumder, S. K., Cultural studies of the pollen population effect and the self-incompatibility inhibition, *Am. J. Bot.* **48,** 457–464 (1961).
31. Campbell, D. R., Pollen and gene dispersal: The influences of competition for pollination, *Evolution* **39,** 418–431 (1985).
32. Charlesworth, D., and Charlesworth, B., Allocation of resources to male and female functions in hermaphrodites, *Biol. J. Linn. Soc.* **15,** 57–74 (1981).
33. Charnov, E. L., Simultaneous hermaphroditism and sexual selection, *Proc. Natl. Acad. Sci. U.S.A.* **76,** 2480–2482 (1979).
34. Charnov, E. L., Sex allocation and local mate competition in barnacles. *Mar. Biol. Lett.* **1,** 269–272 (1980).
35. Charnov, E. L., *The Theory of Sex Allocation.* Princeton Univ, Press, Princeton, 1982.
36. Couvet, D., Henry, J.-P., and Gouyon, P.-H., Sexual selection in hermaphroditic plants: the case of gynodioecy, *Am. Nat.* **126,** 294–299 (1985).
37. Crane, M. B., and Mather, K., The natural cross-pollination of crop plants with particular reference to the radish, *Ann. Appl. Biol.* **30,** 301–308 (1943).
38. Cruden, R. W., Pollen-ovule ratios: A conservative indicator of breeding systems in flowering plants, *Evolution* **31,** 32–46 (1977).
39. Cruden, R. W., and Jensen, K. G., Viscin threads, pollination efficiency and low pollen-ovule ratios, *Am. J. Bot.* **66,** 875–879 (1979).
40. Cruden, R. W., and Lyon, D. L., Patterns of biomass allocation to male and female functions in plants with different mating systems, *Oecologia* **66,** 299–306 (1985).
41. Cruden, R. W., and Miller-Ward, S., Pollen-ovule ratio, pollen size, and the ratio of stigmatic area to the pollen-bearing area of the pollinator: An hypothesis, *Evolution* **35,** 964–974 (1981).
42. Currah, L., Pollen competition in onion *(Allium cepa L.), Euphytica* **30,** 687–669 (1981).
43. Dane, F., and Melton, B., Effect of temperature on self- and cross-compatibility and *in vitro* pollen growth characteristics in alfalfa, *Crop Sci.* **13,** 587–591 (1973).
44. Davis, M. A., The effect of pollinators, predators, and energy constraints on the floral ecology and evolution of *Trillium erectum, Oecologia* **48,** 400–406 (1981).
45. Devlin, B., and Stephenson, A. G., Sex differential floral longevity, nectar secretion, and pollinator foraging in a protandrous species, *Am. J. Bot.* **72,** 303–310 (1985).
46. Dickinson, H. G., and Lawson, J., Pollen tube growth in the stigma of *Oenothera organensis* following compatible and incompatible intraspecific pollinations, *Proc. R. Soc. London, B.* **188,** 327–444 (1975).
47. Dickson, M. H., and Boettger, M. A., Effect of high and low temperatures on pollen germination and seed set in snap beans. *J. Am. Soc. Hortic. Sci.* **109,** 372–374 (1984).
48. Doria, R. S., and Rosario, T. L., Pollen germination and fruit setting in *Mussaenda, Philipp. Agric.* **64,** 203–211 (1981).
49. Eenink, A. H., Compatibility and incompatibility in witloof-chicory *(Cichorium intybus).* 1. The influence of temperature and plant age on pollen germination and seed production, *Euphytica* **30,** 71–76, (1981).
50. Eenink, A. H., Compatibility and incompatibility in witloof-chicory *(Cichorium intybus)* 3. Gametic competition after mixed pollinations and double pollinations, *Euphytica* **31,** 773–886 (1982).

51. Ehrlich, P. R., and Raven, P. H., Differentiation of populations, *Science* **165,** 1228–1232 (1969).
52. Eldridge, K. G., and Griffin, A. R., Selfing effects in *Eucalyptis regnans, Silvae Genet.* **32,** 216–221 (1983).
53. Ellstrand, N. C., Multiple paternity within the fruits of the wild radish, *Raphanus sativus, Am. Nat.* **123,** 819–828 (1984).
54. Ellstrand, N. C., and Marshall, D. L., Interpopulation gene flow by pollen in wild radish, *Raphanus sativus, Am. Nat.* **126,** 606–616 (1985).
55. Fernández-Escobar, R., Gomez-Valledor, G., and Rallo, L., Influence of pistil extract and temperature on *in vitro* pollen germination and pollen tube growth of olive cultivars, *J. Hortic. Sci.* **58,** 219–228 (1983).
56. Fowler, D. P., and Park, Y. S., Population studies of white spruce. I. Effects of self pollination, *Can. J. For. Res.* **13,** 1133–1138, (1983).
57. Frankel, R., and Galun, E., *Pollination Mechanisms, Reproduction and Plant Breeding.* Springer-Verlag, New York, 1977.
58. Freeman, D. C., Harper, K. T., and Charnov, E. L., Sex change in plants: Old and new observations and new hypotheses, *Oecologia* **47,** 222–232 (1980).
59. Ganders, F. R., The biology of heterostyly, *N. Z. J. Bot.* **17,** 607–635 (1979).
60. Gibbs, P. E., Milne, C., and Vargar Carrillo, M., Correlation between the breeding system and recombination index in five species of *Senecio, New Phytol.* **75,** 619–626 (1975).
61. Givnish, T. J., Ecological constraints on the evolution of breeding systems in seed plants: dioecy and dispersal in gymnosperms, *Evolution* **34,** 959–972 (1980).
62. Goldman, D. A., and M. F. Willson, Sex allocation in functionally hermaphroditic plants: A review and critique. *Bot. Rev.* **52,** 157–194 (1986).
63. Guitierrez, M. G., and Sprague, G. F., Randomness of mating in isolated polycross planting of maize, *Genetics* **44,** 1075–1082 (1959).
64. Haig, D., and Westoby, M., Inclusive fitness, seed resources and maternal care, in *Plant Reproductive Ecology: Patterns and Strategies* (J. Lovett Doust and L. Lovett Doust, eds.), Chapter 3. Oxford Univ. Press New York, 1988.
65. Hamilton, W. D., Extraordinary sex ratios, *Science* **156,** 477–488 (1967).
66. Handel, S. N., Contrasting gene flow patterns and genetic subdivision in adjacent populations of *Cucumis sativus* (Cucurbitaceae), *Evolution* **37,** 760–771 (1983).
67. Handel, S. N., Pollination ecology, plant population structure and gene flow, in *Pollination Biology* (L. Real, (ed.), pp. 163–211. Academic Press, New York, 1983.
68. Handel, S. N., and Mishkin, J. L. V., Temporal shifts in gene flow and seed set: Evidence from an experimental population of *Cucumis sativus, Evolution* **38,** 1350–1357 (1984).
69. Hanson, C. H., Longevity of pollen and ovaries of alfalfa, *Crop Sci.* **1,** 114–116 (1961).
70. Harder, L. D., Thomson, J. D., Cruzan, M. B., and Unnasch, R. S., Sexual reproduction and variation in floral morphology in an ephemeral vernal lily, *Erythronium americanum, Oecologia* **67,** 286–291 (1985).
71. Harding, J., and Tucker, C. L., Quantitative studies on mating systems. I. Evidence for the non-randomness of outcrossing in *Phaseolus lunatus, Heredity* **19,** 369–381 (1964).
72. Hayase, H., and Hiraizumi, Y., *Cucurbita* crosses, VI. Relationship between pollen ages and the optimum conditions (saccharose concentrations and pH values) of the artificial media, *Jpn. J. Breed.* **5,** 51–60 (1955).
73. Herpen, M. M. A. van, and Linskens, H. F., Effect of season, plant age and temperature during plant growth on compatible and incompatible pollen tube growth in *Petunia hybrida., Acta. Bot. Neerl.* **30,** 209–218 (1981).
74. Herrero, M., and Dickinson, H. G., Pollen-pistil incompatibility in *Petunia hybrida:* Changes in the pistil following compatible and incompatible intraspecific crosses, *J. Cell Sci.* **36,** 1–18 (1979).
75. Heslop-Harrison, J., Heslop-Harrison, Y., and Shivanna, K. R., The evaluation of pollen quality, and a further appraisal of the fluorochromatic (FCR) test procedure, *Theor. Appl. Genet.* **67,** 367–375 (1984).
76. Hoff, V. J., An analysis of outcrossing in certain complex-heterozygous Euoenotheras. I. Frequency of outcrossing, *Am. J. Bot.* **49,** 715–721 (1962).
77. Hornby, C. A., and Charles, W. B., Pollen germination as affected by variety and number of pollen grains, *Tomato Growers Co-operative* **16,** 11–12 (1966).
78. Horovitz, A., and Harding, J., The concept of male outcrossing in hermaphrodite higher plants, *Heredity* **29,** 223–236 (1972).

79. Hoshino, T., Ujihara, K., and Shikata, S., Time and distance of pollen dispersal in grain sorghum, *Jpn. J. Breed.* **30,** 246–250 (1980).
80. Ibe, R. A., Patterns of pollen deposition around five eastern hemlock trees [*Tsuga canadensis* (L.) Carr], *Bull. Torrey Bot. Club* **110,** 536–541 (1983).
81. Janzen, D. H., DeVries, P., Gladstone, D. E., Higgins, M. L., and Lewinsohn, T. M., Self- and cross-pollination of *Encyclia cordigera* (Orchidaceae) in Santa Rosa National Park, Costa Rica.*Biotropica* **12,** 72–74 (1980).
82. Jimenez, T. J. R., and Nelson, O. E., A new fourth chromosome gametophyte locus in maize, *J. Hered.* **56,** 259–263 (1965).
83. Johnson, C. M., and Mulcahy, D. L., Male gametophyte in maize: II. Pollen vigor in inbred plants, *Theor. Appl. Genet.* **51,** 211–215 (1978).
84. Johnson, C. M., Mulcahy, D. L., and Galinat, W. C., Male gametophyte in maize: Influences of the gametophytic genotype, *Theor. Appl. Genet.* **48,** 299–303 (1976).
85. Johri, B. M., and Vasil, I. K., Physiology of pollen, *Bot. Rev.* **27,** 325–381 (1961).
86. Koncalova, M. N., and Jicinska, D., Ecological factors of flowering and pollen quality in three willow species, *Folia Geobot. Phytotaxon.* **17,** 197–205 (1982).
87. Kress, W. J., Sibling competition and evolution of pollen unit, ovule number, and pollen vector in angiosperms, *Syst. Bot.* **6,** 101–112 (1981).
88. Kuo, C. G., Peng, J. S., and Tsay, J. S., Effect of high temperature on pollen grain germination, pollen tube growth, and seed yield of Chinese cabbage, *HortScience* **16,** 67–68 (1981).
89. Kwack, B. H., and Brewbaker, J. L., The essential role of calcium ion in pollen germination and the population effect [Abstract], *Plant Physiol.* **36** (Suppl 16), XVI (1961).
90. Labarca, C., and Loewus, F., The nutritional role of pistil exudate in pollen tube wall formation in *Lilium longiflorum, Plant Physiol.* **52,** 87–92 (1973).
91. Lavee, S., and Datt, Z., The necessity of cross-pollination for fruit set of Manzanillo olives, *J. Hortic. Sci.* **53,** 261–266 (1978).
92. Lawrence, M. J., Marshall, D. F., Curtis, V. E., and Fearon, C. H., Gametophytic self-incompatibility re-examined: a reply. *Heredity* **54,** 131–138 (1985).
93. Lawrence, M. J., and O'Donnell, S., The population genetics of the self incompatibility polymorphism in *Papaver rhoeas.* III. The number and frequency of S-alleles in two further natural populations (R102 and R104), *Heredity* **46,** 239–252 (1981).
94. Lee, T. D., Patterns of fruit maturation: A gametophyte competition hypothesis, *Am. Nat.* **123,** 427–432 (1984).
95. Lee, T. D., Patterns of fruit and seed production, in *Plant Reproductive Ecology: Patterns and Strategies* (J. Lovett Doust and L. Lovett Doust, eds.), Chapter 9. Oxford Univ. Press, New York, 1988.
96. Lee, T. D., and Bazzaz, F. A., Regulation of fruit maturation pattern in an annual legume, *Cassia fasciculata, Ecology* **63,** 1374–1388 (1982).
97. Lee, T. D., and Hartgerink, A. P., Pollination intensity, fruit maturation pattern, and offspring quality in *Cassia fasciculata* (Leguminosae), in *Biotechnology and Ecology of Pollen* (D. L. Mulcahy et al., eds.), pp. 417–422. Springer-Verlag, New York, 1986.
98. Lemen, C., Allocation of reproductive effort to the male and female strategies in wind-pollinated plants, *Oecologia* **45,** 156–159 (1980).
99. Levin, D. A., Gametophytic selection in *Phlox,* in *Gamete Competition in Plants and Animals* (D. L. Mulcahy, ed.), pp. 207–217. North-Holland Publ. Co., Amsterdam, 1975.
100. Levin, D. A., Pollinator foraging behavior: Genetic implications for plants, in *Topics in Plant Population Biology* (O. T. Solbrig, S. Jain, G. B. Johnson, and P. H. Raven, pp. 131–153. Columbia Univ. Press, New York. 1979.
101. Levin, D. A., Dispersal versus gene flow in plants, *Ann. Missouri Bot. Gard.* **68,** 233–253 (1981).
102. Levin, D. A., Plant parentage: An alternate view of the breeding structure of populations, in *Population Biology* (C. E. King, and P. S. Dawson, eds.), pp. 171–188. Columbia Univ. Press, New York, 1983.
103. Levin, D. A., Inbreeding depression and proximity-dependent crossing success in *Phlox drummondii, Evolution* **38,** 116–127 (1984).
104. Levin, D. A., and Kerster, H. W., Local gene dispersal in *Phlox, Evolution* **22,** 130–139 (1968).
105. Levy, A., Rabinowitch, H. D., and Kedar, N., Morphological and physiological characters affecting flower drop and fruit set of tomatoes at high temperatures, *Euphytica* **27,** 211–218 (1978).
106. Linhart, Y. B., Ecological and behavioral determinants of pollen dispersal in hummingbird-pollinated *Heliconia, Am. Nat.* **107,** 511–523 (1973).

107. Linhart, Y. B., and Feinsinger, P., Plant-hummingbird interactions: Effects of island size and degree of specialization on pollination, *J. Ecol.* **68,** 745–760 (1980).
108. Linskens, H. F., Pollen physiology, *Annu. Rev. Plant Physiol.* **15,** 255–270 (1964).
109. Lloyd, D. G., Breeding systems in *Cotula* L. (Compositae, Anthemideae) I. The array of monoclinous and diclinous systems, *New Phytol.* **71,** 1181–1194 (1972).
110. Lloyd, D. G., Breeding systems in *Cotula* L. (Compositae, Anthemideae), *New Phytol.* **71,** 1195–1202 (1972).
111. Lloyd, D. G., Parental strategies of angiosperms, *N. Z. J. Bot.* **17,** 595–606 (1979).
112. Lloyd, D. G., Gender allocations in outcrossing cosexual plants, in *Principles of Plant Population Ecology* (R. Dirzo, and J. Sarukhan, eds.), pp. 277–300. Sinauer, Sunderland, 1984.
113. Lloyd, D. G., Parallels between sexual strategies and other allocation strategies, *Experientia* **41,** 1277–1285 (1985).
114. Lloyd, D. G., and Webb, C. J., Secondary sex characters in plants, *Bot. Rev.* **43,** 177–216 (1977).
115. Lloyd, D. G., and Yates, J. M. A., Intrasexual selection and the segregation of pollen and stigmas in hermaphrodite plants, exemplified by *Wahlenbergia albomarginata* (Campanulaceae), *Evolution* **36,** 903–913 (1982).
116. Lord, E. M., and Eckard, K. J., Incompatibility between the dimorphic flowers of *Collomia grandiflora*, a cleistogamous species, *Science* **223,** 695–696 (1984).
117. Lovett Doust, J., and Cavers, P. B., Biomass allocation in hermaphrodite flowers, *Can. J. Bot.* **60,** 2530–2534 (1982).
118. Lovett Doust, J., and Harper, J. L., The resource costs of gender and maternal support in an andromonoecious umbellifer, *Smyrnium olusatrum* L., *New Phytol.* **85,** 251–264 (1980).
119. Lund, H. A., Growth hormones in the styles and ovaries of tobacco responsible for fruit development, *Am. J. Bot.* **43,** 562–568 (1956).
120. Lynch, S. P., The floral ecology of *Asclepias solanoana* Woods, *Madroño* **24,** 159–177 (1977).
121. Maisonneuve, B., and Den Nijs, A. P. M., In vitro pollen germination and tube growth of tomato (*Lycopersicon esculentum* Mill.) and its relation with plant growth, *Euphytica* **33,** 833–840 (1984).
122. Mangelsdorf, P. C., and Jones, D. F., The expression of mendelian factors in the gametophyte of maize, *Genetics* **11,** 423–455 (1926).
123. Marshall, D. L., and Ellstrand, N. C., Proximal causes of multiple paternity in wild radish, *Raphanus sativus, Am. Nat.* **126,** 596–605 (1985).
124. Marshall, D. L., and Ellstrand, N. C., Sexual selection in *Raphanus sativus:* Experimental data on nonrandom fertilization, maternal choice, and consequences of multiple paternity, *Am. Nat.* **127,** 446–461 (1986).
125. Martin, F. W., and Brewbaker, J. L., The nature of the stigmatic exudate and its role in pollen germination, in *Pollen: Development and Physiology* (J. Heslop-Harrison, ed.), pp. 262–266. Butterworths, London, 1971.
126. Mascarenhas, J. P., and Altschuler, M., The response of pollen to high temperatures and its potential applications, in *Pollen: Biology and Implications for Plant Breeding* (D. L. Mulcahy and E. Ottaviano, eds.), pp. 3–8. *Elsevier,* New York, 1983.
127. McDade, L. A., Pollination intensity and seed set in *Trichanthera gigantae* (Acanthaceae), *Biotropica* **15,** 122–124 (1983).
128. McDade, L. A., and Davidar, P., Determinants of fruit and seed set in *Pavonia dasypetala* (Malvaceae), *Oecologia* **64,** 61–67 (1984).
129. McKenna, M. A., and Mulcahy, D. L., Gametophytic competition in *Dianthus chinensis:* Effect on sporophytic competitive ability, in *Pollen: Biology and Applications in Plant Breeding* (D. L. Mulcahy, ed.), pp. 419–424. Elsevier, New York, 1983.
130. McKone, M. J., Variation in pollen and seed production within bromegrass (*Bromus* L.) populations. Ph.D. Dissertation, Univ. of Minnesota, Minneapolis, Minnesota, 1985.
131. Meagher, T. R., Analysis of paternity within a natural population of *Chamaelirium luteum.* I. Identification of most-likely male parents, *Am. Nat.* **127,** 199–215 (1986).
132. Miller, M. K., and Schonhorst, M. H., Pollen growth of alfalfa *in vitro* as influenced by grouping of grains on the medium and greenhouse versus field sources, *Crop Sci.* **8,** 525–526 (1968).
133. Morse, D. H., The turnover of milkweed pollinia on bumble bees and implications for outcrossing, *Oecologia* **53,** 187–196 (1982).
134. Mulcahy, D. L., Adaptive significance of gametic competition, in *Fertilization in Higher Plants* (H. F. Linskens, ed.), pp. 27–30. North-Holland Publ. Co., Amsterdam, 1974.

135. Mulcahy, D. L., The rise of the angiosperms: A genecological factor, *Science* 206, 20–23 (1979).
136. Mulcahy, D. L., Curtis, P. S., and Snow A. A., Pollen competition in a natural population, in *Handbook of Experimental Pollination Biology* (C. E. Jones and R. J. Little, eds.), pp. 330–337. Van Nostrand, New York, 1983.
137. Mulcahy, D. L., and Mulcahy, G. B., Gametophytic self-incompatibility reexamined, *Science* **220,** 1247–1251 (1983).
138. Mulcahy, D. L., Mulcahy, G. B., and Ottaviano, E., Sporophytic expression of gametophytic competition on *Petunia hybrida,* in *Gamete Competition in Plants and Animals* (D. L. Mulcahy, ed.), pp. 227–232. North-Holland Publ. Co., Amsterdam, 1975.
139. Mulcahy, G. B., and Mulcahy, D. L., The two phases of growth of *Petunia hybrida* pollen tubes through compatible styles, *J. Palynol.* **18,** 1–3 (1982).
140. Muller-Starck, G., Sexually asymmetric fertility selection and partial self-fertilization. 2. Clonal gametic contributions to the offspring of a Scots pine seed orchard, *Silvae Fennica* **16,** 99–106 (1982).
141. Muller-Starck, G., and Ziehe, M., Reproductive systems in conifer seed orchards. 3. Female and male fitnesses of individual clones realized in seeds of *Pinus sylvestris* L., *Theor. Appl. Genet.* **69,** 173–177 (1984).
142. Neale, D. B., Population genetic structure of the shelterwood regeneration system in southwest Oregon. Ph. D. Dissertation, Oregon State Univ., Corvallis, Oregon, 1983.
143. Nettancourt, D. de., *Incompatibility in Angiosperms.* Springer-Verlag, Berlin, 1977.
144. Nunney, L., Female-biased sex ratios: Individual or group selection? *Evolution* **39,** 349–361 (1985).
145. Ockendon, D. J., and Currah, L., Self-pollen reduces the number of cross-pollen tubes in the styles of *Brassica oleracea* L., *New Phytol.* **78,** 675–680 (1977).
146. Ockendon, D. J., and Gates, P. J., Reduced pollen viability in the onion *(Allium cepa), New Phytol.* **76,** 511–517 (1976).
147. Ottaviano, E., Sari Gorla, M., and Mulcahy, D. L., Genetic and intergametophytic influences on pollen tube growth, in *Gamete Competition in Plants and Animals* (D. L. Mulcahy, ed.), pp. 125–130. North-Holland Publ. Co., Amsterdam, 1975.
148. Ottaviano, E., Sari Gorla, M., and Pe, E., Male gametophytic selection in maize, *Theor. Appl. Genet.* **63,** 249–254 (1982).
149. Park, Y. S., and Fowler, D. P., Inbreeding in black spruce (*Picea mariana* [Mill.] B.S.P.): Self-fertility, genetic load, and performance, *Can. J. For. Res.* **14,** 17–21 (1984).
150. Paterniani, E., and Stort, A. C., Effective maize pollen dispersal in the field, *Euphytica* **23,** 129–134 (1974).
151. Pfahler, P. L., Fertilization ability of maize pollen grains. II. Pollen genotype, female sporophyte and pollen storage interactions, *Genetics* **57,** 513–521 (1967).
152. Pfahler, P. L., *In vitro* germination and pollen tube growth of maize *(Zea mays).* III. The effect of pollen genotype and pollen source vigor, *Can. J. Bot.* **48,** 111–115 (1970).
153. Pfahler, P. L., and Linskens, H. F., *In vitro* germination and pollen tube growth of maize (*Zea mays* L.) pollen. VI. Combined effects of storage and the alleles at the waxy (wx), sugary (su_1), and shrunken (sh_2) loci, *Theor. Appl. Genet.* **42,** 136–140 (1972).
154. Pleasants, J. M., and Chapling, S. J., Nectar production rates of *Asclepias quadrifolia:* causes and consequences of individual variation, *Oecologia* **59,** 232–238 (1983).
155. Price, M. V., and Waser, N. M., Pollen dispersal and optimal outcrossing in *Delphinium nelsoni, Nature (London)* **277,** 294–297 (1979).
156. Price, M. V., and Waser, N. M., Experimental studies of pollen carryover: Hummingbirds and *Ipomopsis aggregata, Oecologia* **54,** 353–358 (1982).
157. Queller, D. C., Sexual selection in a hermaphroditic plant, *Nature (London)* **305,** 706–707 (1983).
158. Queller, D. C., Proximate and ultimate causes of flow fruit production in *Asclepias exaltata, Oikos* **44,** 373–381 (1985).
159. Redalen, G., Pollen germination, fruit set, and dropelet set in raspberry progenies, *Gartenbauwissenschaft* **46 (4),** 159–161 (1981).
160. Reuter, D. J., Robson, A. D., Loneragan, J. F., and Tranthim-Fryer, D. J., Copper nutrition of subterranean clover (*Trifolium subterraneum* L. cv. Seaton Park). 1. Effects of copper supply on growth and development, *Aust. J. Agric. Res.* **32,** 257–266 (1981).
161. Robacker, C. D., and Ascher, P. D., Discriminating styles (DS) and pollen-mediated pseudo-self-compatibility (PMPSC) in *Nemesia strumosa* Benth. Part I: Characteristics and inheritance of DS, *Theor. Appl. Genet.* **60,** 297–302 (1981).

162. Robbins, L. E., Within-population variation in levels of pollen limitation in *Amianthium muscaetoxicum* (Liliaceae), [Abstract], *Am. J. Bot.* **72,** 864 (1985).
163. Roberts, I. N., Gaude, T. C., Harrod, G., and Dickinson, H. G., Pollen-stigma interactions in *Brassica oleracea:* A new pollen germination medium and its use in elucidating the mechanism of self incompatibility, *Theor. Appl. Genet.* **65,** 231–238 (1983).
164. Rosen, W. G., Ultrastructure and physiology of pollen, *Annu. Rev. Plant Physiol.* **19,** 435–462 (1968).
165. Ross, M. D., Five evolutionary pathways to subdioecy, *Am. Nat.* **119,** 297–318 (1982).
166. Ross, M. D., and Gregorius, H. -R., Outcrossing and sex function in hermaphrodites: A resource-allocation model, *Am. Nat.* **121,** 204–222 (1983).
167. Russell, S. D., Participation of male cytoplasm during gamete fusion in an angiosperm, *Plumbago zeylandica, Science* **210,** 200–201 (1980).
168. Russell, S. D., Fertilization in *Plumbago zeylanica:* Gametic fusion and fate of the male cytoplasm, *Am. J. Bot.* **70,** 416–434 (1983).
169. Sari Gorla, M., Ottaviano, E., and Faini, D., Genetic variability of gametophyte growth rate in maize, *Theor. Appl. Genet.* **46,** 289–294 (1975).
170. Schaal, B. A., Measurement of gene flow in *Lupinus texensis, Nature (London)* **284,** 450–451 (1980)
171. Schemske, D. W., Evolution of floral display in the orchid *Brassavola nodosa, Evolution* **34,** 489–493 (1980).
172. Schemske, D. W., and Fenster, C., Pollen-grain interactions in a neotropical *Costus:* effects of clump size and competitors, in *Pollen: Biology and Implications for Plant Breeding* (D. L. Mulcahy and E. Ottaviano, eds.), pp. 405–410. Elsevier, Amsterdam, 1983.
173. Schlichting, C.D., Environmental stress reduces pollen quality in *Phlox: Compounding the fitness deficit,* in *Biotechnology and Ecology of Pollen* (D. L. Mulcahy et al., eds.), pp. 483–488. Springer-Verlag, New York, 1986.
174. Schmidtling, R. C., Genetic variation in fruitfullness in a loblolly pine (*Pinus taeda* L.) seed orchard, *Silvae Genet.* **32,** 76–80 (1983).
175. Schmitt, J., Density-dependent pollinator foraging, flowering phenology, and temporal pollen dispersal patterns in *Linanthus bicolor, Evolution* **37,** 1247–1257 (1983).
176. Schoen, D. J., The breeding system of *Gilia achilleifolia:* Variation in floral characteristics and outcrossing rate, *Evolution* **36,** 352–360 (1982).
177. Schoen, D. J., and Clegg, M. T., The influence of flower color on outcrossing rate and male reproductive success in *Ipomoea purpurea, Evolution* **39,** 1242–1249 (1985).
178. Schoen, D. J., and Lloyd, D. G., The selection of cleistogamy and heteromorphic diaspores, *Biol. J. Linn. Soc.* **23,** 303–322 (1984).
179. Schoen, D. J., and Stewart, S. C., Variation in male reproductive investment and male reproductive success in white spruce, *Evolution* **40,** 1109–1120 (1986).
180. Schultz, R. P., Stimulation of flower and seed production in a young slash pine orchard, *U.S. Dept. Agric. For. Serv. Res. Pap. SE-91,* 1971.
181. Schwartz, D., The analysis of a case of cross-sterility in maize, *Proc. Natl. Acad. Sci. U.S.A.* **36,** 719–724 (1950).
182. Shore, J. S., and Barrett, S. C. H., The effect of pollination intensity and incompatible pollen on seed set in *Turnera ulmifolia* (Turneraceae), *Can. J. Bot.* **62,** 1298–1303 (1984).
183. Silander, J. A., and Primack, R. B., Pollination intensity and seed set in the evening primrose *(Oenothera fruticosa), Am. Midl. Nat.* **100,** 213–216 (1978).
184. Singleton, W. R., and Mangelsdorf, P. C., Gametic lethals on the fourth chromosome of maize, *Genetics* **25,** 366–390 (1940).
185. Smith, E. B., Pollen competition and relatedness in *Haplopappus* section *Isopappus, Bot. Gaz.* **129,** 371–373 (1968).
186. Smith, E. B., Pollen competition and relatedness in *Haplopappus section Isopappus* (Compositae). II, *Am. J. Bot.* **57,** 874–880 (1970).
187. Smith, O., and Cochran, H. L., Effect of temperature on pollen germination and tube growth in the tomato, *Cornell Univ. Agric. Exp. Stn. Mem.* **175,** 1–11 (1935).
188. Smith-Huerta, N. L., and Vasek, F. C., Pollen longevity and stigma pre-emption in *Clarkia, Am. J. Bot.* **71,** 1183–1191 (1984).
189. Snow, A. A., Pollination intensity and potential seed set in *Passiflora vitifolia, Oecologia* **55,** 231–237 (1982).

190. Snow, A. A., Pollination dynamics of *Epilobium canum* (Onagraceae): Consequences for gametophytic selection, *Am. J. Bot.* **73,** 139–157 (1986).
191. Sprague. G. F., Pollen tube establishment and the deficiency of waxy seeds in certain maize crosses. *Proc. Natl. Acad. Sci. U.S.A.* **19,** 838–841 (1933).
192. Stanley, R. G., and Linskens, H. F., *Pollen: Biology, Biochemistry, Management.* Springer-Verlag, Berlin, 1974.
193. Stanton, M. L., and Preston, R. E., Pollen allocation in wild radish: Variation in pollen grain size and number, in *Biotechnology and Ecology of Pollen* (D. L. Mulcahy et al., eds.), pp. 461–466. Springer-Verlag, New York, 1986.
194. Stanton, M. L., Snow, A. A., and Handel, S. N., Floral evolution: Attractiveness to pollinators increases male fitness, *Science* **232,** 1625–1627 (1986).
195. Stephens, S. G., The composition of an open pollinated segregating cotton population, *Am. Nat.* **90,** 25–39 (1956).
196. Stephenson, A. G., Flower and fruit abortion: Proximate causes and ultimate function, *Annu. Rev. Ecol. Syst.* **12,** 253–279 (1981).
197. Stephenson, A. G., and Bertin, R. I., Male competition, female choice, and sexual selection in plants, in *Pollination Biology* (L. Real, ed.), pp. 109–149. Academic Press, New York, 1983.
198. Stephenson, A. G., and Winsor, J. A., *Lotus corniculatus* regulates offspring quality through selective fruit abortion, *Evolution* **40,** 453–458 (1986).
199. Stephenson, A. G., Winsor, J. A., and Davis, L. E., Effects of pollen load size on fruit maturation and sporophyte quality in zucchini, in *Biotechnology and Ecology of Pollen* (D. L. Mulcahy et al., eds.), pp. 429–434. Springer-Verlag, New York, 1986.
200. Sutherland, S., Patterns of fruit-set: What controls fruit-flower ratios in plants? *Evolution* **40,** 117–128 (1986).
201. Sutherland, S., and Delph, L. F., On the importance of male fitness in plants: Patterns of fruit-set, *Ecology* **65,** 1093–1104 (1984).
202. Ter-Avanesian, D. V., Significance of pollen amount for fertilization, *Bull, Torrey Bot. Club* **105,** 2–8 (1978).
203. Ter-Avanesian, D. V. The effect of varying the number of pollen grains used in fertilization, *Theor. Appl. Genet.* **52,** 77–79 (1978).
204. Thomson, J. D., and Barrett, S. C. H. Temporal variation of gender in *Aralia hispida* Vent. (Araliaceae), *Evolution* **35,** 1094–1107 (1981).
205. Thomson, J. D., and Plowright, R. C., Pollen carryover, nectar rewards, and pollinator behavior with special reference to *Diervilla lonicera, Oecologia* **46,** 68–74 (1980).
206. Trivers, R. L., and Willard, D. E., Natural selection of paternal ability to vary the sex ratio of offspring, *Science* **179,** 90–92 (1973).
207. Tseng, H. -P., Pollen-tube competition in *Primula sinensis, J. Genet.* **35,** 289–300 (1938).
208. Vander Kloet, S. P., and Tosh, D., Effects of pollen donors on seed production, seed weight, germination and seedling vigor in *Vaccinium corymbosum* L., *Am. Midl. Nat.* **112,** 392–396 (1984).
209. Vasil, I. K., Effect of boron on pollen germination and pollen tube growth, in *Pollen Physiology and Fertilization* (H. F. Linskens, ed.), pp. 107–119. North-Holland Publ. Co., Amsterdam, 1964.
210. Vasil, I. K., The histology and physiology of pollen germination and pollen tube growth on the stigma and in the style, in *Fertilization in Higher Plants* (H. F. Linskens, ed.), 105–118. North-Holland Publ. Co., Amsterdam, 1974.
211. Visser, T., and Marcucci, M. C., Pollen and pollination experiments. IX. The pioneer pollen effect in apple and pear related to the interval between pollinations and the temperature, *Euphytica* **32,** 703–709 (1983).
212. Visser, T., and Verhaegh, J. J., Pollen and pollination experiments: II. The influence of the first pollination on the effectiveness of the second one in apple, *Euphytica* **29,** 385–390 (1980).
213. Visser, T., Verhaegh, J. J., Marcucci, M. C., and Uijetwaal, B. A., Pollen and pollination experiments: 8. The effect of successive pollinations with compatible and self-incompatible pollen in apple *(Malus)* and pear *(Pyrus), Euphytica* **32,** 57–64 (1983).
214. Vuilleumier, B. S., The origin and evolutionary development of heterostyly in the angiosperms, *Evolution* **21,** 210–226 (1967).
215. Waddington, K. D., Factors influencing pollen flow in bumblebee-pollinated *Delphinium virescens, Oikos* **37,** 153–159 (1981).

216. Wade, M. J., and Arnold, S. J., The intensity of sexual selection in relation to male sexual behavior, female choice and sperm precedence, *Anim. Behav.* **28,** 446–461 (1980).
217. Waser, N. M., and Price, M. V., A comparison of pollen and fluorescent dye carry-over by natural pollinators of *Ipomopsis aggregata* (Polemoniaceae), *Ecology* **63,** 1168–1172 (1982).
218. Waser, N. M., and Price, M. V., Optimal and actual outcrossing in plants, and the nature of plant-pollinator interaction, in *Handbook of Experimental Pollination Biology* (C. E. Jones and R. J. Little, eds.), pp. 341–359. Van Nostrand Reinhold, New York, 1983.
219. Waser, N. M., Price, M. V., Montalvo, A. M., and Gray, R. N., Female mate choice in a perennial herbaceous wildflower, *Delphinium nelsonii, Evol. Trends Plants* **1,** 29–33 (1987).
220. Webb, C. J., and Bawa, K. S., Pollen dispersal by hummingbirds and butterflies: A comparative study of two lowland tropical plants, *Evolution* **37,** 1258–1270 (1983).
221. Webster, S. R., Nutrition of seed orchard pine in Virginia. Ph.D. Thesis, North Carolina State Univ., Raleigh, 1974.
222. Weinbaum, S. A., Parfitt, D. E., and Polito, V. S., Differential cold sensitivity of pollen grain germination in two *Prunus* species, *Euphytica* **33,** 419–416 (1984).
223. Willson, M. F., Sexual selection in plants, *Am. Nat.* **113,** 777–790 (1979).
224. Wilson, M. F., and Burley, N., *Mate Choice in Plants: Tactics, Mechanisms and Consequences.* Princeton Univ. Press, Princeton, 1983.
226. Willson, M. F., and Price, P. W., The evolution of inflorescence size in *Asclepias* (Asclepiadaceae), *Evolution* **31,** 495–511 (1977).
227. Willson, M. F., and Rathke, B. J., Adaptive design of the floral display in *Asclepias syriaca* L., *Am. Midl. Nat.* **92,** 47–57 (1974).
228. Wyatt, R., The reproductive biology of *Asclepias tuberosa:* I. Flower number, arrangement, and fruit set, *New Phytol* **85,** 119–131 (1980).
229. Wyatt, R., Pollinator-plant interactions and the evolution of breeding systems, in *Pollination Biology* (L. Real, ed.), pp. 51–95. Academic Press, New York, 1983.
230. Zmair, D., Tanksley, D. S., and Jones, R. A., Low temperature effect on selective fertilization by pollen mixtures of wild and cultivated tomato species, *Theor. Appl. Genet.* **59,** 235–238 (1981).
231. Zeven, A. C., Kapok tree *Ceiba pentandra* Gaertn, in *Outlines of Perennial Crop Breeding in the Tropics* (F. P. Ferwerda and F. Wit, eds.). Veenam and Zonen, Wageningen, Netherlands, 1969.

3

Inclusive Fitness, Seed Resources, and Maternal Care

DAVID HAIG and MARK WESTOBY

Natural selection should favor that allocation of resources among all the activities of a plant that maximizes the plant's fitness.[43] However, we wish here to consider only the allocation of a given maternal investment *MI* among alternative expenditures within a single season. The seeds produced by a mother in a single season will be referred to as a "brood." Maternal investment is not restricted to seed provisioning but includes all costs associated with the production, protection, and dispersal of seeds. If resources limit female reproductive success, then direct trade-offs exist among these costs. Our aim is to consider how a given *MI* can be used so as to maximize the maternal contribution to fitness.

Recent reviews[124,153] have concluded that resource limitation of seed production is far more prevalent than pollen limitation, though others have emphasized cases of the latter.[8] The distinction between resource and pollen limitation is not always clear-cut. If nutrition of fruits is local within a plant[62,123,143] then some flowers could be pollen limited and other flowers on the same plant could be resource limited. Increased yield due to hand-pollination could deplete resources for reproduction in subsequent seasons;[64] hand-pollination of only some flowers on a plant could increase their yield at the expense of other flowers on the plant;[124] and closely related species or even different individuals within a population could differ as to limiting factor.[52]

Just as extra pollen does not usually increase the quantity of seed produced, most plants under natural pollination produce more fertilized ovules than develop into mature seeds.[124,149,153] In this sense we can say that a surplus of fertilized ovules is created. In angiosperms, this is possible without enormous expenditure of resources because the major commitment of resources to propagules occurs after fertilization, and indeed after the abortion of the surplus.[148]

In the first part of this chapter we will consider the allocation of resources among functions and structures within maternal effort. Particular attention will be paid to the trade-off between seed size and number. In the second part, conflict between mother and offspring over allocation to individual seeds will be discussed, and new explanations suggested for some features of angiosperm embryology. Finally, we will consider the factors favoring diversified germination behavior. Except where otherwise stated, discussion will be of angiosperms.

MATERNAL ALLOCATION

Optimal Seed Size

Assume (following Smith and Fretwell,[116] MacNair and Parker,[77] Westoby and Rice[148]) the probability s that an individual will survive and reproduce is a function of the provisions m supplied to it by its mother, and the mother makes a fixed total investment MI in a group of offspring. MI and m should be measured in units of the resource that limits reproduction. Further assume that there is a minimum m below which s is zero and that s rises with m but with diminishing returns at some higher level of investment (Fig. 3.1a). Then the tangent to the curve from the origin gives m^*, the maternal investment in an individual offspring at which s/m is maximal. Maternal fitness is maximized by investing m^* in each of MI/m^* offspring and aborting the rest. This model predicts that the effect of varying MI should be to change the number of seeds which are produced, rather than their size. However, it should be noticed that this is only true for a particular case of the function. Different mothers within a year, or the same mother in different years, might very well be subject to different versions of the function (Fig. 3.1b), and in that case the model would predict different optimal seed sizes between mothers or between years. The actual location of the function has not been estimated for any single case, much less compared for different species or for mothers in different locations or different years. This model has assumed that MI is a global quantity that can be freely partitioned among all of a mother's offspring. The model is basically unchanged if MI is separately defined for each local unit (e.g., branch, inflorescence, or fruit) at which allocation to individual offspring is determined.[143]

The inclusive fitness[49,54,55] of an organism or tissue performing some act is the sum of the increments in individual fitness of all extant relatives (including self) weighted by their relatedness r to the organism. Relatedness r is the probability that a randomly chosen allele in the organism is present by common descent in the relative (exact formulations of r in Michod and Hamilton[86]). A mother will be selected to abort all offspring except those that can be supplied with m^* provisions. Investment in individual offspring should, therefore, be terminated either (1) before major resource commitment (early abortion) or (2) after a seed has received m^*. This allocation pattern need not maximize the inclusive fitness of individual offspring because full-sibs only share half their alleles by common descent, and half-sibs only share a quarter. A seed that avoided abortion would increase its individual fitness but would reduce the amount of resources available for other seeds. Maternal fitness would suffer because of the costs of suppressing offspring or because the increased number of offspring receive less than m^*. Similarly, a provisioned offspring could benefit at the mother's (and sibs') expense by obtaining greater than m^* provisions. Consider the effect of increasing maternal investment in one offspring among a family of half-sibs. Additional investment is of benefit B to the offspring's *individual* fitness and of cost C to its half-sibs. Because of diminishing returns, B/C should decrease with increasing m. Since the mother is equally related to all offspring, her fitness is greatest at m^* when $B/C = 1$ but the *inclusive* fitness of the individual offspring is greatest at m_C when $B/C = \frac{1}{4}$ (Fig. 3.2). The costs to the mother of unequal resource distribution among offspring must be weighed against the cost to the mother of suppressing sib competition.

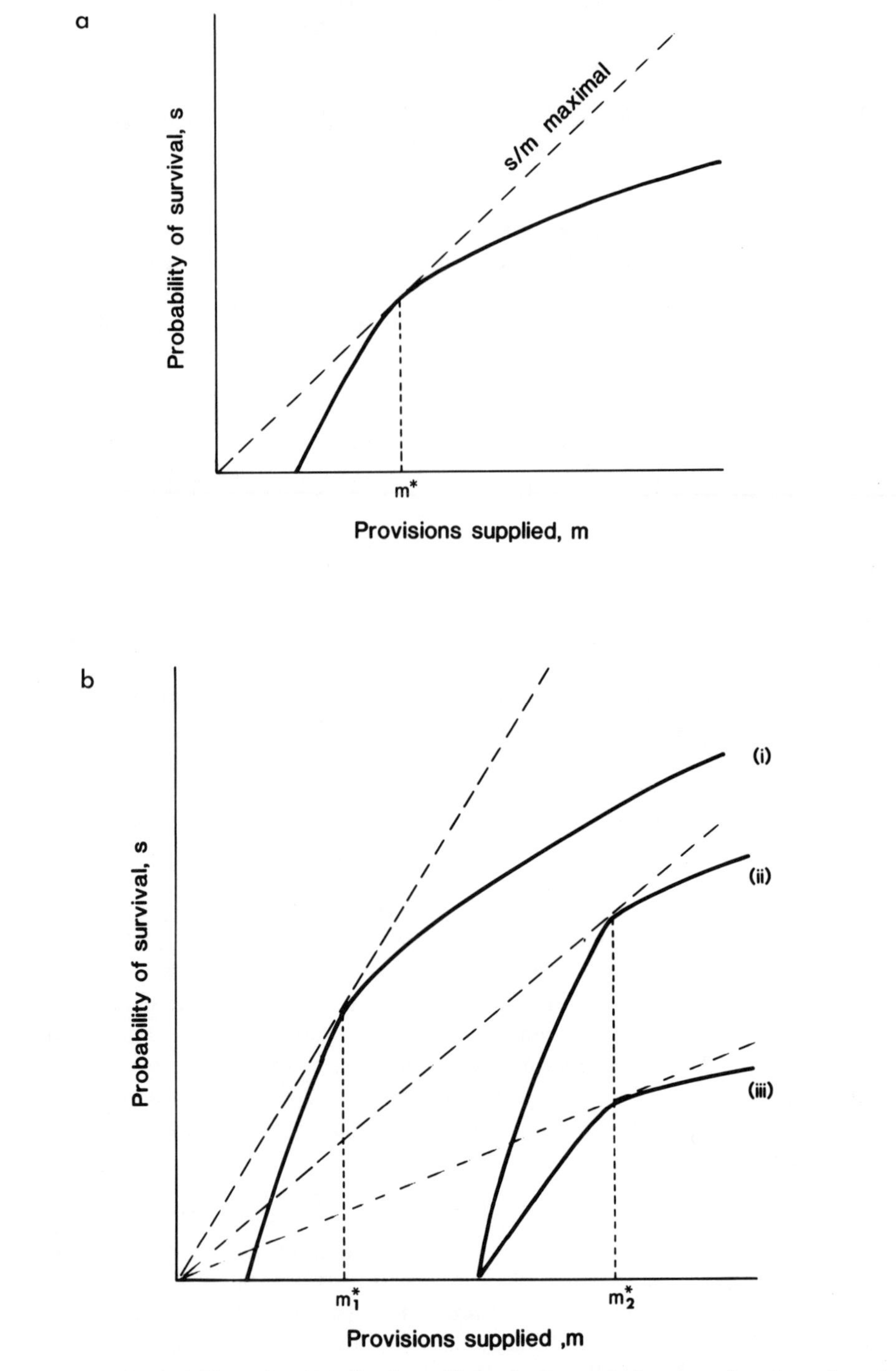

Fig. 3.1. (a) Probability s that an offspring will survive to reproduce as a function of m, the provisions supplied by the mother. The mother's fitness is greatest when m^* provided to each of MI/m^* offspring, where MI is the total maternal investment. (After Smith and Fretwell.[116]) (b) Different functions of m may give different optimal seed sizes. Functions (i) and (ii) give optima of m_1^* and m_2^*, respectively. Functions (ii) and (iii) give the same maternal optimum, m_2^*.

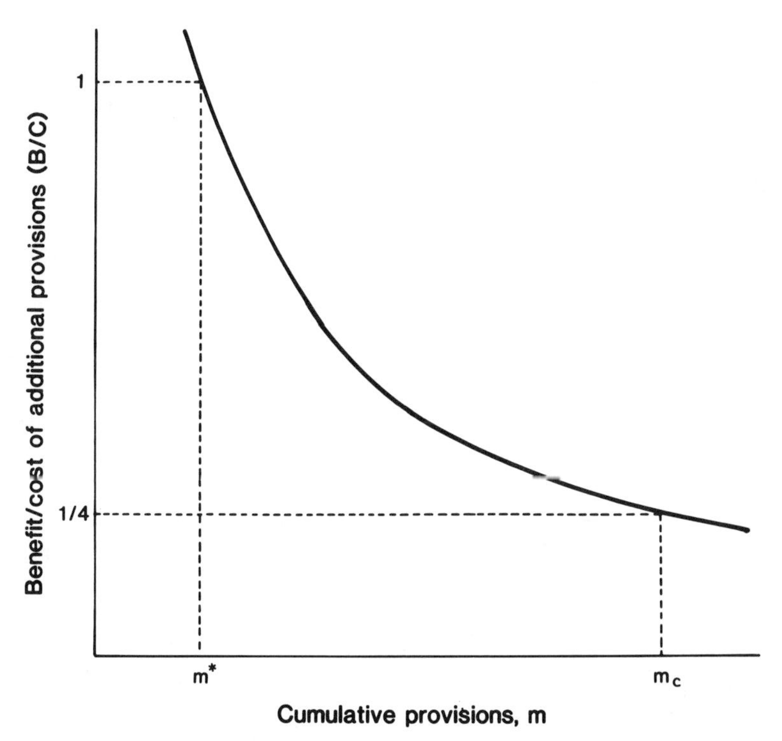

Fig. 3.2. Benefit/cost of additional investment in an individual offspring where B is the benefit to the offspring and C is the cost to its half-sibs. The mother favors termination of investment at m^* ($B/C = 1$), the offspring at m_c ($B/C = \frac{1}{4}$). Mother's and offspring's interests are in conflict between these values. (After Trivers.[136])

The use of coefficients of relatedness to describe parent/offspring conflict is quantitatively accurate provided the costs of increased provisioning of an offspring are experienced by its siblings independently of their genotype at those loci determining overconsumption. This would be the case if overconsumption reduces *MI* available for future broods but not if costs are experienced by current brood members. This is because, in broods containing both under- and overconsumers, the costs of a seed acquiring resources in excess of the maternal optimum are experienced by underconsumers and not spread evenly among siblings.[103,104] When costs fall upon underconsumers, the conflict between the interests of a mother and of her individual offspring is greater than if the costs are experienced by future siblings.

For simplicity, the discussion of parent/offspring conflict has assumed that there is no inbreeding. However, inbreeding does not affect the model's qualitative predictions except in the case of obligate self-fertilization.

Under the view of seed provisioning presented above we should expect:

1. The size of mature seeds within broods should be relatively constant
2. Changes in resource supply should change seed number rather than seed size

3. Within broods, larger seeds should have greater fitness
4. "Surplus" offspring should be aborted before major resource commitment
5. Evidence should exist of conflict between mother and offspring.

The next section will consider evidence on the first three points. It will be followed by a brief discussion of ovule abortion. Evidence for conflict between mother and offspring will be treated later, in the section Angiosperm Embryology.

Variation in Seed Size

Harper, Lovell, and Moore[58] emphasized the relative constancy of seed weight within species. Seed size, they claimed, is "the least plastic of all components of reproductive yield." However, in many of their examples the observed constancy was in the *mean* seed weight from pooled samples, and any variation in seed weight among mothers or within broods was obscured. The model developed in Fig. 3.1 is essentially one of constancy within broods. Comparatively few studies allow this component of variation to be determined.[38,59,63,79,99,119,128,156] In these studies seed weight varies significantly but by less than an order of magnitude. Several studies have shown variation in seed size between different positions on a plant or within a fruit.[20,59,82,112,119,140,158] In general, larger seeds occur closer to resource supplies. In a study of eight weedy species, mean seed weight declined over the season in all species with a maximum decrease of 25% in *Melilotus alba.*[21]

Variation among mothers in mean seed weight need not invalidate the model of optimum seed size. This variation may be environmental or genetic. If environmental, it is possible that the variation is an adaptive response to changes in the seed provisions/survival relationship (Fig. 3.1b). Variation among the broods of a mother could have a similar explanation. *Prunella vulgaris* grown in woodland produces fewer but larger seeds than plants grown on old fields. Transplant experiments indicate this is a strong environmental effect rather than ecotypic differentiation.[155] Similarly, *Hakea sericea* grown in nutrient-enriched soil produced smaller seeds than plants grown in the same soil without added nutrients,[3] and natural populations of this species on fertile soils produced smaller seeds than populations at less fertile sites.[47] As mineral nutrients are thought to be limiting growth and reproduction, nitrogen or phosphorus content may be more appropriate as a measure of maternal expenditure than seed weight. On richer soils, seeds may accumulate nutrients faster and the smaller seeds may represent an equivalent maternal investment. Alternatively, seedling survival may be less dependent on stored resources, and the trade-off between seed size and number may have shifted toward smaller seed reserves. Genetic differences in seed size between populations may reflect divergent natural selection; however, genetic differences within populations are harder to explain if the selective importance of a particular seed size is to be defended. Seed size heritability is rather high in crop species[58] and significant heritabilities have been reported in wild populations of *Raphanus raphanistrum*[119] and *Lupinus texensis.*[112]

The model of optimal seed size predicts plants will respond to changes in resource supply by a change in seed numbers rather than size. The constancy of seed size over a wide range of parental densities[58] and much of the evidence for resource-limited seed production support this prediction. In *Asclepias verticillata* seed number was more sensitive than seed size to changes in resource level.[154] Defoliation prior to anthesis of *Abutilon theophrasti*[73] and *Chelone* spp.[118] reduced seed number but had little effect on seed size, whereas the opposite was true in *Rumex crispus*.[81] The timing of changes in resource supply should be crucial. If resource levels are changed after seed number is determined, a response in seed size might be expected. In *Gymnocladus dioicus* defoliation just after anthesis greatly reduced seed number whereas later defoliation did not. Seed size, however, was reduced in both treatments. Because of its highly toxic leaves this species is not subject to natural defoliation by herbivores and may lack adaptive mechanisms for buffering seed size.[62] Lloyd has argued that the more predictable the resources to be available at the time of seed provisioning the earlier, relative to investment, should seed number be determined.[75]

The model assumes larger seeds have greater fitness, and numerous studies show an advantage within a species to seedlings from larger seeds in terms of emergence or early growth,[9–12,51,59,61,62,67,90,99,112,125,141,144,155,121] though these advantages may not be maintained in later growth except under competitive conditions.[25,38,50,94,120,145,159] Swards of *Trifolium subterraneum* established from plantings of small, large, or mixed-size seeds had similar biomasses but, in the mixed swards, self-thinning mortality was confined to plants from small seeds. The early growth disadvantage allowed these plants to be outcompeted for light.[11] Larger seeds may also be at an advantage in emergence from greater depth in the soil.[9,145,159]

There may be other disadvantages to large seed size apart from reduced seed number.[19,122] However, in the Smith–Fretwell model of the relationship between seedling survival and maternal investment (Fig. 3.1), m^* will always occur on a rising section of the curve, even though seed fitness may decrease at some higher level of investment. Therefore, natural selection on seed size is always likely to be operating in situations where an increase in maternal investment will result in a *net* increase in a seed's individual fitness.

The evidence on variation in seed size gives equivocal support to the model. Large seed size has selective advantages and seed size is less variable than many other components of reproductive yield. It is sterile to argue whether the evidence proves or disproves the prediction of uniform seed size unless the level of variability that is consistent with the prediction is defined. A more productive approach is to consider the factors that could contribute to variation within broods. Temme[127] has shown that Smith and Fretwell's model predicts variation in seed size if mothers are able to detect differences in offspring quality. Lloyd[76] has shown that optimal seed size may vary if the availability of resources is unpredictable at the time of offspring initiation. Alternatively, variation within broods may be adaptive in diversifying seed germination or dispersal, may reflect conflict between the different optima of mother and offspring, or may be a consequence of imperfect regulatory mechanisms.

Seed Abortion

Many plants only produce mature fruits from a fraction of their pollinated flowers and abort both flowers and immature fruits. If resources are limiting, the production of "surplus" flowers would appear to reduce potential fruit production.[124] In *Catalpa speciosa* the costs due to aborted fruits significantly reduced the seed weight of maturing fruits on the same inflorescence.[125] These costs may be small, as fruits are usually aborted before major resource commitment.[123,124] Aborted flowers and pods of *Asclepias speciosa* contained less than 4% as much nitrogen, phosphorus, potassium, and magnesium as mature pods.[13]

Stephenson has identified three kinds of advantage that have been proposed to compensate for these costs[124]:

1. "Surplus" flowers could provide a reserve in case of losses due to unpredictable environmental stress, predation, or disease, and would allow fruit production to be adjusted to fluctuations in resource levels.[75]
2. Extra flowers could be favored by selection for greater male contribution to fitness; see Bertin[5] this volume.
3. Selective abortion could improve fruit/seed quality.

The potential for mothers to improve the quality of their provisioned offspring by abortion of seeds or fruits is reviewed by Willson and Burley,[153] Stephenson and Bertin,[126] and Lee[72] (this volume).

"Nonseed" Maternal Effort

Maternal costs other than seed food reserves may limit total seed number. Defining these costs may be difficult as plant organs may serve more than one function. Nectar rewards of hermaphroditic flowers serve both male and female fitness because pollinators not only bring pollen but remove it. Photosynthesis by flowers and fruits may satisfy their own energy requirement,[2,143] but the photosynthetic structures enabling self-sufficiency in carbon metabolism contain nitrogen that must be diverted from other locations.[88] This illustrates the difficulties that arise if the limiting currency of seed production is unknown. In *Hakea undulata,* seeds contained only 2% of the fruit's dry weight but 53% of its nitrogen and 76% of its phosphorus.[60] In eight other species of Proteaceae, seeds had 30–500 times the phosphorus concentration of the fruit's woody parts.[68] Space does not permit a review of the diversity of "nonseed" maternal effort. Pollinator attraction, protection from predation, and seed dispersal are discussed by Willson.[152]

ANGIOSPERM EMBRYOLOGY

Background

In most angiosperms, the megaspore mother cell undergoes meiotic division to produce a single functional megaspore (monosporic development). Mitotic divisions of this megaspore produce an eight-nucleate embryo sac consisting of an egg cell flanked by two synergids at the micropylar end, three antipodal cells at the opposite (chalazal) end, and two polar nuclei in a large central cell. This is known as a *Polygonum*-type embryo sac and is found in over 70% of angiosperms studied. The other types of development differ principally in the number of megaspores producing the embryo sac and in the number and origin of the polar nuclei. *Oenothera*-type embryo sacs are monosporic and four-nucleate with a single polar nucleus. In bisporic development (*Allium*-type), the embryo sac is formed by two megaspores from the same side of the first meiotic division. An eight-nucleate embryo sac is formed with each megaspore contributing a polar nucleus. (The two megaspores have been treated as genetically identical[71,102,153] but this is not necessarily the case as they may differ at loci distal to points of crossing over). In the various forms of tetrasporic development, all four megaspores produce the embryo sac and there are from 2 to 14 polar nuclei.[78]

Fertilization is double. A pollen tube enters the embryo sac and discharges two genetically identical sperm nuclei. One nucleus fertilizes the egg cell to produce the zygote and the other fuses with the polar nuclei to produce the primary endosperm nucleus. The endosperm that develops from this nucleus is a tissue specialized to obtain resources from the mother for the nourishment of the embryo.[78,105,139] The developing angiosperm ovule is, therefore, an amalgam of four genetically distinct tissues. They are (1) maternal tissues (the nucellus, and the integuments which surround the offspring and form the seed coat in the mature seed); (2) the remnants of the haploid gametophyte; (3) the zygote/embryo, combining maternal and paternal genes; and (4) the endosperm. In the *Oenothera*-type, embryo and endosperm are genetically identical but in all other types the endosperm contains a higher dosage of maternal than paternal genes.

The relatedness of different tissues to their own embryo and to an embryo in another ovule on the same plant is given in Table 3.1 for *Polygonum*-type development. Also given is the value of B/C (sensu Fig. 3.2) at which each tissue will favor termination of investment in its own ovule. By relatedness arguments, the tissues represent a series of increasing "preference" for investment in their own seed in the order: maternal tissues, gametophyte, endosperm, embryo. This ordering among offspring tissues is maintained as long as some embryos have different fathers.[102,104,148] Bulmer gives coefficients of relatedness for non-*Polygonum* endosperms.[18] Queller defines the conditions under which models based on relatedness are accurate.[103]

As we have seen, maternal tissues favor termination of investment in an ovule either before substantial investment (i.e., early abortion) or when seed provisions have reached the maternal optimum (m^*). However, the other tissues of the ovule will favor further investment and may attempt to acquire these resources against maternal interests. Therefore, we should expect the rapid erection of maternal barriers to nutrient transfer once the "decision" to abort an ovule is taken. Alternatively, preexisting barriers could be removed for those offspring to be provisioned. For provisioned offspring, maternally imposed blocks to translocation would be expected once seed

Table 3.1. Coefficients of Relatedness for Different Ovular Tissues[a]

Tissue	Relatedness to: Associated embryo (r_1)	Relatedness to: Other embryo (r_2)	B/C
Maternal tissues	½	½	1
Female gametophyte	1	½	½
Endosperm	1	⅓	⅓
Embryo	1	¼	¼

[a] r_1 is the probability that a randomly chosen allele in the tissue is also present by common descent in the ovule's embryo; r_2 is the probability that the allele is present in the embryo of another ovule on the same mother; r_2/r_1 gives the benefit to cost ratio (*B/C*; sensu Fig. 3.2) of additional maternal investment at which termination maximizes each tissue's inclusive fitness. Calculations are for *Polygonum*-type embryo sacs and half-sib families.

provisions reach m^*. We would also expect adaptations of offspring tissues (i.e., gametophyte, endosperm, and embryo) to circumvent maternal barriers.

Maternal Tissues

Two maternal tissues, the hypostase and the endothelium (see Fig. 3.3), are of particular interest in a discussion of seed provisioning. Neither tissue is present in all angiosperms. The hypostase is a specialized group of cells found opposite the chalazal end of the gametophyte. Typically, the cells have little cytoplasm and thick, partially lignified or suberized walls.[14,78,80] The endothelium, or integumentary tapetum, differentiates from the innermost layer of the integuments and is usually in direct contact with the embryo sac. As the embryo approaches maturity the *inner* surface of the endothelium often becomes cutinized and the tissue has been considered to exchange a nutritive for a protective function.[14,66,78,80] We believe some of the confusion[14] in the literature as to the function of these tissues is due to their dual role of both facilitating and restricting nutrient transfer to the seed.

Seed provisions must pass from maternal symplast to apoplast to offspring sym-

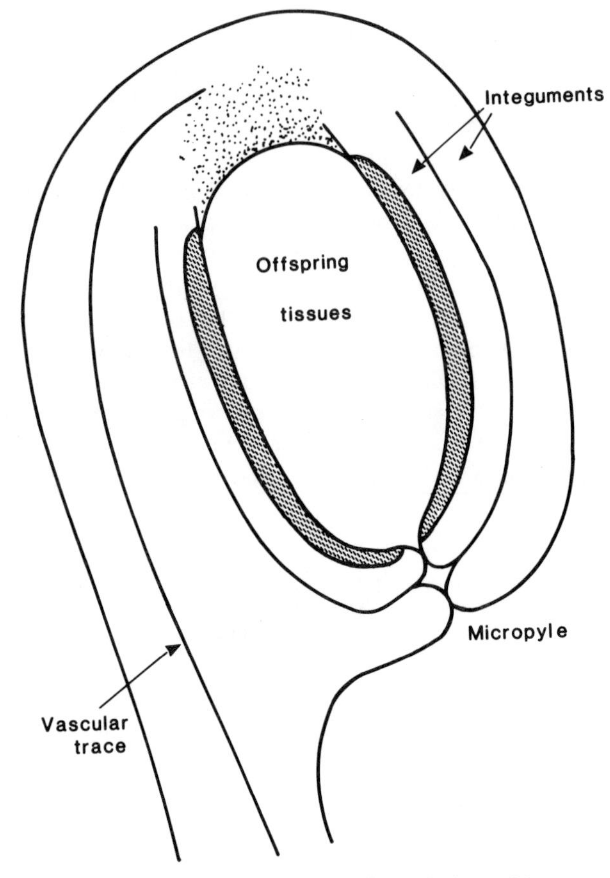

Fig. 3.3. Schematic cross-section of an ovule. Differentiation of hypostase and endothelium occurs in the spotted and hatched regions, respectively. "Offspring tissues" refers to the female gametophyte, embryo, or endosperm, depending on stage of development.

plast.[42] Plasmodesmata connect the megaspore or megagametophyte and the nucellus during the earliest stages of offspring development,[109,150,151] but, during later development, plasmodesmata are absent between maternal and offspring tissues.[37,45,100,130,132,133,157] Similarly, there are no vascular links between mother and offspring. Vascular traces usually terminate at the base of the integuments, upstream from the hypostase.[80,87,131,132,134] When the integuments are vascularized there is no vascular connection to the embryo sac.[82,129,131] The lack of direct connections between mother and offspring may give the mother greater control over resource movement. In *Zea mays* the rate-limiting step in sugar movement appears to be the unloading into the apoplast from maternal tissues rather than the uptake by endosperm transfer cells.[45]

Hypostase and endothelium have been assigned nutritive functions[14] but they are characterized at various stages of their development by the deposition of substances that would seem to block translocation. In *Petunia hybrida,* pollination induced callose deposition in the walls of the endothelium. After fertilization callose gradually disappeared in developing ovules.[41] It would be interesting to know if callose remained intact in aborting ovules. In *Pisum sativum,* callose is found in the hypostase of aborting ovules but not of developing ovules.[15] Tannin is deposited in the hypostase and/or integuments of *Ananas comosus,*[107] *Trochodendron aralioides,*[87] and *Daphniphyllum himalayense.*[6] In *Iberis amara* and *Alyssum maritimum* endothelial cells accumulate tannins, except at the micropylar end, which is active in seed nutrition. At seed maturity the endothelium is completely tanniniferous.[100] In *Torenia fournieri,* hypostase cells are apparently devoid of cytoplasm, with their cell walls composed predominantly of cellulose and callose. Callose is present from the beginning of hypostase differentiation and "effectively hinder[s] the flow of substances."[134] In *Zephyranthes drummondii,* radioactive tracers showed that nutrients moving from maternal vascular tissue to the embryo sac were forced to pass around the hypostase.[27] The protective role of the mature integuments has been emphasized[66,78] but we argue here that they also seal off the seed from maternal resources. This would explain why "protective" layers often develop first in the inner rather than outer integuments and why it is the *inner* wall of the endothelium that is first cutinized.[78]

A consideration of inclusive fitnesses suggests that the abortion of potentially viable seeds is most likely to be initiated by the action of the maternal genome. Numerous reports have associated embryo abortion with proliferation of the endothelium, though the causal relationships are disputed.[66] Brink and Cooper found embryo abortion was preceded by slow endosperm growth and increased growth of the ovule's maternal tissues. Increased growth of maternal tissues was proposed to be a consequence of reduced nutrient demand of the endosperm.[16] It is at least as plausible that maternal proliferation is a cause of slow endosperm growth. In aborting seeds of *Medicago sativa,* integuments continued to grow while their contents degenerated.[34] In *Asclepias syriaca,* thickening of the integuments preceded embryo sac degeneration in early-aborting ovules. Later abortion was accompanied by collapse of the inner layers of the integuments.[89] In incompatible crosses of *Datura,* endothelial cells mutiplied inward, absorbing the contents of the embryo sac.[110] Similarly, abortion of fertilized ovules of *Cichorium intybus* was associated with centripetal proliferation of the endothelium.[24] Abnormal endothelial growth is associated with embryo sac degeneration in *Helianthus annuus*[111] and aborting almond ovules are surrounded by callose, blocking translocation.[98]

Offspring Tissues (Gametophyte, Embryo, and Endosperm)

Usually, early growth of the endosperm is more rapid than that of the embryo.[16,139] Embryo growth may be postponed until just before maximum ovule size, and take place at the expense of endosperm reserves (e.g., *Ananas comosus*[107]). Alternatively, embryos may be active in obtaining resources from their mother (e.g., *Alyssum maritimum*[101]) and, in later development, can completely supersede the endosperm in this role (e.g., *Glycine max*[129]). Rapid resource accumulation may be important to offspring because of competition among endosperms (the "bird in the hand" principle) and because the greater the resources acquired, the greater the cost to the mother of abortion. If the mother uses endosperm vigor as a criterion by which the offspring to be provisioned are determined, rapid early growth is at a premium.[148] Many endosperms display coenocytic development in early growth,[16,78] perhaps saving time and resources by temporarily dispensing with cell walls. Nuclear doubling times in the Triticeae increase as endosperm becomes cellular.[4]

The dependence of the embryo on endosperm appears to be relaxed in apomicts.[113] Unlike the sexual species *Taraxacum kok-saghyz,* in apomictic *T. officinale* the embryo could draw directly on food stored in maternal tissues and normal development could proceed in spite of very limited endosperm growth.[33] In apomicts, there is no genetic conflict over provisioning as mother and embryos are genetically identical.[102,148]

Offspring tissues might be expected to evolve methods of counteracting maternal barriers to nutrient supply. Haustorial outgrowths are known from the megaspore, gametophyte, endosperm, and suspensor (part of the embryo)[65,80,139] but, to our knowledge, are unknown from maternal seed tissues. These haustoria are often described as "aggressive." A nutritive function for endosperm haustoria has been confirmed in *Linaria bipartita.*[7] In *Lobelia dunnii*[135] and *Pisum sativum,*[15] endosperm haustoria penetrate the integuments crushing maternal cells. Synergid haustoria are found in both apomictic and sexual species of *Cortaderia,*[95,96] so haustoria are not always associated with genetic conflict. Different tissues take the primary nutritive role in different taxonomic groups. Endosperm haustoria are generally absent where antipodal cells are active in embryo sac nutrition[7] and species with massive suspensors usually have reduced endosperm.[36,161] Suspensor haustoria are extensively developed in the Orchidaceae, Trapaceae, and Podostemaceae, families in which endosperm is lacking or poorly developed.[80]

Origin of Endosperm

Endosperm is one of the distinctive features of angiosperms. In what is inferred to be the primitive condition,[35] endosperm is a food storage tissue from which the embryo absorbs nutrients during germination. In other groups, the embryo absorbs the endosperm's food reserves prior to seed maturity or obtains most of its food reserves direct from maternal tissues. Endosperm formation is suppressed in the Orchidaceae, in which the minute seeds lack substantial food reserves; in the Podostemaceae, in which the reserves are stored in maternal seed tissues; in the Trapaceae; and in some apomicts. These exceptions are rare and are presumed to be derived conditions.[35,78,106]

Recently, a number of authors have attempted to use parent–offspring conflict and relatedness arguments to explain the adaptive significance of endosperm. Central to these theories is the belief that the unique genetic constitution of endosperm affects the process of conflict between mother and offspring over seed provisioning. Westoby and Rice[148] and Queller[102] independently derived the relevant coefficients of relatedness (Table 3.1). Queller, developing an idea of Charnov,[22] suggested double fertilization shifted the interests of the nutritive tissue (the female gametophyte in gymnosperms) in favor of the embryo.[102] Westoby and Rice emphasized that endosperm is only present where investment is deferred till after fertilization. They argued that provisioning the endosperm allows the mother to control resource allocation for a smaller cost than is possible with direct provisioning of the embryo.[148]

Subsequently, more formal models have investigated the level of seed provisioning favored by endosperm as compared to that favored by mother and embryo.[18,71,103,104] If the costs of increased provisioning of an offspring are experienced by future siblings rather than current brood members, the interests of endosperm tend to be intermediate to those of mother and embryo.[18,104] If the costs are experienced by current brood members,the interests of monosporic endosperm are similar to those of the embryo.[71] A characteristic of all models is that the outcome of selection on alleles expressed in endosperm depends critically on whether gene expression is additive, dominant/recessive, or has threshold effects.

The genetic constitution of endosperm is determined by (1) fertilization with a sperm nucleus and (2) fusion of the polar nuclei. Phylogenetically, the fertilization could antedate the fusion of polar nuclei or vice versa. In the following discussion, we will assume that monosporic development is the ancestral condition among angiosperms. This assumption is in accord with general opinion[35] but should be open to question. Its important consequence is that the nuclei of the female gametophyte (including the egg and polar nuclei) are genetically identical.

If fertilization were the earlier event, the endosperm precursor would have been diploid and genetically identical to the embryo. An evolutionary model would need to define conditions favoring the specialization of one "embryo" as a nutritive tissue for the benefit of its identical twin, and the conditions favoring the addition of a second female nucleus. However, we would not necessarily have to hypothesize a novel fertilization because polyembryony due to fertilization of multiple eggs within a female gametophyte is a common feature of gymnosperms, and male gametophytes of some species are capable of fertilizing more than one egg.[44] If polar fusion preceded fertilization, the endosperm precursor would have been a diploid version of the female gametophyte (assuming two polar nuclei). An evolutionary model would need to explain the advantages of increased ploidy of the female gametophyte and then define the conditions favoring fertilization of this diploid tissue. The models would be more complex if a tetrasporic angiosperm ancestor was assumed. This is only a sketch of ideas we intend to explore elsewhere.

DIVERSIFIED SEED BEHAVIOR

Frequency Dependence and Risk Spreading

Diversified seed behavior will here refer to all plant properties that spread the germination of a brood in space (dispersal) and time. Specifically we are concerned with somatic (i.e., nonheritable) variation among brood members. Because of dispersal, brood members germinate in different locations. Microsite variation may contribute to temporal spread in germination. Seed release may be gradual and prolonged.[70] Given continuously favorable conditions, germination time within broods is variable. Within morphologically uniform broods some seeds may be immediately germinable and others not.[57] Other species produce two or more seed morphs that may differ in germination requirements and/or mode of dispersal.[1,23,26,46,67,83,84,97,117,138] Here we will be concerned with the selective forces that operate on the end products of brood development, rather than with the mechanisms of development themselves (see Silvertown[115] for a discussion of mechanisms).

Two broad classes of models can explain the evolution of diversified seed behavior. Frequency-dependent selection can favor diversification if a seed's probability of success is negatively related to the number of other brood members germinating at the same time and place. Alternatively, diversified germination could be favored by risk-spreading in unpredictable environments. Not all variation within broods need benefit maternal fitness: some may be a consequence of environmental and developmental stochasticity, nonadaptive but economically unavoidable.

Diversified germination due to frequency-dependent selection increases the *average* reproductive success from a single brood and does not require unpredictable environments. A number of models have considered the optimal dispersal fraction in stable habitats and have found that some dispersal is advantageous even when risks are high.[31,32,56,91–93,114] In all of these models, the number of adults able to occupy a site is limited. Thus, density dependence at the parental site favors attempted colonization of new sites whether they are empty or occupied by individuals with alternative alleles.[31] (Note that the relative fitness of dispersed and undispersed seeds depends on their frequency.)

Diversified germination due to risk spreading does not increase the average reproductive success from a single brood, but does require unpredictable environments.[17,28–30,39,48,74,85,108,137] It is expected where conditions suitable for the establishment and ultimate reproduction of a germinating seed cannot be predicted at the time when, or in the place where, the "decision" to disperse or germinate is made. Intuitively, the probability that some seeds, at least, will germinate under favorable conditions should be increased by spreading the time and place of germination. Mathematically, the advantage of diversified germination arises because of a reduction in the variance of the number of successful offspring in each generation. This has the consequence of increasing the geometric mean of successful offspring number, which is a long-term estimate of fitness. A decreased probability of total failure in any one generation is likely to be the dominant factor, as a single generation with no survivors negates any amount of reproductive success in other years. Diversified germination in time should be most pronounced in broods of semelparous species because iteroparous species can spread the risk of failure over successive broods.

Frequency dependence and risk spreading can provide alternative explanations of the same phenomena (compare Zeide[160] and Schoen and Lloyd[114] on amphicarpic annuals), but they are clearly not mutually exclusive (Table 3.2).

Parent–Offspring Conflict and Control of Diversification

Diversified germination may produce seeds with different expected fitnesses (e.g., dispersed seeds may be less fit than nondispersed seeds[56]). Therefore, one might expect

Table 3.2. Properties of Models in which Diversified Germination Is Due to Frequency-Dependent Selection or Risk Spreading

	Frequency-dependent selection	Risk spreading
Diversified germination increases single generation average fitness	Yes	No
Evolution of diversified germination requires variable environments	No	Yes
Parent and offspring are in conflict over diversified germination	Yes	No?

selection on the offspring genome to ensure membership of the most-fit seed class. Frequency-dependent dispersal models predict a lower dispersal fraction if the probability of dispersing is determined by offspring rather than parent.[32,56,92,93] Thus, a mother may favor greater diversification of seed germination than offspring.[40] In unpredictable environments, offspring should benefit from some spreading of risk within sibships because if the individual fails to leave offspring, its alleles may still be transmitted through siblings. Ellner has shown that, under certain genetic assumptions, the optimal germination fraction is identical for mother and offspring.[40]

Diversified germination is usually determined by properties of maternal tissues. (1) Seed release is controlled by maternal tissues;[69,142] (2) dispersal structures (wings, awns, pappi, fleshy fruits, etc.) are genotypically maternal; and (3) properties of the genotypically maternal seed coat usually determine variation in germination requirements within broods.[53,146,147] Westoby[147] believed selection for risk spreading could only operate on the mother, whereas frequency-dependent selection could operate on either mother or offspring. He therefore argued that, because germination-diversifying machinery is located in maternal tissues, risk spreading was a more important selective force than frequency dependence in explaining diversified germination. He was wrong, because risk spreading within broods can also be in an offspring's interest (see above). Maternal control of diversification may allow maternal interests to prevail in parent–offspring conflict due to frequency dependence, or control may reside in maternal tissues because it is these tissues that actually communicate with the outside environment and with the other members of a brood.

ACKNOWLEDGMENTS

We would like to thank R. I. Bertin, M. G. Bulmer, S. Ellner, D. G. Lloyd, and L. Venable for access to unpublished manuscripts. We would also like to thank the above people and R. Law, J. and L. Lovett Doust, D. Queller, P. Werner, and M. Willson for helpful comments on various drafts of the manuscript.

REFERENCES

1. Baker, G. A., and O'Dowd, D. J., Effects of parent plant density on the production of achene types in the annual *Hypochoeris glabra, J. Ecol.* **70,** 201–215 (1982).
2. Bazzaz, F. A., Carlson, R. W., and Harper, J. L., Contribution to reproductive effort by photosynthesis of flowers and fruits, *Nature, (London)* **279,** 554–555 (1979).

3. Beadle, N. C. W., Some aspects of the ecology and physiology of Australian xeromorphic plants, *Aust. J. Sci.* **30,** 348–355 (1968).
4. Bennett, M. D., Smith, J. B., and Barclay, I., Early seed development in the Triticeae, *Philos. Trans. R. Soc. London, B* **272,** 199–227 (1975).
5. Bertin, R. I., Paternity in plants, in *Plant Reproductive Ecology: Patterns and Strategies* (J. Lovett Doust and L. Lovett Doust eds.), Chapter 2. Oxford Univ. Press, New York, 1988.
6. Bhatnagar, A. K., and Kapil, R. N., Seed development in *Daphniphyllum himalayense* with a discussion on taxonomic position of Daphniphyllaceae, *Phytomorphology* **32,** 66–81 (1982).
7. Bhatnagar, S. P., and Kallarackal, J., Cytochemical studies on the endosperm of *Linaria bipartita* (Vent.) Willd. with a note on the role of endosperm haustoria, *Cytologia* **45,** 247–256 (1980).
8. Bierzychudek, P., Pollinator limitation of plant reproductive effort, *Am. Nat.* **117,** 838–840 (1981)
9. Black, J. N., The influence of seed size and depth of sowing on pre-emergence and early vegetative growth of subterranean clover (*Trifolium subterraneum* L.), *Aust. J. Agric. Res.* **7,** 98–109 (1956).
10. Black, J. N., The early vegetative growth of three strains of subterranean clover (*Trifolium subterraneum* L.) in relation to size of seed. *Aust. J. Agric. Res.* **8,** 1–14 (1957).
11. Black, J. N., Competition between plants of different initial seed size in swards of subterranean clover (*Trifolium subterraneum* L.) with particular reference to leaf area and microclimate, *Aust. J. Agric. Res.* **9,** 299–318 (1958).
12. Black, J. N., Seed size in herbage legumes, *Herb. Abstr.* **29,** 235–241 (1959).
13. Bookman, S. S., Costs and benefits of flower abscission and fruit abortion in *Asclepias speciosa, Ecology* **64,** 264–273 (1983).
14. Bouman, F., The ovule, in *Embryology of Angiosperms* (B. M. Johri, ed.), Chapter 3. Springer-Verlag, Berlin, 1984.
15. Briggs, C. L., Westoby, M., Selkirk, P. M., and Oldfield, R. J., Embryology of early abortion due to limited maternal resources in *Pisum sativum* L., *Ann. Bot. (London)* **59,** 611–619 (1987).
16. Brink, R.A., and Cooper, D. C., The endosperm in seed development, *Bot. Rev.* **13,** 423–541 (1947).
17. Bulmer, M. G., Delayed germination of seeds: Cohen's model revisited, *Theor. Popul. Biol.* **26,** 367–377 (1984).
18. Bulmer, M. G., Genetic models of endosperm evolution in higher plants, *Proc. Int. Workshop Conf. Evol. Process Theory* (S. Karlin and E. Nevo, eds.). Israel, 1985.
19. Capinera, J. L., Qualitative variation in plants and insects: effect of propagule size on ecological plasticity, *Am. Nat.* **114,** 350–361 (1979).
20. Cavers, P. B., and Harper J. L., Germination polymorphism in *Rumex crispus* and *Rumex obtusifolius, J. Ecol.* **54,** 367–382 (1966).
21. Cavers, P. B, and Steel, M. G., Patterns of change in seed weight over time on individual plants, *Am. Nat.* **124,** 324–335 (1984).
22. Charnov, E. L., Simultaneous hermaphroditism and sexual selection, *Proc. Natl. Acad. Sci.* U.S.A. **76,** 2480–2484 (1979).
23. Cheplick, G. P., and Quinn, J. A., The shift in aerial/subterranean fruit ratio in *Amphicarpum purshii:* Causes and significance, *Oecologia* **57,** 374–379 (1983).
24. Cichan, M. A. and Palser, B. F., Development of normal and seedless achenes in *Cichorium intybus* (Compositae), *Am. J. Bot.* **69,** 885–895 (1982).
25. Cideciyan, M. A., and Malloch, A. J. C., Effects of seed size on the germination, growth and competitive ability of *Rumex crispus* and *Rumex obtusifolia, J. Ecol.* **70,** 227–232 (1982).
26. Clay, K., The differential establishment of seedlings from chasmogamous and cleistogamous flowers in natural populations of the grass *Danthonia spicata* (L.) Beauv., *Oecologia* **57,** 183–188 (1983).
27. Coe, G. E., Distribution of carbon 14 in ovules of *Zephyranthes drummondii, Bot. Gaz.* **115,** 342–346 (1954).
28. Cohen, D., Optimizing reproduction in a randomly varying environment, *J. Theor. Biol.* **12,** 119–129 (1966).
29. Cohen, D., Optimizing reproduction in a randomly varying environment when a correlation may exist between the conditions at the time a choice has to be made and the subsequent outcome, *J. Theor. Biol.* **16,** 1–14 (1967).
30. Cohen, D., A general model of optimal reproduction in a randomly varying environment, *J. Ecol.* **56,** 219–228 (1968).

31. Comins, H. N., Evolutionarily stable strategies for localized dispersal in two dimensions, *J. Theor. Biol.* **94,** 579–606 (1982).
32. Comins, H. N., Hamilton, W. D., and May R. M., Evolutionarily stable dispersal strategies, *J. Theor. Biol.* **82,** 205–230 (1980).
33. Cooper, D. C., and Brink, R. A., The endosperm-embryo relationship in an autonomous apomict, *Taraxacum officinale, Bot. Gaz.* **111,** 139–153 (1949).
34. Cooper, D. C., Brink, R. A., and Albrecht, H. R., Embryo mortality in relation to seed formation in alfalfa *(Medicago sativa), Am. J. Bot.* **24,** 203–213 (1937).
35. Cronquist, A., *The Evolution and Classification of Flowering Plants,* Chapter 2. Houghton Mifflin, Boston, 1968.
36. D'Amato, F., Role of polyploidy in reproductive organs and tissues, in *Embryology of Angiosperms* (B. M. Johri, ed.), Chapter 11. Springer-Verlag, Berlin, 1984.
37. Diboll, A. G., and Larson, D. A., An electron microscopic study of the mature megagametophyte in *Zea mays, Am. J. Bot.,* **53,** 391–402 (1966).
38. Dolan, R. W., The effect of seed size and maternal source on individual size in a population of *Ludwigia leptocarpa* (Onagraceae), *Am. J. Bot.* **71,** 1302–1307 (1984).
39. Ellner, S., ESS germination strategies in randomly varying environments. I. Logistic type models, *Theor. Popul. Biol.* **28,** 50–79 (1985).
40. Ellner, S., Germination dimorphisms and parent-offspring conflict in seed germination, *J. Theor. Biol.* **123,** 173–185 (1986).
41. Esser, K., cited in Kapil, R. N., and Tiwari, S. C., Plant embryological investigations and fluorescence microscopy, *Int. Rev. Cytol.* **53,** 291–331 (1978).
42. Evenari, M., Seed physiology: From ovule to maturing seed, *Bot. Rev.* **50,** 143–170 (1984).
43. Evenson, W. E., Experimental studies of reproductive energy allocation in plants, in *Handbook of Experimental Pollination Biology* (C. E. Jones and R. J. Little, eds.), pp. 249–276. Van Nostrand Reinhold, New York, 1983.
44. Favre-Duchartre, M., Homologies and phylogeny, in *Embryology of Angiosperms* (B. M. Johri, ed.), Chapter 14. Springer-Verlag, Berlin, 1984.
45. Felker, F. C., and Sharmon, J. C., Movement of ^{14}C-labelled assimilates into kernels of *Zea mays* L., *Plant Physiol.* **65,** 864–870 (1980).
46. Flint, S. D., and Palmblad, I. G., Germination dimorphism and developmental flexibility in the ruderal weed *Heterotheca grandiflora, Oecologia* **36,** 33–43 (1978).
47. Gill, A. M., *Acacia cyclops* and *Hakea sericea* at home and abroad, in *Proc. 4th Int. Conf. Mediterranean Ecosyst.* (B. Dell, ed.), p. 57. University of Western Australia, Perth, 1984.
48. Gillespie, J. H., Natural selection for variances in offspring numbers: a new evolutionary principle, *Am. Nat.* **111,** 1010–1014 (1977).
49. Grafen, A., Natural selection, kin selection and group selection, in *Behavioural Ecology* (J. R. Krebs and N. B. Davies, eds.), 2nd Ed. Chapter 3. Blackwell Scientific, Oxford, 1984.
50. Gross, K. L., Effects of seed size and growth form on seedling establishment of six monocarpic perennial plants, *J. Ecol.* **72,** 369–387 (1984).
51. Gross, K. L., and Soule, J. D., Differences in biomass allocation to reproductive and vegetative structures of male and female plants of a dioecious, perennial herb, *Silene alba* (Miller) Krause, *Am. J. Bot.* **68,** 801–807 (1981).
52. Gross, R. S., and Werner, P. A., Relationships among flowering phenology, insect visitors, and seed-set of individuals: Experimental studies on four co-occurring species of golden rod (*Solidago:* Compositae) *Ecol. Monogr.* **53,** 95–117 (1983).
53. Gutterman, Y., Influences on seed germinability. Phenotypic maternal effects during seed maturation, *Isr. J.Bot.* **29,** 105–117 (1980/81).
54. Hamilton, W. D., The evolution of altruistic behavior, *Am. Nat.* **97,** 354–356 (1963).
55. Hamilton, W. D., The genetical evolution of social behavior, I and II, *J. Theor. Biol.* **7,** 1–52 (1964).
56. Hamilton, W. D., and May, R. M., Dispersal in stable habitats, *Nature (London)* **269,** 578–581 (1977).
57. Harper, J. L., *Population Biology of Plants.* Academic Press, London, 1977.
58. Harper, J. L., Lovell, P. H., and Moore, K. G., The shapes and sizes of seeds, *Annu. Rev. Ecol. Syst.* **1,** 327–356 (1970).
59. Hendrix, S. D., Variation in seed weight and its effects on germination in *Pastinaca sativa* L. (Umbelliferae), *Am. J. Bot.* **71,** 795–802 (1984).

60. Hocking, P. J., The nutrition of fruits of two proteaceous shrubs, *Grevillea wilsonii* and *Hakea undulata* from south-western Australia, *Aust. J. Bot.* **30,** 219–230 (1982).
61. Howe, H. F., and Richter, W. M., Effects of seed size on seedling size in *Virola surinamensis;* a within and between tree analysis, *Oecologia* **53,** 347–351 (1982).
62. Janzen, D. H., Effect of defoliation on fruit-bearing branches of the Kentucky coffee tree, *Gymnocladus dioicus* (Leguminosae), *Am. Midl. Nat.* **95,** 474–478 (1976).
63. Janzen, D. H., Variation in seed size within a crop of a Costa Rican *Mucuna andreana* (Leguminosae), *Am. J. Bot.* **64,** 347–349 (1977).
64. Janzen, D. H., de Vries, P., Gladstone, D. E., Higgins, M. L., and Lewinsohn, T. M., Self- and cross-pollination of *Encyclia cordigera* (Orchidaceae) in Santa Rosa National Park, Costa Rica, *Biotropica* **12,** 72–74 (1980).
65. Johri, B. M., and Ambegaokar, K. B., Embryology: Then and now, in *Embryology of Angiosperms* (B. M. Johri, ed.,) Chapter 1. Springer-Verlag, Berlin, 1984.
66. Kapil, R. N., and Tiwari, S. C., The integumentary tapetum, *Bot. Rev.* **44,** 457–490 (1978).
67. Khan, M.A., and Ungar, I. A., Seed polymorphism and germination responses to salinity stress in *Atriplex triangularis* Willd., *Bot. Gaz.* **145,** 487–494 (1984).
68. Kuo, J., Hocking, P. J., and Pate, J. S., Nutrient reserves in seeds of selected proteaceous species from south-western Australia, *Aust. J. Bot.* **30,** 231–249 (1982).
69. Lacey, E. P., The influence of hygroscopic movement on seed dispersal in *Daucus carota* (Apiaceae), *Oecologia* **47,** 110–114 (1980).
70. Lacey, E. P., Timing of seed dispersal in *Daucus carota, Oikos* **39,** 83–91 (1982).
71. Law, R., and Cannings, C., Genetic analysis of conflicts arising during development of seeds in the Angiospermophyta, *Proc. R. Soc. London, B.* **221,** 53–70 (1984).
72. Lee, T. D., Patterns of fruit and seed production, in *Plant Reproductive Ecology: Patterns and Strategies* (J. Lovett Doust and L. Lovett Doust, eds.), Chapter 9. Oxford University Press, Oxford, 1988.
73. Lee, T. D., and Bazzazz, F. A., Effects of defoliation and competition on growth and reproduction in the annual plant *Abutilon theophrasti, J. Ecol.* **68,** 813–821 (1980).
74. Levin, S. A., Cohen, D., and Hastings, A., Dispersal strategies in patchy environments, *Theor. Popul. Biol.* **26,** 165–191 (1984).
75. Lloyd, D. G., Sexual strategies in plants. I. An hypothesis of serial adjustment of maternal investment during one reproductive session, *New Phytol.* **86,** 69–79 (1980).
76. Lloyd, D. G., Selection of offspring size at independence and other size-versus-number strategies, *Am. Nat.* **129,** 800–817 (1987).
77. MacNair, M. R., and Parker, G. A., Models of parent-offspring conflict. III. Intrabrood conflict, *Anim. Behav.* **27,** 1202–1209 (1979).
78. Maheshwari, P., *An Introduction to the Embryology of Angiosperms.* McGraw-Hill, New York, 1950.
79. Marshall, D. L., Fowler, N. L, and Levin, D. A., Plasticity in yield components in natural populations of three species of *Sesbania, Ecology* **66,** 753–761 (1985).
80. Masand, P., and Kapil, R. N., Nutrition of the embryo sac and embryo—a morphological approach, *Phytomorphology* **16,** 158–175 (1966).
81. Maun, M. A., and Cavers, P. B., Seed production and dormancy in *Rumex crispus.* 1. The effects of removal of cauline leaves at anthesis, *Can. J. Bot.* **49,** 1123–1130 (1971).
82. Maun, M. A., and Cavers, P. B., Seed production and dormancy in *Rumex crispus* II. The effects of removal of various proportions of flowers at anthesis, *Can. J. Bot.* **49,** 1841–1848 (1971).
83. Maurya, A. N., and Ambasht, R. S., Significance of seed dimorphism in *Alysicarpus monilifer,* D. C., *J. Ecol.* **61,** 213–217 (1973).
84. McEvoy, P. B., Dormancy and dispersal in dimorphic achenes of tansy ragwort, *Senecio jacobaea* L. (Compositae), *Oecologia* **61,** 160–168 (1984).
85. Metz, J. A. J., de Jong, T. J. and Klinkhamer, P. G. L., What are the advantages of dispersing; a paper by Kuno explained and extended, *Oecologia* **57,** 166–169 (1983).
86. Michod, R. E., and Hamilton, W. D., Coefficients of relatedness in sociobiology, *Nature (London)* **288,** 694–697 (1980).
87. Mohana Rao, P. R., Seed and fruit anatomy of *Trochodendron aralioides, Phytomorphology* **31,** 18–23 (1981).
88. Mooney, H. A., and Chiariello, N. R., The study of plant function—the plant as a balanced system, in *Perspectives on Plant Population Ecology* (R. Dirzo and J. Sarukhan, eds.), Chapter 15. Sinauer, Sunderland, Massachusetts, 1984.

89. Moore, R. J., Investigations on rubber-bearing plants. III. Development of normal and aborting seeds in *Asclepias syriaca* L., *Can. J. Res.* **24,** 55–65 (1946).
90. Morse, D. H., and Schmitt, J., Propagule size, dispersal ability and seedling performance in *Asclepias syriaca, Oecologia* **67,** 372–379 (1985).
91. Motro, U., Optimal rates of dispersal I. Haploid populations, *Theor. Popul. Biol.* **21,** 394–411 (1982).
92. Motro, U., Optimal rates of dispersal II. Diploid populations. *Theor. Popul. Biol.* **21,** 412–429 (1982).
93. Motro, U., Optimal rates of dispersal III. Parent-offspring conflict, *Theor. Popul. Biol.* **23,** 159–168 (1983).
94. Parrish, J. A. D., and Bazzaz, F. A., Nutrient content of *Abutilon theophrasti* seeds and the competitive ability of the resulting plants, *Oecologia* **65,** 247–251 (1985).
95. Philipson, M. N., Haustorial synergids in *Cortaderia* (Gramineae), *N.Z. J. Bot.* **15,** 777-778 (1977).
96. Philipson, M.N., Apomixis in *Cortaderia jubata* (Gramineae), *N.Z. J. Bot.* **16,** 45–59 (1978).
97. Philipupillai, J., and Ungar, I. A., The effect of seed dimorphism on the germination and survival of *Salicornia europaea* L. populations, *Am. J. Bot.* **71,** 542–549 (1984).
98. Pimienta, E., and Polito, V. S., Ovule abortion in 'Nonpareil' almond (*Prunus dulcis* [Mill.] D. A. Webb), *Am. J. Bot.* **69,** 913–920 (1982).
99. Pitelka, L. F., Thayer, M. E., and Hansen, S. B., Variation in achene weight in *Aster acuminatus, Can. J. Bot.* **61,** 1415–1420 (1983).
100. Prabhakar, K., and Vijayaraghavan, M. R., Endothelium in *Iberis amara* and *Alyssum maritimum*—its histochemistry and ultrastructure, *Phytomorphology* **32,** 28–36 (1982).
101. Prabhakar, K., and Vijayaraghavan, M. R., Histochemistry and ultrastructure of suspensor cells in *Alyssum maritimum, Cytologia* **48,** 389–402 (1983).
102. Queller, D. C. Kin selection and conflict in seed maturation, *J. Theor. Biol.* **100,** 153–172 (1983).
103. Queller, D. C. Kin selection and frequency dependence: A game theoretic approach, *Biol. J. Linn. Soc.* **23,** 133–143 (1984).
104. Queller, D. C., Models of kin selection on seed provisioning, *Heredity* **53,** 151–165 (1984).
105. Raghavan, V., *Experimental Embryogenesis in Vascular Plants,* Chapter 5. Academic Press, London, 1976.
106. Ram, M. Floral morphology and embryology of *Trapa bispinosa* Roxb. with a discussion of the systematic position of the genus, *Phytomorphology* **6,** 312–323 (1956).
107. Rao, A. N., and Wee, Y. C. Embryology of the pineapple, *Ananas comosus* (L.) Merr., *New Phytol.* **83,** 485–497 (1979).
108. Real, L.A., Fitness, uncertainty, and the role of diversification in evolution and behavior, *Am. Nat.* **115,** 623–638 (1980).
109. Russell, S. D., Fine structure of megagametophyte development in *Zea mays, Can. J. Bot.* **57,** 1093–1110 (1979).
110. Satina, S., Rappaport, J., and Blakeslee, A. F., Ovular tumors connected with incompatible crosses in *Datura, Am. J. Bot.* **37,** 576–586 (1950).
111. Savchenko, M. I., Anomalies in the structure of angiosperm ovules, *Dokl. Bot. Sci.(Engl. Transl.)* **130,** 15–17 (1960).
112. Schaal, B. A., Reproductive capacity and seed size in *Lupinus texensis, Am. J. Bot.* **67,** 703–709 (1980).
113. Schnarf, K., cited in Cooper and Brink (Ref. 31).
114. Schoen, D. J., and Lloyd, D. G., The selection of cleistogamy and heteromorphic diaspores, *Biol. J. Linn. Soc.* **23,** 303–322 (1984).
115. Silvertown, J. W., Phenotypic variety in seed germination behavior: The ontogeny and evolution of somatic polymorphism in seeds, *Am. Nat.* **124,** 1–16 (1984).
116. Smith, C. C., and Fretwell, S. D., The optimal balance between size and number of offspring, *Am. Nat.* **108,** 499–506 (1974).
117. Sorensen, A. E., Somatic polymorphism and seed dispersal, *Nature (London)* **276,** 174–176 (1978).
118. Stamp, N. E., Effect of defoliation by checkerspot caterpillars *(Euphydryas phaeton)* and sawfly larvae (*Macrophya nigra* and *Tenthredo grandis*) on their host plants (*Chelone* spp), *Oecologia* **63,** 275–280 (1984).
119. Stanton, M. L., Developmental and genetic sources of seed weight variation in *Raphanus raphanistrum* L. (Brassicaceae), *Am. J. Bot.* **71,** 1090–1098 (1984).
120. Stanton, M. L., Seed variation in wild radish: Effect of seed size on components of seedling and adult fitness, *Ecology* **65,** 1105–1112 (1984).

121. Stanton, M. L., Seed size and emergence time within a stand of wild radish (*Raphanus raphanistrum* L.): The establishment of a fitness hierarchy, *Oecologia* **67,** 524–531 (1985).
122. Stebbins, G. L., Adaptive radiation of reproductive characteristics in angiosperms II: Seeds and seedlings, *Annu. Rev. Ecol. Syst.* **2,** 237–260 (1971).
123. Stephenson, A. G., Fruit set, herbivory, fruit reduction, and the fruiting strategy of *Catalpa speciosa* (Bignoniaceae), *Ecology* **61,** 57–64 (1980).
124. Stephenson, A. G., Flower and fruit abortion: Proximate causes and ultimate functions, *Annu. Rev. Ecol. Syst.* **12,** 253–279 (1981).
125. Stephenson, A. G., The cost of over-initiating fruit, *Am. Midl. Nat.* **112,** 379–386 (1984).
126. Stephenson, A. G., and Bertin, R. I., Male competition, female choice and sexual selection in plants, in *Pollination Biology* (L. Real, ed.), Chapter 6. Academic Press, Orlando, Florida, 1983.
127. Temme, D. H., Seed size variability: A consequence of variable genetic quality among offspring? *Evolution* **40,** 414–417 (1986).
128. Thompson, J. N., Variation among individual seed masses in *Lomatium grayi* (Umbelliferae) under controlled conditions: Magnitude and partitioning of the variance, *Ecology* **65,** 626–631 (1984).
129. Thorne, J. H., Kinetics of ^{14}C-photosynthate uptake by developing soybean fruit, *Plant Physiol.* **65,** 975–979 (1980).
130. Thorne, J. H., Morphology and ultrastructure of maternal seed tissues of soybean in relation to the import of photosynthate, *Plant Physiol.* **67,** 1016–1025 (1981).
131. Tilton, V. R., and Lersten, N. R., Ovule development in *Ornithogalum caudatum* (Liliaceae) with a review of selected papers on angiosperm reproduction. I. Integuments, funiculus and vascular tissue, *New Phytol.* **88,** 439–457 (1981).
132. Tilton, V. R., and Mogensen, H. L., Ultrastructural aspects of the ovule of *Agave parryi* before fertilization, *Phytomorphology* **29,** 338–350 (1979).
133. Tilton, V. R., Wilcox, L. W., and Palmer, R. G., Postfertilization Wandlabrinthe formation and function in the central cell of soybean, *Glycine max* (L.) Merr. (Leguminosae), *Bot. Gaz.* **145,** 334–339 (1984).
134. Tiwari, S. C., The hypostase in *Torenia fournieri* Lind.: A histochemical study of the cell walls, *Ann. Bot. (London)* **51,** 17–26 (1983).
135. Torosian, C. D., Ultrastructural study of endosperm haustorial cells of *Lobelia dunnii* Greene (Campanulaceae, Lobelioideae), *Am. J. Bot.* **58,** (Abstr.) 456 (1971).
136. Trivers, R. L., Parent-offspring conflict, *Am. Zool.* **14,** 249–264 (1974).
137. Venable, D. L., The evolutionary ecology of seed heteromorphism, *Am. Nat.* **126,** 577–595 (1985).
138. Venable, D. L., and Levin, D. A., Ecology of achene dimorphism in *Heterotheca latifolia.* I-III., *J. Ecol.* **73,** 133–145 (1985).
139. Vijayaraghavan, M. R., and Prabhakar, K., The endosperm, in *Embryology of Angiosperms* (B. M. Johri, ed.) Chapter 7. Springer-Verlag, Berlin, 1984.
140. Waller, D. M. Factors influencing seed weight in *Impatiens capensis* (Balsaminaceae), *Am. J. Bot.* **69,** 1470–1475 (1982).
141. Waller, D. M., The genesis of size hierarchies in seedling populations of *Impatiens capensis* Meerb., *New Phytol.* **100,** 243–260 (1985).
142. Wardrop, A. B., The opening mechanism of follicles of some species of *Banksia, Aust. J. Bot.* **31,** 485–500 (1983).
143. Watson, M. A., and Casper, B. B., Morphogenetic constraints on patterns of carbon distribution in plants, *Annu. Rev. Ecol. Syst.* **15,** 233–258 (1984).
144. Weis, I. M., The effects of propagule size on germination and seedling growth in *Mirabilis hirsuta, Can. J. Bot.* **60,** 1868–1874 (1982).
145. Weller, S. G., Establishment of *Lithospermum caroliniense* on sand dunes: the role of nutlet mass, *Ecology* **66,** 1893–1901 (1985).
146. Werker, E., Seed dormancy as explained by the anatomy of embryo envelopes, *Isr. J. Bot.* **29,** 22–44 (1980/81).
147. Westoby, M., How diversified seed germination behavior is selected, *Am. Nat.* **118,** 882–885 (1981).
148. Westoby, M., and Rice, B., Evolution of the seed plants and inclusive fitness of plant tissues. *Evolution* **36,** 713–724 (1982).
149. Wiens, D., Ovule survivorship, brood size, life history, breeding systems, and reproductive success in plants, *Oecologia* **64,** 47–53 (1984).

150. Willemse, M. T. M., and Bednara, J., Polarity during megasporogenesis in *Gasteria verrucosa, Phytomorphology* **29,** 156–165 (1979).
151. Willemse, M. T. M., and van Went, J. L., The female gametophyte, in *Embryology of Angiosperms* (B. M. Johri, ed.), Chapter 4. Springer-Verlag, Berlin, 1984.
152. Willson, M. F., *Plant Reproductive Ecology.* J.Wiley & Sons, New York, 1983.
153. Willson, M. F., and Burley, N., *Mate Choice in Plants: Tactics, Mechanisms and Consequences.* Princeton University Press, Princeton, 1983.
154. Willson, M. F. and Price, P. W., Resource limitation of fruit and seed production in some *Asclepias* species, *Can. J. Bot.* **58,** 2229–2233 (1980).
155. Winn, A. A., Effects of seed size and microsite on seedling emergence of *Prunella vulgaris* in four habitats, *J. Ecol.* **73,** 831–840 (1985).
156. Wolf, L. L., Hainsworth, F. R., Mercier, T., and Benjamin, R., Seed size variation and pollinator uncertainty in *Ipomopsis aggregata* (Polemoniaceae), *J. Ecol.* **74,** 361–371 (1986).
157. Woodcock, C. L. F., and Bell, P. R., Features of the ultrastructure of the female gametophyte of *Myosurus minimus, J. Ultrastruct. Res.* **22,** 546–563 (1968).
158. Wulff, R. D., Seed size variation in *Desmodium paniculatum.* I. Factors affecting seed size, *J. Ecol.* **74,** 87–97 (1986).
159. Wulff, R. D., Seed size variation in *Desmodium paniculatum.* II. Effects on seedling growth and physiological performance, *J. Ecol.* **74,** 99–114 (1986).
160. Zeide, B., Reproductive behaviour of plants in time, *Am. Nat.* **112,** 636–639 (1978).
161. Zinger, N. V., and Poddubnaya-Arnoldi, V. A., Application of histochemical techniques to the study of embryonic processes in certain orchids, *Phytomorphology* **16,** 111–124 (1966).

4

Monomorphic and Dimorphic Sexual Strategies: A Modular Approach

PAUL ALAN COX

INTRODUCTION: SEXUALITY IN MODULAR ORGANISMS

A seed plant has two possible pathways to transmit its genes to the next generation: pollen and/or ovules. The precise combination of pollen and ovules used to transmit genes depends on the types of sexual organs found within a single type of flower, the different types of flowers produced by a single individual, and the different types of individuals occurring within a population. Thus, compared to animals, plants have a wealth of means of varying their sexual expression. This inherent sexual plasticity results from an open manner of growth where major growth increments result from the addition of more modules as well as the expansion of previously formed parts. Plants can therefore be viewed as metapopulations[126] in the sense that an individual plant comprises a population of individual modules. One consequence of the modular nature of plants[45] is that plant taxonomy is based not upon the comparison of entire genetic individuals but rather on the comparative study of plant parts such as flowers, fruits, or leaves.

The modularity of plants occurs at a variety of levels. For example, in green algae the reproductive module can be a single cell (see DeWreede and Klinger,[49] this volume), while in angiosperms the reproductive module can be a single stamen, a flower, an inflorescence, a monopodial shoot terminated by an inflorescence, or an entire genetic individual, depending on the interests of the observer. Most animals lack such modularity at different levels, hence their outer surfaces are usually portrayed as closed curves (e.g., the silhouette of a giraffe) while plants (e.g., a bracken clone) are best represented mathematically by fractal curves[87] that detail repetitive self-similar structures at different levels of organization.

At any given level of organization a plant may be constructed of different types of modules; indeed, the architecture of many trees depends upon the relationships of different type of modules.[62] For example, at the level of the shoot in the cycad species *Cycas circinalis* (Cycadaceae) two different sorts of module occur.[62] One consists of a hapaxanthic (determinate), shoot terminated by a staminate cone. The other type of module consists of a pleonanthic (indeterminate) shoot that produces ovules on lateral sporophylls. In *Cycas circinalis* these two types of modules never occur on the same individual and therefore the species is dioecious. However, one could easily imagine an hermaphroditic individual that was constructed from both types of modules. From an evolutionary perspective we may ponder why such individuals do not exist in nature. As I will suggest later in this chapter, a more useful task would be to predict their likelihood of successfully invading a dioecious population if they did exist.

By the very nature of their modular construction, plants seem to have an inherent plasticity in sexual expression at a variety of levels of morphological complexity. Such plasticity may not be as important to most animals as it is to plants, since plants are sedentary and cannot move to more favorable patches or populations for their reproductive activities. However, the central mystery in plant reproductive biology is not why plants are so plastic in their sexual expression (as this seems to follow directly from their modular construction), but rather why some plants have surrendered this plasticity to different degrees and at different modular levels. Some plants, for example, exhibit little sexual plasticity, yet can be shown to have evolved from sexually plastic ancestors. Loss of sexual plasticity has independently evolved in a variety of phylogenetic lineages in a variety of habitats.[125] Why should plants give up an advantageous ability?

Perhaps the nature of the problem can best be illustrated by reference to a single taxon, *Freycinetia* (Pandanaceae), a genus of dioecious lianas found throughout the South Pacific and tropical southeast Asia. In all known species of *Freycinetia* a minimum of four levels of modular construction can be recognized. For example, at one modular level, entire genetic individuals (Fig. 4.1a) of *Freycinetia reineckei* may be considered as modules, with entire plants being either male or female. A second modular level consists of hapaxanthic axes terminated by an inflorescence (Fig. 4.1b). A renewal shoot then develops from an axillary bud beneath the inflorescence and hence growth is sympodial. These hapaxanthic axes can be either male, with the terminal inflorescence consisting of an involucre of male spikes and fleshy bracts, or female, with the inflorescence consisting of female spikes and fleshy bracts. A third modular level is found within the inflorescence, namely each spike together with its subtending bract (Fig. 4.1c). The fourth modular level is found within the spike itself, which is made up of small modules consisting of a flower subtended by a tiny bract (Fig. 4.1d). These tiny modules are male or female, depending upon the sex of the flower, although in early floral organogenesis all of the developing flowers possess both androecial and gynoecial primorida.

Thus a single *Freycinetia* individual conceivably could be hermaphroditic at each of these four modular levels. A single plant could (1) have both male and female axes, or (2) produce inflorescences containing both male and female spikes, or (3) produce spikes containing both male and female flowers, or (4) produce flowers containing both stamens and pistils. Although the genus and indeed the entire Pandanaceae have long been believed to be strictly dioecious,[70] recent research has revealed some individuals of *Freycinetia reineckei* in Samoa to be hermaphroditic at level 3 by producing hermaphroditic spikes.[32] Some individuals of *Freycinetia scandens* in Australia are hermaphroditic at level 2 by producing both male and female axes.[40] The question of interest is not why such exceptions to dioecism occur, but rather why the genus in general is dimorphic in gender at four different modular levels. It is interesting to consider what the reproductive success would be of individuals that were monomorphic rather than dimorphic at one or more levels of morphological complexity.

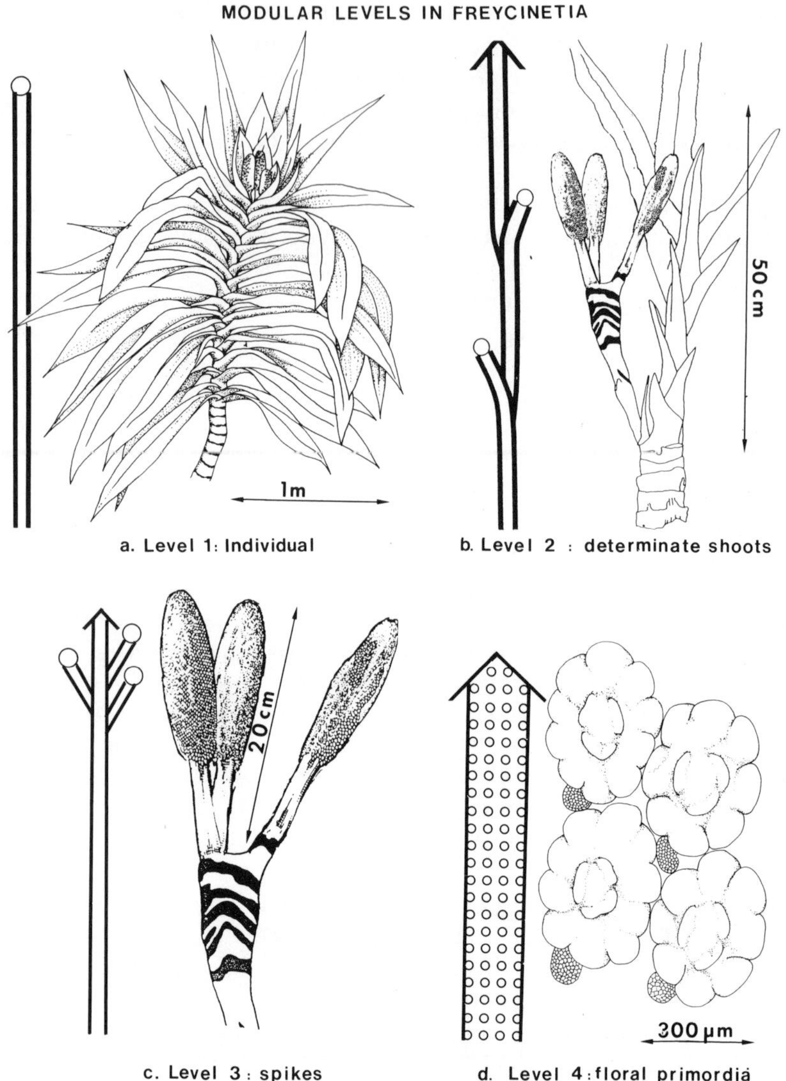

Fig. 4.1. Illustration of the different modular levels in *Freycinetia reineckei* at which various dimorphic or monomorphic reproductive strategies could possibly be expressed; the floral primordia in (d) occur early during organogenesis; crowding on the mature axis obscures individual floral units, particularly on staminate spikes.

SEXUAL STRATEGIES IN PLANTS

To the extent that such differences in sexual expression at different levels of morphological complexity reflect the genotypic program of the organism, they may be considered to represent different reproductive strategies.[65] Since the evolutionary game is always retrospective and never anticipatory, and since plants lack neural systems, the use of the term "strategy" in plant ecology cannot impute directed, goal-seeking behaviors to plants. The term instead provides a useful contrast to the term "tactics,"[65] which describes plastic responses to prevailing conditions. Although use of the term "strategy" is relatively recent in plant biology, the idea that different lineages of plants have consistently different forms of sexual expression was used in the early days of plant taxonomy as a means of artificially classifying the plant kingdom.

The many different morphological levels at which sexuality is expressed in plants has resulted in a variety of typologies to categorize the resultant different sexual systems. Linnaeus[75] developed an artificial system of plant classification based upon whether stamens and pistils occur in the same or different flowers, whether staminate and pistillate flowers are mixed with hermaphroditic flowers, and whether staminate and pistillate flowers occur on the same or different plants. His resultant division of the plant kingdom into hermaphrodite, monoecious, dioecious, and polygamous species was subsequently elaborated upon by Darwin in his classic *The Different Forms of Flowers on Plants of the Same Species.*[44]

Although the strength of the Linnaean system has been its simplicity of use, many botanists, particularly those interested in plant populations, have long been aware of its deficiencies. As Darwin admitted,[44] "the classification is artificial, and the groups often pass into one another" (p. 1). A major problem with use of the Linnaean system is the existence of sexual intergrades at a variety of morphological levels, and the possibility that some plants may even change sex and thus be sexually labile.[104] Partly in response to the presence of such diverse intergrades between the major sexual classes, Lloyd[81] has devised quantitative methods for describing plant gender. This system has been developed[83] to estimate both "phenotypic gender," or the plant's total investment in male and female functions, and the "functional gender," or relative success of a plant as a maternal and paternal parent. The concept of "functional gender" treats sexuality at the level of the population, as it recognizes that the reproductive success of a plant depends not only on its own sexual expression, but on the sexual composition of the surrounding population. It also allows quantification of reproductive success in populations where sexual phenotype may be unrelated to an individual's morphological gender. For example, individuals of the species *Discaria toumatou* (Rhamnaceae) have morphologically hermaphroditic flowers that vary in male and female function.[94] Between 0 and 44% of flowers develop into fruits while some plants produce pollen but no seeds. Such a situation was foreseen by earlier workers[68,69,78] who argued that hermaphroditic plants should only rarely prove to be equally effective as male and female parents.

These new quantitative measures of gender were developed as supplements to, rather than replacements for, the traditional Linnaean system (Lloyd, personal communication). These measures of gender can be applied to both plants and animals since they depend only on measuring the total investment of an individual in male and female reproductive functions or the total success of an individual as a maternal or paternal parent relative to other individuals in the population. Also of interest are measures of gender specialization that quantify a plant's enchanced ability to serve as a pollen donor at the expense of its ability to serve as a pollen recipient, without reference to the performance of other individuals in the population.[96] These new concepts of gender thus potentially provide a powerful tool in studying sexuality in both plants and animals.

Measures of phenotypic and function gender, as currently formulated, do not, however, explicitly recognize the different morphological levels at which plant sexuality can be expressed. For example, the phenotypic maleness[83] of a rhizomatous plant that produces separate staminate and pistillate shoots would be the same as that of a plant that produces hermaphroditic flowers on all shoots as long as the pollen/ovule ratios of both plants are equal. Similarly, the two plants would have equal functional genders if they have equivalent successes as paternal parents and equivalent successes as maternal parents. To determine why plants are inherently more plastic than animals in their sexual expression, and why they express this plasticity to different degrees and at different levels of morphological complexity, requires a melding of concepts from plant morphology as well as sex allocation theory.

MODULAR CONSTRUCTION AND SEX ALLOCATION IN PLANTS

At any particular level of morphological complexity, reproductive modules may be either monomorphic, i.e., all of a single type, or they may be reproductively dimorphic, i.e., of two types (the possible significance of polymorphic individuals and modules is the subject of current debate and is briefly discussed below). The effectiveness of monomorphic versus dimorphic strategies at any level can be analyzed through the use of evolutionary game theory, which is intuitively simple, but mathematically powerful. As articulated by Maynard Smith,[88] an "evolutionary stable strategy" (ESS) is a strategy such that if all members of a population adopted it, then no individual with a different strategy could invade the population through the forces of natural selection. Charnov,[24] in perhaps the most important work written on this topic, has extensively used these techniques to analyze the problem of sex allocation in plants and animals.

As developed by Maynard Smith, Charnov, Bull,[25,27] and others, evolutionary game theory can prove extremely useful in studying problems of sexuality in plants, particularly at the modular level. For example, even at the level of the individual gamete, species of plants have different sexual strategies since some produce monomorphic (isogamous) gametes while other species produce dimorphic (anisogamous) gametes. Since anisogamy and isogamy are found within certain phylogenetic lines (all higher plants, bryophytes, and most algae are anisogamous), they can be taken to represent different reproductive strategies. The effectiveness of isogamy and anisogamy can be compared[38,39] by considering a hypothetical population of gametangia with equal reproductive masses producing type A gametes, all of equal size and volume (and hence isogamous). What is the vulnerability of this population to invasion by a mutant gametangium of equal reproductive mass which can be divided into N type B gametes?

By use of search theory or, alternatively, through numerical simulations, it can be shown[38,39] that an isogamous population can successfully be invaded by an anisogamous mutant as long as the gametes it produces are of greatly different size. However, if only slight size differences exist between the anisogamous gametes and the isogamous gametes, the invasion will fail. Thus a low adaptive peak exists for isogamy while a much higher adaptive peak exists for anisogamy. It is therefore likely that stochastic forces such as drift are important in driving a population across the fitness saddle[130] from isogamy to anisogamy. Anisogamy can be shown to be an ESS since anisogamous populations, once established, are invulnerable to reinvasion by isogamous mutants.[39]

The theory of sex allocation can also be used to analyze sexuality at higher levels of morphological complexity. For example, analyis of *Freycinetia reineckei* populations in Samoa at the level of the shoot and terminal inflorescence indicates populations of plants possessing male and female shoots to be invulnerable to invasion by plants producing hermaphroditic shoots since the fitness of hermaphroditic shoots is significantly lowered by the actions of the large flying foxes and birds that function as pollinators.[33,35] During pollination, any pollen-bearing spikes are destroyed. This destruction does not reduce the fitness of males, which transmit their genes via the pollen grains on the pollinator's face, nor does it reduce the fitness of females, whose spikes do not produce pollen and are not damaged. The fitness of hermaphrodites, however, whose spikes produce pollen as well as pistils, is significantly reduced since any investment made in female structures is lost due to pollinator damage.

That dioecism is an ESS in *Freycinetia reineckei* can be shown through examination of the fitness set obtained by graphing the relative fitnesses of males, females, and bisexuals.[27,33] Let us assume that an hermaphrodite disperses some fraction m of the pollen dispersed by a single male, and produces some fraction f of seed produced by a single female. Thus as m approaches 1, the pollen dispersed by an hermaphrodite approaches that dispersed by a male; conversely, as f approaches 1, the seed set by an hermaphrodite approaches that set by a female. It can be shown[27] that an hermaphroditic population can be invaded by dioecious mutants only if $m + f < 1$. Similarly, if $m + f > 1$, the population is resistant to invasion. By graphing m versus f a fitness set can be constructed. (Fig. 4.2), which, if concave, indicates $m + f < 1$ and, if con-

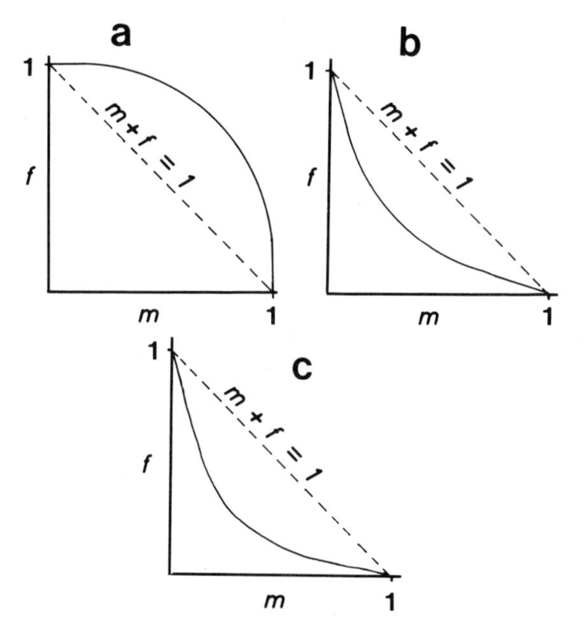

Fig. 4.2. Fitness sets representing different ESS where m is relative male fitness and f is relative female fitness. (a) The fitness set is convex, indicating the monomorphic population to be invulnerable to invasion by dimorphic mutants; thus a monomorphic stategy is the ESS. (b) The fitness set is concave, indicating that the monomorphic population can be successfully invaded by dimorphic mutants; thus a dimorphic strategy is an ESS. (c) Empirically calculated fitness set for *Freycinetia reineckei* lianas in Samoa; its concavity indicates a dimorphic strategy to be the ESS at modular level 1.

vex, indicates $m + f > 1$. The fitness set calculated[33] for *Freycinetia reineckei* (Fig. 4.2) is clearly concave, showing dioecism to be the ESS.

Since *Freycinetia reineckei* plants produce only a single shoot per year, the large flying foxes and birds that pollinate it[33,35] maintain the population in a dioecious condition. If, however, these plants could produce multiple shoots per year, the same selective pressure could produce, as Willson[129] points out, a monoecious system. Monoecism has recently been discovered[40] in the chiropterophilous species *Freycinetia scandens,* where some lateral shoots off a single primary axis produce male inflorescences while others produce female inflorescences.

The genus *Freycinetia* therefore illustrates how a single factor (e.g., the feeding behavior of pollinators) can select for different breeding systems (e.g., dioecism versus monoecism) in closely related plants possessing slightly different architecture. In both cases, selection favors a dimorphic sexual strategy, but this dimorphism is expressed at different modular levels due to the different morphologies of the species. The case of anisogamy illustrates, conversely, that selective pressures favoring dimorphism at one modular level (male and female gametes) may have little to do with selection for dimorphic or monomorphic strategies at other levels of morphological complexity. In both cases, however, the analysis of ESS allows us to explore the evolution of sexual strategies at different modular levels. Fitness sets calculated for each modular level of a plant can be examined for convexity, which would indicate a monomorphic strategy to be the ESS at that level, or for concavity, which would indicate a dimorphic strategy to be the ESS at that level. An alternative but equivalent algebraic technique to determine the ESS is to use the product theorem[24] that states that the ESS values of m and f are those that maximize the product mf. Most, if not all, current problems in plant breeding system evolution may be usefully analyzed in these terms. In short, one can ask at a particular morphological level, are the modules monomorphic or dimorphic, and is the population as a whole resistant to invasion by mutant individuals possessing modules of a different strategy?

MONOMORPHIC AND DIMORPHIC SEXUAL STRATEGIES IN PLANTS: A REVIEW BY MODULAR LEVEL

Although studies have been made of plant sexuality at various modular levels[33,51,85,94,112] the modular approach has yet to be applied rigorously to more than a few species of plants. Therefore this review deals with only three modular levels, namely, the flower, the inflorescence, and the individual, although a number of other modular levels can be identified and could be studied in many taxa. In the following discussion it should be noted that evolution of a monomorphic or dimorphic strategy at a particular modular level may be driven by selection operating at a higher level of morphological complexity. Thus selection for dimorphism at the level of the individual (dioecism, etc.) may of necessity drive the evolution of monomorphic flowers and inflorescences. It should also be noted that phylogenetic constraints may limit evolution at one morphological level, but not at another.[125]

Flowers

The majority of angiosperm species (though no greater than 75%, as indicated by Yampolsky and Yampolsky[132]) produce monomorphic flowers. Plants producing monomorphic flowers, when such a strategy is an ESS, may have a number of advantages,[24,82] including facilitation of pollination in self-compatible species (particularly in cleistogamous flowers[28]) and efficiency of pollinator attraction (since both male and female functions share the same floral display),[11,26] although the optimal number of flowers to perform both male and female reproductive functions may vary.[110] Indeed, in some entomophilous plants such as species of *Solanum* (Solanaceae) where pollen is the only pollinator reward,[111] monomorphic flowers may be necessary for stigmas to be pollinated, unless the plant relies on some system of "mistake" pollination.[8] Male and female reproductive effort can be controlled in monomorphic flowers by varying the relative allocation to stamens and pistils within flowers. Thus the first flowers produced in *Muntingia calabura* (Elaeocarpaceae) have as few as 10 stamens and large pistils, while those produced later in the season have up to 100 stamens, but small pistils.[18] In *Gilia achilleifolia* (Polemoniaceae), allocation to male reproductive effort decreases with increased rate of selfing.[103] Given the advantages of cost-sharing for male and female functions, the ease of self fertilization, and the possibility that temporal separation of male and female activity, if necessary, can be facilitated through dichogamy, it is understandable why monomorphic flowers predominate in angiosperms. Recently, two other conditions, differential timing of reproduction through pollen and ovules and differing limiting resources for pollen and ovules, have been suggested as leading to an ESS for monomorphic strategies at the level of the flower.[58]

On the other hand, dimorphic flowers may become an ESS for a variety of reasons. Preferential feeding on one type of sexual organ by predators or pollinators may favor dimorphic flowers.[17,33] In anemophilous pollination systems, requirements for pollen dispersal from a flower may be incompatible with those for pollen receipt. For example, in the dioecious ephydrophilous seagrass, *Halodule pinnifolia* (Cymodoceaceae), the stamens are erect and exposed to the air at low tide, allowing dehiscence to occur while the long, female stigmas remain flaccid so they can float on the surface as the tide rises and remain oriented in the direction of the current.[37] Another example of selective pressures favoring dimorphic flowers may be found in species where optimal flowering phenologies differ significantly from optimal fruiting phenologies. In these cases the production of different sexes of flowers may provide a means of uncoupling these two phenomena. For example, if selection by frugivores favored a single flush of fruit to be produced, while selection by trap-lining pollinators favored a long flowering period, these two disparate phenological requirements could be met by having a long period of staminate flowering, but a short flowering period of separate pistillate flowers. Complex flowering phenologies in monoecious plants are well known, e.g., in *Cupania quatemalensis,* plants first bear staminate, then pistillate, then finally more staminate flowers.[13] Cruden and Hermann-Parker[42] argue that such temporal differences in sexual expression are adaptive in preventing geitonogamy while lacking some of the disadvantages associated with true dioecism. Andromonoecious species (i.e., those producing both perfect and staminate flowers on the same individual) such as *Solanum carolinense* (Solanaceae) may increase their male fitness component by increasing the number of male flowers without the ovule wastage that would occur in a monomorphic species if male contribution could only be increased through an increase in the number of hermaphroditic flowers.[107]

Production of dimorphic flowers also permits differential placement or positioning on the plant itself. For example, in many monocotyledonous families such as the Araceae or Cyclanthaceae, female flowers occur at the base of the inflorescence and male flowers at the distal end.[46] The existence of dimorphic flowers may allow subtle fine-tuning of the sexual expression of the plant to the requirements and opportunites of the environment. Of course, dimorphic flowers may also become an ESS through selection at higher levels of morphological complexity, such as selection for dimorphic inflorescences or even dimorphic individuals.

Perhaps the most puzzling cases of dimorphic flowers and those in need of greatest study from an evolutionary perspective involve cases of cryptic polymorphism. In the monoecious species *Cupania guatemalensis* (Sapindaceae), pistillate flowers have well developed stamens with anthers containing pollen grains, but the anthers do not dehisce.[13] Similarly, the fiddlewood tree *Citharexylum fruticosum* (Verbenaceae) in Florida produces superficially monomorphic flowers containing both stamens and pistils, but closer examination reveals that the stamens in female flowers never produce pollen or dehisce.[114]

Even more cryptic is the case of *Solanum appendiculatum* where both male and female flowers produce pistils, stamens, and well-formed pollen, but the pollen produced by females does not germinate to produce pollen tubes.[2,3] Similar situations occur in the morphologically androdioecious (but functionally dioecious) genus *Saurauia*[61] (Actinidiaceae) and in the related dioecious genus *Actinidia* (Actinidaceae).[102] In each of these three cases, morphologically monomorphic flowers may be important for facilitating "mistake pollination"[8] since pollen is a major pollinator reward. The functional dimorphisms in these cases may relate to selection for outbreeding potential.

Inflorescences

Many of the arguments advanced in favor of monomorphic flowers being an ESS, such as facilitation of fertilization, efficiency of pollinator attraction, etc., apply to monomorphic inflorescences. Even pollinator movement, and thus amount of outcrossing, can also be controlled by monomorphic inflorescences[131] with mixtures of cleistogamous and chasmogamous flowers[51,101,119,127] or with mixtures of dichogamous flowers, e.g., those of the protandrous species *Lobelia cardinalis* (Campanulaceae), whose phenology can in turn be affected by removal of pollen from flowers by pollinators.[48]

However, some conditions can lead to dimorphic inflorescences being an ESS. As previously mentioned, vertebrates that damage male flowers more frequently than female flowers during pollination cause dimorphic inflorescences to be the ESS in *Freycinetia,* as well as in other paleotropic genera, such as *Collospermum* or *Astelia* in the Liliaceae with similar pollination systems.[33,43] Anemophilous pollination systems may also select for dimorphic inflorescences, particularly where aerodynamic requirements for pollen dispersal differ from those for pollen reception.[90,91] Wind tunnel experiments[36] indicate that the tristichous pistillate inflorescences of *Pandanus tectorius* (Pandanaceae) function superbly as pollen collection devices. They show hydrodynamic form that is similar to some marine invertebrate filter feeders, while the pendulous staminate inflorescences seem well-adapted to pollen dispersal. Such instances of aerodynamic divergence between male and female functions may partially explain some of the correlations between anemophily and sexually dimorphic breeding systems.[14,57,66,93]

Dimorphic inflorescences also permit a degree of control of male/female reproductive effort by a single plant. For example, in andromonoecious umbellifers, the proportion of male flowers increases in umbels produced later in the season, in effect creating protogyny at the level of the inflorescence, although the earlier-produced hermaphroditic flowers are themselves protandrous. This allows avoidance of competition for resources between male and female functions as well as allowing careful regulation of self- and out-crossed offspring.[85,86] Outcrossing can also be controlled by degree of aggregation of flowers within an umbel. In *Thaspium trifoliatum* (Umbelliferae) and *Zizia trifoliata* (Umbelliferae) umbels of flowers with receptive stigmas are compact, encouraging pollinator movements within the inflorescence, while inflorescences with dehiscing anthers are more open and encourage movement between inflorescences.[74] Finally, the benefits of dichogamy are accrued through multiple flowering cycles within a single season in species such as *Aralia hispida* (Araliaceae), which produces synchronized cycles of protandry as successive umbel orders flower; in this case periods of anthesis of successive orders do not overlap.[112] Details of dimorphic inflorescences as a factor controlling male and female reproductive effort may be found elsewhere in this book.[120]

Individuals

Perhaps the greatest amount of study on plant breeding systems has been made on the level of the genetic individual. Sexual polymorphism at this level includes several different breeding systems as recognized by the Linnaean system, including dioecism (separate male and female individuals), gynodioecism (separate hermaphroditic and female individuals), androdioecism (separate hermaphroditic and male individuals), and polygamodioecism (separate male, female, and hermaphroditic individuals), although in practice many plant populations are found to have breeding systems intermediate between these various extremes. Recently, Lloyd and Bawa[83] questioned whether polymorphic (trioecious, polygamodioecious, etc.) populations exist in nature, arguing that "departures from strict unisexuality occur in many male and female morphs" and therefore separating bifunctional individuals into a separate class lumps together "inconsistent males and inconstant females of many species into a separate class." Here again, sexual allocation theory can be useful in determining if sexual polymorphism is due only to statistical aberrations, or if it can indeed, as has been argued by some workers,[55] represent an ESS in some situations.

Of these dimorphic breeding systems, dioecism has received the most attention and has been the subject of several excellent reviews.[14,24,84,129] A number of population genetic models have been made to outline potential pathways to dioecism[21,22,77,78,80,97–100] and a number of different selective forces have been proposed to drive the evolution of dioecism from various starting points.[24] One of the long-established suggestions is that selection for outbreeding could lead to the evolution of dioecism. Darwin[44] (p. 279) struggled with this explanation, but eventually rejected it arguing, "There would be no such conversion, unless pollen was already carried regularly by insects or by the wind from one individual to the other; for otherwise every step towards dioeciousness would lead towards sterility."

During the last decade, a variety of different ecological factors have been proposed as driving the evolution of dioecism, although some workers believe selection for outbreeding to predominate.[113] Clearly, in plants with large clone sizes but small pollen shadows, such as the dioecious seagrasses,[34] convincing arguments can be made for selection for outbreeding as the major force in the evolution of dioecism. In *Aralia* species, for example, with synchronized dichogamy, development of large clone sizes may disrupt synchrony of flowering, greatly increasing the odds of inbreeding.[12] Some recent population genetics models consider deleterious effects of selfing and inbreeding depression.[21–23,76,79] Even though there are many other mechanisms that can promote outbreeding, they may not be as easily derived as dioecisms,[9] nor may they function as well in populations with large clones. Similarly, the presumed advantages of self-incompatibility systems over dioecism as an outbreeding mechanism disappear in plants of large clone size at the edge of their range, since the number of potential mates having a compatible genotype is likely to be low.[4]

All plausible population genetics models for the evolution of gynodioecism and androdioecism that do not involve selfing[100] indicate that males or females can be maintained in such populations if they have greater pollen or ovule fertility than hermaphrodites. This prediction, which was first demonstrated by Darwin,[44] has been recently confirmed for several gynodioecious species such as *Plantago lanceolata*[73] (Plantaginaceae) in California, *Carpodetus serratus*[105] (Escalloniaceae), and a variety of umbellifers[123,124] in New Zealand, where females consistently achieve higher seed set than hermaphrodites. In *Iris douglasiana* (Iridaceae), hermaphrodites, which flower later than male steriles, lose a greater number of seeds to larval predation.[115] However, fitness differences between male-steriles and hermaphrodites in *Limnanthes douglasii* (Limnanthaceae) are inadequate to explain the observed maintenance of the nucleo-cytoplasmic male sterility.[72]

It has been suggested that gynodioecism allows more efficient coupling of reproductive effort to environmental resources.[47] Temporal differences in phenology can occur; in the gynodioecious species *Gingidia decipiens* (Umbelliferae), hermaphrodites and females begin flowering at approximately the same time, but hermaphrodites reach peak flowering and finish flowering later than females.[122]

Models of gynodioecy involving the effects of selfing postulate that females are maintained in the population because their offspring are obligately outcrossed and hence more fit than the progeny of hermaphrodites resulting from self-fertilization. Obviously this mechanism can work only if female reproduction is not unduly constrained by limited pollination. Observed correlations between selfing rates and frequency of female plants in gynodioecious populations of *Bidens* (Compositae)[109] lend plausibility to this outcross-advantage model, although the specific rates of selfing could not by themselves completely account for the frequency of females. Clearly a combination of factors, including superiority of outcrossed progeny, increased ovule production by females, differential predation on the sexes,[33] and differential adult survival,[116] could be responsible for the maintenance of females in gynodioecious populations.

Dimorphic individuals also permit differences between male and female phenologies. For example, although the majority of *Ficus* (Moraceae) species are monoecious, *Ficus carica* is gynodioecious, with staminate trees producing 3 crops of figs (all containing short-styled sterile pistillate flowers). The staminate trees are responsible for maintenance of the pollinating wasp, while the pistillate trees produce only one crop of viable fruits per year (an earlier flush of figs falls before pollination). Since the wasps cannot lay their eggs on the long-styled flowers of the pistillate synconia (the pseudocarp or hollow receptacle that bears the flowers), pollination of the pistillate trees depends on "mistake pollination."[8,117] A similar case of one sex maintaining the pollinator occurs in the polygamodioecious species *Fuchsia lyciodes* (Onagraceae), where hermaphrodites produce up to six times the nectar of female plants and feed the hummingbird pollinators for a much longer period than female plants.[6] "Mistake pollination" has also been found in *Rubus chamaemorus* (Rosaceae), which is pollinated by syrphids and bumblebees.[1]

Another possible advantage for sexual polymorphism is sexual niche partitioning. Although this idea has a long history,[31] it was brought to recent prominence through work in North America[27,56] and North Wales.[31,92,95] Under this "Jack Sprat" scenario, if certain environmental patches are better suited to male than female reproductive functions, then dioecious mutants may be able to successfully invade a monomorphic population if there is a tight coupling of different reproductive requirements to the different types of patches. This requires, however, different patterns of resource allocation to male and female reproductive activities.[67] A number of cases of sexual niche partitioning in both temperate regions and the tropics have been documented in recent years,[31,50,54,56,60,92,121] although sexual niche partitioning is by no means ubiquitous in dimorphic species.[16,63,64,89,118] An alternative explanation for niche differences in dioecious plants is that they evolve subsequent to the establishment of dioecism in response to deleterious intersexual competition (as has been suggested for some animal species).[31] However, empirically distinguishing between these two alternative explanations for niche differences has proved to be exceedingly difficult. Recent investigations[133] indicate that observed niche differences between male and female plants mirror distinct physiological differences between the sexes, but the question why some, but not all, dioecious species partition their niche remains unanswered.

A possible disadvantage of dimorphic strategies at the level of the individual is the reduced ability to colonize islands. "Baker's Law"[108] suggests that self-compatible taxa should be favored in long-distance dispersal, but Bawa has suggested that dioecious taxa may have been disproportionately successful in colonizing islands.[15] Analyses of numerous oceanic island floras[10] indicate that in this circumstance dioecious taxa do not do better in long-distance dispersal than self-compatible hermaphrodites, but neither do they fare worse. This result was unexpected and was not predicted by extensive numerical simulations.[36]

One factor mitigating the disadvantages of dioecism in island colonization, however, is the existence of "leaky dioecy"[10,36] or occasional departures from strict dioecism. Although numerous cases from island floras have been described,[10] one of the best documented mainland cases is that of the strawberry *Fragaria chiloensis* (Rosaceae) where, in 12 separate populations, polygamodioecy as well as hermaphroditism was discovered.[63] An even more important and yet surprisingly little-studied factor mitigating the deleterious effects of dioecism on colonization ability is apomixis. Recent studies[36] in Tahiti and Hawaii reveal *Pandanus tectorius* to be facultatively apomictic, with somatic polyembryony resulting in viable seed if pollination does not occur. Facultative apomixis has allowed *Pandanus tectorius* to colonize numerous islands and atolls throughout the Pacific and Indian oceans despite its dioecious condition. Facultative apomixis is not limited, however, solely to island taxa. In the nineteenth century, Kerner von Marilaun, curious about the absence of male plants in coastal populations of the dioecious aquatic thallophyte *Chara crinata* (Characeae), found it, a variety of dioecious mosses, and the flowering species *Mercurialis annua*

(Euphorbiaceae), *Gnaphalium alpinum* (Compositae), and *Coelobogyne ilcifolia* (Euphorbiaceae) to be apomictic.[71] It is likely that further investigations will reveal facultative apomixis to be widespread in other dioecious taxa.

A perennial habit as well as fleshy fruits may also assist dioecious plants in establishment during long-distance dispersal.[19,36] Analyses of various floras indicate distinct correlations between perennation, fleshy fruits, and dioecism.[14,29,41,53,57,59,106] Clearly fleshy fruits can be seen as one of a set of dispersal strategies in dioecious plants [e.g., floating fused syncarps in *Pandanus* or tumbling glomerules in *Spinacea* (Chenopodiaceae)] that ensures dissemination of entire breeding populations rather than just single isolated seeds. However, both Bawa[14] and Givnish[59] suggest that correlations between fleshy fruits and dioecism evidences selection pressure for dioecism in fleshy-fruited hermaphroditic populations by frugivores that prefer large fruit displays.

Sexual selection has also been suggested[128,129] as a possible factor in the evolution of sexual polymorphism, particularly where competition for pollinators gives disproportionately high paternity to plants capable of reallocating female reproductive resources to male functions. Pollinator response to floral dimorphisms has been studied in the androdioecious rain forest species *Xerospermum intermedium*[5] (Sapindaceae) as well as in the temperate dioecious species *Silene dioica* (Caryophyllaceae).[52] Some support for sexual selection in plants comes from the finding that sexual selection is probably the primary determinant of relative flower number in the gynodioecious species *Thymus vulgaris* (Labiatae).[30] In a similar vein, Beach[19] has suggested that temporal differences in pollinator movement can cause some tropical trees to have high success at male reproductive functions and others high success at female reproductive functions, thus creating disruptive selection favoring dioecism. In *Mussaenda* (Rubiaceae), dioecy appears to have evolved from heterostyly because of the rarity of cross-pollination from low anthers to short styles.[7]

Although the evolution of plant breeding systems has long been the object of much interest, it is only within the last several years that the appropriate tools have been created to study it in sufficient detail. These tools include careful analysis of the population genetics controlling different breeding systems (see Bertin,[20] this volume), and the use of evolutionary game theory to arrive at a quantitative theory of sex allocation.[24,27,88] If these theoretical tools are combined with an appreciation of plant morphology and the modular construction of plants, significant advances can be made in understanding not only the evolution of plant breeding systems, but larger questions concerning the evolution of sex and sexuality in all organisms.

ACKNOWLEDGMENTS

I thank Herbert Baker, Spencer Barrett, Ric Charnov, Tom Elmqvist, John Harper, Kim Harper, Josephine Kenrick, Bruce Knox, David Lloyd, Jon Lovett Doust, Lesley Lovett Doust, Doug Schemske, Barry Tomlinson, and Don Waller for criticisms of an earlier version of this chapter and T. Hough for assistance with the illustrations. This work was supported by a University of Melbourne Research Fellowship and a National Science Foundation Presidential Young Investigator Award BSR-84 52090.

REFERENCES

1. Agren, J., Elmqvist, T., and Tunlid, A., Pollination by deceit, floral sex ratios and seed set in dioecious *Rubus chamaemorus* L., *Oecologia* **70,** 332–338 (1986).

2. Anderson, G. J., Dioecious *Solanum* species of hermaphroditic origin is an example of broad convergence, *Nature (London)* **282,** 837–838 (1979).
3. Anderson, G. J., and Levine, D. A., Three taxa constitute the sexes of a single dioecious species of *Solanum, Taxon* **31,** 667–672 (1982).
4. Anderson, G. J., and Stebbins, G. L., Dioecy versus gametophytic self-incompatibility: A test. *Am. Nat.* **124,** 423–428 (1984).
5. Appanah, S., Pollination of androdioecious *Xerospermum intermedium* Radlk. (Sapindaceae) in a rain forest, *Biol. J. Linn. Soc.* **18,** 11–34 (1982).
6. Atsatt, P. R., and Rundel, P. W., Pollinator maintenance vs. fruit production: Partitioned reproductive effort in subdioecious *Fuchsia lycioides, Ann. Missouri Bot. Gard.* **69,** 199–208 (1982).
7. Baker, H. G., Reproductive methods as factors in speciation in flowering plants, *Cold Spring Harbor Symp. Quant. Biol.* **24,** 177–191 (1959).
8. Baker, H. G, "Mistake" pollination as a reproductive system, with special reference to the Caricaceae, in *Tropical Trees: Variation, Breeding, and Conservation* (J. Burley and B. T. Styles, eds.), pp. 161–170. Academic Press, London, 1976.
9. Baker, H. G., Some functions of dioecy in seed plants, *Am. Nat.* **124,** 149–158 (1984).
10. Baker, H. G., and Cox, P. A., Further thoughts on islands and dioecism, *Ann. Missouri Bot. Gard.* **71,** 230–239 (1984).
11. Baker, H. G., and Hurd, P. D., Intrafloral ecology *Annu. Rev. Ecol. Syst.* **13,** 385–414 (1968).
12. Barrett, S. C. H., Variation in floral sexuality of diclinous *Aralia, Ann. Missouri Bot. Gard.* **71,** 278–288 (1984).
13. Bawa, K. S., The reproductive biology of *Cupania guatemalensis* (Sapindaceae), *Evolution* **31,** 52–63 (1977).
14. Bawa, K. S., Evolution of dioecy in flowering plants, *Annu. Rev. Ecol. Syst.* **11,** 15–39 (1980).
15. Bawa, K. S., Outcrossing and the incidence of dioecism in island floras, Am. Nat. **119,** 866–871 (1982).
16. Bawa, K. S., and Opler, P. A., Spatial relationships between staminate and pistillate plants of dioecious tropical forest trees, *Evolution* **31,** 64–68 (1977).
17. Bawa, K. S., and Opler, P. A., Why are pistillate inflorescences of *Simarouba glauca* eaten less than staminate inflorescences? *Evolution* **32,** 673–767 (1978).
18. Bawa, K. S., and Webb, C. J., Floral variation and sexual differentiation in *Muntingia calabura* (Elaeocarpaceae), a species with hermaphrodite flowers, *Evolution* **37,** 1271–1282 (1983).
19. Beach, J., Pollinator foraging and the evolution of dioecy, *Am. Nat.* **118,** 572–577 (1981).
20. Bertin, R., Paternity in plants, in *Plant Reproductive Ecology: Patterns and Strategies* (J. Lovett Doust and L. Lovett Doust, eds.) Chapter 3. Oxford Univ. Press, New York, 1988.
21. Charlesworth, B., and Charlesworth, D., Population genetics of partial male sterility and the evolution of monoecy and dioecy, *Heredity* **41,** 137–153 (1978).
22. Charlesworth, B., and Charlesworth, D., A model for the evolution of dioecy and gyndioecy, *Am. Nat.* **112,** 975–997 (1978).
23. Charlesworth, B., and Charlesworth, D., The evolutionary genetics of sexual systems in flowering plants, *Proc. R. Soc. London, B.,* **205,** 513–530 (1979).
24. Charnov, E. L., *The Theory of Sex Allocation.* Princeton University Press, Princeton, 1982.
25. Charnov, E. L., and Bull, J. J., When is sex environmentally determined? *Nature (London)* **266,** 828–830 (1977).
26. Charnov, E. L., and Bull, J. J., Sex allocation, pollinator attraction, and fruit dispersal in cosexual plants, *J. Theor. Biol.* **118,** 321–325 (1986).
27. Charnov, E. L., Maynard Smith, J., and Bull, J., Why be an hermaphrodite? *Nature (London)* **289,** 27–33 (1976).
28. Clay, K., Environmental and genetic determinants of cleistogamy in a natural population of the grass *Danthonia spicata, Evolution* **31,** 32–46 (1982).
29. Conn, J. S., Wentworth, T. R., and Blum, U., Patterns of dioecism in the flora of the Carolinas USA, *Am. Midl. Nat.* **103,** 310–315 (1980).
30. Couvet, D., Henry, J., and Gouyon, P., Sexual selection in hermaphroditic plants: The case of gynodioecy, *Am. Nat.* **126,** 22–299 (1985).
31. Cox, P. A., Niche partitioning between sexes of dioecious plants, *Am. Nat.* **117,** 295–307 (1981).
32. Cox, P. A., Bisexuality in the Pandanaceae: New findings in the genus *Freycinetia, Biotropica* **13,** 195–198 (1981).

33. Cox, P. A., Vertebrate pollination and the maintenance of dioecism in *Freycinetia, Am. Nat.* **120,** 65–80 (1982).
34. Cox, P. A., Search theory, random motion, and the convergent evolution of pollen and spore morphologies in aquatic plants, *Am. Nat.* **121,** 9–31 (1983).
35. Cox, P. A., Chiropterophily and ornithophily in *Freycinetia* in Samoa, *Plant Syst. Evol.* **144,** 277–290 (1984).
36. Cox, P. A., Islands and dioecism: Insights from the reproductive ecology of *Pandanus tectorius* in Polynesia, in *Studies on Plant Demography: a Festschrift for John L. Harper* (J. White, ed.), pp. 359–372. Academic Press, London, 1985.
37. Cox, P. A., and Knox, R. B., Pollination postulates and two-dimensional pollination in hydrophilous monocotyledons, *Ann. Missouri Bot. Gard.,* in press.
38. Cox, P. A., and Sethian, J., Search, encounter rates, and the evolution of anisogamy, *Proc. Natl. Acad. U.S.A* **81,** 6078–6079 (1984).
39. Cox, P. A., and Sethian, J., Gamete motion, search, and the evolution of anisogamy, oogamy, and chemotaxis, *Am. Nat.* **125,** 74–101 (1984).
40. Cox, P. A., Wallace, B., and Baker, I., Monoecism in the genus *Freycinetia, Biotropica* **16,** 313–314 (1984).
41. Croat, T. B., The sexuality of the Barro Colorado Island flora (Panama), *Phytologia* **42,** 312–348 (1979).
42. Cruden, R. W., and Hermann-Parker, S. M., Temporal dioecism: An alternative to dioecism? *Evolution* **31,** 863–866 (1977).
43. Daniel, M., Feeding by the short-tailed bat *(Mystacina tuberculata)* on fruit and possibly nectar, *N.Z. J. Zool.* **3,** 391–398 (1976).
44. Darwin, C., *The Different Forms of Flowers on Plants of the Same Species.* John Murray, London, 1877.
45. Darwin, E., *Phytologia.* Johnson, London, 1800.
46. De Guevara, L. C., Unisexuality in monocotyledons in families of the Venezuelan flora, *Acta Bot. Venez.* **10,** 19–86 (1975).
47. Delannay, X., Gynodioecism in angiosperms, *Naturalistes Belges* **59,** 223–237 (1978).
48. Devlin, B., and Stephenson, A. G., Sex differential floral longevity, nectar secretion, and pollinator foraging in a portandrous species, *Am. J. Bot.,* **72,** 303–310 (1985).
49. DeWreede, R., and Klinger, T., Reproductive strategies in algae, in *Plant Reproductive Ecology: Patterns and Strategies* (J. Lovett Doust and L. Lovett Doust, eds.), Chapter 12. Oxford Univ. Press, New York, 1988.
50. Dommee, B., Assouad, M. W., and Valderyon, G., Natural selection and gynodioecy in *Thymus vulgaris, Bot. J. Linn. Soc.* **77,** 17–28 (1978).
51. Ellstrand, N. C., Lord, E. M., and Eckard, K. J., The inflorescence as a metapopulation of flowers: Position-dependent differences in function and form in the cleistogamous species *Collomia grandiflora* Doug. ex Lindl. (Polemoniaceae), *Bot. Gaz.* **145,** 329–333 (1984).
52. Elmqvist, T., Tunlid, A., and Agren, J., Bumble bee foraging and sexual dimorphisms in *Silene dioica* L., manuscript in preparation.
53. Flores, S., and Schemske, D. W., Dioecy and monoecy in the flora of Puerto Rico and the Virgin Islands: Ecological correlates, *Biotropica* **16,** 132–139 (1984).
54. Fox, J. F., and Harrison, A. T., Habitat assortment of sexes and water balance in a dioecious grass, *Oecologia* **49,** 233–235 (1981).
55. Freeman, D. C., and Vitale, J. J., The influence of environment on the sex ratio and fitness of spinach, *Bot. Gaz.* **146,** 137–142 (1985).
56. Freeman, D. C., Klikoff, L. G., and Harper, K. T., Differential resource utilization by the sexes of dioecious plants, *Science* **193,** 597–599 (1976).
57. Freeman, D. C., Harper, K. T., and Ostler, W. K., Ecology of plant dioecy in the intermountain region of North America and California, USA, *Oecologia* **44,** 410–417 (1980).
58. Geber, M. A., and Charnov, E. L., Sex allocation in hermaphrodites with partial overlap in male/female resource inputs, *J. Theor. Biol.* **118,** 33–43 (1986).
59. Givnish, T. J., Ecological constraints on the evolution of breeding systems in seed plants: Dioecy and dispersal in gymnosperms, *Evolution* **34,** 959–972 (1980).
60. Grant, M. C., and Mitton, J., Elevational gradients in adult sex ratios and sexual differentiation in vegetative growth rates of *Populus tremuloides* Michx., *Evolution* **33,** 914–918 (1979).
61. Haber, W. A., and Bawa, K. S., Evolution of dioecy in *Saurauia* (Dilleniaceae), *Ann. Missouri Bot. Gard.* **71,** 289–293 (1984).

62. Halle, F., Oldemann, R., and Tomlinson, P. B., *Tropical Trees and Forests.* Springer-Verlag, Berlin, 1978.
63. Hancock, J. F., and Bringhurst, R. S., Hermaphroditism in predominantly dioecious populations of *Fragaria chiloensis, Bull. Torrey Bot. Club* **106,** 229–231 (1979).
64. Hancock, J. F., and Bringhurst, R. S., Sexual dimorphism in the strawberry *Fragaria chiloensis, Evolution* **34,** 762–768 (1980).
65. Harper, J. L, and Ogden, J., The reproductive strategy of higher plants I. The concept of strategy with special reference to *Senecio vulgaris* L., *J. Ecol.* **58,** 681–698 (1970).
66. Heslop-Harrison, J., Sexuality in angiosperms, *Plant Physiol.* **6C,** 133–289 (1972).
67. Hoffman, A. J., and Alliende, M. C., Interactions in the patterns of vegetative growth and reproduction in woody dioecious plants, *Oecologia* **61,** 109–114 (1984).
68. Horovitz, A., Is the hermaphrodite flowering plant equisexual? *Am. J. Bot.* **65,** 485–486 (1978).
69. Horovitz, A., and Harding, J., The concept of male outcrossing in hermaphroditic higher plants, *Heredity* **29,** 223–236 (1972).
70. Hutchinson, J., *The Families of Flowering Plants,* 3rd Ed. Clarendon, Oxford, 1973.
71. Kerner Von Marilaun, A., *The Natural History of Plants* (F. W. Oliver, translator), Vol. II. Gersham, London, 1895.
72. Kesseli, R., and Jain, S. K., An ecological genetic study of *Limnanthes douglasii* (Limnanthaceae), *Am. J. Bot.* **71,** 775–786 (1984).
73. Krohne, D. T., Baker, I., and Baker, H. G., The maintenance of the gynodioecious breeding system in *Plantago lanceolata, Am. Midl. Nat.* **103,** 269–279 (1980).
74. Lindsey, A. H., and Bell, C. R., Reproductive biology of Apiaceae. II. Cryptic specialization and floral evolution in *Thaspium* and *Zizia, Am. J. Bot.* **73,** 231–247 (1985).
75. Linnaeus, C., *Genera Plantarum, Editio Quinta.* Laurentii Salvii, Holmiae, 1754.
76. Lloyd, D. G., The genetic contributions of individual males and females in dioecious and gynodioecious angiosperms, *Heredity* **32,** 45–51 (1974).
77. Lloyd, D. G., The maintenance of gynodioecy and androdioecy in angiosperms, *Genetica* **45,** 325–339 (1975).
78. Lloyd, D. G., The transmission of genes via pollen and ovules in gynodioecious angiosperms, *Theor. Popul. Biol.* **9,** 229–316 (1976).
79. Lloyd, D. G., Evolution toward dioecy in heterostylous populations, *Plant Syst. Evol.* **131,** 71–80 (1979).
80. Lloyd, D. G., Parental strategies of angiosperms, *N.Z.J. Bot.* **17,** 595–606 (1979).
81. Lloyd, D. G., Sexual strategies in plants III: A quantitative method for describing the gender of plants, *N. Z. J. Bot.* **18,** 103–108 (1980).
82. Lloyd, D. G., Selection of combined versus separate sexes in seed plants, *Am. Nat.* **120,** 571–585 (1982).
83. Lloyd, D. G., and Bawa, K. S., Modification of the gender of seed plants in varying conditions, *Evol. Biol.* **17,** 255–338 (1984).
84. Lloyd, D. G., and Webb, C. J., Secondary sex characters in plants, *Bot. Rev.* **43,** 177–216 (1977).
85. Lovett Doust, J., Floral sex ratios in andromonoecious Umbelliferae, *New Phytol.* **85,** 265–273 (1980).
86. Lovett Doust, J., and Lovett Doust, L., Life-history patterns in British Umbelliferae: A review, *Bot. J. Linn. Soc.* **85,** 179–194 (1982).
87. Mandelbrot, B. B., *Fractals: Form, Chance and Dimension.* Freeman, San Francisco, 1977.
88. Maynard Smith, J., *Evolution and the Theory of Games.* Cambridge Univ. Press, Cambridge, 1982.
89. Melampy, M. N., and Howe, H. F., Sex ratio in the tropical tree *Triplaris americana* (Polygonaceae), *Evolution* **31,** 867–872 (1977).
90. Niklas, K. J., Pollination and airflow patterns around conifer ovulate cones, *Science* **217,** 442–444 (1982).
91. Niklas, K. J., The motion of windborne pollen grains around conifer ovulate cones: Implications on wind pollination, *Am J. Bot.* **71,** 356–374 (1984).
92. Onyekwelu, S., and Harper, J. L., Sex ratio and niche differentiation of spinach (*Spinacia oleracea* L.), *Nature (London)* **282,** 609–611 (1979).
93. Pijl, L. van der, *The Principles of Pollination Ecology.* Pergamon Press, Oxford, 1979.
94. Primack, R. B., and Lloyd, D. G., Sexual strategies in plants IV. The distribution of gender in two monomorphic shrub populations, *N.Z. J. Bot.* **18,** 109–11 (1980).
95. Putwain, P. D., and Harper, J. L., Studies in the dynamics of plant populations, V., Mechanisms governing the sex ratios in *Rumex acetosa* and *R. acetosella, J. Ecol.* **60,** 113–129 (1972).

96. Robbins, L., and Travis, J., Examining the relationship between functional gender and gender specialization in hermaphroditic plants, *Am. Nat.* **128,** 409–415 (1986).
97. Ross, M. D., Evolution of dioecy from gynodioecy, *Evolution* **24,** 827–828 (1970).
98. Ross, M. D., The evolution of gynodioecy and subdioecy, *Evolution* **32,** 174–188 (1978).
99. Ross, M. D., The evolution and decay of overdominance during the evolution of gynodioecy, subdioecy, and dioecy, *Am. Nat.* **116,** 607–620 (1980).
100. Ross, M. D., and Weir, B. S., Maintenance of males and females in hermaphrodite populations and the evolution of dioecy, *Evolution* **30,** 425–441 (1976).
101. Schemske, D. W., Evolution of reproductive characters in *Impatiens* (Balsaminaceae): The significance of cleistogamy and chasmogamy, *Ecology* **59,** 596–613 (1978).
102. Schmid, R., Reproductive anatomy of *Actinidia chinensis* (Actinidiaceae), *Bot. Jahrb. Syst. Pflanzengesch. Pflanzengeogr.* **100,** 149–195 (1978).
103. Schoen, D. J., Male reproductive effort and breeding system in an hermaphroditic plant, *Oecologia* **53,** 255–257 (1982).
104. Schlessman, M., Gender diphasy ("sex choice"), in *Plant Reproductive Ecology: Patterns and Strategies* (J. Lovett Doust and L. Lovett Doust, eds.), Chapter 7. Oxford Univ. Press, New York, 1988.
105. Shore, B. F., Breeding systems in *Carpodetus serratus, N.Z. J. Bot.* **16,** 179–184 (1978).
106. Sobrevila, C., and Arroyo, M. T. K., Breeding systems in a montane tropical cloud forest in Venezuela, *Plant Syst. Evol.* **140,** 19–38 (1982).
107. Solomon, B. P., Sexual allocation and andromonoecy: Resource investment in male and hermaphroditic flowers of *Solanum carolinense* (Solanaceae), *Am. J. Bot.* **73,** 1215–1221 (1986).
108. Stebbins, G. L., Self-fertilization and variability in the higher plants, *Am. Nat.* **41,** 337–354 (1957).
109. Sun, M., and Ganders, F. R., Female frequencies in gynodioecious populations correlated with selfing rates in hermaphrodites, *Am. J. Bot.* **73,** 1645–1648 (1986).
110. Sutherland, S., and Delph, L. F., On the importance of male fitness in plants: Patterns of fruit set, *Ecology* **65,** 1093–1104 (1984).
111. Symon, D. E., Sex forms in *Solanum* (Solanaceae) and the role of pollen collecting insects, in *The Biology and Taxonomy of the Solanaceae* (G. J. Hawker, R. N. Lester, and A. D. Skelding, eds.), pp. 385–397. Academic Press, London, 1979.
112. Thomson, J. D., and Barrett, S. C. H., Temporal variation of gender in *Aralia hispida* Vent. (Araliaceae), *Evolution* **35,** 102–1107 (1981).
113. Thomson, J. D., and Barrett, S. C. H., Selection for outcrossing, sexual selection, and the evolution of dioecy in plants, *Am. Nat.* **118,** 443–449 (1981).
114. Tomlinson, P. B., and Fawcett, P., Dioecism in *Citharexylum* (Verbenaceae), *J. Arnold Arbor. Harv. Univ.* **53,** 386–389 (1972).
115. Uno, G. E., Comparative reproductive biology of hermaphroditic and male sterile *Iris douglasiana* Herb. (Iridaceae), *Am J. Bot.* **69,** 818–823 (1982).
116. Van Damme, J. M. M., and Van Damme, R., On the maintenance of gynodioecy: Lewis' result explained, *J. Theor. Biol.* **121,** 339–350 (1986).
117. Valdeyron, G., and Lloyd, D. G., Sex differences and flowering phenology in the common fig *Ficus carica, Evolution* **33,** 673–685 (1979).
118. Wallace, C. S., and Rundel, P. W., Sexual dimorphism and resource allocation in male and female shrubs of *Simmondsia chinensis, Oecologia* **44,** 34–39 (1979).
119. Waller, D. M., Environmental determinants of outcrossing in *Impatiens capensis* (Balsaminaceae), *Evolution* **34,** 747–761 (1980).
120. Waller, D. M., Plant morphology and reproduction, in *Plant Reproductive Ecology: Patterns and Strategies* (J. Lovett Doust and L. Lovett Doust, eds.), Chapter 10. Oxford Univ. Press, New York, 1988.
121. Waser, N. M., Sex ratio variation in populations of a dioecious desert perennial, *Simmondsia chinensis, Oikos* **42,** 343–348 (1984).
122. Webb, C. J., Flowering periods in the gynodioecious species *Gingidia decipiens* (Umbelliferae), *N.Z. J. Bot.* **14,** 207–210 (1976).
123. Webb, C. J., Sex ratios in the New Zealand apioid Umbelliferae, *N.Z. J. Bot.* **18,** 121–126 (1979).
124. Webb, C. J., Test of a model predicting equilibrium frequencies of females in populations of gynodioecious angiosperms, *Heredity* **46,** 397–405 (1981).
125. Webb, C. J., Constraints on the evolution of plant breeding systems and their relevance to systematics, in *Plant systematics,* (W. F. Grant, ed.), Academic Press, Orlando, Florida, 1984.
126. White, J., The plant as a metapopulation, *Annu. Rev. Ecol. Syst.* **10,** 109–145 (1979).

127. Wilken, D. H., The balance between chasmogamy and cleistogamy in *Collomia grandiflora* (Polemoniaceae), *Am. J. Bot.* **69,** 1326–1333 (1982).
128. Willson, M. F., Sexual selection in plants, *Am. Nat.* **113,** 777–790 (1979).
129. Willson, M. F., *Plant Reproductive Ecology.* Wiley, New York, 1983.
130. Wright, S., *Evolution and the Genetics of Populations: Experimental Results and Evolutionary Deductions,* Vol. 3. Univ. of Chicago Press, Chicago, 1977.
131. Wyatt, R., Inflorescence architecture: How flower number, arrangement, and phenology affect pollination and fruit-set, *Am. J. Bot.* **69,** 585–594 (1982).
132. Yampolsky, E., and Yampolsky, H., Distribution of sex forms in phanerogamic flora, *Bibl. Genet.* **3,** 1–62 (1922).
133. Zimmerman, J. K., and Lechowicz, M. J., Responses to moisture stress in male and female plants of *Rumex acetosella* L. (Polygonaceae), *Oecologia* **53,** 305–309 (1982).

5

The Evolution, Maintenance, and Loss of Self-Incompatibility Systems

SPENCER C. H. BARRETT

Self-incompatibility, the inability of a fertile hermaphrodite plant to produce viable seeds upon self-pollination, is the principal and most effective mechanism preventing self-fertilization in flowering plants. While its manifestations are diverse, in all cases, the major effect is to promote outcrossing between genetically different individuals of the same species. Discrimination between self and nonself is not usually the result of incompatibility between the gametes themselves, or between gametophytes. Instead the siphonogamous habit (with pollen tubes) of angiosperms enables direct interaction between the male gametophyte and the female-acting sporophyte, the parent of the female gametophyte.[57] Systems of self-incompatability are widely distributed among flowering plant taxa and are reported from at least 19 orders and 71 families.[6,22] These include both dicotyledons and monocotyledons, plants from all geographical regions, and virtually all life forms.

Recently, several new hypotheses have been proposed to explain the selective forces involved in the evolution and maintenance of plant breeding systems.[33,128] Sexual selection, the optimal allocation of resources to maternal and paternal function, and strategies for coping with environmental uncertainty have all been invoked to explain the evolution of different reproductive modes. Self-incompatibility systems have remained relatively immune from these considerations. The traditional view of inbreeding avoidance as an explanation for the evolution of self-incompatibility has been largely unchallenged, since the evidence in support of the role of self-incompatibility systems as outbreeding devices is strong. This is not to imply that no difficulties exist in explaining the functional significance of different systems of incompatibility and their evolutionary relationships with one another. As the floral biology of a broader range of plant species has been investigated, it has become apparent that some revision of our concepts of self-incompatibility may be in order, since several types of self-incompatibility that have been recently discovered do not readily fit into existing classifications.

In this chapter, I review current research on the evolution and genetics of self-incompatibility. The discussion is organized into three main topics: (1) evolutionary relationships among incompatibility systems, (2) maintenance of self-incompatibility in natural populations, and (3) genetic modifications and evolutionary loss of self-incompatibility systems. Since the literature on self-incompatibility is vast, I have made no attempt to be comprehensive and in many cases have cited only recent references on a particular topic. Initially, a brief review of the major classes of self-incompatibility and their general properties is given. For more detailed treatments of self-incompatibility, the reader is referred to general reviews of the topic.[2,47,71,85]

TYPES OF INCOMPATIBILITY

Self-incompatibility systems can be divided into two distinct groups: gametophytic self-incompatibility, in which the incompatibility phenotype of the pollen is determined by its own haploid genotype, and sporophytic self-incompatibility, in which the incompatibility phenotype is governed by the genotype of the pollen-producing parent. The difference may arise from the time of *S* gene action, which in sporophytic systems appears to be premeiotic (or at the latest meiotic) before individualization in the tetrads, but in gametophytic systems occurs after the first metaphase of meiosis in pollen mother cells.[85]

Whereas all mating types in gametophytic self-incompatibility systems are morphologically similar (homomorphic), sporophytic self-incompatibility can be further subdivided into homomorphic and heteromorphic systems on the basis of whether or not the mating types are morphologically alike. Two classes of heteromorphic incompatibility are known (distyly and tristyly), depending on whether there are two or three mating groups. The mating groups usually differ in style length, anther height, pollen size, pollen production, and incompatibility behavior. The reader is referred to general reviews of heteromorphic self-incompatibility systems.[51,114]

Within the major types of self-incompatibility there occurs a variety of systems that differ from one another largely in their genetic basis (Table 5.1). Typically, in homomorphic systems a single locus (*S*) with multiple alleles controls incompatibility, although in recent years systems involving two to four loci and multiple alleles have been demonstrated,[78] and several cases of the polygenic control of self-incompatibility have been claimed.[38,113] In heteromorphic systems, distyly is controlled by a single locus with two alleles and tristyly by two loci each with two alleles and epistasis operating between the loci.[51]

Despite the variation in patterns of inheritance within gametophytic and sporophytic systems, each possesses distinctive cytological and physiological characteristics. For example, with few exceptions, in gametophytic systems pollen is binucleate and pollen tubes are inhibited in the style, whereas in sporophytic systems pollen is trinucleate and the rejection response is on the stigmatic surface. These differences may arise because of the contrast in the timing of *S* gene action. Although these differences break down in heteromorphic systems, where pollen can be bi- or trinucleate and inhibition stigmatic or stylar, the relationship between pollen cytology and site of inhibition within individual heterostylous species appears to be maintained.[92]

As research on self-incompatibility continues several "anomalous" systems have been identified, forcing us to reconsider and perhaps revise our views on the classification and overall properties of the different types of self-incompatibility. While it seems unlikely that the general dichotomy between sporophytic and gametophytic systems will be affected by recent discoveries, we may need to alter our thinking about the evolutionary origins and relationships among the major classes of self-incompatibility. Three examples of pollen–pistil interactions that are not readily interpretable

Table 5.1. Systems of Self-Incompatibility in Flowering Plants[a]

Major Types	Genes	Alleles	Allelic interaction	Stage of inhibition	Effect of polyploidy	Selected families
HOMOMORPHIC						
1 Gametophytic	1	Many	Codominant	Style	Breakdown	30 Rosaceae, Leguminosae, Solanaceae
2 Gametophytic	2	Many	Codominant	Style	None	1 Gramineae
3 Gametophytic	4	Many	Codominant	Style	None	2 Ranunculaceae, Chenopodiaceae
4 Sporophytic	1	Many	Dominant	Pollen germination	None	20 Cruciferae, Compositae, Rubiaceae
5 Sporophytic	3–4	Many	Dominant codominant	Pollen germination and stigma penetration	None	1 Cruciferae *(Eruca sativa)*
6 Sporophytic/ gametophytic[b]	1	Many	—	Ovary	—	1 Sterculiaceae *(Theobroma cacao)* (perhaps many)
7 Polygenic[b]	many	?	—	Ovary	—	1 Boraginaceae *(Borago officinalis)*
HETEROMORPHIC						
8 Distyly	1	2	Dominant	Pollen germination, stigma penetration, and style	None	23 Primulaceae, Linaceae, Turneraceae
9 Tristyly	2	2	Dominant	Style and ovary	None	3 Lythraceae, Oxalidaceae, Pontederiaceae
10 Distyly "anomalous"[b]	1	2 floral many SI	Dominant —	Ovary	None	2 Boraginaceae *(Anchusa)* Amaryllidaceae *(Narcissus)*
11 Enantiostyly[b]	1?	2?	—	—	—	2 Haemodoraceae *(Wachendorfia)* ? Tecophilaeaceae *(Cyanella)*

[a]Modified from Lewis.[71]

[b]These reports of self-incompatibility are not well understood and require further study (see text for references).

within the traditional classification schemes will serve to illustrate this point. In each case workers examining these phenomena have suggested that self-incompatibility mechanisms are involved, but a critical appraisal of the nature of these systems is required.

Late-Acting Self-Incompatibility

Early studies of pollen tube growth in self-incompatible species suggested that inhibition of self-pollen tubes in the ovary occurred only rarely in flowering plants. Ovarian inhibition of self-pollinations, either pre- or postzygotically, was generally considered an aberrant or maladaptive condition since, in some cases (e.g., *Theobroma cacao*[34]), ovules were apparently irreversibly sterilized by selfs. Numerous reports of ovarian inhibition have appeared in the literature recently, suggesting that late-acting incompatibility systems are more widespread than had been previously thought. Seavey and Bawa[103] have reviewed the subject and discuss the nature, occurrence, and functional significance of various ovarian phenomena. They distinguish three types of response: (1) ovarian inhibition of self-pollen tubes before ovules are reached, (2) prefertilization inhibition in the ovules, and (3) postzygotic rejection. Most workers have not considered postzygotic effects as involving a true self-incompatibility system, the former being usually reserved for prezygotic interactions between the pollen and pistil. The postzygotic rejection of selfs is often excluded from definitions of self-incompatibility owing to the difficulty of distinguishing such an effect from inbreeding influences. Embryo abortion due to the action of recessive lethals uncovered by selfing may be similar in appearance to a true self-rejection reaction. Seavey and Bawa[103] discuss these difficulties and outline how the effects of inbreeding may be distinguished from late-acting self-incompatibility.

Models of genetic load do not anticipate levels of deleterious recessives sufficient to obtain zero or low levels of seed set upon selfing (although see Sorensen[108]). In addition, it seems unlikely that the expression of inbreeding depression upon selfing would be concentrated in only the early embryonic phase of the life history. Thus measures of seed set upon selfing and the subsequent evaluation of growth of selfed progeny might be useful in distinguishing between inbreeding depression and a true self-rejection response, since inbreeding effects would be manifest at a variety of different developmental stages. In contrast, a late-acting self-incompatibility system would be expected to operate at a specific developmental period of embryo growth. Thus, detailed studies of embryo development might distinguish such effects.

If the genetic basis of late-acting self-incompatibility is similar to gametophytic and sporophytic systems in the possession of mating types, then these should be detectable as cross-incompatible matings. Such mating groups would not be anticipated from inbreeding unless consanguineous matings were involved. The distinction here may be particularly difficult if late-acting self-incompatibility phenomena are polygenically based, since the expectations for both involve quantitative variation as opposed to clear segregation of seed set values.

Cryptic Self-Incompatibility

The distinction between self-incompatibility phenomena and inbreeding effects is also relevant to plant species exhibiting cryptic self-incompatibility systems. In families with both homomorphic and heteromorphic incompatibility, pollen tube growth is often significantly faster in cross-pollen compared to self-pollen (prepotency) in self-compatible relatives. This effect was studied by Darwin[41] and has been termed cryptic self-incompatibility by Bateman.[16] It may be more widespread in angiosperms than previously thought, since the usual method of testing for self-incompatibility will not reveal its presence. Where this type of behavior occurs, differential fertilization of selfs and crosses can reflect the presence of weak self-incompatibility or it may be a reflection of inbreeding effects. In the latter case it is not always clear whether pre- or post-zygotic influences are at work, unless direct observations of differential pollen tube growth are made.

Using controlled pollen mixtures and the style length locus as a genetic marker in self-compatible, distylous *Amsinckia grandiflora,* Weller and Ornduff[124] showed that self and intramorph pollen were at a competitive disadvantage to intermorph pollen. Hence the cryptic self-incompatibility system found in this species resembles a weaker version of that occurring in related distylous species of the family, which show inhibition of *cross*-pollen among individuals of the same floral morph. The existence of cryptic self-incompatibility in *Amsinckia* has been questioned by Carey and Ganders (unpublished data cited in Ganders[51]) who failed to find differences in pollen tube growth in *Amsinckia grandiflora* or in any other distylous species in the genus. This has led to the suggestion that selective abortion of embryos may occur in *Amsinckia.*[26,27] However, Weller (unpublished data) has recently repeated the pollen tube experiments on *Amsinckia grandiflora* with results similar to his earlier findings and hence there seems little doubt that the species exhibits a true cryptic dimorphic-incompatibility system.

A different phenomenon appears to operate in self-compatible tristylous *Eichhornia paniculata* where inbreeding effects seem to be more important in regulating the parentage of offspring. Using pollen mixtures and an isozyme marker locus (GOT-3) Glover and Barrett[52] observed an approximately twofold advantage to cross-pollen over self-pollen. The treatments involved both intramorph and intermorph cross-pollen. In both, a similar advantage to cross-pollen was observed, a result not expected if a weak trimorphic incompatibility system was functioning. An additional treatment also compared the competitive ability of both legitimate (between anthers and stigmas of equivalent level) and illegitimate (between anthers and stigmas at different levels) cross-pollen. The two classes of pollen were equally competitive, again an outcome not expected in a conventional trimorphic incompatibility system.

While in this study it seems unlikely that cryptic trimorphic incompatibility is responsible for the observed advantage to cross-pollen, it is by no means clear at what stage the advantage to the cross treatment is manifest. While prepotency of cross-pollen may occur, it is possible that there is some selective elimination of selfed zygotes through embryo abortion. Here, as in putative cases of late-acting self-incompatibility, detailed dissection of the complete reproductive cycle from pollination to seed set is required to distinguish between the possibility of incompatibility phenomena and inbreeding effects. In the former case, we are dealing with a mechanism operated by the maternal parent and controlled by its genotype and that of the pollen donor. In contrast, inbreeding depression is a process acting in the progeny zygote determined by its own genotype. While it is conceptually straightforward to distinguish between these factors, in practice it may not always be clear which of these processes is responsible for the reduced seed set after selfing compared with outcrossing.

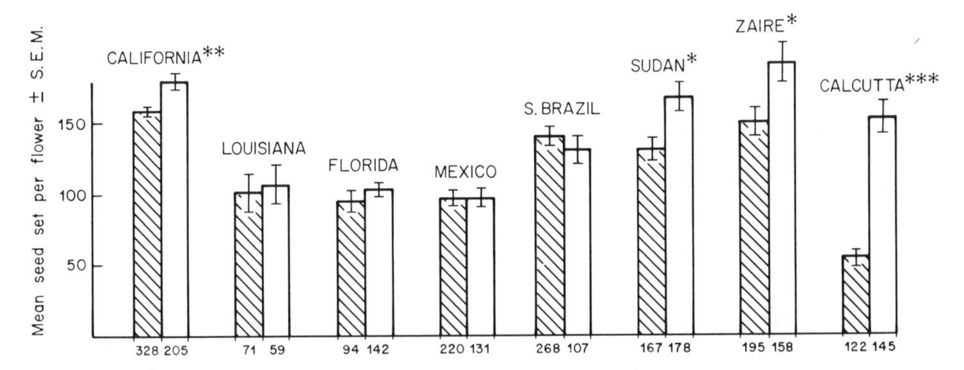

Fig. 5.1. Seed production following controlled self- and cross-pollinations of *Eichhornia crassipes* clones. All cross-pollinations involved a single clone from Costa Rica. Sample sizes are the number of flowers pollinated. ▧ = self-pollinations, □ = cross-pollinations, *$p < .025$, **$p < .01$, ***$p < .001$. Significant differences in seed set between self- and cross-pollination may result from weak self-incompatibility and/or inbreeding depression. (After Barrett.[7])

This problem is often encountered in interpreting results of controlled pollination studies. Figure 5.1 illustrates such a difficulty from the results of controlled self- and legitimate cross-pollinations of tristylous *Eichhornia crassipes.*[24]

Anomalous Heteromorphic Incompatibility Systems

A final example in which the incompatibility phenomena described do not readily fit into conventional schemes involves two distinctive types of floral heteromorphism. In both cases, controlled pollinations indicate reduced seed set on selfing, but it is by no means clear what mechanism is operating, whether or not inbreeding depression is involved, and how the systems are related to typical heteromorphic incompatibility.

In the Boraginaceae both self-incompatible and self-compatible distylous taxa are known.[26] Experimental studies on *Anchusa officinalis*[93,101,102] and *Anchusa hybrida*[44] have revealed a distinct and unusual form of floral heteromorphism, which may also occur in *Narcissus.*[43] In both *Anchusa* species there is considerable variation in style length in natural populations, although the ratio of style length to anther height shows a clear bimodal distribution. In *Anchusa officinalis,* surveys of morph ratio indicate that in all populations the long-styled morph is far in excess of the short-styled morph.[93] Yet curiously, genetic studies of inheritance of style length are suggestive of the common pattern for distylous plants with a single diallelic locus governing style length variation.[102] High frequencies of the long-styled morph might occur if this morph experienced a high degree of self-fertilization, but controlled self-pollinations of both morphs yield little to no seed.[93] Of particular interest is the observation that both intramorph and intermorph pollinations are compatible (Table 5.2). Because of this finding, workers studying these species have concluded that *Anchusa* possesses a multiallelic incompatibility system unlinked to the locus governing floral dimorphism. Observations of pollen tube growth in selfs indicate that pollen tubes reach the ovary and enter the micropyle, suggesting that the recognition reaction resides in the ovules.[101] Detailed studies of the genetic basis of the incompatibility system of *Anchusa* are hindered by the generally low female fertility of crosses. Despite this, it would seem worthwhile to investigate further the cause of self rejection in the species,

Table 5.2. Seed Set of Intrafamilial Pollinations in *Anchusa officinalis.*[a,b]

	Plant number									
Plant number	S 1	S 2	S 3	L 4	L 5	S 6	S 7	L 8	S 9	S 10
S 1	0	6	38	**0**	5	31	6	19	**25**	0
S 2	10	0	**10**	13	**0**	25	13	**50**	0	19
S 3	5	**0**	0	38	20	44	**19**	50	25	25
L 4	**38**	44	63	0	31	38	31	6	31	**50**
L 5	31	**13**	19	31	0	13	56	50	50	19
S 6	45	6	6	25	25	0	19	75	31	25
S 7	35	25	**0**	38	25	25	0	19	63	25
L 8	20	**0**	25	10	56	50	63	0	44	44
S 9	**0**	0	6	19	15	6	25	38	0	13
S 10	0	25	25	**0**	15	25	13	6	63	0

[a]After Schou and Philipp.[102]

[b]The numbers are the percentage of maximum seed set, boldface numbers indicate the seven cases in which a difference in compatibility is manifested in a complete absence of seeds in one cross. All plants are from a single family.

particularly since the apparent absence of clear-cut mating groups and the fact that ovarian phenomena are involved suggest that inbreeding depression may also be a factor.

The data for *Anchusa* resemble those obtained by Crowe[38] for the related nonheterostylous *Borago officinalis.* She argued that in this species self-incompatibility is polygenically controlled and is expressed postzygotically. Evidence to support this claim was obtained from pollen chase experiments (prior application of self-pollen before cross-pollen) in which a sterilization effect was observed from self-pollinations. However, prezygotic rejection mechanisms operating in the nucellus or micropyle may also block subsequent compatible pollen tubes and thus it may be premature to conclude that postzygotic mechanisms are at work in this species.

One mechanism by which incompatibility could operate postzygotically involves the postponement of the rejection response relative to recognition. This would require labeling of the developing zygote by a product synthesized during the recognition period. No such chemical has yet been demonstrated in species in which postzygotic incompatibility has been claimed and, therefore, it may be more satisfactory to reserve the term incompatibility for prezygotic interactions.

A second distinctive form of heteromorphic incompatibility involves differences among mating types in style orientation and has been observed in two monocotyledonous families (Haemodoraceae, Tecophilaeaceae).[46,89,90] In the genera *Wachendorfia* and *Cyanella,* some plants have styles that are sharply deflected to the right, while in others the style bends to the left. This condition is referred to as enantiostyly and is interpreted as an outbreeding mechanism promoting pollination between floral morphs in a manner similar to heterostyly.

In *Wachendorfia paniculata,* controlled self- and cross-pollinations within and between plants with right- and left-bending styles suggest that intermorph crosses produce more seeds than self- or intramorph pollinations. On the basis of these results, Ornduff and Dulberger[90] concluded that a weakly developed self- and intramorph incompatibility system is present in the species. The occurrence of 1 : 1 ratios of the two morphs in four *Wachendorfia* populations[38] suggests that intermorph matings may predominate under field conditions and that the mechanism of inheritance of floral enantiostyly may be similar to that found in heterostylous plants. Since no relatives of enantiostylous plants are heterostylous, it seems unlikely that the polymorphisms are related in any way, except in as much as they may represent distinctive and independent responses to selection favoring outcrossing. More detailed genetic and ecological studies of these curious polymorphisms are required before any firm conclusions on their adaptive significance can be reached. In addition, controlled selfs and crosses among the morphs combined with observations of pollen tube growth are required to firmly establish the presence of a self-incompatibility system in the species.

Heterosis Model of Self-Incompatibility

The examples reviewed above indicate some of the difficulties in distinguishing the various forms of self-incompatibility from the influences of inbreeding depression. Recently, Mulcahy and Mulcahy[83] have attempted to extend the significance of inbreeding effects to encompass typical style-mediated gametophytic self-incompatibility systems. They have questioned the conventional genetic model of gametophytic self-incompatibility by one or a few multiallelic loci with oppositional effects and have instead argued that many loci, which are spread throughout the genome with complementary effects, govern the incompatibility response. According to this view, gametophytic self-incompatibility is simply an expression of genetic load mediated via extensive pollen style interactions. This model, called the "heterosis model" of gametophytic self-incompatibility, is based on the assumption that if the pollen and style carry dissimilar allelic combinations, there will be heterotic interactions between them, resulting in increased pollen tube growth rates. In contrast, if both the pollen and style share the same deleterious recessive alleles, pollen tube growth will be reduced accordingly. The actual growth rate of the pollen tube will be the sum of all pollen–style interactions, and incompatible pollinations are due not to specific inhibitory molecules (oppositional model) but rather to the growth of pollen tubes being too slow to allow fertilization.

The heterosis model and evidence used to support it have been strongly criticized by Lawrence *et al.*,[66] who argue that much of the evidence used by the Mulcahys against the oppositional hypothesis is either not relevant or not inconsistent with it. They point out difficulties concerned with the genetic and biochemical basis of the model, the most serious of which is that it is not capable of providing an explanation for the compatibility relationships observed in either single locus or multilocus systems, unless in the latter case it is assumed that the constituent loci of the proposed supergenes which govern self-incompatibility are very tightly or completely linked.

While the Mulcahys' model may be inconsistent with available information for gametophytic systems of self-incompatibility, it may help explain other facets of pollen–pistil interactions such as those involved with pollen prepotency, optimal outcrossing, and extraneous pollen advantage in interpopulation crosses.[69,115] Observations of pollen germination, pollen tube growth, and the fertility of crosses within and between subpopulations at different spatial scales would be useful in assessing whether or not the genetic relatedness of sexual partners can influence pollen–pistil interactions in ways that mimic incompatibility phenomena.

EVOLUTION OF SELF-INCOMPATIBILITY SYSTEMS

Two contrasting views on the evolutionary origins of self-incompatibility systems are evident in the literature. The first, originally proposed by Whitehouse,[125] suggests that self-incompatibility arose once in association with the origin of flowering plants. Following this interpretation, the present range of self-recognition systems is fundamentally similar because of the presence of an ancient, but strictly conserved, *S* locus in all families. Variation among systems arises from superimpositions on the basic mechanism underlying self-rejection. An alternative view follows Bateman,[15] who argued against the monophyletic origin of self-incompatibility systems and suggested that it was more probable that weak polygenic incompatibility had arisen de novo several times and that progressive genetic modifications had taken place to give the range of systems observed today. Modifications involved either selection of nonspecific modifiers influencing all loci or specific modifiers increasing the effectiveness of one or two loci at the expense of the rest.

Each view on the evolutionary origin(s) of self-incompatibility has its supporters, but until more information on the taxonomic distribution, genetic basis, and physiological properties of incompatibility systems is available, the question is likely to remain unresolved. The problem may eventually be solved by molecular characterization of the *S* gene from species with different systems of incompatibility.

Primitive Systems of Incompatibility

While controversy exists over the phylogenetic relationships between the different types of sporophytic self-incompatibility (see below), there is general consensus that the primitive system of self-incompatibility in flowering plants is gametophytic. In addition to the single locus form of control, more complex systems with three, four, and perhaps even more loci are known.[78,91] Since these occur in species from relatively unspecialized families (Ranunculaceae, Chenopodiaceae), it is possible that they may be similar to the original forms of self-incompatibility with the common one-gene system derived by progressive homozygosis or deactivation (silencing) of all but one of the genes. The observation of ovarian self-incompatibility in the primitive *Pseudowintera colorata* (Winteraceae) by Godley and Smith[54] is also of interest, and raises the possibility that unspecialized forms of polygenic self-incompatibility, with rejection mechanisms residing in the ovary, may have evolved first in the angiosperms, and that, later in evolution, progression to stylar and finally stigmatic inhibition with monogenic control occurred.

Detailed genetic data from species with late-acting (ovarian) self-incompatibility systems are badly needed to enable an assessment of their relationships to sporophytic and gametophytic systems. Unfortunately, since many of the plants in which these systems have been observed are tropical woody species, this may be some time in coming. The only data available for a species with this type of self-incompatibility system *(Theobroma cacao)* are difficult to interpret and suggest that genetic control is gametophytic for the pollen and sporophytic for the ovules.[34] A similar system may also operate in the related *Sterculia chicha,*[111] where ovarian inhibition has been observed.

The view that gametophytic self-incompatibility is phylogenetically primitive whereas sporophytic self-incompatibility is derived has recently been challenged by Zavada[134] and Zavada and Taylor[135] on the basis of fossil evidence. Studies of early Cretaceous angiosperm pollen indicate that many taxa possess reticulate exine sculpturing, a feature of extant plants with sporophytic self-incompatibility. In addition, current fossil evidence indicates that the style did not evolve until the Lower Cretaceous or lower Upper Cretaceous, thus postdating the occurrence of pollen types indicative of sporophytic systems. Since gametophytic self-incompatibility depends primarily on interactions between the pollen tube and style, this observation is difficult to reconcile with the view that gametophytic self-incompatibility is ancestral, unless the early plants with this system possessed stigmatic recognition mechanisms such as those that occur in *Papaver.* Zavada and Taylor[135] suggest that early angiosperm self-incompatibility may have involved a system similar to that found in *Theobroma cacao,* with stigmatic recognition but with the rejection response resulting in abortion of the carpel. According to this view, the subsequent development of pollen tube inhibition, without the accompanying abortion of reproductive structures as a result of incompatible pollinations, provided energetic advantages as well as opportunities for prezygotic mate assessment.[128]

Relationships Between Homomorphic and Heteromorphic Self-Incompatibility

Current information on the distribution of homomorphic self-incompatibility systems is fragmentary but suggests that not only are sporophytic and gametophytic systems found in different families but heteromorphic incompatibility occurs in yet another group of families distinct from these.[30,31] Despite contrary views,[131] there are no convincing genetic data indicating that homomorphic and heteromorphic systems of sporophytic incompatibility co-occur in the same family, with the exception of the large family Rubiaceae. This point is of relevance to ideas on the evolution of dimorphic incompatibility.[80] Following the view of a unitary, strictly conserved *S*-locus in flowering plants, Muenchow[81] has developed a theoretical model for the evolution of distyly by loss of alleles from an existing multiallelic sporophytic system. Rather than invoking genetic drift,[131] Muenchow's model suggests that selection for maximal cross-incompatibility can, under rather restricted conditions, remove incompatibility alleles in such a way that remaining alleles display the pattern of dominance and recessiveness found in distylous groups. Until closely related taxa with both homomorphic and heteromorphic systems of sporophytic incompatibility are discovered, however, the model may have no more than theoretical value.

The physiological and biochemical properties of incompatibility systems are still relatively poorly understood, but available data hardly support the view of a unitary *S* gene for sporophytic systems. It is possible that the recognition factors normally associated with the tapetum in homomorphic systems have no role to play in the incompatibility systems of heterostylous plants and that physiological differences between pollen tubes and the pistil mediate incompatibility.[110] In this connection, it is worth noting that inhibition sites in heterostylous species can involve the stigma, style, or ovary.[1,17,96,104] Charlesworth[30] has suggested that if the general properties of heteromorphic incompatibility turn out to be fundamentally different from homomorphic systems, the conventional use of the term *S* gene should probably not be applied to the incompatibility locus in heterostylous plants.

Selective Forces

Few workers have considered the selective forces that have given rise to two distinctly different types of incompatibility in flowering plants, namely the gametophytic and sporophytic systems. Beach and Kress[19] suggest that the answer may stem from the conflict created by the contrasting reproductive behaviors of the sporophytic and gametophytic generations. In order for the male gametophyte to be evolutionarily successful, it must fertilize an egg, or none of the gametophyte's genes will be transmitted to the next generation. The quality of the resultant zygote is not open to choice since the male gametophyte is already "committed." In contrast the female sporophyte does not benefit by indiscriminate male gametophyte success but rather by inhibiting self-pollen and promoting cross-pollen. Beach and Kress[19] propose that the development of sporophytic incompatibility from gametophytic incompatibility may represent an evolutionary response by sporophytes that is due to opportunities available to "committed" gametophytes for circumventing the inhibition mechanisms of gametophytic systems. Sporophytic systems can be viewed as more effective in discriminating against committed male gametophytes since they operate before the haploid genome is expressed, as a result of the biochemical labeling of pollen with sporophytically derived products in the exine.[63] Willson[127] considers other aspects of conflict between male and female function in self-incompatible plants. While these ideas are both novel and plausible, they provide little opportunity for experimental analysis and as a result the hypotheses are unfortunately largely untestable.

MAINTENANCE OF SELF-INCOMPATIBILITY SYSTEMS

Our understanding of the evolutionary development of incompatibility systems is largely speculative and based on an imperfect knowledge of their distribution and general characteristics. A rich literature has, however, developed on their maintenance and function in contemporary plant populations. Much of this work is theoretical and there is considerable scope for experimental field studies on the ecology and population genetics of self-incompatible species to assess the validity and predictions of the theoretical models.

The major selective force proposed to explain the maintenance of incompatibility systems is substantial inbreeding depression in the fitness of selfed progeny due to the expression of largely recessive deleterious mutations in homozygotes. Virtually every natural outbreeding plant and animal population that has been examined displays the complementary effects of inbreeding depression and heterosis.[130] The total inbreeding depression, in normally outcrossing species, that results from selfing is frequently greater than 50%, and the average individual is typically heterozygous for one or more recessive lethal factors.[39,64]

Unfortunately, the magnitude and quantitative patterns of inbreeding depression have not been examined in natural populations of many self-incompatible species, although numerous reports of the deleterious effects of inbreeding in cultivated self-incompatible species are available in the agricultural and horticultural literature. A paucity of data for natural populations of self-incompatible species may in part be a consequence of the difficulties in obtaining selfed seed. Bud pollinations and other techniques can be employed to circumvent this problem, but these approaches are frequently time-consuming and technically difficult.[85] For example, by using bud pollinations in self-incompatible *Turnera ulmifolia,* substantial inbreeding depression has been demonstrated for vegetative and reproductive traits in several diploid populations (J. S. Shore and S. C. H. Barrett, unpublished data). However, the yield from bud selfs differs between the style morphs and the amount of seed obtained is generally low. An alternative approach for examining inbreeding depression in self-incompatible plants involves the use of sib-matings. This could be of particular interest in species with contrasting incompatibility systems since the control of sib-mating differs markedly between them.[71] Unfortunately, since different systems of incompatibility rarely, if ever, occur within related taxonomic groups, it seems likely that other factors (e.g., life history, dispersal mechanism, population size) would overwhelm effects on inbreeding that could be ascribed to the system of mating alone.

Olmstead[86] has recently considered the relationship between the breeding system of self-incompatible species and the level of inbreeding in populations. He proposes that the evolution and maintenance of self-incompatibility may have been largely independent of the level of inbreeding in the population as a whole. This is because the avoidance of selfing, the primary outcome of all self-incompatibility systems, has a negligible influence on the level of inbreeding in comparison with population size effects. Since many flowering plants are characterized by small effective population sizes and considerable genetic substructure, they are likely to experience considerable inbreeding. Olmstead argues that inbreeding has beneficial effects (reduced cost of meiosis, maintenance of coadapted gene complexes), and an optimal level exists in plant populations. Following this view, the maintenance of self-incompatibility primarily results from differences in the relative fitness of selfed and outcrossed progeny, not from any positive influence brought about by increased outbreeding.

Number and Frequency of S *Alleles*

The number and frequency of *S* alleles that can be maintained in finite populations of self-incompatible plants with multiallelic systems has been the subject of extensive theoretical treatment[133] but little empirical work. Until the recent studies by Lawrence and co-workers on the field poppy, *Papaver rhoeas,*[65] the sum total of our knowledge was based on Emerson's pioneering work on *Oenothera organensis*[48,49] and the less detailed studies of *Trifolium repens*[3] and *Trifolium pratense.*[129]

Work on *Papaver rhoeas*[24,25,65] is sufficiently detailed so that the data can be compared validly with those of Emerson. The first point is that in both studies similar numbers of *S* alleles were found within populations of the two species. However, while in *Oenothera organensis* the frequency of *S* alleles was not significantly different, in *Papaver rhoeas* large differences in frequency were evident in each of three populations examined (Fig. 5.2). Two hypotheses could account for the unequal frequencies of *S* alleles in *Papaver rhoeas* populations. The first proposes that the alleles are subject to selection unrelated to that associated with the incompatibility system, either directly or via close linkage with other genes. The alternative hypothesis invokes random genetic drift, associated with repeated colonizing episodes, and assumes that populations were not at equilibrium when sampled. While the first hypothesis predicts

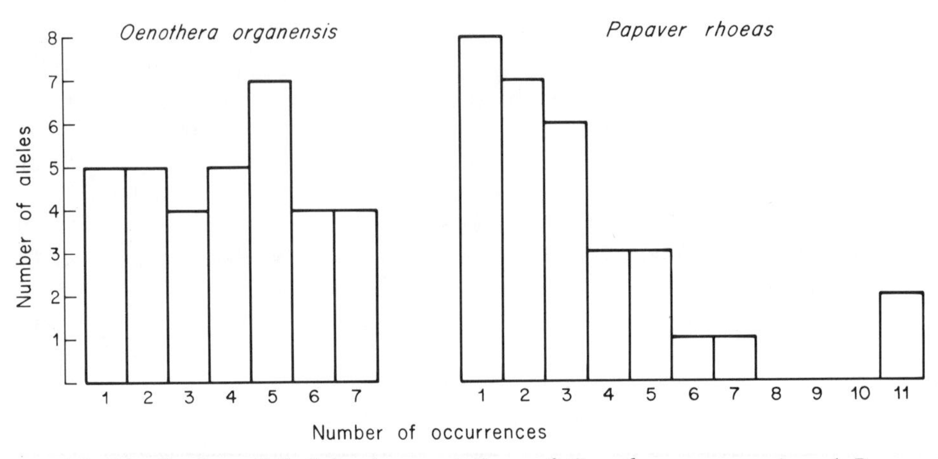

Fig. 5.2. Distribution of *S*-alleles in populations of *Oenothera organensis* and *Papaver rhoeas.* (After Emerson[49] and Campbell and Lawrence.[25])

that the same alleles will occur at high frequency in different populations, the second does not. The fact that *Papaver rhoeas* is a weed of arable land and disturbed sites certainly appears to favor the nonequilibrium hypothesis since weed species are usually subject to repeated colonizing episodes. However, the number of *S* alleles found in each of the three populations is large,[26,30,31] suggesting that genetic bottlenecks may be of less importance than might be indicated from a consideration of the population ecology of the species.

Lawrence and O'Donnell[65] believe that, despite its weedy tendencies, *Papaver rhoeas* is a permanent and stable member of arable weed communities and that the large dormant seed bank found in the species may buffer populations against the chance effects associated with restrictions of population size. The cause(s) of the differences in *S* allele frequencies within populations, therefore, still remains unresolved. Cross-classification of *S* alleles among populations is necessary to determine if the same alleles predominate in different populations. If this turns out to be the case, differential selection among heterozygotes may be involved.

Mating Groups in Heterostylous Species

Identification of mating types in species with homomorphic incompatibility can be ascertained only by extensive pollination programs, and this probably accounts for the paucity of data from natural populations. In contrast the frequencies of mating types in species with heteromorphic incompatibility can be readily obtained by visual inspection of plants within populations. In addition, equilibrium genotype frequencies at the heterostyly loci can also be determined, although in tristylous species this involves progeny testing and complex mathematical analysis.[61]

There is considerable information on the population structure of both distylous and tristylous species. Survey data from distylous populations typically indicate that the long- and short-styled morphs are equally frequent (isoplethy), although in some species unequal morph frequencies (anisoplethy) are a feature of populations.[68,89] Owing to the rarity of floral trimorphism, there are fewer observations of tristylous species, and both isoplethic and anisoplethic population structures are reported.[14] Studies of style morph frequency in heterostylous plants are of special interest because they can provide information on the dynamics of selection at the loci controlling mating system.[8,28]

Heuch[58,59] has shown theoretically that, provided no fitness differences among the style morphs occur, an isoplethic equilibrium is the only possible condition in large populations with disassortative mating. This outcome follows from the genetic systems that govern heterostyly. Where unequal morph frequencies prevail, several possible factors may be involved. These can include founder effects and clonal propagation,[11] mating asymmetries among the style morphs,[8,94] differential selfing owing to relaxation of self-incompatibility,[29] or modification and breakdown of heterostyly.[8,28,117]

Of interest to problems concerned with the maintenance of heteromorphic incompatibility is a consideration of the minimum population size required for the polymorphisms to remain stable. This issue is relevant mainly to tristylous species because of their complex systems of inheritance. In a study of 16 populations of tristylous *Lythrum salicaria* on Finnish islands, Halkka and Halkka[56] found that the three style morphs were present in all populations, despite their small size. They concluded that gene flow between the islands must be frequent in order for populations to remain tristylous. However, as Heuch[60] has shown theoretically, the genetic system governing tristyly in *Lythrum salicaria* can remain stable in isolated populations consisting of as few as 20 plants. Loss of style morphs occurs with regularity in populations below this size, and when this happens the short-styled morph is lost more frequently, since the dominant *S* allele governing this phenotype is only carried by short-styled plants. Fluctuations in population size, associated with colonizing episodes and drought, are postulated as the major factor leading to deficiency and loss of this morph from populations of tristylous *Eichhornia* species.[8,11]

Function of Floral Polymorphisms in Heterostylous Species

Although mating types in self-incompatible species are maintained in populations by frequency-dependent selection, it is by no means clear what selective forces are responsible for the evolution and maintenance of the complementary set of floral polymorphisms that is associated with the incompatibility groups in heterostylous species.[51,132] The most widely accepted explanation of the functional significance of floral heteromorphism was originally formulated by Darwin,[42] who hypothesized that the reciprocal placement of stamens and styles in the floral morphs is a mechanical device to promote insect-mediated cross-pollination among morphs with anthers and stigmas at equivalent levels (legitimate pollination). Although statistically significant levels of legitimate pollination have been demonstrated in both distylous[50] and tristylous[12] species, in many studies heterostyly appears to have little effect on pollination patterns. With random pollination, however, sufficient numbers of compatible pollen grains are usually deposited on naturally pollinated stigmas of heterostylous plants to ensure maximum seed set.[72]

Observations of random pollination in heterostylous species have led to the development of several alternative hypotheses to explain the maintenance of heterostyly. These hypotheses view heterostyly as a floral mechanism that (1) reduces self-pollination,[82] (2) is maintained by sexual selection and the optimal allocation of sexual resources,[27,112,126] (3) avoids mutual pollen–stigma interference and stigmatic clogging,[77,116] and (4) enhances pollen carryover.[115] A major challenge will be to devise experimental tests to distinguish among these hypotheses. It is possible that in some heterostylous species the floral polymorphisms are selectively neutral under contemporary conditions and are maintained because of a close developmental association with the incompatibility system. More information on the developmental genetics of heterostyly is required to assess this possibility. Dulberger[45] and Richards and Barrett[97] discuss the developmental relationships between the floral polymorphisms and incompatibility in heterostylous species.

MODIFICATION AND LOSS OF INCOMPATIBILITY

Comparative studies of closely related taxa with contrasting breeding systems provide strong evidence for the repeated loss of self-incompatibility in flowering plants.[109] The tendency of incompatibility loci to mutate toward increased self-compatibility has been demonstrated in both homomorphic and heteromorphic systems.[70,85,107] Various types of genetic modification leading to self-compatibility occur. These include (1) mutation of the incompatibility gene(s), (2) alteration of the genetic background in which *S* alleles function, (3) occurrence of polyploidy in gametophytic systems (excluding *Ranunculus, Beta,* and monocotyledons), and (4) homostyle formation in distylous species as a result of crossing-over in the supergene controlling the heterostylous syndrome. Whether or not self-compatible variants establish and spread is dependent on their ability to compete with their outbreeding progenitors or establish in novel environments.[62,76] Inbreeding depression is likely to be the major factor restricting spread, particularly if population sizes are large in the outcrossing progenitor, resulting in high genetic loads. Sporadic pollinator failure in zoophilous species and population bottlenecks on a time scale of less than 100 generations can, however, promote selection for a highly self-fertilizing mode of reproduction since these processes reduce genetic load and hence the magnitude of inbreeding depression.[64] Of course, mutations at incompatibility loci do not necessarily mean that self-compatible individuals are self-fertilizing. The degree of selfing will depend on a range of factors of which floral morphology and the abundance of pollen vectors are usually the most important.[100]

Homomorphic Incompatibility

Among homomorphic systems, loss of self-incompatibility has been particularly well documented in *Leavenworthia,* in which several species (e.g., *Leavenworthia crassa* and *Leavenworthia alabamica*) exhibit both self-incompatible and self-compatible populations.[99] Self-compatible populations have developed adaptations (e.g., introrse anthers, small flower size) that increase the efficiency of self-pollination. These have been documented in detail by Lloyd.[74] In some cases, loss of self-incompatibility may be associated with speciation, as has been proposed for *Stephanomeria malheurensis*[55] and *Lasthenia maritima.*[35] In both cases, it appears that genetic modifications at loci governing sporophytic incompatibility have initiated the events leading to reproductive isolation. In neither case is the genetic basis of the change in incompatibility behavior known. The genetic basis of self-compatibility in *Stephanomeria* has recently been investigated (see Brauner and Gottlieb [21a]).

Most of the detailed information on genetic modifications at incompatibility loci in homomorphic systems is based on studies of agricultural and horticultural plants.[85] Plant breeders have endeavored to select for self-compatibility to facilitate production of homozygous lines. Unfortunately, there is relatively little information on the variation in expression of self-incompatibility in populations of most wild species. Occasional self-compatible individuals in normally self-incompatible species (pseudocompatibility) have been studied in detail by Ascher.[73,98] The extent of this variation in natural populations, how it is maintained, and its influence on the mating system of populations are largely unknown.

Breakdown of Dimorphic Incompatibility

While the evolution of heteromorphic incompatibility systems presents a complex problem that is still poorly understood,[29,32] breakdown of these genetic polymorphisms has been documented in many heterostylous families.[28,51] Modifications include replacement of one type of outcrossing mechanism by another, such as the evolution of distyly from tristyly (see below) and the origin of dioecism from distyly.[18,75] More frequently, heterostylous systems break down in the direction of increased self-fertilization by the formation of homostylous population systems. Two recent studies of this shift in breeding system illustrate how similar genetic pathways can result in different outcomes with regard to the mating system.

The breakdown of distyly to homostyly in *Primula* is one of the classic examples of the evolution of self-fertilization in flowering plants. Homostyles are interpreted as products of crossing-over within the supergene that controls heterostyly. The product is an allelic combination and phenotype, which combines the style length and compatibility group of one morph with the stamen length and compatibility of the alternate morph. Homostyles are thus self-pollinating, due to the close proximity of sexual organs, and self-compatible.

Whether or not homostylous variants will spread following their origin depends on several factors, including the mating system of morphs, the relative fitness of their progeny, and the availability of pollinating agents.[28] A controversy exists concerning the presence of locally high frequencies of homostylous variants in populations of *Primula vulgaris* in two regions of England (Somerset and the Chilterns). Crosby's early studies[36,37] predicted that homostylous variants would increase in frequency and eventually replace the distylous morphs as a result of their high selfing rates. This view was disputed by Bodmer,[20,21] who suggested, based on garden studies, that homostyles were up to 80% outcrossed as a result of marked protogyny. Two recent studies have clarified some of these issues. Using isozyme loci as genetic markers, Piper *et al.*[94,95] have shown in several populations that the homostylous morph is highly self-fertilizing ($s = 0.92$) while, as expected, the long- and short-styled morphs are highly outcrossed (and see Cahalan and Gliddon[23]). Comparison of several components of fitness in natural populations (e.g., flower production, seeds per capsule, total seed production) demonstrated that homostyles were significantly more fertile than the other morphs. However, this difference varied in both time and space, owing to fluctuations in pollinator service due to differences in rainfall. Although surveys of *Primula* populations in Somerset have been conducted over a 25- to 40-year period,[40] they indicate only small changes in morph frequency, preventing any firm conclusion about whether a stable equilibrium has been reached, or whether homostyles are slowly replacing the

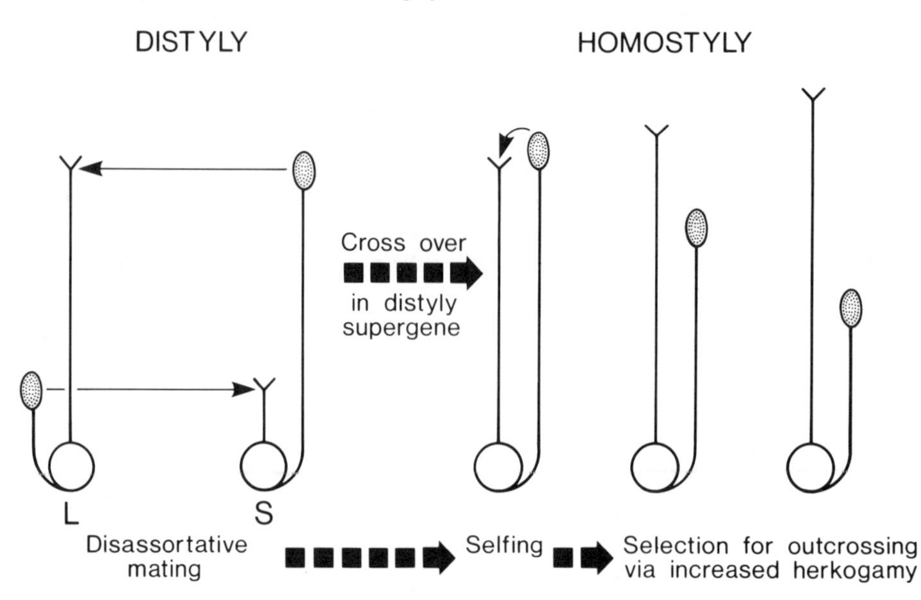

Fig. 5.3. Breeding system evolution in the *Turnera ulmifolia* complex. (For details see Barrett and Shore.[13])

distylous morphs. Clearly, without long-term demographic work, it is extremely difficult to provide conclusive evidence about the net direction of selection on the mating system, particularly in long-lived perennial plants.

There are many cases in which the close relatives of heterostylous taxa are homostylous. This suggests that the shift in breeding system from outcrossing to selfing may be frequently associated with speciation events. Homostylous taxa are often found at the geographical margins of the progenitor's range, raising the possibility that reduced pollinator service may have favored their establishment and spread. This geographical pattern is evident in *Turnera ulmifolia,* a Neotropical polyploid complex of perennial weeds. Our studies of this group[4,13,105–107] have revealed the striking lability of breeding systems and cast doubt on the frequently held view that the evolution of selfing involves a unidirectional change (Fig. 5.3).

The *Turnera ulmifolia* complex is composed of diploid, tetraploid, and hexaploid varieties. Diploids and tetraploids exhibit typical dimorphic incompatibility, whereas hexaploids are self-compatible and homostylous.[13,105] The three homostylous varieties of *Turnera ulmifolia* that we have studied experimentally are differentiated for morphological traits and isozyme patterns as well as being intersterile. They occur at different margins of the range of the species complex, indicating that dimorphic incompatibility has broken down to homostyly on at least three separate occasions in the complex, always in association with the hexaploid condition. The reason for the association between homostyly and hexaploidy is unclear. Hexaploids synthesized using colchicine remain distylous, indicating that at its inception hexaploidy per se does not cause homostyle formation.[107]

Cytological studies indicate that while tetraploid varieties in the complex form quadrivalents and appear to be autoploids, hexaploids form bivalents and are therefore likely to be allopolyploids. Evidence to support this comes from isozyme studies, which indicate that tetraploids exhibit tetrasomic inheritance for enzyme loci, whereas hexaploids display considerable fixed heterozygosity (J. S. Shore and S. C. H. Barrett, unpublished data). This raises the possibility that, following their origin, homostyles might spread more easily in hexaploid populations as a result of a reduction in the magnitude of inbreeding depression associated with allopolyploidy. Lande and Schemske[64] consider the influence of polyploidy on inbreeding depression.

The patterns of floral variation in *Turnera ulmifolia* are particularly complex in the Caribbean region. On large islands (e.g., Greater Antilles) populations are either tetraploid and distylous or hexaploid and homostylous. However, on smaller islands (e.g., Bahamas) only homostyles occur. Presumably, repeated colonizing episodes and the facility for establishment after long-distance dispersal favor homostyles over the self-incompatible distylous morphs in island colonization. On Jamaica, populations are uniformly hexaploid and self-compatible but display a range of floral phenotypes from long homostyle (long stamens and long styles) to plants with flowers resembling those of the typical long-styled morph from distylous populations. Barrett and Shore[13] interpret these latter phenotypes as resulting from selection for outcrossing in homostylous colonists (Fig. 5.4). This may be more readily achieved by the development of herkogamy (spatial separation of stigmas and anthers) in homostylous stocks, through selection on polygenic variation, than by the de novo development of alternative outbreeding mechanisms.

To test the hypothesis that the range of homostylous floral variants on Jamaica is secondarily derived from distylous ancestors, crosses between homostylous and distylous forms were conducted.[13] The predicted crossing relationships from the cross-over model for the origin of homostyly were revealed in all floral phenotypes (see Table 4 in Ref. 51). Hence, despite possessing "short-level" anthers, the herkogamous populations exhibit the residual incompatibility reaction of long-level anthers of the short-styled morph. It is remarkable that, despite the absence of distylous populations on Jamaica, both the pistils and pollen of homostylous forms retain their ancestral incompatibility behavior. Unlike unilateral interspecific incompatibility,[85] the incompatibility behavior expressed in crosses between heterostylous and homostylous forms is usually reciprocal in nature.

Although there is no evidence of changes in floral traits owing to selection for outcrossing in homostylous variants of *Primula vulgaris,* this may have occurred in other taxa in the genus. Many monomorphic relatives of heterostylous *Primula* species are known that possess large flowers and outcrossing adaptations. Similar patterns are also evident in *Linum.*[79] Whether homostyles maintain selfing or redevelop outcrossing adaptations may depend in part on the capacity of other components of the genetic system to influence recombination, as well as local selection pressures favoring outcrossing.

In both of the above examples, the breakdown of dimorphic incompatibility arises as a result of recombination in the supergene that controls distyly. This may not be the only genetic pathway by which incompatibility can be modified, as a number of distylous taxa are known in which the style morphs are highly self-compatible.[51] Since it seems unlikely that floral dimorphism can evolve in the absence of incompatibility,[32] these taxa have most likely secondarily lost their incompatibility systems. The genetic basis of self-compatibility in these taxa has not been studied in detail. Shore and Barrett[107] have examined the inheritance of a range of self-compatible variants in

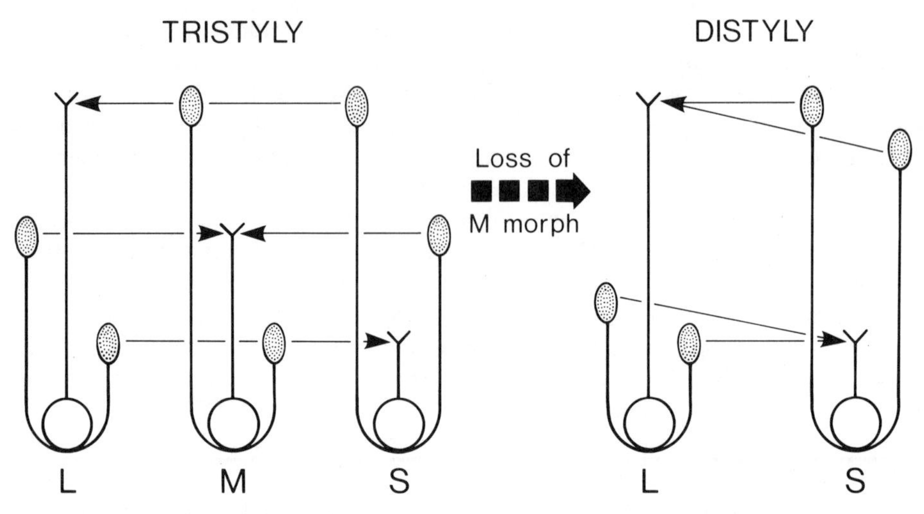

Fig. 5.4. Evolution of distyly from tristyly in *Oxalis alpina.* (For details see Weller.[117])

distylous *Turnera ulmifolia* populations. The variants display varying degrees of self-compatibility as a result of either aberrant style or pollen behavior and the genetic control of compatibility behavior is polygenic in nature.

Modification and Loss of Trimorphic Incompatibility

Modification and loss of trimorphic incompatibility have been reported from each of the three tristylous families.[29] Current work on two cases illustrates the complexity of these systems; the first involves the multiple origins of distyly from tristyly in *Oxalis* and the second the evolution of selfing in *Eichhornia.*

In both the Lythraceae and Oxalidaceae, distyly is derived from tristyly by loss of one of the style morphs. The most detailed investigations of this change in breeding system are those of Weller[117–123] on *Oxalis alpina* (Fig. 5.4). In populations of this species from southeast Arizona, the mid-styled morph ranges in frequency from 0–46%. Where populations exhibit high frequencies of this morph, the floral architecture and incompatibility relationships of the morphs are typical of most taxa with trimorphic incompatibility. However, in populations in which the mid-styled morph is rare or absent the reproductive morphology and incompatibility behavior of the long- and short-styled morphs are typical of distylous species.[117] Crossing studies[120] among populations with the two breeding systems indicate that distylous populations have diverged more substantially from one another than have tristylous populations. This pattern is consistent with the view that contrasting selection pressures in populations have resulted in the evolution of distyly in some and the retention of tristyly in others.

The difficulty arises in trying to determine the selective forces responsible for loss of the mid-styled morph from populations. Weller has examined a number of hypotheses, and several have been clearly falsified. These include preferential foraging by pollinators on the style morphs[121] and differences in clonal propagation and ovule and seed fertility of the morphs.[127] The most likely hypothesis concerns the loss of incompatibility differentiation in mid-level stamens of the long- and short-styled morphs.[117] This could favor these forms as male parents, since pollen capable of fertilizing their ovules would be more likely derived from these morphs than from the mid-styled morph. However, detailed progeny tests conducted over a 3-year period that had been designed specifically to evaluate this hypothesis gave unexpected results.[123] The mid-styled morph was disproportionately represented in families derived from this morph, and there was no clear evidence of its reduced male fertility as anticipated. The progeny test results also indicated large deficiencies of the short-styled morph in mid-styled families and suggested that anomalous transmission of alleles at the *S* and *M* loci may occur. To detect the differential transmission of alleles by gametophytic selection during megasporogenesis or through embryo abortion, controlled crosses among known genotypes and progeny analysis will be required. At this time it is too early to evaluate whether or not these phenomena are involved in the origin of distyly, but it is difficult to believe that the loss of incompatibility differentiation in tristylous populations has no role to play. Several other cases of genetic modification of trimorphic incompatibility in the genus *Oxalis* are equally difficult to interpret.[67,87]

Breakdown of trimorphic incompatibility in the Pontederiaceae involves the repeated shift to semihomostyly and selfing,[8] rather than the evolution of distyly or other outcrossing systems. These changes may or may not be associated with speciation events. In species in which incompatibility is maintained, as in the genus *Pontederia,* it is variable in expression, with the mid-styled morph displaying a high level of self-compatibility in comparison with the long- and short-styled morphs.[10] Barrett and Anderson[10] have proposed a developmental model to explain the weak expression of self-incompatibility in the mid-styled morph and discuss its implications for the breakdown of tristyly.

In *Eichhornia,* floral trimorphism is associated with high levels of self-compatibility and the occurrence of autogamous semihomostylous variants in each of the tristylous species.[5,6,9] The breakdown process has been studied in detail in *Eichhornia paniculata,* in which populations exhibit modifications ranging from complete tristyly in northeast Brazil to semihomostyly on the island of Jamaica. Proposed stages in the breakdown process are illustrated in Fig. 5.5. Critical events involve loss of the *S* allele (and hence the short-styled morph) through stochastic influences on population size

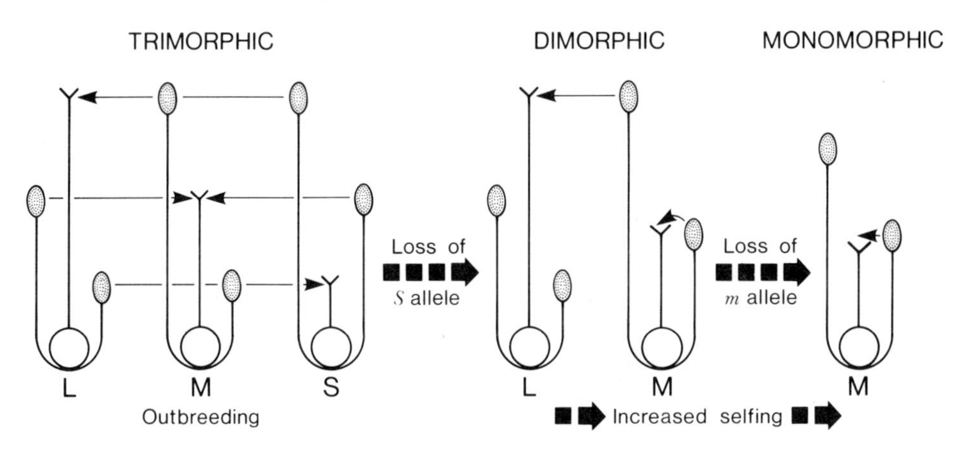

Fig. 5.5. Evolutionary breakdown of tristyly to semihomostyly in *Eichhornia paniculata.* (For details see Barrett[8] and Glover and Barrett.[52])

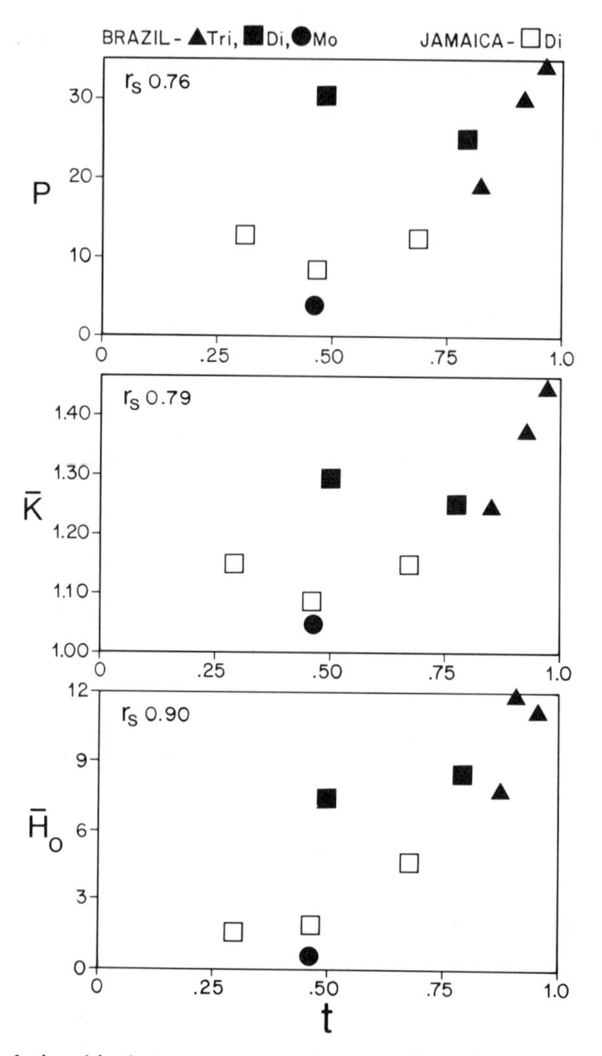

Fig. 5.6. The relationship between outcrossing rate (t) and several population genetic parameters (P = proportion of loci polymorphic, $\overline{K}$ = mean number of alleles per locus, $\overline{H}_0$ = mean observed heterozygosity) in populations of *Eichhornia paniculata.* (After Glover and Barrett.[53])

and loss of the *m* allele (and hence the long-styled morph) in association with the automatic selection of genes modifying the position of short-level stamens in the mid-styled morph.[8] All monomorphic populations so far examined are composed exclusively of semihomostylous mid-styled individuals. These populations are frequently small, suggesting that the facility for self-pollination is advantageous at low density.

The breakdown of tristyly in *Eichhornia paniculata* involves a shift in mating system from predominant outcrossing to high levels of selfing. This has been verified by multilocus estimates of outcrossing rate, using isozyme loci as genetic markers.[52] Lack of electrophoretically detectable isozyme variation in semihomostylous populations from Jamaica precluded quantitative estimates of their mating systems. However, the lack of variation and the highly autogamous behavior suggest that they are predominantly selfing. Associated with the evolutionary change in mating system of *Eichhornia paniculata* is a reduction in levels of genetic variation and heterozygosity (Fig. 5.6). Tristylous populations are significantly more variable than dimorphic or monomorphic populations.[53] The breakdown of tristyly in *Eichhornia* species depends in large part on the initial relaxation of self-incompatibility. Self-compatible populations are likely to be more sensitive to demographic and ecological factors (e.g., plant density, pollinator levels) that influence mating patterns. While it is evident that populations of species with trimorphic incompatibility contain considerable genetic variation for self-compatibility,[10] it is not obvious how this variation is maintained and what selective factors lead to its eventual loss independently of changes in floral form.

CONCLUSIONS

Self-incompatibility systems in flowering plants can be classified according to different criteria including the time of gene action, the inhibition site of self-pollen tubes, the association with floral polymorphism, and the number of loci and alleles governing the incompatibility reaction. Future research on this diversity is likely to benefit considerably from recent advances in molecular biology. Molecular characterization of incompatibility systems through the use of recombinant DNA technologies[84] and comparison of gene homologies in contrasting systems (sporophytic versus gametophytic and homomorphic versus heteromorphic) by hybridization techniques should enable a more rigorous assessment of phylogenetic relationships. Other topics that are likely to provide promising avenues for research include (1) clarification of the general properties (genetics, inhibition mechanisms) of "late-acting" and "cryptic" self-incompatibility systems, (2) evaluation of the role of inbreeding depression in the maintenance of self-incompatibility, (3) estimation of mating system parameters (e.g., levels of inbreeding through sib-mating) in populations of self-incompatible species, and (4) ecological, demographic, and life history correlates of different self-incompatibility systems. It will be of particular interest to see whether the classical view of self-incompatibility as an outbreeding mechanism survives the challenge of alternative hypotheses that will undoubtedly be formulated in the coming years.

ACKNOWLEDGMENTS

I thank Kamal Bawa, Robert Bertin, Deborah and Brian Charlesworth, Deborah Glover, Richard Olmstead, John Piper, Joel Shore, and Stephen Weller for providing unpublished manuscripts and information; Brenda Missen for editorial advice and typing; and the Natural Sciences and Engineering Research Council of Canada for supporting my own research on heteromorphic incompatibility systems.

REFERENCES

1. Anderson, J. M., and Barrett, S. C. H., Pollen tube growth in tristylous *Pontederia cordata* L. (Pontederiaceae), *Can. J. Bot.* **64,** 2602–2607 (1986).
2. Arasu, N. N., Self-incompatibility in angiosperms: A review, *Genetica* **39,** 1–24 (1968).
3. Atwood, S. S., Oppositional alleles in natural populations of *Trifolium repens, Genetics* **29,** 428–435 (1944).

4. Barrett, S. C. H., Heterostyly in a tropical weed: The reproductive biology of the *Turnera ulmifolia* complex (Turneraceae), *Can. J. Bot.* **56,** 1713–1725 (1978).
5. Barrett, S. C. H., The floral biology of *Eichhornia azurea* (Swartz) Kunth (Pontederiaceae), *Aquat. Bot.* **5,** 217–228 (1978).
6. Barrett, S. C. H., The evolutionary breakdown of tristyly in *Eichhornia crassipes* (Mart.) Solms. (water hyacinth), *Evolution* **33,** 499–510 (1979).
7. Barrett, S. C. H., Sexual reproduction in *Eichhornia crassipes* (Water Hyacinth) I. Fertility of clones from diverse regions, *J. Appl. Ecol.* **17,** 101–112 (1980).
8. Barrett, S. C. H., Ecological genetics of breakdown in tristyly, in *Structure and Functioning of Plant Populations. 2. Phenotypic and Genotypic Variation in Plant Populations* (I. Haeck and J. W. Woldendrop, eds.), pp. 267–275. North-Holland Publ. Co., Amsterdam, 1985.
9. Barrett, S. C. H., Floral trimorphism and monomorphism in continental and island populations of *Eichhornia paniculata* (Spreng.) Solms. (Pontederiaceae), *Biol. J. Linn. Soc.* **25,** 41–60 (1985).
10. Barrett, S. C. H., and Anderson, J. M., Variation in expression of trimorphic incompatibility in *Pontederia cordata* L. (Pontederiaceae), *Theor. Appl. Genet.* **70,** 355–362 (1985).
11. Barrett, S. C. H., and Forno, I. W., Style morph distribution in New World populations of *Eichhornia crassipes* (Mart.) Solms-Laubach (Water Hyacinth), *Aquat. Bot.* **13,** 299–306 (1982).
12. Barrett, S. C. H., and Glover, D. E., On the Darwinian hypothesis of the adaptive significance of tristyly, *Evolution* **39,** 766–774 (1985).
13. Barrett, S. C. H., and Shore, J. S., Variation and evolution of breeding systems in the *Turnera ulmifolia* complex (Turneraceae), *Evolution* **41,** 340–354 (1987).
14. Barrett, S. C. H., Price, S. D., and Shore, J. S., Male fertility and anisoplethic population structure in tristylous *Pontederia cordata* (Pontederiaceae), *Evolution* **37,** 745–759 (1983).
15. Bateman, A. J., Self-incompatibility systems in angiosperms, I: Theory, *Heredity* **6,** 285–310 (1952).
16. Bateman, A. J., Cryptic self-incompatibility in the wallflower: *Cheiranthus cheiri* L., *Heredity* **10,** 257–261 (1956).
17. Bawa, K. S., and Beach, J. H., Self-incompatibility systems in the *Rubiaceae* of a tropical lowland forest, *Am. J. Bot.* **70,** 1281–1288 (1983).
18. Beach, J. H., Pollinator foraging and the evolution of dioecy, *Am. Nat.* **118,** 572–577 (1981).
19. Beach, J. H., and Kress, W. J., Sporophyte versus gametophyte: A note on the origin of self-incompatibility in flowering plants, *Syst. Bot.* **5,** 1–5 (1980).
20. Bodmer, W. F., Natural crossing between homostyle plants of *Primula vulgaris, Heredity* **12,** 363–370 (1958).
21. Bodmer, W. F., Genetics of homostyly in populations of *Primula vulgaris, Phil. Trans. R. Soc. London,* B **242,** 517–549 (1960).
21a. Brauner, S., and Gottlieb, L. D., A self-incompatible plant of *Stephanomeria exigua* sub sp. *coronaria* (Asteracea) and its relevance to the origin of its self-pollinating derivative *F. malheurensis, Systematic Botany* **12,** 299–304 (1987).
22. Brewbaker, J. L., Pollen cytology and incompatibility systems in plants, *J. Hered.* **48,** 217–277 (1957).
23. Cahalan, C. M., and Gliddon, C., Genetic neighbourhood sizes in *Primula vulgaris, Heredity* **54,** 65–70 (1985).
24. Campbell, J. M., and Lawrence, M. J., The population genetics of the self-incompatibility polymorphism in *Papaver rhoeas.* I. The number and distribution of S-alleles in families from three localities, *Heredity* **46,** 69–79 (1981).
25. Campbell, J. M., and Lawrence, M. J., The population genetics of the self-incompatibility polymorphism in *Papaver rhoeas.* II. The number and frequency of S-alleles in a natural population (R106), *Heredity* **46,** 81–90 (1981).
26. Casper, B. B., Self-compatibility in distylous *Cryptantha flava* (Boraginaceae), *New Phytol.* **99,** 149–154 (1985).
27. Casper, B. B., and Charnov, E. L., Sex allocation in heterostylous plants, *J. Theor. Biol.* **96,** 143–149 (1982).
28. Charlesworth, B., and Charlesworth, D., The maintenance and breakdown of heterostyly, *Am. Nat.* **114,** 499–513 (1979).
29. Charlesworth, D., The evolution and breakdown of tristyly, *Evolution* **33,** 486–498 (1979).
30. Charlesworth, D., On the nature of the self-incompatibility locus in homomorphic and heteromorphic systems, *Am. Nat.* **119,** 732–735 (1982).
31. Charlesworth, D., Distribution of dioecy and self-incompatibility in angiosperms, in *Evolution—Essays*

in Honour of John Maynard Smith (J. J. Greenwood and M. Slatkin, eds.), p. 237. Cambridge Univ. Press, Cambridge, 1985.
32. Charlesworth, D., and Charlesworth, B., A model for the evolution of heterostyly, *Am. Nat.* **114,** 467–498 (1979).
33. Charnov, E. L., *The Theory of Sex Allocation, Monographs in Population Biology* No. 18. Princeton Univ. Press, Princeton, 1982.
34. Cope, F. W., The mechanisms of pollen incompatibility in *Theobroma cacao* L., *Heredity* **17,** 157–182 (1962).
35. Crawford, D. J., Ornduff, R., and Vasey, M. C., Allozyme variation within and between *Lasthenia minor* and its derivative species, *L. maritima* (Asteraceae), *Am. J. Bot.* **72,** 1177–1184 (1985).
36. Crosby, J. L., Selection of an unfavourable gene-complex, *Evolution* **3,** 212–230 (1949).
37. Crosby, J. L., Outcrossing on homostyle primroses, *Heredity* **13,** 127 (1959).
38. Crowe, L. K., The polygenic control of outbreeding in *Borago officinalis, Heredity* **27,** 111–118 (1971).
39. Crumpacker, D. W., Genetic loads in maize (*Zea mays* L.) and other cross-fertilized plants and animals, *Evol. Biol.* **1,** 1–131 (1967).
40. Curtis, J., and Curtis, C. F., Homostyle primrose re-visited. I. Variation in time and space, *Heredity* **54,** 227–234 (1985).
41. Darwin, C., *The Effects of Cross and Self Fertilization in the Vegetable Kingdom.* Appleton, New York, 1876.
42. Darwin, C., *The Different Forms of Flowers on Plants on the Same Species.* Murray, London, 1877.
43. Dulberger, R., Flower dimorphism and self-incompatibility in *Narcissus tazetta* L., *Evolution* **18,** 361–363 (1964).
44. Dulberger, R., Floral dimorphism in *Anchusa hybrida* Ten., *Isr. J. Bot.* **19,** 37–41 (1970).
45. Dulberger, R., *S*-gene action and the significance of characters in the heterostylous syndrome, *Heredity* **35,** 407–415 (1975).
46. Dulberger, R., and Ornduff, R., Floral morphology and reproductive biology of four species of *Cyanella* (Tecophilaeaceae), *New Phytol.* **86,** 45–56 (1980).
47. East, E. M., The distribution of self-sterility in flowering plants, *Proc. Am. Philos. Soc.* **82,** 449–518 (1940).
48. Emerson, S., The genetics of self-incompatibility in *Oenothera organensis, Genetics* **23,** 190–202 (1938).
49. Emerson, S., A preliminary survey of the *Oenothera organensis* population, *Genetics* **24,** 524–537 (1939).
50. Ganders, F. R., Disassortative pollination in the distylous plant *Jepsonia heterandra., Can. J. Bot.* **52,** 2401–2406 (1974).
51. Ganders, F. R., The biology of heterostyly, *N.Z. J. Bot.* **17,** 607–635 (1979).
52. Glover, D. E., and Barrett, S. C. H., Variation in the mating system of *Eichhornia paniculata* (Spreng.) Solms. (Pontederiaceae), *Evolution* **40,** 1122–1131 (1986).
53. Glover, D. E., and Barrett, S. C. H., Genetic variation in continental and island populations of *Eichhornia paniculata* (Pontederiaceae), *Heredity* **59,** 7–17 (1987).
54. Godley, E. J., and Smith, D. H., Breeding systems in New Zealand plants 5. *Pseudowintera colorata* (Winteraceae), *N.Z. J. Bot.* **19,** 151–156 (1981).
55. Gottlieb, L. D., Genetic differentiation, sympatric speciation, and the origin of a diploid species of *Stephanomeria, Am. J. Bot.* **60,** 545–553 (1973).
56. Halkka, O., and Halkka, L., Polymorphic balance in small island populations of *Lythrum salicaria, Ann. Bot. Fenn.* **11,** 267–270 (1974).
57. Heslop-Harrison, J., Self-incompatibility: Phenomenology and physiology, *Proc. R. Soc. London,* B **218,** 371–395 (1983).
58. Heuch, I., Equilibrium populations of heterostylous plants, *Theor. Popul. Biol.* **15,** 43–57 (1979a).
59. Heuch, I., The effect of partial self-fertilization on type frequencies in heterostylous plants, *Ann. Bot. (London)* **44,** 611–616 (1979b).
60. Heuch, I., Loss of incompatibility types in finite populations of the heterostylous plant, *Lythrum salicaria, Hereditas* **92,** 53–57 (1980).
61. Heuch, I., and Lie, R. T., Genotype frequencies associated with incompatibility systems in tristylous plants, *Theor. Popul. Biol.* **27,** 318–336 (1985).
62. Jain, S. K., The evolution of inbreeding in plants, *Annu. Rev. Ecol. Syst.* **7,** 469–495 (1976).
63. Knox, R. B., Heslop-Harrison, J., and Heslop-Harrison, Y., Pollen-wall proteins: Localization and characterization of gametophytic and sporophytic fractions, in *The Biology of the Male Gamete* (J. G. Duckett and P. A. Racey, eds.), p. 177. *Biol. J. Linn. Soc.* (1975).

64. Lande, R., and Schemske, D. W., The evolution of self-fertilization and inbreeding depression in plants. I. Genetic models, *Evolution* **39,** 24–40 (1985).
65. Lawrence, M. J., and O'Donnell, S., The population genetics of the self-incompatibility polymorphism in *Papaver rhoeas.* III. The number and frequency of *S*-alleles in two further natural populations (R102 and R104), *Heredity* **47,** 53–61 (1981).
66. Lawrence, M. J., Marshall, D. F., Curtis, V. E., and Fearon, C. H., Gametophytic self-incompatibility re-examined: A reply, *Heredity* **54,** 131–138 (1985).
67. Leach, C. R., Fluctuations in heteromorphic self-incompatibility systems, *Theor. Appl. Genet.* **66,** 307–312 (1983).
68. Levin, D. A., Plant density, cleistogamy and self-fertilization in natural populations of *Lithospermum caroliniense, Am. J. Bot.* **59,** 71–77 (1972).
69. Levin, D. A., and Clay, K., Extraneous pollen advantage in *Phlox cuspidata, Heredity* **54,** 145–148 (1985).
70. Lewis, D., Structure of the incompatibility gene. III. Types of spontaneous and induced mutation, *Heredity* **5,** 399–414 (1951).
71. Lewis, D., *Sexual Incompatibility in Plants.* Edward Arnold, London, 1979.
72. Lewis, D., Incompatibility, stamen movement and pollen economy in a heterostyled tropical forest tree, *Cratoxylum formosum* (Guttiferae), *Proc. R. Soc. London,* B **214,** 273–283 (1982).
73. Litzow, M. E., and Ascher, P. D., The inheritance of pseudo-self compatibility (PSC) in *Raphanus sativus* L., *Euphytica* **32,** 9–15 (1983).
74. Lloyd, D. G., Evolution of self-compatibility and racial differentiation in *Leavenworthia* (Cruciferae), *Contrib. Gray. Herb.* **195,** 3–134 (1965).
75. Lloyd, D. G., Evolution towards dioecy in heterostylous populations, *Plant Syst. Evol.* **131,** 71–80 (1979).
76. Lloyd, D. G., Demographic factors and mating patterns in Angiosperms, in *Demography and Evolution of Plant Populations* (O. T. Solbrig, ed.), p. 67. Blackwell, Oxford, 1980.
77. Lloyd, D. G., and Yates, J. M. A., Intrasexual selection and the segregation of pollen and stigmas in hermaphrodite plants, exemplified by *Wahlenbergia albomarginata* (Campanulaceae), *Evolution* **36,** 903–913 (1982).
78. Lundquist, A., Complex self-incompatibility systems in angiosperms, *Proc. R. Soc. London,* B **188,** 235–245 (1975).
79. Mosquin, T., Biosystematic studies in the North American species of *Linum,* section *Adenolinum* (Linaceae), *Can. J. Bot.* **49,** 1379–1388 (1971).
80. Muenchow, G., An *S*-locus model for the distyly supergene, *Am. Nat.* **118,** 756–760 (1981).
81. Muenchow, G., A loss-of-alleles model for the evolution of distyly, *Heredity* **49,** 81–93 (1982).
82. Mulcahy, D. L., and Caporello, D., Pollen flow within a tristylous species: *Lythrum salicaria, Am. J. Bot.* **57,** 1027–1030 (1970).
83. Mulcahy, D. L., and Mulcahy, G. B., Gametophytic self-incompatibility reexamined, *Science* **220,** 1247–1251 (1983).
84. Nasrallah, J. B., Kao, T. H., Goldberg, M. L., and Nasrallah, M. E., A cDNA clone encoding an *S*-locus-specific glycoprotein from *Brassica oleracea, Nature (London)* **318,** 263–267 (1985).
85. Nettancourt, D. de, *Incompatibility in Angiosperms.* Springer-Verlag, Berlin, 1977.
86. Olmstead, R. G., Self-incompatibility in light of population structure and inbreeding, in *Biotechnology and Ecology of Pollen* (D. L. Mulcahy et al., eds.), p. 239. Springer-Verlag, New York, 1986.
87. Ornduff, R., The breeding system of Oxalis suksdorfii, *Am. J. Bot.* **51,** 307–314 (1964).
88. Ornduff, R., Heterostyly in South African flowering plants: A conspectus, *J. South Afr. Bot.* **40,** 169–187 (1974).
89. Ornduff, R., Heterostyly, population composition and pollen flow in *Hedyotis caerulea, Am. J. Bot.* **67,** 95–103 (1980).
90. Ornduff, R., and Dulberger, R., Floral enantiomorphy and the reproductive system of *Wachendorfia paniculata* (Haemodoraceae), *New Phytol.* **80,** 427–434 (1978).
91. Osterbye, U., Self-incompatibility in *Ranunculus acris* L. I. Genetic interpretation and evolutionary aspects, *Hereditas* **80,** 91–112 (1975).
92. Pandey, K. K., Overcoming incompatibility and promoting genetic recombination in flowering plants, *N.Z. J. Bot.* **17,** 645–663 (1979).
93. Philipp, M., and Schou, O., An unusual heteromorphic incompatibility system: Distyly, self-incompatibility, pollen load and fecundity in *Anchusa officinalis* (Boraginaceae), *New Phytol.* **89,** 693–703 (1981).

94. Piper, J. G., Charlesworth, B., and Charlesworth, D., Breeding system evolution in *Primula vulgaris* and the role of reproductive assurance, *Heredity* **56,** 207–217 (1986).
95. Piper, J. G., Charlesworth, B., and Charlesworth, D., A high rate of self-fertilization and increased seed fertility of homostyle primroses, *Nature (London)* **310,** 50–51 (1984).
96. Richards, A. J., and Ibrahim, H. B. T., The breeding system in *Primula veris* L. II. Pollen tube growth and seed set, *New Phytol.* **90,** 305–314 (1982).
97. Richards, J. H., and Barrett, S. C. H., The developmental basis of tristyly in *Eichhornia paniculata* (Pontederiaceae), *Am. J. Bot.* **71,** 1347–1363 (1984).
98. Robacker, C. D., and Ascher, P. D., Effect of selection for pseudo-self-compatibility in advanced inbred generations of *Nemesia strumosa* Benth, *Euphytica* **31,** 591–610 (1982).
99. Rollins, R. C., The evolution and systematics of *Leavenworthia* (Cruciferae), *Contrib. Gray. Herb.* **192,** 3–98 (1963).
100. Schemske, D. W., and Lande, R., The evolution of self-fertilization and inbreeding depression in plants. II. Empirical observations, *Evolution* **39,** 41–52 (1985).
101. Schou, O., and Philipp, M., An unusual heteromorphic incompatibility system. 2. Pollen tube growth and seed sets following compatible and incompatible crossings within *Anchusa officinalis* L. (Boraginaceae), in *Pollen: Biology and Implications for Plant Breeding* (D. L. Mulcahy, ed.), pp. 219–227. Elsevier, New York, 1983.
102. Schou, O., and Philipp, M., An unusual heteromorphic incompatibility system. 3. On the genetic control of distyly and self-incompatibility in *Anchusa officinalis* L. (Boraginaceae), *Theor. Appl. Genet.* **68,** 139–144 (1984).
103. Seavey, S. R., and Bawa, K. S., Late-acting self-incompatibility in angiosperms, *Bot. Rev.,* **52,** 195–219 (1986).
104. Shivanna, K. R., Heslop-Harrison, J., and Heslop-Harrison, Y., Heterostyly in *Primula* 2. Sites of pollen inhibition, and effects of pistil constituents on compatible and incompatible pollen tube growth, *Protoplasma* **107,** 319–337 (1981).
105. Shore, J. S., and Barrett, S. C. H., Morphological differentiation and crossability among populations of the *Turnera ulmifolia* L. complex (Turneraceae), *Syst. Bot.* **10,** 308–321 (1985).
106. Shore, J. S., and Barrett, S. C. H., The genetics of distyly and homostyly in *Turnera ulmifolia* L. (Turneraceae), *Heredity* **55,** 167–174 (1985).
107. Shore, J. S., and Barrett, S. C. H., Genetic modifications of dimorphic incompatibility in the *Turnera ulmifolia* L. complex (Turneraceae), *Can. J. Genet. Cytol.* **28,** 796–807 (1986).
108. Sorensen, F., Embryonic genetic load in coastal Douglas-fir, *Pseudotsuga menziesii* var. *menziessi, Am. Nat.* **103,** 389–398 (1969).
109. Stebbins, G. L., *Flowering Plant Evolution above the Species Level.* Belknap Press of Harvard University Press, Cambridge, 1974.
110. Stevens, V. A. M., and Murray, B. G., Studies on heteromorphic self-incompatibility systems: Physiological aspects of the incompatibility system of *Primula obconica, Theor. Appl. Genet.* **61,** 245–256 (1982).
111. Taroda, T., and Gibbs, P. E., Floral biology and breeding system of *Sterculia chicha* St. Hil., *New Phytol.* **90,** 735–743 (1982).
112. Taylor, P. D., Evolutionarily stable reproductive allocations in heterostylous plants, *Evolution* **38,** 408–416 (1984).
113. Varopoulos, A., Breeding systems in *Myosotis scorpioides* (Boraginaceae) I self-incompatibility, *Heredity* **42,** 149–157 (1979).
114. Vuilleumier, B. S., The origin and evolutionary development of heterostyly in the angiosperms, *Evolution* **21,** 210–226 (1967).
115. Waser, N. M., and Price, M. V., Optimal and actual outcrossing in plants, and the nature of plant-pollinator interaction, in *Handbook of Experimental Pollination Biology* (C. E. Jones and R. J. Little, eds.), Chapter 17. Van Nostrand Reinhold Company Inc., New York, 1983.
116. Webb, C. J., and Lloyd, D. G., Selection to avoid interference between the presentation of pollen and stigmas in angiosperms. II Herkogamy, *N.Z. J. Bot.* **24,** 163–178 (1986).
117. Weller, S. G., Breeding system polymorphism in a heterostylous species, *Evolution* **30,** 442–454 (1976).
118. Weller, S. G., The genetic control of tristyly in *Oxalis* section *Ionoxalis, Heredity* **37,** 387–393 (1976).
119. Weller, S. G., Dispersal patterns and the evolution of distyly in *Oxalis alpina, Syst. Bot.* **3,** 115–126 (1978).

120. Weller, S. G., Variation in heterostylous reproductive systems among populations of *Oxalis alpina* in southeastern Arizona, *Syst. Bot.* **4,** 57–71 (1979).
121. Weller, S. G., Pollination biology of heteromorphic populations of *Oxalis alpina* (Rose) Knuth (Oxalidaceae) in southeastern Arizona, *Bot. J. Linn. Soc.* **83,** 189–198 (1981).
122. Weller, S. G., Fecundity in populations of *Oxalis alpina* in southeastern Arizona, *Evolution* **35,** 197–200 (1981b).
123. Weller, S. G., Factors influencing frequency of the mid-styled morph in tristylous populations of *Oxalis alpina, Evolution* **40,** 279–289 (1986).
124. Weller, S. G., and Ornduff, R., Cryptic self-incompatibility in *Amsinckia grandiflora, Evolution* **31,** 47–51 (1977).
125. Whitehouse, H. L. K., Multiple-allelomorph incompatibility of pollen and style in the evolution of the angiosperms, *Ann. Bot. (London)* **14,** 198–216 (1950).
126. Willson, M. F., Sexual selection in plants, *Am. Nat.* **113,** 777–790 (1979).
127. Willson, M. F., *Plant Reproductive Ecology.* John Wiley and Sons, New York, 1983.
128. Willson, M. F., and Burley, N., *Mate Choice in Plants: Tactics, Mechanisms, and Consequences, Monographs in Population Biology No. 19.* Princeton Univ. Press, Princeton, 1983.
129. Williams, R. D., and Williams, W., Genetics of red clover (*Trifolium pratense* L.) compatibility, *J. Genet.* **48,** 67–79 (1947).
130. Wright, S., *Evolution and the Genetics of Populations, Vol. 3 Experimental Results and Evolutionary Deductions.* Univ. of Chicago Press, Chicago, 1977.
131. Wyatt, R., Pollinator-plant interactions and the evolution of breeding systems, in *Pollination Biology* (L. Real, ed.), Chapter 4. Academic Press, Orlando, Florida, 1983.
132. Yeo, P. F., Some aspects of heterostyly, *New Phytol.* **75,** 147–153 (1975).
133. Yokoyama, S., and Hetherington, L. E., The expected number of self-incompatibility alleles in finite plant populations, *Heredity* **48,** 299–303 (1982).
134. Zavada, M. S., The relation between pollen exine sculpturing and self-incompatibility mechanisms. *Plant Syst. Evol.* **147,** 63–78 (1984).
135. Zavada, M. S., and Taylor, T. N., The role of self-incompatibility and sexual selection in the gymnosperm-angiosperm transition: A hypothesis. *Am. Nat.* **128,** 538–550 (1986).

6

Sex Determination in Plants

THOMAS R. MEAGHER

Sex determination refers to the balance between expression of female versus male sexuality within a plant. The mechanism by which sex is determined has a profound impact on the reproductive function of a plant. Population level consequences of differential male and female function or allocation have been recently reviewed by Charnov,[16] Charlesworth and Charlesworth,[13] and Lloyd and Bawa.[57] The evolutionary implications of this process are most striking in dioecious species, where individual sex expression is limited to either maleness or femaleness. Many plant species also exhibit male sterility such that populations consist of females and hermaphrodites. In this chapter, the phenomenon of sex determination will be viewed from a broad perspective, including physiological, ecological, and genetic influences on plant sexuality.

A wide diversity of sexual states is exhibited within flowering plant species, ranging from discrete sex morphs, as in dioecy or monoecy, to more quantitative variation in which individual plants vary in their degree of maleness or femaleness. This diversity has in turn led to a diverse array of terminology to describe the sexuality of different species. Lloyd has suggested that such terminology should be replaced by a numerical scale that would more adequately represent the quantitative nature of plant sexuality.[56] Although such a scale is in many cases probably more appropriate, the older terminology has been adhered to here because it was used in the work that will be discussed.

The evolutionary genetics of sex-determining mechanisms has recently been treated by Bull,[9] who emphasized zoological systems. Sex determination in certain plant groups, notably bryophytes, has also been the focus of several recent reviews (see Mishler,[68] this volume, and Refs. 77 and 93). On the other hand, the last major review of sex determination that dealt primarily with flowering plants was conducted in 1958 by Westergaard.[91] Thus this chapter will focus primarily on flowering plants. Westergaard placed strong emphasis on sex chromosomes and the overall genetics of sex determination. This article will draw on his review for much of the work prior to 1958 and will emphasize areas in which significant progress has been made since that time.

PHYSIOLOGICAL ASPECTS OF SEX DETERMINATION

Physiological differences associated with sexuality in dioecious species can be viewed on two levels. On the one hand, since males and females perform very different reproductive functions that in turn impose very different resource demands on the plant,[54,55,61–63,66] many aspects of differentiation can be regarded as a consequence of sexuality.[1,66] On the other hand, certain physiological differences between the sexes are thought to be involved as a cause of plant sexuality. This latter type of differentiation is the focus of the present discussion.

At the time of Westergaard's 1958 review, very little was known about the physiology of sex determination.[91] Although there is still much that is not known about this phenomenon, this area has received more attention since that time.[12,23,32,33] There are two principal aspects of the physiology of sex expression that have been studied in dioecious plant species: influences of plant growth substances and differential enzyme activity.

In the course of investigation of the physiology of sex expression, the influence of almost every conceivable category of physiologically active substances on sex expression has been explored by exogenous application. In particular, studies involving the exogenous application of plant growth substances have been widespread,[12,17,32,44,74,75,83] and many plant growth substances have been found to influence plant sexuality. As a general, but not universal, rule, gibberellins (GA_3) are thought to enhance male expression.[12,23,27] Recent studies on endogenous levels of GA_3 in male and female plants of several species provide confirmation of the masculinization effect of these growth substances.[29,48] Auxins [e.g., indoleacetic acid (IAA)], on the other hand, have been shown to enhance female expression.[12,23,27]

Cytokinins (CK) have also been shown to play a role in sex expression, although their effects are more variable than those of IAA and GA_3.[12,23,27,32,60] In *Mercurialis annua,* exogenous application of CK results in the masculinization of female plants.[23] Male and female plants of this species have also been found to differ with respect to endogenous CK levels, which were elevated in males.[21] Differences were found between the sexes in nonflowering individuals, and these differences became more extreme once plants had begun to flower.

In conjunction with studies of plant growth substances, there has been increasing attention directed toward the role of different enzyme activity levels or differential expression of isozymes in male and female plants.[27,34,38,40,42,45,67,88] Most of this work has focussed on peroxidases, a broadly defined group of enzymes that are known to vary in degree of expression at different developmental stages. Levels of peroxidase activity have been shown to differ between the sexes across a range of species, including *Actinidia chinensis,*[34] *Mercurialis annua,*[42] *Carica papaya,*[67] *Morus nigra,*[38] and *Phoenix dactylifera.*[88] In those cases that have been examined in more detail, these changes are accompanied by differential expression of peroxidase isozymes as detected by electrophoresis.[34,38] Although sex specific banding patterns may not be evident in all whole plant tissue, differences between male and female plants are stable even in callus tissue culture.[34] The significance of peroxidase in sex determination is that these enzymes are believed to play a role in the regulation of plant growth substances. In dioecious crop species where there is an economic incentive for being able to distinguish male from female plants early in the life cycle (e.g., before flowering), peroxidase isozymes have been studied as a possible diagnostic tool.[67,88]

In addition to dioecious species, the influence of physiological factors on sex determination has also been studied in hermaphroditic species.[23,32,33] For these latter species, the object of investigation is the degree of maleness or femaleness within perfect flowers[12,32,33] or the localization of male- and female-determining influences within a monoecious plant.[41] As in dioecious species, the balance of IAA, GA_3, and CK plays a significant role with very similar effects. Also, localization of differential expression of peroxidase activity within male versus female flower primordia has been observed in various species of the Cucurbitaceae[41] as well as in *Ricinus communis.*[39]

Although there is considerable variation in the physiological correlates of sex determination across taxa, there does appear to be an underlying homology in the effects of such factors in dioecious and hermaphroditic species. A great deal is now known about possible effects of various enzymes and plant growth regulators on sex expression, but most of the work discussed above is based on observation of correlated events. There is still much research needed to identify the mechanisms and interactions by which these substances determine sexuality.

ENVIRONMENTAL ASPECTS OF SEX DETERMINATION

Environmental factors are known to influence sex determination in a number of plant species.[27,57,82] Since environmental conditions are not necessarily stable over time, the sex of a plant subject to environmental influences may change from season to season.[16,57] Such an effect would lead to sex lability, a phenomenon which is discussed in detail by Schlessman,[81] this volume. In this chapter, discussion of environmental effects will be limited to instances where such effects have contributed to our understanding of sex expression.

Freeman et al.[27] provide an extensive list of species in which environmental factors have been shown to influence sexuality, and a number of specific cases are discussed in detail by Lloyd and Bawa.[57] Basically, conditions favorable for growth (high CO_2, mild temperatures, moist soil, high light intensity, and fertilization) appear to enhance female sex expression, whereas less favorable conditions are associated with male sex expression.[22,28,63] This overall observation is in keeping with the observation that female sexuality is costlier than male sexuality within any given plant due to additional costs of seed and fruit production.[55,61,62]

A number of environmental effects are also known to influence the metabolism of plant growth regulators associated with sex expression, as discussed in the previous section. For example, nitrogen availability is thought to influence IAA metabolism. In addition, photoperiod, which also influences sexuality (long days promote maleness; short days promote femaleness),[27] directly influences the levels of various growth regulators.[12,32] Thus the particular balance of growth substances within a plant or, more specifically, within developing flower primordia may be reflecting the environmental background. Consequently, those plants which express appropriate sexuality in response to the growth substances reflecting either male-suitable or female-suitable backgrounds will have a selective advantage. Therefore the patterns of sexuality associated with specific growth regulators may be the outcome of a selective advantage of maleness or femaleness under different environmental conditions.

GENETIC ASPECTS OF SEX DETERMINATION

The existence of variation in genes that influence sexuality is a necessary prerequisite to the evolution of a genetic sex determination mechanism. Consequently, whereas studies of physiological and environmental effects on sex expression focus on individual plants, genetic analyses of sex expression also encompass population level phenomena. Thus the influence of genetic effects on sexuality will be considered not only in terms of the existence of different variants, but also in terms of the evolutionary behavior of these variants within populations.

There are three basic types of genetic effects involved in plant sex expression: (1) genes determining male expression, (2) genes determining female expression, and (3) genes that modify sex expression. As in much of genetics, most of our knowledge of the effects of genes controlling male or female sexuality is based on the study of cases where particular genes are not functioning properly, giving rise to male or female sterility, respectively.

Male Sterility

The presence of a male-sterile morph results in gynodioecious populations. This is a very widespread phenomenon among flowering plants.[25,37,49] There are two recognized causal bases for male sterility: cytoplasmic effects and nuclear genetic effects. Cytoplasmic male sterility occurs when extrachromosomal effects within the particular cytoplasm of an individual suppress male function. In such a case, male sterility is maternally inherited so that all of a female's progeny will also be female, although there are nuclear genes that restore male function in spite of the cytoplasmic effect.[35,47,60] Within populations in which cytoplasmic male sterility occurs, hermaphroditic individuals (plants that express both male and female sexuality) set seed that will all give rise to hermaphroditic plants in addition to the genes that they contribute as pollen parents to male-sterile individuals. In the case of nuclear genetic male sterility, genes that result in male sterility are inherited in a normal Mendelian fashion.

In order for male sterility to persist within a population, male-sterile (female) individuals must exceed male-fertile individuals in their seed production to offset the loss of genetic contribution through male gametes. There has been extensive theoretical and mathematical discussion of the threshold conditions for the evolution and persistence of male-sterility factors in plant populations.[13,49,53] For cytoplasmic effects, male-sterile plants need to set at least as many seeds as male-fertile plants in order to persist. In the case of nuclear genetic effects, male-sterile plants need to set at least twice as many seeds as male-fertile individuals in order to be maintained in a population.[49] In addition, for populations in which cytoplasmic male sterility is already established, the fitness of females resulting from nuclear male-sterility genes is diminished even further by the presence of cytoplasmic male-sterile individuals. As a result of this phenomenon, the simultaneous occurrence of both forms of male sterility within natural populations has been shown on theoretical grounds to be evolutionarily unstable.[15]

In addition to work on natural plant populations, male-sterility effects have received a lot of attention in crop species.[43] This is largely because of the usefulness of male sterility in plant breeding programs.[24] For example, cytoplasmic male sterility is often incorporated into one of the parental lines in the production of hybrid seed to ensure that none of the seed produced are the result of selfing. In corn alone, there are at least three forms of cytoplasmic male sterility[8] as well as 19 nuclear genetic loci[71] that are known to influence male sex expression.

Female Sterility

Female-sterility factors, which would result in androdioecious populations, are almost unknown in natural populations. The only genera in which this author could find androdioecy reported for natural populations were *Solanum*[2] and *Fuchsia.*[30] Just as the widespread occurrence of male sterility has attracted a fair amount of scientific attention, so has the rarity of female sterility.[13,53] It is easy to understand why cytoplasmic effects on female fertility would be strongly selected against; all of a cytoplasmically female-sterile's progeny would be normal, so that such an effect would be lost after only one generation. Similarly, mathematical studies have shown that the conditions under which nuclear genes for female sterility may become established within a population are much more restricted than in the case of male-sterility genes.[13,53] Such restrictions are based on the observation that under partial selfing, female-sterile mutants are not only losing their potential female gamete contribution, but also the male gamete contribution that would have resulted from selfed progeny.

Genes that confer female sterility are well known in the crop literature.[11,43,90] As in the case of male sterility, genes that result in female sterility have applied significance for plant breeding. Such genes also play a role in genetic studies of the regulation of female fecundity, which for many crop species, notably seed crops, is the essence of their productivity.

Sex Modifiers

Genetic effects that modify sex expression include loci that shift the balance between male and female sex expression. Although such loci would include the sterility effects discussed above, they would also include effects such as developmental "decisions" between production of male or female flowers in monoecious species. The best-documented cases of the latter type of effect occur in the Cucurbitaceae,[41,78] a family that has been thoroughly studied genetically due to its worldwide economic importance.

In the species *Cucumis sativus* and *Cucumis melo,* there is a locus that directly influences sex expression,[78] in addition to loci that influence male fertility. In *Cucumis sativus,* this locus carries four alleles: *F* enhances female sex expression and is dominant to *m* which enhances male sex expression, *a* results in an androecious (male) individual with no female flowers, and *Tr* results in a monoecious plant with some perfect flowers. In a corresponding locus in *Cucumis melo,* there is a pair of codominant alleles with effects similar to *F* and *m*. Other loci in this genus influence the location of male and female flower initiation along the shoot so that femaleness (*st*, more female; st^{+}, normal) or maleness (*a*, more male; *A*, normal) is enhanced.[50] These mechanisms lead to a wide array of sex phenotypes.[41]

The genetic effects associated with sex determination in *Cucumis* have also been noted in other species. A secondary sex-modifying locus has been proposed for *Asparagus* that is thought to behave very similarly to the *A*/*a* locus in *Cucumis.*[26] In a series of crossing experiments, Kubicki[46] noted that both males and females could be homogametic or heterogametic. To explain this seemingly contradictory result, he postulated a two-locus model along the lines of the genic effects in *Cucumis,* in which the first locus contains a dominant allele for female sex expression and a recessive allele for male sex expression, while the second locus contains a dominant allele for male expression that overrides the female allele at the first locus.

A two-locus model for sex determination in *Vitis* has recently been discredited in favor of one locus with three alleles.[4] A first allele, *M,* causes male expression and is dominant to *H* for hermaphrodite expression and *F* for female expression; *H* is in turn dominant to *F.*

The genus *Cotula* (Compositae) contains an array of species that vary in their sex expression, from monoecious (equivalent numbers of male and female florets within flower heads) to dioecious.[51] The types of genetic effects thought to be involved in sex expression in this genus control the timing of initiation of male and female flower primordia,[52] similar in effect to the secondary loci (st/st^+, A/a) in *Cucumis.*

EVOLUTIONARY MODELS OF SEX DETERMINATION

Theoretical investigations of sex determination have encompassed three major areas. First, the concept of evolutionarily stable strategies has been used to study the fitness and stability of different sex morphs within populations.[16] In such studies, precise details of the sex determination mechanism are only of secondary interest. Second, polygenic models in which sex is determined on a quantitative basis have been explored primarily in reference to sex determination in animals, particularly reptiles.[9,10] Finally, one- and two-locus genetic models in which sex is determined on a qualitative basis have been developed. This last area has received the most emphasis in the context of flowering plants. In fact, the evolutionary behavior of sex-determining genes on plant populations has recently been explored and reviewed by Bawa,[6] Charlesworth and Charlesworth,[13] and Lloyd[53]; consequently the following discussion will be limited to a brief overview.

The establishment of a stable sex determination mechanism involves the combined effects of several loci with different influences on plant sexuality. In the case of dioecious species, sex determination usually involves closely linked loci that affect male fertility and female fertility. The evolution of dioecy in plants is thought to occur primarily with gynodioecy as an intermediate step.[6,13,50,54,79] There are two reasons why this view is widely held. From an empirical standpoint, close relatives of dioecious species are often gynodioecious, which suggests that gynodioecy may have been ancestral to dioecy in those instances. Second, it is intuitively straightforward that the establishment of a dioecious state from an hermaphroditic origin requires the establishment of two types of genes: those that suppress male function in some individuals and those that suppress female function in others. Population genetics models for such evolutionary sequences have been extensively investigated.[13] As noted above, the evolutionary conditions under which genes conferring male sterility may become established are much less restrictive than the conditions under which genes conferring female sterility may become established. However, once a male-sterility gene is widespread in a population, linked genes conferring female sterility, leading to dioecy, are themselves more likely to spread. If the initial male sterility allele is dominant, males will be heterogametic; if it is recessive, female heterogamy will result. Generally, dominant male-sterility alleles have a higher probability of establishment, which may explain why male heterogamy is more common than female heterogamy in plants.

CHROMOSOMAL ASPECTS OF SEX DETERMINATION

Evidence

Any species in which a stable genetic sex determination mechanism has evolved can be said to have sex chromosomes. These sex chromosomes may be distinguishable as an heteromorphic pair, or, as is the case in many plant species, may only be recognized by their effects on inheritance of sex. On the simplest level, sex chromosomes (or the sex-determining loci) are homozygous (XX) in one sex and heterozygous (XY) in the other, with the sex of the offspring being determined by the gametes (.5X and .5Y) from the heterozygous, or heterogametic, parent.

The first step toward interpreting the chromosomal basis of sex determination in plants is the identification of the heterogametic sex. Of the six methods outlined by Westergaard[91] for distinguishing heterogamy, the following three will be discussed here:

1. Cytogenetic identification of heteromorphic sex chromosomes,
2. Crossing experiments in which the sex distribution of progeny are studied, and
3. Analysis of sex phenotypes of aneuploid or polyploid derivatives of dioecious species.

Both male and female heterogamy have been found in plants, and these phenomena are discussed separately below following a discussion of evidence relating to sex chromosomes.

Cytology

Westergaard[91] outlined three criteria for the identification of heteromorphic sex chromosomes: (1) demonstration of an unequal X–Y pair in meiotic configurations in both sexes, (2) absence of unequal pairing in the homogametic sex, and (3) recognition of the sex chromosomes in mitotic configurations in both sexes. Applying these criteria, he identified only 13 well-established cases of sex chromosomes in flowering plant species, 7 of these cases were in the genus *Rumex.* Westergaard also cited 22 cases, based primarily on work done in the 1920s and 1930s, where sex chromosomes had been reported but not adequately demonstrated.

Although there was strong interest in cytological investigation of heteromorphic sex chromosomes at the time of Westergaard's review, this area has quiesced substantially since the 1960s. Most of the work done during that period involved the genus *Rumex.*[64,84,86,87,94,95] There are two sections of that genus that contain dioecious species in which sex chromosomes have been reported[64]: Section *Acetosella* encompasses a polyploid series with a basic chromosome number of $2n = 12 + XX$ in females and $2n = 12 + XY$ in males. Section *Acetosa* consists of a number of species with $2n = 12 + XX$ in females and $2n = 12 + XY_1Y_2$ in males. One species in section *Acetosa, Rumex hastatulus,* has a reduced chromosome count of $2n = 6 + XY_1Y_2$ in males.

In spite of the apparent similarity of the tripartite (XY_1Y_2) sex chromosomes of *Rumex acetosa* and *Rumex hastatulus,* sex determination in these species is believed to have evolved independently, which, in conjunction with the *Acetosella* complex and *Rumex paucifolius,* gives four separate dioecious lineages in this genus.[85]

Another genus in which the cytogenetics of sex determination has been studied in detail is *Silene* (= *Melandrium* or *Lychnis*)[72,91]. In addition to *Silene alba* and *Silene rubra,* in which females are XX and males are XY, sex chromosomes have recently been reported in *Silene diclinus.*[73] In *Silene alba,* inspection of the sex morphs of plants carrying various deletions on the Y chromosome has enabled the physical location of sex-determining genes to be identified.[91] Thus three different genetic effects on sexuality were found on the nonpairing segment of the Y chromosome in *Silene alba:* (1) genes that suppress female function, (2) genes that initiate anther development, and (3) genes that enable formation of viable pollen.

Spinacea oleracea is one of the species listed by Westergaard in which sex chromosomes were reported as present by some authors and as absent by others. This controversy still rages on! Iizuka and Janick[36] have reported that sex chromosomes are present in spinach, although they vary in different lineages. Meanwhile, Ramanna[76] has shown that different chromosome pairs in spinach show heteromorphism as an artifact of different preparation protocols. Using trisomic analysis, Loptien[59] localized sex determination to a particular chromosome pair that was not heteromorphic in his material.

Finally, sex determination has been associated with translocation heterozygosity in the mistletoe *Viscum fischeri.*[92] In this species, there are effectively nine sex chromosomes, four X and five Y, that pair normally in females and form a nonavalent complex in males, the heterogametic sex.

Crossing Experiments

In cases where dioecious species are interfertile with related hermaphroditic taxa, the heterogametic sex can be identified by examining the segregation of sex forms in interspecific crosses involving males and females. In such crosses, segregation of different sex morphs in the progeny of male × hermaphrodite crosses indicates male heterogamy and, conversely, segregation in female × hermaphrodite crosses indicates female heterogamy. Westergaard[91] reported in detail on the use of this approach to demonstrate male heterogamy in dioecious species of *Bryonia, Acnida, Ecballium,* and *Thalictrum* as well as female heterogamy in *Fragaria.* Reciprocal crosses between dioecious species and related hermaphroditic species have also been used more recently to demonstrate female heterogamy in *Potentilla*[31] and *Cotula.*[52] A similar approach involving crosses between sex morphs within species has shown that female plants of *Fuchsia procumbens* are heterogametic.[30]

Variation in progeny sex ratios of intraspecific crosses may also shed light on the genetics of sex determination. For example, in *Dioscorea floribunda,* progeny sex ratios are either 1 : 1 or 3 : 1 (males : females), depending on the genotype of the male parent.[65] This observation has lead to the suggestion that males of this tetraploid species are heterogametic with an active-Y mechanism of sex determination (see next section); thus there are two male genotypes (XXYY and XXXY). This mechanism of sex determination would explain the male bias observed in seed progenies of open-pollinated *Dioscorea* species,[65] and may also explain a similar male sex ratio bias in

Solanum,[2] in which variation in progeny sex ratios has also been observed (G. Anderson, personal communication).

Experimental crosses may also be useful in identifying cryptic dioecy in which male and female plants differ in sex fertility but not in morphological appearance. For example, in the few known dioecious species of *Solanum,* the two sexes differ primarily in style length, with both sexes appearing to be hermaphroditic. As a consequence, the two sexes of *Solanum appendiculatum* had actually been named as different species.[3] In such a case, the only way to demonstrate that a species is in fact dioecious is to test the intra- and interfertility of different floral morphs through test crosses.[2,3,5] The phenomenon of cryptic dioecy has also been recognized in *Citharexylum*[89] as well as in a number of tropical tree species.[7]

Polyploidy and Aneuploidy

There are two basic forms of chromosomal sex determination that are widely recognized among species with male heterogamy.[87,91] In the first form, the sex of a plant is determined by the ratio of X chromosomes to autosomes (X/A balance), with the Y chromosome(s) playing little or no role in sex determination. In the second form, it is the presence of the Y chromosome that determines maleness; this form is referred to as the X/Y or active-Y mechanism. In order to establish which form of sex determination prevails within particular species, extensive use has been made of artificially generated polyploids.[87,91,94,95] Polyploidy wreaks havoc with the X/A mechanism because the X/A balance is unstable at higher ploidy levels. On the other hand, it is possible to retain a stable dioecious condition at higher ploidy levels under the active-Y mechanism, as demonstrated in the naturally occurring polyploid series of species of *Rumex* section *Acetosella.*[64]

Aneuploidy is also a valuable tool for analyzing chromosomal determination. By analyzing crosses involving individuals with trisomic configurations for different chromosomes, Loptien was able to associate sex determination with specific homomorphic chromosome pairs in *Asparagus officinalis*[58] and *Spinacea oleracea.*[59]

Male Heterogamy

Among dioecious species in which sex determination has been investigated, male heterogamy is by far the most prevalent mechanism of sex determination.[91] Many of the properties of male heterogamy have already been discussed. However, an additional peculiarity of this mechanism of sex determination is that the progeny sex ratio in some species is strongly dependent on the level of abundance of pollen on the stigmatic surface.[14,18,19,69,70,80] This phenomenon was first noted by Correns,[19] who coined the term "certation" to refer to this effect, and it has been reported repeatedly in dioecious species of *Rumex*[18,19,80] and *Silene.*[19,69,70] Certation results in an excess of females under abundant pollen deposition, an effect that has been attributed to the superior competitive ability of X-bearing pollen. Y chromosomes are believed to accumulate deleterious mutations in their nonhomologous regions.[14] However, in the haploid phase of the life cycle of flowering plants, e.g., the pollen tube, these deleterious effects might be exposed, an effect which would give X-bearing pollen a competitive edge. It has even been argued that male heterogamy itself might be selectively advantageous because it provides a mechanism for facultative adjustment of the sex ratio to favor female offspring in instances where pollen is abundant.[69,70] However, such a mechanism would only be selectively advantageous in populations subject to competition for mates among close relatives, giving rise to a situation generally known in studies of sex ratio evolution as local mate competition.[16]

Female Heterogamy

There are very few instances of female heterogamy in plants. The two primary taxa in which the females are heterogametic are *Fragaria*[20] and *Potentilla,*[31] closely related genera in the family Rosaceae. These two groups are also unusual in that the dioecious species in both genera are polyploid whereas their hermaphroditic congeners are diploid. The only other genera in which female heterogamy in plants has been documented are *Cotula*[52] and *Fuchsia.*[30]

CONCLUSION

Separation of the sexes, resulting in dioecy, has evolved many times in flowering plants. Therefore, there is no reason to expect any overall genetic homology in the determination of sex among the diversity of extant dioecious taxa. However, physiological attributes associated with sex expression, which are several levels removed from precise genetic phenomena, do show some overall trends related to maleness and femaleness, both in dioecious and monoecious species.

The overall development of a plant, particularly with respect to the distribution and abundance of plant growth regulators in different tissues, is likely to be strongly affected by local environmental conditions in combination with the physiological background of the plant itself. Thus, if the growth status of a plant is such that it would be best served by performance as one sex or the other, then any appropriate differential sex expression in response to the growth regulators (which in turn are influenced by the resource or environmental status of that plant) would be favored by natural selection. If the variation in sexual opportunity that a population is experiencing is strictly due to environmental variation, one might expect some form of sexual lability to evolve. If, on the other hand, there are deterministic properties inherent to the plants that lead to variation in success as a maternal or paternal parent, one might expect separate sex morphs to become associated with that variation.

There is growing evidence that even within hermaphroditic populations there is variation in maternal versus paternal success among plants. There are particular growth trends associated with femaleness in both hermaphroditic and dioecious populations. For example, female parents tend to be larger and/or have faster growth rates.[66] Thus these secondary characteristics may be a cause rather than a consequence of the evolution of the physiological basis of sex determination.

The genetic bases of sex determination are quite diverse, ranging from very specific nuclear genes and cytoplasmic effects to fully developed differentiation of sex chromosomes. The mechanistic link between sex determination on a genetic level and sex phenotype has yet to be elucidated. Even very specific gene effects, such as male sterility, are heterogeneous over different species and, thus far, have yet to be attributed to particular biochemical mechanisms.

The manner in which genic effects are put together to give rise to sex chromosomes is also only poorly understood in plants. Theoretical and phylogenetic studies into the evolution of dioecy have provided us with hypotheses concerning the combination of genetic loci that might be involved in dioecy in particular, and of the likelihood of occurrence of these various possibilities, but empirical genetic evidence is still lacking. Except for cytological aspects of homologous versus nonhomologous pairing and early work on the localization of very general genetic effects of sex determination in *Silene alba,* work on the genetic structure of sex chromosomes in plants has been virtually at a standstill for about 20 years.

There have, however, been considerable advances in the development and understanding of model genetic systems, as well as developmental processes underlying sex expression. We are now in a better position to know what to look for in terms of types of genetic effects and, building on some of the earlier work that has established our knowledge of the chromosomal basis of sex determination in particular taxa, we also know where to look. Thus we may be optimistic that it will not be another 20 years before significant progress in the study of sex determination in plants is made.

REFERENCES

1. Akoroda, M. O., Wilson, J. E., and Chheda, H. R., The association of sexuality with plant traits and tuber yield in white yam, *Euphytica* **33,** 435–442 (1984).
2. Anderson, G. J., Dioecious *Solanum* species of hermaphroditic origin is an example of broad convergence, *Nature (London)* **282,** 836–838 (1979).
3. Anderson, G. J., and Levine, D. A., Three taxa constitute the sexes of a single dioecious species of *Solanum, Taxon* **31,** 667–672 (1982).
4. Antcliff, A. J., Inheritance of sex in *Vitis, Amelior. Plantes* **30,** 113–122 (1980).
5. Baksh, S., Iqbal, M., and Jamal, A., Breeding system of *Solanum integrifolium* Poir. with an emphasis on sex potential and intercrossability, *Euphytica* **27,** 811–815 (1978).
6. Bawa, K. S., Evolution of dioecy in flowering plants, *Annu. Rev. Ecol. Syst.* **11,** 15–40 (1980).
7. Bawa, K. S., and Opler, P. A., Dioecism in tropical forest trees, *Evolution* **29,** 167–179 (1975).
8. Becket, J. B., Classification of male sterile cytoplasms in maize (*Zea mays* L.), *Crop Sci.* **11,** 724–727 (1971).
9. Bull, J. J., *Evolution of sex determining mechanisms,* Benjamin/Cummings Publishing Company, Inc., Menlo Park, California, 1983.
10. Bulmer, M. G., and Bull, J. J., Models of polygenic sex determination and sex ratio evolution, *Evolution* **36,** 13–26 (1982).
11. Carbonneau, A., Sterilities male et femelle dans le genre *Vitis.* II. Consequences en genetique et selection, *Agronomie (Paris)* **3,** 645–649 (1983).
12. Chailakhyan, M. Kh., Genetic and hormonal regulation of growth, flowering, and sex expression in plants, *Am. J. Bot.* **66,** 717–736 (1979).
13. Charlesworth, B., and Charlesworth, D., A model for the evolution of dioecy and gynodioecy, *Am. Nat.* **112,** 975–997 (1978).
14. Charlesworth, B., Model for evolution of Y chromosomes and dosage compensation, *Proc. Natl. Acad. Sci.* U.S.A. **75,** 5618–5622 (1978).
15. Charlesworth, D., and Ganders, F. R., The population genetics of gynodioecy with cytoplasmic-genic male-sterility, *Heredity* **43,** 213–218 (1979).
16. Charnov, E. L., *The Theory of Sex Allocation.* Princeton Univ. Press, Princeton, New Jersey, 1982.
17. Christopher, D. A., and Lory, J. B., Influence of foliarly applied growth regulators on sex expression in watermelon, *J. Am. Soc. Hortic. Sci.* **107,** 401–404 (1982).
18. Conn, J. S., and Blum, U., Sex ratios of *Rumex hastatulus:* The effect of environmental factors and certation, *Evolution* **35,** 1108–1116 (1981).

19. Correns, C., Bestimmung, Vererbung und Verteilung des Geschlechtes bei den hoheren Pflanzen, *Handb. Vererbungsw.* **2,** 1–138 (1928).
20. Darrow, G. M., *The Strawberry: History, Breeding and Physiology.* Holt, Rinehart and Winston, New York, 1966.
21. Dauphin-Guerin, B., Teller, G., and Durand, B., Different endogenous cytokinins between male and female *Mercurialis annua* L., *Planta* **148,** 124–129 (1980).
22. Davey, A. J., and Gibson, C. M., Note on the distribution of sexes in *Myrica gale, New Phytol.* **16,** 147–151 (1917).
23. Durand, R., and Durand, B., Sexual differentiation in higher plants, *Physiol. Plant.* **60,** 267–274 (1984).
24. Duvick, D. N., The use of cytoplasmic male sterility in hybrid seed production, *Econ. Bot.* **13,** 167–195 (1959).
25. Edwardson, J. R., Cytoplasmic male sterility, *Bot. Rev.* **86,** 341–420 (1970).
26. Franken, A. A., Sex characteristics and inheritance of sex in asparagus (*Asparagus officinalis* L.), *Euphytica* **19,** 277–287 (1970).
27. Freeman, D. C., Harper, K. T., and Charnov, E. L., Sex change in plants: Old and new observations and new hypotheses, *Oecologia* **47,** 222–232 (1980).
28. Freeman, D. C., and Vitale, J. J., The influence of environment on the sex ratio and fitness of spinach, *Bot. Gaz.* **146,** 137–142 (1985).
29. Friedlander, M., Atsmon, D., and Galun, E., Sexual differentiation in cucumber: Abscissic acid and gibberellic acid contents of various sex genotypes, *Plant Cell Physiol.* **18,** 681–691 (1977).
30. Godley, E. J., Breeding systems in New Zealand plants. 2. Genetics of the sex forms in *Fuchsia procumbens, N.Z. J. Bot.* **1,** 48–52 (1963).
31. Grewal, M. S., and Ellis, J. R., Sex determination in *Potentilla fruticosa, Heredity* **29,** 359–362 (1972).
32. Heslop-Harrison, J., The experimental modification of sex expression in flowering plants, *Biol. Rev. Cambridge Philos. Soc.* **32,** 38–90 (1956).
33. Heslop-Harrison, J., Sex expression in flowering plants, *Brookhaven Symp. Biol.* **16,** 109–122, 1963.
34. Hirsch, A. M., and Fortune, D., Peroxidase activity and isoperoxidase composition in cultured stem tissue, callus, and cell suspensions of *Actinidia chinensis, Z. Pflanzenphysiol.* **113,** 129–139 (1984).
35. Hughes, W. G., and Bodden, J. J., Single gene restoration of cytoplasmic male sterility in wheat and its implications in the breeding of restorer lines, *Theor. Appl. Genet.* **50,** 129–136 (1977).
36. Iizuka, M., and Janick, J., Sex chromosome variation in *Spinacia oleracea* L., *J. Hered.* **62,** 349–352 (1971).
37. Jain, S. K., Male sterility in flowering plants, *Bibliogr. Genet.* **18,** 101–166 (1959).
38. Jaiswal, V. S., and Kumar, A., Activity and isozymes of peroxidase in *Morus nigra* L. during sex differentiation, *Z. Pflanzenphysiol.* **102,** 299–302 (1981).
39. Jaiswal, V. S., and Kumar, A., Peroxide activity and its isozymes in relation to flower sex-expression in *Ricinus communis* L., *Curr. Sci.* **52,** 368–369 (1983).
40. Jaiswal, V. S., Narayan, P., and Lal, M., Activities of acid and alkaline phosphatases in relation to sex differentiation in *Carica papaya* L., *Biochem. Physiol. Pflanz.* **179,** 799–801 (1984).
41. Jeffrey, C., A review of the Cucurbitaceae, *Bot. J. Linn. Soc.* **81,** 233–247 (1980).
42. Kahlem, G., Isolation and localization by histoimmunology of isoperoxidases specific for male flowers of the dioecious species *Mercurialis annua* L., *Dev. Biol.* **50,** 58–67 (1976).
43. King, R. C., *Handbook of Genetics,* Vol. 2. Plenum Press, New York, 1974.
44. Khryanin, V. N., Kovaleva, L. V., and Chailakhyan, M. Kh., The specific characteristics of cytokinins and proteins in female and male plants of hemp, *Ref. Zh.* 1.65.29 (1981).
45. Kovaleva, L. V., Khryanin, V. N., and Chailakhyan, M. Kh., Specificity of proteins in the expression of sex in hemp plants, *Dokl. Akad. Nauk SSSR (Dokl. Bot. Sci. Engl. transl.)* **252,** 45–47 (1980).
46. Kubicki, B., The mechanism of sex determination in flowering plants, *Genet. Pol.* **13,** 53–66 (1972).
47. Laughnam, J. R., and Gobay, S. J., Mutations leading to nuclear restoration of fertility in S male sterile cytoplasm in maize, *Theor. Appl. Genet.* **43,** 109–116 (1973).
48. Leshem, Y., and Oplin, D., Differences in endogenous levels of giberrellin activity in male and female partners of two dioecious tree species, *Ann. Bot. (London)* **41,** 375–379 (1977).
49. Lewis, D., Male sterility in natural populations of hermaphrodite plants, *New Phytol.* **40,** 56–63 (1941).
50. Lewis, D., The evolution of sex in flowering plants, *Biol. Rev. Cambridge Philos. Soc.* **17,** 46–67 (1942).
51. Lloyd, D. G., Breeding systems in *Cotula* L. (Compositae, Anthemideae). I. The array of monoclinous and diclinous systems, *New Phytol.* **71,** 1181–1194 (1972).
52. Lloyd, D. G., Breeding systems in *Cotula* L. III. Dioecious populations, *New Phytol.* **74,** 109–123 (1975).

53. Lloyd, D. G., The maintenance of gynodioecy and androdioecy in angiosperms, *Genetica* **45,** 325–339 (1975).
54. Lloyd, D. G., Parental strategies of angiosperms, *N.Z. J. Bot.* **17,** 595–606 (1979).
55. Lloyd, D. G., Sexual strategies in plants. I. An hypothesis of serial adjustment of maternal investment during one reproductive session, *New Phytol.* **86,** 67–79 (1980).
56. Lloyd, D. G., Sexual strategies in plants. III. A quantative method for describing the gender of plants, *N.Z. J. Bot.* **18,** 103–108 (1980).
57. Lloyd, D. G., and Bawa, K. S., Modification of the gender of seed plants in varying conditions, *Evol. Biol.* **17,** 255–338 (1984).
58. Loptien, H., Identification of the sex chromosome pair in asparagus (*Asparagus officinalis* L.), *Z. Pflanzenzuecht.* **82,** 162–173 (1979).
59. Loptien, H., Untersuchungen zur Bestimmung der Geschlechtschromosomen beim Spinat (*Spinacia oleracea* L.), *Z. Pflanzenzuecht.* **82,** 90–92 (1979).
60. Louis, J. P., and Durand, B., Studies with the dioecious angiosperm *Mercurialis annua* L. (2n = 16): Correlation between genic and cytoplasmic male sterility, sex segregation and feminizing hormones (cytokinins), *Mol. Gen. Genet.* **165,** 309–322 (1978).
61. Lovett Doust, J., and Cavers, P. B., Biomass allocation in hermaphroditic flowers, *Can. J. Bot.* **60,** 2530–2534 (1982).
62. Lovett Doust, J., and Harper, J. L., The resource costs of gender and maternal support in an andromonoecious umbellifer, *Smyrnium olusatrum* L., *New Phytol.* **85,** 251–264 (1980).
63. Lovett Doust, J., O'Brien, G., and Lovett Doust, L., Effect of density on secondary sex characteristics and sex ratio in *Silene alba* (Caryophyllaceae), *Am. J. Bot.* **74,** 40–46 (1987).
64. Love, A., and Kapoor, B. M., A chromosome atlas of the collective genus *Rumex, Cytologia* **32,** 328–342 (1967).
65. Martin, F. W., Sex ratio and sex determination in *Dioscorea, J. Hered.* **57,** 95–99 (1966).
66. Meagher, T. R., Sexual dimorphism and ecological differentiation of male and female plants, *Ann. Missouri Bot. Gard.* **71,** 254–264 (1984).
67. Minoz, S., Lima, H., Perez, M., and Rodriguez, O. L., Use of the peroxidase enzyme system for the identification of sex in *Carica papaya, Cienc. Tec. Agric. Citricios Otros Frutales* **5,** 39–48 (1982).
68. Mishler, B. D., Reproductive ecology of bryophytes, in *Plant Reproductive Ecology: Patterns and Strategies,* (J. Lovett Doust and L. Lovett Doust, eds.) Chapter 10. Oxford Univ. Press, New York, 1988.
69. Mulcahy, D. L., Optimal sex ratio in *Silene alba, Heredity* **22,** 411–423 (1967).
70. Mulcahy, D. L., The selective advantage of staminate heterogamy, *Taxon* **16,** 280–283 (1967).
71. Neuffer, M. G., and Coe, G. H., Jr., Corn (maize), in *Handbook of Genetics* (R. C. King, ed.), Vol. 2, pp. 3–30. Plenum Press, New York, 1974.
72. Nigtevecht, G. van, Genetic studies in dioecious *Melandrium.* II. Sex determination in *Melandrium album* and *Melandrium dioicum, Genetica* **37,** 307–344 (1966).
73. Nigtevecht, G. van, and Prentice, H. C., A note on the sex chromosomes of the Valencian endemic, *Silene diclinis* (Caryophyllaceae), *An. Jard. Bot. Madr.* **41,** 267–270 (1985).
74. Ram, H. Y. M., and Sett, R., Modification of growth and sex expression in *Cannabis sativa* by aminoethoxyvinylglycine and ethephon, *Z. Pflanzenphysiol.* **105,** 165–172 (1982).
75. Ram, H. Y. M., and Sett, R., Reversal of ethephon-induced feminization in male plants of *Cannabis sativa* by ethylene antagonists, *Z. Pflanzenphysiol.* **107,** 85–89 (1982).
76. Ramanna, M. S., Are there heteromorphic sex chromosomes in spinach (*Spinacia oleracea* L.)? *Euphytica* **25,** 277–284 (1976).
77. Ramsay, H. P., and Berrie, G. K., Sex determination in bryophytes, *J. Hattori Bot. Lab.* **52,** 255–274 (1982).
78. Robinson, R. W., and Whitaker, T. W., *Cucumis,* in *Handbook of Genetics* (R. C. King, ed.), Vol. 2, pp. 145–150. Plenum Press, New York, 1974.
79. Ross, M. D., Evolution of dioecy from gynodioecy, *Evolution* **24,** 827–828 (1970).
80. Rychlewski, J., and Zarzycki, K., Sex ratio in seeds of *Rumex acetosa* L. as a result of sparse or abundant pollination, *Acta Biol. Cracov. Ser. Bot.* **18,** 101–114 (1975).
81. Schlessman, M. A., Gender diphasy ("sex choice"), in *Plant Reproductive Ecology: Patterns and Strategies,* (J. Lovett Doust and L. Lovett Doust, eds.), Chapter 8. Oxford Univ. Press, New York, 1988.
82. Schwabe, W. W., Physiology of vegetative reproductive and flowering, in *Plant Physiology: A Treatise,* (F. C. Steward, ed.), pp. 233–412. Academic Press, New York, 1971.
83. Shanna, C. P., Jyotishi, R. P., and Agrawal, G. P., Studies on sex expression, sex ratio, and fruit set as

affected by nitrogen levels and growth regulator sprays in *Lagemaria siceraria,* (Mol.) Standl. *JNKVV Res. J.* **14,** 47–52 (1980).
84. Smith, B. W., Cytogeography and cytotaxonomic relationships of *Rumex paucifolius, Am. J. Bot.* **55,** 673–683 (1968).
85. Smith, B. W., Evolution of sex determining mechanisms in *Rumex, Chromosomes Today* **2,** 172–182 (1967).
86. Smith, B. W., The evolving karyotype of *Rumex hastatulus, Evolution* **18,** 93–104 (1964).
87. Smith, B. W., The mechanism of sex determination in *Rumex hastatulus, Genetics* **48,** 1265–1288 (1963).
88. Suganuma, H., and Iwasaki, F., Sex identification of dioecious plants by the isozyme method. Date (*Phoenix dactylifera* L.), *Jpn. J. Trop. Agric.* **27,** 75–78 (1983).
89. Tomlinson, P. B., and Fawcett, P., Dioecism in *Citherexylum* (Verbenaceae), *J. Arnold Arbor, Harv. Univ.* **53,** 386–389 (1972).
90. Tsujimoto, H., and Tsunewaki, K., Gametocidal genes in wheat and its relatives. I. Genetic analyses in common wheat of a gametocidal gene derived from *Aegilops speltoides, Can. J. Genet. Cytol.* **26,** 78–84 (1984).
91. Westergaard, M., The mechanism of sex determination in dioecious flowering plants, *Adv. Genet.* **9,** 217–281 (1958).
92. Wiens, D., and Barlow, B. A., Permanent translocation heterozygosity and sex determination in East African mistletoes, *Science* **187,** 1208–1209 (1975).
93. Wyatt, R., and Anderson, L. E., Breeding systems in bryophytes, in *The Experimental Biology of Bryophytes* (A. F. Dyer and J. G. Duckett, eds.), pp. 39–64. Academic Press, Orlando, Florida, 1984.
94. Zuk, J., Function of Y chromosomes in *Rumex thyrsiflorus, Theor. Appl. Genet.* **40,** 124–129 (1970).
95. Zuk, J., Y-chromosome hyperploidy in *Rumex, Theor. Appl. Genet.* **40,** 147–154 (1970).

7

Gender Diphasy ("Sex Choice")

MARK A. SCHLESSMAN

This chapter is concerned with diphasy, a sexual system in which " . . . individuals belong to a single genetic class but choose their sexual mode in any season according to circumstances."[37] In well-documented cases, the female phase produces ovules and seeds while the male phase does not. The female phase may bear only female (pistillate) flowers, only hermaphroditic flowers, a mixture of male (staminate) and female flowers, or both unisexual and hermaphroditic flowers (Table 7.1). The essential feature of diphasy, then, is that a developmental "decision" regarding production of ovules is made before flowers mature. Diphasy involves a "choice" between two different modes of prefertilization investment in femaleness versus maleness. More gradual, continuous gender adjustments and postfertilization phenomena (i.e., abortion of developing seeds) are not considered here (see Lee,[31] this volume, and Ref. 37).

The terms "sex changing," "labile sexuality," "sex reversal," "sex choice," "phase choice," "alternative gender," "sequential gender," "sequential cosexuality," and "sequential hermaphroditism" have all been used in reference to diphasy.[4,10,19,34,37,39,53,65] Lloyd has argued that since the vast majority of plants are capable of reproducing as both males and females at the same time, it is best to approach most questions of sex allocation in plants using the concept of a plant's gender, i.e., its "maleness or femaleness as a parent of adults of the next generation," rather than its "sex."[34,35,37] The concept of gender is important here because the ovule-bearing phases of diphasic plants often produce pollen as well. I refer to the ovule-bearing phases of all diphasic species as females, even though they may produce pollen and obtain some of their fitness via male function. Policansky discussed the inappropriate application of "hermaphroditism," in its zoological connotation, to plants.[53] "Sequential" also has a well-established meaning in reference to certain animals that normally change sex only once. The term sequential should not be applied to diphasic plants because it obscures the fact that they may change gender several times.[6,39,52,58,59,61,63,65]

Lloyd and Bawa clarified a useful distinction between diphasy in sexually monomorphic populations and the possibility that members of either sex of a dimorphic (e.g., dioecious, gynodioceious) species might be capable of phase change.[37] They termed the latter phenomenon "dimorphism with phase choices."

In this chapter, the theory for diphasy, the evidence for gender choice, and the extent to which available data support various aspects of theoretical models are reviewed.

Table 7.1. Plants Cited as Examples of Gender Choice[a]

Plant	Gender phases[b]	Nature of evidence[c]	References
Arisaema (Araceae)			
A. triphyllum	♂, ♀ (♂ ♀)	A, B, C, D	1, 4–8, 18, 25, 39, 51, 52, 58–60, 65
A. japonica	♂, ♀ (♂ ♀)	A, B, C, D	30, 45
A. dracontium	♂, ♂ ♀	B, C, D	40, 58
Six other spp.	♂, ♀ (♂ ♀)	D	30
Guraniinae (Cucurbitaceae)	♂, ♀	A, B, D	13, 14
Gurania spp.			
Psiguria spp.			
Panax trifolium (Araliaceae)	♂, ⚥	A, B, D	50, 61, 63
Elaeis guinensis (Arecaceae)	♂, ♀ (♂ ♀)	E	28, 70
Catasetinae (Orchidaceae)		B, C, D	17, 26, 27
Catasetum spp.	♂, ♀ (♂ ♀)		
Cycnoches spp.	♂, ♀		
Mormodes spp.	♂, ♂ ♀		

Atriplex spp. (Chenopodiaceae)[d]	(♂ ♀) ♂, ♀ (♂ ♀)	B, C	21, 22, 46, 47
Spinacea oleracea (Chenopodiaceae)	(♂ ♀) ♂, ♀ (♂ ♀)	C	23
Acer (Aceraceae)			
A. pensylvanicum[e]	(♂ ♀) ♂, ♀	A	29
A. grandidentatum[f]	♂, ♂ ♀	A, B	2
Juniperus (Cupressaceae)		A	66
J. australis[d]	(♂ ♀) ♂, ♀ (♂ ♀)		
J. osteosperma[f]	(♂) ♂ ♀ (♀)		

[a]Compiled from Refs. 10, 23, and 37.

[b]♂, staminate flowers (or male cones) only, ♀, pistillate flowers (or female cones) only; ♂ ♀, both staminate and pistillate flowers; ⚥, hermaphroditic flowers only. Symbols in parentheses denote variation within a gender phase or a morph.

[c]A, gender dynamics of marked individuals in natural populations; B, interpopulational variation in gender ratios; C, experimental manipulation of gender; D, females larger than males in natural populations; E, gender dynamics of cultivated plants.

[d]♂ ♀ individuals may be inconstant members of genetically determined morphs.

[e]♂ ♀ individuals are probably inconstant males.

[f]Unisexual individuals may represent extreme gender adjustment rather than distinct gender phases.

THEORY FOR DIPHASY

The Size Advantage Hypothesis

The theory for diphasy is based on the size advantage hypothesis advanced by Ghiselin to explain sequential hermaphroditism in animals.[24] Ghiselin stated that if efficiency of reproduction via female versus male function varies with age or size, the fitness of an individual that "assumes" the sex most "advantageous" to its current status will be greater than that of one that remains the same sex throughout its life. The hypothesis was elaborated and quantified by Charnov, Leigh, Robertson, and Warner.[9,32,67,68] These quantitative models were presented in terms of fitness curves, i.e., relationships of the fitnesses of females and males to age or size. Warner noted that if the fitness curves for females and males do differ they should intersect, and that fitness is maximized when individuals change sex at the age (or size) corresponding to the intersection of the two curves.[67] Leigh et al. noted that "individuals should be born into the sex where the penalty of youth [smallness] is less . . . , and change later to the sex where age [larger size] is more advantageous," and they concluded that "if one sex gains fertility even slightly faster with age [size] than the other, then selection will favor sex change."[32]

Freeman et al. suggested that environmental heterogeneity might be the fundamental cause of diphasy.[19] They proposed that if gender expression is correlated with "resource state," it would be "advantageous" for gender "to fluctuate in synchrony with variable environments." This version of the size advantage hypothesis has been elaborated by Charnov, Freeman, McArthur, and Harper, and has come to be known as the "patchy environment model."[12,20–23,47]

In 1982, Charnov summarized the evidence for "labile sexuality" in plants and provided a provocative discussion of gender choice in the broader context of sex allocation. He concluded that diphasy is, at least in part, "a form of sequential hermaphroditism, where an individual changes sex as it grows larger (but see Ref. 10a)."

Constraints on the Evolution of Diphasy

Three essential requirements for gender modification of any sort are (1) that individuals cannot control the conditions that they experience, (2) that all parts of the individual experience similar conditions, and (3) that individuals can somehow "assess" their prospects for successful maternity and paternity. The first requirement is met by all plants,[11] but the conditions experienced by different flowers may vary, and the abilities of plants to assess conditions are limited. Failure to meet the latter two requirements may be the primary reasons why the magnitude of reproductive effort and gender are uncorrelated in most plants.[37]

If the above requirements are fulfilled and some aspect of a plant's condition (e.g., size) affects male and female fitness differently, some form of gender modification should result. Differing fitness curves alone, however, are insufficient for the evolution of phase choice. This is because most plants are simultaneous hermaphrodites that can adjust their relative allocations to male and female reproductive functions. As long as some fitness can be achieved via both sexual functions, both pollen and ovules will be produced.

Lloyd and Bawa listed four factors that generally favor continuous gender adjustment rather than a choice between two distinct gender phases.[37] Two of these are consequences of the open, modular system of growth that is characteristic of plants. First, if the rate of gain in either female or male fitness decreases at higher allocations (i.e., if either fitness curve decelerates), adjustments will confer higher fitness than phase choice. Second, conditions within a plant (e.g., on different branches) may vary, making it advantageous for an individual to adjust its gender allocation accordingly. The third factor, locally biased gender ratios, was also discussed by Bull,[7] Charnov,[10] and Bierzychudek.[4] As more of an individual's neighbors express the same gender phase, the fitness of a gender chooser is reduced, and adjustment becomes the selected strategy. The fourth factor is the nature of environmental variability. It is difficult to imagine the kinds of spatial and temporal variation that would account for the gender dynamics of individuals that change gender several times during their lives.

Lloyd and Bawa investigated optimal allocations to paternal and maternal functions under a variety of circumstances.[37] Their approach differed from earlier work in that it assumed continuous variation in gender, whereas previous models assumed that at any given time all reproductive individuals were functionally unisexual. Specifically, they considered situations in which (1) reproductive status affects both pollen and seed production, (2) offspring produced via male and female function have different viabilities, (3) reproductive status affects seed production only, and (4) varying rates of pollen removal affect male fitness. They found that diphasy is likely to evolve only in the last three situations, and then only when competition for mates and competition among offspring are minimal.

EVIDENCE FOR DIPHASY

The strongest evidence for diphasy comes from observations of the gender dynamics of individuals in natural populations. Quantitative assessments of gender are often necessary to separate gender adjustments from phase changes. Since the sexes of many dioecious species cannot be distinguished cytologically (see Refs. 33 and 69, and Meagher,[48] this volume), observation of gender dynamics may be necessary not only to detect diphasy but also to distinguish it from variable gender expression ("inconstancy" or "leaky dioecism") of genetically determined morphs. Such conclusive data are not always available for species that have been considered diphasic (Table 7.1).

Freeman et al. listed more than 50 "dioecious and subdioecious" species for which some individuals were reported to display different "sexual states" or to produce "hermaphroditic" offspring.[19] In most cases, the literature they cited provides only suggestive evidence of gender variation. Table 7.1 lists the plants that have been specifically cited as established cases of phase choice.[10,23,37]

The evidence for diphasy in *Arisaema triphyllum* (jack-in-the-pulpit; Araceae) is voluminous and incontrovertible.[1,4–6,8,18,25,39,43,51,52,58–60,65] Maekawa conducted an extensive study of *Arisaema japonica,* and 10 other congeners are reported to be diphasic.[45,65] Recently, Kinoshita investigated the relationships between gender and size in *Arisaema japonica* and five other species in Japan.[30] Schaffner induced gender change in *Arisaema dracontium* (green dragon),[59] but gender change has not been documented for this species in nature.

Studies of plants in natural populations have established diphasy in *Gurania* and

Psiguria (Cucurbitaceae)[13,14] and *Panax trifolium* (dwarf ginseng; Araliaceae).[61,63] Gender dynamics of the oil palm (*Elaeis guinensis;* Arecaceae) have been documented for plants in cultivation.[28,70]

Although it is widely accepted that orchids of the subtribe Catasetinae (*Cycnoches, Mormodes,* and most spp. of *Catasetum*) are diphasic, there are no documented observations of gender dynamics in natural populations. The evidence for phase choice is based entirely on reports of variation in gender ratios and work with transplants to greenhouses or other experimental conditions.[17,26,27] The only documented case of gender change involved plants that were moved from a natural population to an experimental plot.[17]

Barker et al. recorded the gender expression of 46 canyon maples *(Acer grandidentatum)* over two consecutive reproductive seasons.[2] In the first season, 29 trees produced both staminate and pistillate flowers and 17 produced only staminate flowers. In the following season, nine of the male trees bore some pistillate flowers as well as staminate ones, and three previously ambisexual individuals bore only staminate flowers. Barker et al. referred to these changes as "sex conversion" and Charnov[10] cited them as an example of "sex choice." Without determination of the proportions of staminate and pistillate flowers on ambisexual plants, it is impossible to say whether staminate individuals represent a distinct gender phase or simply one extreme of a more or less continual gradient of gender expression. Given that gender adjustments are known for several monoecious maples,[16] the latter possibility seems most likely.[62]

Hibbs and Fisher recorded gender expression for a population of striped maples *(Acer pensylvanicum)* for the years 1976 and 1977.[29] Plants were scored as male (staminate flowers only), female (pistillate flowers only), or ambisexual. Most transitions were between the ambisexual and the male and female states, and a few plants changed from male to female. Hibbs and Fisher took this as evidence for sex change, and Lloyd and Bawa[37] discussed striped maple as an example of diphasy.

Of the 243 striped maples marked in 1976, 199 were male, 31 were female, and 13 were ambisexual. In 1977, 2 males, 1 ambisexual, and 21 females had died, and 4 males did not flower. Of the remaining males, 7 became ambisexual, 8 became female, and 178 stayed male. Two ambisexuals became male, 6 became female, and 4 remained ambisexual. Of the females, 1 became ambisexual and 9 remained female (see Refs. 29 and 37 or 62 for tabulations of these data).

Considering the possibility that striped maple might be sexually dimorphic, I reanalyzed the data assuming that: (1) individuals that showed transitions between the ambisexual and the male or female states were inconstant males or females, and therefore (2) only transitions between the male and female states represented phase changes (see Ref. 37 and the discussion of *Atriplex* below). Under these assumptions, only 2% of the trees (the ambisexual that died and the four that stayed ambisexual) could not be classified as males or females, and only 3.3% (the eight that changed from male to female) exhibited phase choice.[62]

Although they did not count them, Hibbs and Fisher reported that ambisexual trees had many more staminate flowers than pistillate ones. If all ambisexuals are regarded as inconstant males, the estimated percentage of plants exhibiting gender choice rises to six (eight plants changing from male to female, six from ambisexual to female, and one from female to ambisexual). In either case, gender change is largely unidirectional (male to female) and infrequent. Exclusion of the trees that died or did not flower in 1977 would not significantly alter the estimated frequency of gender change.[62]

Given this low incidence of observed phase changes, the case for gender choice in striped maple must be considered inconclusive. However, the highly male-biased gender ratios for striped maple are inconsistent with dimorphism. It is more likely that striped maple is cosexual, i.e., that all individuals belong to a single genetic class. If the data of Hibbs and Fisher are representative, female striped maples suffer higher mortality than males. Hibbs and Fisher noted that females were more likely to exhibit indications of poor health than were males. From these observations they hypothesized that gender change occurs relatively late in the life of an individual. Lloyd and Bawa elaborated this hypothesis.[37] They noted that delaying expression of the female phase may assure maximum availability of resources for maturation of seeds, and that a large expenditure late in life would have minimal impact on future growth and reproduction.

Additional data on gender dynamics, preferably from two or more populations observed for at least 3 years, would firmly establish the significance of gender choice in striped maple. Estimates of the ages of males and females (e.g., by growth ring analysis) would also be informative.

Freeman and McArthur have compiled extensive data on the gender ratios and dynamics of several species of *Atriplex* (Chenopodiaceae).[21,22,46,47] The species under investigation have been traditionally classified as dioecious, and in some cases genetic dimorphism has been demonstrated.[47] The question is whether or not gender choice contributes to variation in the observed "sex ratio" or, in other terms, whether or not these species exhibit dimorphism with phase choices. In the original studies, individuals were classified as male (staminate flowers only), female (pistillate flowers only), or ambisexual ("monoecious," both staminate and pistillate flowers). Usually, no quantitative data on the gender of ambisexuals were obtained. "Sex change" was defined as a transition between any of these states, and proportions of individuals exhibiting gender choice were reported to be as high as 40%.[10]

In a reanalysis of data for *Atriplex,* Lloyd and Bawa assumed that ambisexuals were probably inconstant males and females rather than a distinct gender phase.[37] They classified individuals exhibiting transitions between the ambisexual and the male or female states as males or females. Only transitions between the male and female states were scored as gender change.

Specifically, Lloyd and Bawa reanalyzed data from a half-sib family of the perennial, *Atriplex canescens,* for which gender states had been recorded over 7 years (Table 3 in Refs. 22 and 46; Table VIII in Ref. 37). McArthur and Freeman originally concluded that, of the approximately 660 plants in their study, 51% of those that were initially female, 15% of those that were initially male, and all that were initially ambisexual were capable of gender change. In contrast, Lloyd and Bawa estimated that at most, only 5% of the plants changed gender. Thus McArthur and Freeman considered phase choice an important component of the reproductive strategy of *Atriplex cancescens* (see also Ref. 10), while Lloyd and Bawa considered it "at most a subsidiary element" (but see Ref. 3).

Freeman et al. elaborated their conclusion regarding *Atriplex canescens* using additional data on fruit set and a more quantitative view of gender.[22] When 80% or more of an individual's flowers were of a sex "not previously displayed" (e.g., female to ambisexual with 85% staminate and 15% female flowers) they scored the transition as "change in the primary sexual state." By extrapolation from comparisons of the weights of fruit crops and subjective ratings of the proportions of male and female flowers on ambisexual plants, they estimated that 21% of the plants "changed sexual state." In a later paper, Freeman and McArthur reported that on average, 71% of year-to-year "state changes" were due to "sex changes."[21] In that paper, the "states" included mortality and nonflowering as well as male, female, and ambisexual.

Given that individuals of many dioecious and subdioecious species exhibit inconstancies,[33,37,69] a priori classification of ambisexuals as a distinct gender class is unjustified. Because weights of fruit crops, rather than numbers of flowers or ovules, were used to estimate female allocation, and because so many extrapolations were involved, the data of Freeman et al.[22] are difficult to interpret. Further elucidation of the role of gender choice in *Atriplex* will require new, quantitative descriptions of the proportions of staminate and pistillate flowers on ambisexual plants.

Some evidence has been obtained for phase choice in cultivated spinach, (*Spinacea oleracea;* Chenopodiaceae). Since this species is a dioecious annual, detection of gender choice requires experimental manipulation of gender ratios. Onyekwelu and Harper observed greater proportions of males in populations grown at high densities than at lower ones, but the differences were all attributable to variation in rates of germination and flowering.[49] There was no evidence for phase choice. However, Freeman and Vitale reported that males were disproportionally abundant in dry versus wet experimental environments.[23] In at least one of their three experiments, the excess males could not be accounted for by variable germination, survival, or flowering of the sexual morphs.

The gender dynamics of *Juniperus australis* and *Juniperus osteosperma* (Cupressaceae) were studied by Vasek.[66] Again, plants were scored as male (staminate cones only), female (ovulate cones only), or ambisexual, and transitions among the states were reported as sex change.[10] For *Juniperus australis,* all transitions were between male and ambisexual with only a few female cones. These changes probably represent inconstancy of males rather than phase choice.[37] In a sample of 50 *Juniperus osteosperma,* only two trees exhibited both the male and female states, but these trees were also ambisexual in at least 1 year. All other transitions were between the ambisexual and the female or male states. Thus, there is no real evidence for phase choice as opposed to gender adjustment.[37]

SUPPORT FOR THE SIZE ADVANTAGE HYPOTHESIS

Difficulties of Testing Quantitative Models

There are several theoretical and practical difficulties associated with the collection and evaluation of data that would provide critical tests of quantitative models of the size advantage hypothesis. First, the models generally assume constant population sizes and stable distribution of plant size (age in the zoological models). Bierzychudek pointed out that since these conditions may not obtain in nature, an optimal timing for phase change may not exist.[5]

Second, there are two theoretical reasons why fitness curves for male and female gender phases should not intersect. If individuals make accurate gender choices, the largest male will be smaller than the smallest female. In the case of strictly staminate and pistillate phases, if the fitness curves have different shapes they should approach each other, but they should not intersect. If data yielding intersecting curves are obtained, we must conclude that the estimates of size, fitness, or both are inaccurate. If the male gender phase is strictly staminate and the female phase is ambisexual, the mean fitness of females will be higher than that of males, and the fitness curves may not even approach each other. In any case, estimates of size and fitness for individuals in natural populations provide only portions of the desired fitness curves.[5]

It is important to distinguish a plant's reproductive status (e.g., its size at the end of a growing season) from the gender allocations it makes on the basis of that status (e.g., ovule and pollen production in the next season), and, if possible, to distinguish these from actual reproductive success (RS). Since the fates of individual pollen grains cannot be determined, male RS is often assumed to be proportional to pollen production. Counts of fruits or seeds may reflect the results of inadequate pollination or abortion of fertilized ovules, thus obscuring the strategy of phase choice.

Costs of Maternity and Paternity

Sexual reproduction involves different kinds of costs, which may be distinguished as the immediate expenditure of resources and the consequences of that expenditure in terms of future growth and reproduction. The size advantage hypothesis predicts that small individuals should express the gender that incurs the lowest costs. In many plants, the total expenditure of metabolic energy and mineral nutrients on ovules, seeds, and fruit exceeds expenditure on pollen.[15,36,37,38,41,42,56,71,72] It is reasonable to expect that, in diphasic species, the reproductive effort of males will be less than that of females. Determinations of biomass (dry weight) allocations to sexual reproduction have been made for three diphasic species.

For *Arisaema triphyllum*, J. Lovett Doust and Cavers reported that, on average, males allocate 17% of their dry weight to inflorescences, and Bierzychudek reported 11%.[5,40] In contrast, females with mature fruit have allocated approximately 44% of their biomass to reproduction.[40] Allocations for males and fruiting females of *Arisaema dracontium* were 10 and 19%,[41] and for *Panax trifolium* the respective allocations were 2 and 16%.[63] Bierzychudek calculated that, for *Arisaema triphyllum,* plants with leaf areas less than 147 cm^2 would be incapable of maturing a single seed. This is very close to the upper size limit for males in the populations she studied.[5] I found that in *Panax trifolium,* the absolute allocation of biomass to a male inflorescence (2.11 mg) was slightly less than that to a single mature seed (2.47 mg). Although the difference is not statistically significant,[63] it does suggest that males lack sufficient resources to mature more than one seed.

In *Arisaema, Panax trifolium,* and orchids of the Catasetinae, storage organs (corms, roots, or pseudobulbs) of females were larger than those of males.[1,4–6,26,27,30,39,40,45,58,59,61,62,63] Decreasing the size of corms and partial defoliation of *Arisaema* females promoted switching to the male phase.[1,6,58,59] Expression of the female phase in *Elais quinensis* was correlated with environmental conditions that favor accumulation of photosynthetic reserves.[28,70] Here too, defoliation promoted switching to the male phase. The female phases of *Gurania* and *Psiquria* were preceded by an increase in stem diameter, but not by an increase in total biomass. Condon and Gilbert suggested that, in these climbing vines, large stem diameter is a correlate of reaching a position in the forest canopy that is favorable for photosynthesis.[14]

Gender ratios in natural populations of diphasic species were usually male-biased.[4,14,17,26,39,51,52,58,63] For *Arisaema triphyllum* and *Panax trifolium,* populations in which mean plant size is large also had relatively high proportions of females.[39,63] Taken together, data on biomass allocations, gender ratios, and the relationship of gender to size indicate that the resource expenditure required for successful maternity imposes a size threshold for expression of the female phase.

In natural populations of *Arisaema triphyllum* and *Panax trifolium,* females were generally more likely to become vegetative or male than to remain female in the next reproductive season.[6,61,63,64] Similar results were obtained for experimental populations of *Arisaema triphyllum.*[39,65] In both species, there were apparently no differences in mortality between females and males.[4,5,63] In *Acer pensylvanicum,* however, females experienced much higher mortality than males[29] (see previous section, Evidence for Diphasy).

Fitness Curves

Relationships among reproductive status (size), investment (RI), and success (RS) have been evaluated for natural populations of *Arisaema triphyllum* and *Panax trifolium.*[5,43,52,61,63] For both species, the average RS of females was greater than that of males. For *Arisaema triphyllum,* Policansky found no relationship between the size of a male and the number of flowers it produced (RI),[52] but Bierzychudek found a weak correlation in one of two populations.[5] Bierzychudek made direct estimates of male RS by marking male–female pairs and removing all other male inflorescences within 5 m of the females. She assumed that this procedure would restrict pollination of each female to its paired male. There was no correlation between the number of seeds set by females and the size of paired males. Bierzychudek felt that inadequate pollination was the primary reason for the lack of correlation between male size and RS (see also Ref. 65). Lloyd and Bawa speculated that, for mechanical reasons, increasing the size of male inflorescences beyond a certain point might not increase dispersal of pollen.[37]

In two populations of *Panax trifolium,* there was no relationship between root weight and number of flowers for males.[64] Male-phase RS was estimated using phenological observations to determine the number of receptive stigmas that could have been pollinated by each male. Reproductive success was not correlated with size and only weakly correlated with RI (flower number). The timing of pollen presentation relative to stigmatic receptivity and other stochastic influences on the transport of pollen appear to have greater influence on male RS than the amount of pollen produced.

In the two populations studied by Bierzychudek, there was no correlation between the size of a female *Arisaema triphyllum* and the number of seeds it matured. Seed set of hand-pollinated plants, however, was significantly correlated with size.[5] Policansky found a significant correlation for naturally pollinated females that did mature seed, but "about half" of the females in his study did not set fruit and were excluded from his analysis.[52] In a third study, L. Lovett-Doust et al. found that larger females produced more flowers, fruits, and seeds than smaller ones. The percentage of ovules and ovaries maturing to seeds and fruits and the mean seed weight also increased with size.[43] In summary, the data for *Arisaema triphyllum* indicate that females invest (produce ovules) according to their size, but that inadequate pollination may occasionally obscure this pattern.

In *Panax trifolium,* neither the number of ovules produced (RI) nor the number of seeds matured (RS) is strongly correlated with root weight at the end of the season.[64] However, numbers of ovules and seeds are strongly correlated (mean *r* for seven samples = 0.79), so the lack of a relationship between size and RS is primarily due to the lack of correlation between size and RI. It may be that size and RI were not correlated because root weights were determined at the end, rather than the beginning, of the growing season. A second factor may be the way that ovules are "packaged" in *Panax trifolium.* Females produce, on average, five to nine flowers, each containing 3 ovules. A large female with reserves sufficient to mature 25 seeds would have to produce nine flowers (27 ovules) in order to assure optimal utilization of resources. If such a plant produced only eight flowers (24 ovules), it would forfeit the opportunity to produce 1 seed, or 4% of its potential maternal RS. Smaller females would forfeit larger proportions of their potential seeds by underproduction of flowers and ovules.

Bierzychudek concluded that, since the slopes of observable regions of the fitness curves for *Arisaema triphyllum* were not significantly different from zero, they could not be used to predict the size at which plants should change gender.[5] I have reached a similar conclusion for *Panax trifolium.*[63] By excluding females with no seed set, Policansky was able to obtain a significant linear regression of female RS on size, and intersecting fitness curves.[52] From this he was able to show that the size distributions of male and seed-bearing female *Arisaema triphyllum* in the population he studied conformed fairly well to those expected if gender choice maximizes lifetime RS.

Gender Ratios and the Environment

The available evidence indicates that environmental conditions account for at least some variation in gender ratios of diphasic populations. Early reports suggested that moist soils favored expression of the female phase of *Arisaema triphyllum.*[51,58] Treiber found that females were much more prevalent on a floodplain than on the adjacent upland.[65] However, J. Lovett Doust and Cavers recorded the lowest proportion of females at their wettest site, and they found that females were more prevalent at sites with high levels of light intensity, soil nutrients, and pH.[39] Bierzychudek found that a single application of fertilizer neither increased size nor promoted gender change in a natural population.[6]

Dodson found higher proportions of female Catasetinae orchids in sunlit sites than in shaded ones, and Gregg demonstrated experimentally that sunlight promotes femaleness in the Catasetinae.[17,26] Femaleness in *Elaeis guinensis* is promoted by adequate moisture and sunlight.[28]

Gender ratios of natural populations have been determined for *Arisaema triphyllum, Arisaema dracontium, Catasetum macroglossum, Catasetum macrocarpum, Cycnoches lehmanii,* and *Panax trifolium.*[5,17,26,40,50,52,58,63,65] As noted above (previous section on Costs of Maternity and Paternity), the vast majority of these ratios are clearly male-biased. Male-biased ratios are to be expected because (1) young, small plants will express the male phase, and (2) having expressed the female phase, a plant is often more likely to become male or vegetative than to remain female.

Female-biased ratios have been reported for three populations of *Arisaema tri-*

phyllum and one population of *Catesetum macrocarpum.*[26,39,58] The population of *Catasetum* was growing in "full sun," a condition that has been shown to promote femaleness in the Catasetinae. Given that populations of large plants tend to have higher proportions of females than those composed of small plants (see Costs of Maternity and Paternity), female-biased gender ratios appear to result from usually favorable conditions that allow most individuals to surpass the threshold size for expression of the female phase.

CONCLUSIONS

The study of gender modification in plants, and plant reproductive ecology in general, is in danger of becoming a field in which theory proliferates in inverse proportion to available data.[44] Gender diphasy in nature has been established for only about half of the species reported to exhibit sex choice. It is instructive to note that there are no data on natural gender dynamics of the Catasetinae orchids, one of the most widely cited cases. Evidence for phase choice in other groups *(Acer, Atriplex, Juniperus)* is weak. However, studies of *Catasetum ochraceum* and *Acer pensylvanicum* are in progress.[54,57]

Much of the data on natural populations that has been published are equivocal because they were not collected with a quantitative concept of plant gender in mind. The notion that male, female, and ambisexual states always represent distinct phases of gender expression must be abandoned. Future studies should incorporate the concept that gender may vary within and among individuals, even when individuals are genetically predisposed toward femaleness or maleness.[34,35,37,62] Quantitative data on the gender expression of individuals should be collected over several reproductive seasons. A recent study of the red maple by Primack and McCall is a good example of the approach that is needed.[55]

Diphasic plants conform to the general predictions of the size advantage hypothesis as applied to plants. A physiological threshold for expression of the female phase imposed by the costs of maternity appears to be the primary factor determining gender dynamics. There are several theoretical and practical problems with application of quantitative models of the size advantage hypothesis, and attempts to test them have produced mixed results. Lack of correlations among reproductive status, investment, and success is largely due to factors such as pollinator efficiency, which affects success but cannot be regulated by adjusting investment. Future studies should carefully distinguish the relationship between status and investment, which represents the gender allocation strategy, from that between status and success, which may represent the combined effects of strategy and essentially stochastic environmental factors.

The potential conflict between the advantages of size and a female-biased gender ratio has little significance. Female-biased ratios do occur, but they are rare and there is no evidence for a mechanism that would allow a plant to respond to the gender choices of its neighbors.

While the patchy environment model may help to explain interpopulational variation in gender ratios and observed gender ratios in dimorphic species with phase choice, it is largely irrelevant to the gender dynamics of cosexual, perennial species that frequently change gender.[39] Plants in favorable sites may be prone to express the female phase earlier and more often than those in less favorable sites, but the spatial and temporal scales of patchiness that would be necessary to account for a specific individual's sequence of gender changes are unlikely to occur.

ACKNOWLEDGMENTS

I wish to thank M. Condon, E. Kinoshita, J. and L. Lovett Doust, C. McCall, R. Primack, G. Romero, and K. Turi for sharing unpublished information. I also wish to thank Vassar College for a generous sabbatical leave that provided the leisure to contribute this chapter. J. and L. Lovett Doust and D. G. Lloyd provided many helpful comments on an earlier draft. The Department of Plant and Microbial Sciences, University of Canterbury, provided secretarial help. My research on diphasy in dwarf ginseng has been supported by Vassar College and by a Cottrell College Science Grant from the Research Corporation.

REFERENCES

1. Atkinson, G. F., Experiments on the morphology of *Arisaema triphyllum, Bot. Gaz.* **25,** 114 (1898).
2. Barker, P., Freeman, D. C., and Harper, K. T., Variation in the breeding system of *Acer grandidentatum, For. Sci.* **28,** 563–572 (1982).
3. Bawa, K. S., The evolution of dioecy—concluding remarks, *Ann. Missouri Bot. Gard.* **71,** 294–296 (1984).
4. Bierzychudek, P., The demography of jack-in-the-pulpit, a forest perennial that changes sex, *Ecol. Monogr.* **52,** 335–351 (1982).
5. Bierzychudek, P., Assessing "optimal" life histories in a fluctuating environment: The evolution of sex-changing by jack-in-the-pulpit, *Am. Nat.* **123,** 829–840 (1984).
6. Bierzychudek, P., Determinants of gender in jack-in-the-pulpit: The influence of plant size and reproductive history, *Oecologia* **65,** 14–18 (1984).
7. Bull, J. J., Sex determination in reptiles, *Q. Rev. Biol.* **55,** 3–21 (1980).
8. Camp, W. H., Sex in *Arisaema triphyllum, Ohio J. Sci.* **32,** 147–151 (1932).
9. Charnov, E. L., Natural selection and sex change in pandalid shrimp: test of a life-history theory, *Am. Nat.* **113,** 715–734 (1979).
10. Charnov, E. L., *The Theory of Sex Allocation,* Chapter 13. Princeton Univ. Press, Princeton, 1982.

10a. Charnov, E. L., Size advantage may not always favor sex change, *J. Theor. Biol.* **119,** 283–285 (1986).

11. Charnov, E. L., and Bull, J. J., When is sex environmentally determined? *Nature (London)* **266,** 828–830 (1979).
12. Charnov, E. L., and Bull, J. J., Sex allocation in a patchy environment: A marginal value theorem, *J. Theor. Biol.* **115,** 619–624 (1985).
13. Condon, M. A., Reproductive biology, demography, and natural history of neotropical vines *Gurania* and *Psiguria* (Guraniinae, Cucurbitaceae): A study of the adaptive significance of size related sex change. Ph.D. dissertation, Univ. of Texas, Austin, 1984.
14. Condon, M. A., and Gilbert, L. E., Reproductive biology and natural history of neotropical vines *Gurania* and *Psiguria* (*Anguria*), in *Biology and Chemistry of Cucurbitaceae* (R. W. Robinson and C. Jeffry, eds.). Cornell Univ. Press, Ithaca, in press.
15. Cruden, R. W., and Lyon, D. L., Patterns of biomass allocation to male and female functions in plants with different mating systems, *Oecologia* **66,** 299–306 (1985).
16. De Jong, P. C., Flowering and sex expression in *Acer* L. A biosystematic study, *Meded. Landbouwhogesch. Wageningen* **76,** 1–201 (1976).
17. Dodson, C. H., Pollination and variation in the subtribe Catasetinae (Orchidaceae), *Ann. Missouri Bot. Gard.* **49,** 35–56 (1962).
18. Ewing, J. W., and Klein, R. M., Sex expression in jack-in-the-pulpit, *Bull. Torrey Bot. Club* **109,** 47–50 (1982).
19. Freeman, D. C., Harper, K. T., and Charnov, E. L., Sex change in plants: Old and new observations and new hypotheses, *Oecologia* **47,** 222–232 (1980).
20. Freeman, D. C., McArthur, E. D., Harper, K. T., and Blauer, A. C., Influence of environment on the floral sex ratio of monoecious plants, *Evolution* **35,** 194–197 (1981).
21. Freeman, D. C., and McArthur, E. D., The relative influences of mortality, nonflowering, and sex change on the sex ratios of six *Atriplex* species, *Bot. Gaz.* **145,** 385–394 (1984).

22. Freeman, D. C., McArthur, E. D., and Harper, K. T., The adaptive significance of sexual lability in plants using *Atriplex canescens* as a principal example, *Ann. Missouri Bot. Gard.* **71,** 265–277 (1984).
23. Freeman, D. C., and Vitale, J. J., The influence of environment on the sex ratio and fitness of spinach, *Bot. Gaz.* **146,** 137–142 (1985).
24. Ghiselin, M. T., The evolution of hermaphroditism among animals, *Q. Rev. Biol.* **44,** 189–208 (1969).
25. Gow, J. E., Observations on the morphology of the aroids, *Bot. Gaz.* **56,** 127–142 (1913).
26. Gregg, K. B., The effects of light intensity on sex expression in species of *Cycnoches* and *Catasetum* (Orchidaceae), *Selbyana* **1,** 101–112 (1975).
27. Gregg, K. B., The interaction of light intensity, plant size, and nutrition in sex expression in *Cycnoches* (Orchidaceae), *Selbyana* **2,** 212–223 (1978).
28. Hartley, C. W. S., Some environmental factors affecting flowering and fruiting in the oil palm, in *Physiology of Tree Crops* (L. C. Luckwill and C. V. Cutting, eds.), pp. 269–285. Academic Press, London, 1970.
29. Hibbs, D. E., and Fischer, B. C., Sexual and vegetative reproduction of striped maple *(Acer pensylvanicum), Bull. Torrey Bot. Club* **106,** 222–226 (1979).
30. Kinoshita, E., Size-sex relationships and sexual dimorphism in Japanese *Arisaema* (Araceae), *Ecol. Res.* **1,** 157–172 (1986).
31. Lee, T., Patterns of fruit and seed production, in *Plant Reproductive Ecology: Patterns and Strategies* (J. Lovett Doust and L. Lovett Doust, eds.), Chapter 9. Oxford Univ. Press, New York, 1988.
32. Leigh, E. G., Charnov, E. L., and Warner, R. R., Sex ratio, sex change, and natural selection, *Proc. Natl. Acad. Sci. U.S.A.* **73,** 3656–3660 (1976).
33. Lewis, D., The evolution of sex in flowering plants, *Biol. Rev.* **17,** 46–67 (1942).
34. Lloyd, D. G., Parental strategies of angiosperms, *N.Z. J. Bot.* **17,** 595–606 (1979).
35. Lloyd, D. G., Sexual strategies in plants III. A quantitative method for describing the gender of plants, *N.Z. J. Bot.* **18,** 103–108 (1980).
36. Lloyd, D. G., Evolutionarily stable sex ratios and sex allocations, *J. Theor. Biol.* **105,** 525–539 (1983).
37. Lloyd, D. G., and Bawa, K. S., Modification of the gender of seed plants in varying conditions, *Evol. Biol.* **17,** 255–338 (1984).
38. Lloyd, D. G., and Webb, C. J., Secondary sex characteristics in seed plants, *Bot. Rev.* **43,** 177–216 (1977).
39. Lovett Doust, J., and Cavers, P. B., Sex and gender dynamics in jack-in-the-pulpit, *Arisaema triphyllum* (Araceae), *Ecology* **63,** 797–807 (1982).
40. Lovett Doust, J., and Cavers, P. B., Resource allocation and gender in the green dragon, *Arisaema dracontium* (Araceae), *Am. Midl. Nat.* **108,** 144–148 (1982).
41. Lovett Doust, J., and Harper, J. L., The resource costs of gender and maternal support in an andromonoecious umbellifer, *Smyrnium olusatrum* L., *New Phytol.* **85,** 251–264 (1980).
42. Lovett Doust, J., and Lovett Doust, L., Parental strategy: Gender and maternity in higher plants, *BioScience* **33,** 180–186 (1983).
43. Lovett Doust, L., Lovett Doust, J., and Turi, K., Fecundity and size relationships in jack-in-the-pulpit, *Arisaema triphyllum* (Araceae), *Am. J. Bot.* **73,** 489–494 (1986).
44. Macior, L. W., Pollination biology (book review), *Bull. Torrey Bot. Club* **112,** 200 (1985).
45. Maekawa, T., On the phenomena of sex transition in *Arisaema japonica, J. Coll. Agric. Hokkaido Imp. Univ.* **13,** 217–305 (1924).
46. McArthur, E. D., Environmentally induced changes of sex expression in *Atriplex canescens, Heredity* **38,** 91–103 (1977).
47. McArthur, E. D., and Freeman, D. C., Sex expression in *Atriplex canescens:* Genetics and environment, *Bot. Gaz.* **143,** 476–482 (1982).
48. Meagher, T., Sex determination in plants, in *Plant Reproductive Ecology: Patterns and Strategies* (J. Lovett Doust and L. Lovett Doust, Chapter 6. Oxford Univ. Press, New York, 1988.
49. Onyekwelu, S. S., and Harper, J. L., Sex ratio and niche differentiation in spinach (*Spinacea oleracea* L.), *Nature (London)* **282,** 609–611 (1979).
50. Philbrick, C. T., Contributions to the reproductive biology of *Panax trifolium* L. (Araliaceae), *Rhodora* **85,** 97–113 (1983).
51. Pickett, F. L., A contribution to our knowledge of *Arisaema triphyllum, Mem. Torrey Bot. Club* **16,** 1–55 (1915).
52. Policansky, D., Sex choice and the size advantage model in jack-in-the-pulpit *(Arisaema triphyllum), Proc. Natl. Acad. Sci. U.S.A.* **78,** 1306–1308 (1981).
53. Policansky, D., Sex change in plants and animals, *Annu. Rev. Ecol. Syst.* **13,** 471–496 (1982).

54. Primack, R. B., personal communication, 1986.
55. Primack, R. B., and McCall, C., Gender variation in the red maple: a seven-year study of a "polygamodioecious" species, *Am. J. Bot.* **73,** 1239–1248 (1986).
56. Putwain, P. D., and Harper, J. L., Studies in the dynamics of plant populations. V. Mechanisms governing sex ratio in *Rumex acetosa* and *R. acetosella, J. Ecol.* **60,** 113–129 (1972).
57. Romero, G. A., personal communication, 1986.
58. Schaffner, J. H., Control of the sexual state in *Arisaema triphyllum* and *Arisaema dracontium, Am. J. Bot.* **9,** 72–78 (1922).
59. Schaffner, J. H., Experiments with various plants to produce change of sex in the individual, *Bull. Torrey Bot. Club* **52,** 35–47 (1925).
60. Schaffner, J. H., Siamese twins of *Arisaema triphyllum* of opposite sex experimentally induced, *Ohio J. Sci.* **26,** 276–280 (1926).
61. Schlessman, M. A., Gender modification in North American ginseng: Dichotomous choice versus adjustment, *BioScience* **37,** 469–475 (1987).
62. Schlessman, M. A., Interpretation of evidence for gender choice in plants, *Am. Nat.* **128,** 416–420 (1986).
63. Schlessman, M. A., Gender diphasy in dwarf ginseng, *Panax trifolium* (Araliaceae), manuscript in preparation.
64. Schlessman, M. A., unpublished data, 1987.
65. Treiber, M., Biosystematics of the *Arisaema triphyllum* complex. Ph.D. dissertation, Univ. of North Carolina, Chapel Hill, 1980.
66. Vasek, F. C., The distribution and taxonomy of three western junipers, *Brittonia* **18,** 350–372 (1966).
67. Warner, R. R., The adaptive significance of sequential hermaphroditism in animals, *Am. Nat.* **109,** 61–82 (1975).
68. Warner, R. R., Robertson, D. R., and Leigh, E. G., Sex change and sexual selection, *Science* **190,** 633–638 (1975).
69. Westergard, M., The mechanisms of sex determination in dioecious flowering plants, *Adv. Genet.* **9,** 217–281 (1958).
70. Williams, C. N., and Thomas, R. L., Observations on sex differentiation in the oil palm, *Elaeis guinensis* L., *Ann. Bot. (London)* **34,** 957–963 (1970).
71. Willson, M. F., *Plant Reproductive Ecology.* John Wiley & Sons, New York, 1983.
72. Willson, M. F., and Ruppell, K. P., Resource allocation and floral sex ratios in *Zizania aquatica, Can. J. Bot.* **62,** 799–805 (1984).

II

Ecological Forces

8

Nectar Production, Flowering Phenology, and Strategies for Pollination

MICHAEL ZIMMERMAN

In this chapter I focus on one general question exemplifying the interactions between plants and their biotic pollinators: Can plants manipulate their pollinators to their own advantage? In other words, do plants possess phenotypic traits that are heritable, variable, and capable of influencing foraging decisions? Furthermore, can the new foraging behavior affect gene flow in the plant population so that selective pressure is put on plants to optimize certain traits? Both plants and pollinators are under pressure to maximize fitness but this comes about via different proximate mechanisms. Individual plants, for example, need to produce the largest number of the highest quality seeds possible. Pollinators, on the other hand, enhance their fitness by maximizing their net rate of energy intake while foraging.[146,148,164] Conceivably, conflicts might arise between the pressures on plants and pollinators. For this reason it is important to understand how pollinators react to particular plant traits.

Even addressing such a specific question in its entirety in a single chapter is impossible. This chapter, therefore, deals with only two of the plant traits that might prove able to influence pollinator behavior: rate of nectar production and flowering phenology. Because the work addressing this general question is not very complete, in addition to reviewing the literature on each of these characteristics, I outline the type of data that must be acquired for this sort of question to be answered. Rate of nectar production is used as a model because of the volume of data already gathered relating this trait to pollinator behavior. Although a number of workers[134,191,196] have suggested that it is unlikely that plants are able to influence where they send their pollen grains, I believe that such an assumption is premature. Rather than actually answering the question, can plants manipulate pollinators to their own advantage, I hope in the present chapter, to demonstrate how best to pose it.

Two types of analyses must be performed if the effect of alterations in any plant trait on plant fitness is to be determined. First, it is necessary to decide whether seed set is limited by pollen or other resources. The female component of reproduction, for example, responds differently to increases in pollen depending on what limits seed set. If resources limit seed production, additional pollen might enhance seed quality but not number while, if pollen is in short supply and resources are not, increased pollen deposition might increase seed number but not necessarily quality. Furthermore, what limits seed production is of evolutionary importance because, if reproduction is limited by resources other than pollen, and if excess pollen is available, then pollen grains may compete for ovules[116] or plants may select the pollen donors of their offspring by making judicious use of seed and/or fruit abortion (see Bertin,[14] Haig and Westoby,[64] and Lee, [94] all this volume, and Ref. 86). The importance of any particular pollen flow pattern is thus dependent on the particular factors that limit seed set in the taxon in question.

Second, the variability of the trait in question must be assessed and related to pollinator behavior. The influence of various behaviors on pollen flow and ultimately on plant fitness must then be determined. Fig. 8.1 presents a flow diagram exemplifying this chain of events for rate of floral nectar production, a trait so central to plant–pollinator interactions that it is ideal for this sort of analysis. The left side of the figure represents the benefits of a particular rate of nectar production while the right depicts the costs. Any benefits arising from floral nectar production are due to the actions of nectar-gathering pollinators and thus the standing crop of nectar (i.e., the amount of reward that pollinators actually encounter) is of greater proximate significance than is the rate of production. Pollinator behavior in turn influences the pattern of standing crop in a population.[130,131,137,210,219] The amount of standing crop of nectar in blossoms has the potential to affect three kinds of decisions made by pollinators: (1) whether to visit a particular plant; (2) how long to remain there; and (3) which plant to visit next. All of these decisions may then produce marked alterations in the patterns of pollen movement in the population. The changes may be brought about either by new patterns of movement between plants by pollinators or by modifications in the dynamics of pollen turnover on pollinators' bodies. A complete assessment of fitness will consider both the number and quality of seeds produced. Similarly, the costs of nectar production are ultimately measured in terms of the decrease in the number and/or quality of seeds produced by a plant in its lifetime because of the energy spent on nectar. A complete determination of the ways in which plants may manipulate their pollinators via the amount of reward that they offer will be possible only when the full costs and benefits of various nectar production rates are calculated for a single species. Although parts of Fig. 8.1 have been examined for many species, the entire analysis has not yet been completed for a single taxon. Similarly, a comparable analysis has not yet been performed for any other plant trait.

DIFFICULTIES DETERMINING THE FACTORS LIMITING SEED SET

On first thought, it seems that it should be easy to determine whether pollen or other resources limit seed set. As Bierzychudek[18] says, "if hand-pollinated plants produce more seeds than naturally pollinated controls, then reproduction is being limited, not by energy levels, but by pollinator activity." In practice, the situation is not always this simple. If additional pollen yields no increase in seed production relative to control blossoms, then seed number must be resource rather than pollen limited. As Zimmerman and Pyke[221] point out, however, the reverse is not necessarily the case. If flowers receiving additional pollen set more seed than control blossoms, the situation is ambiguous because plants may compensate for increased seed set in some flowers by allocating fewer resources to untreated flowers that bloom simultaneously. Also,

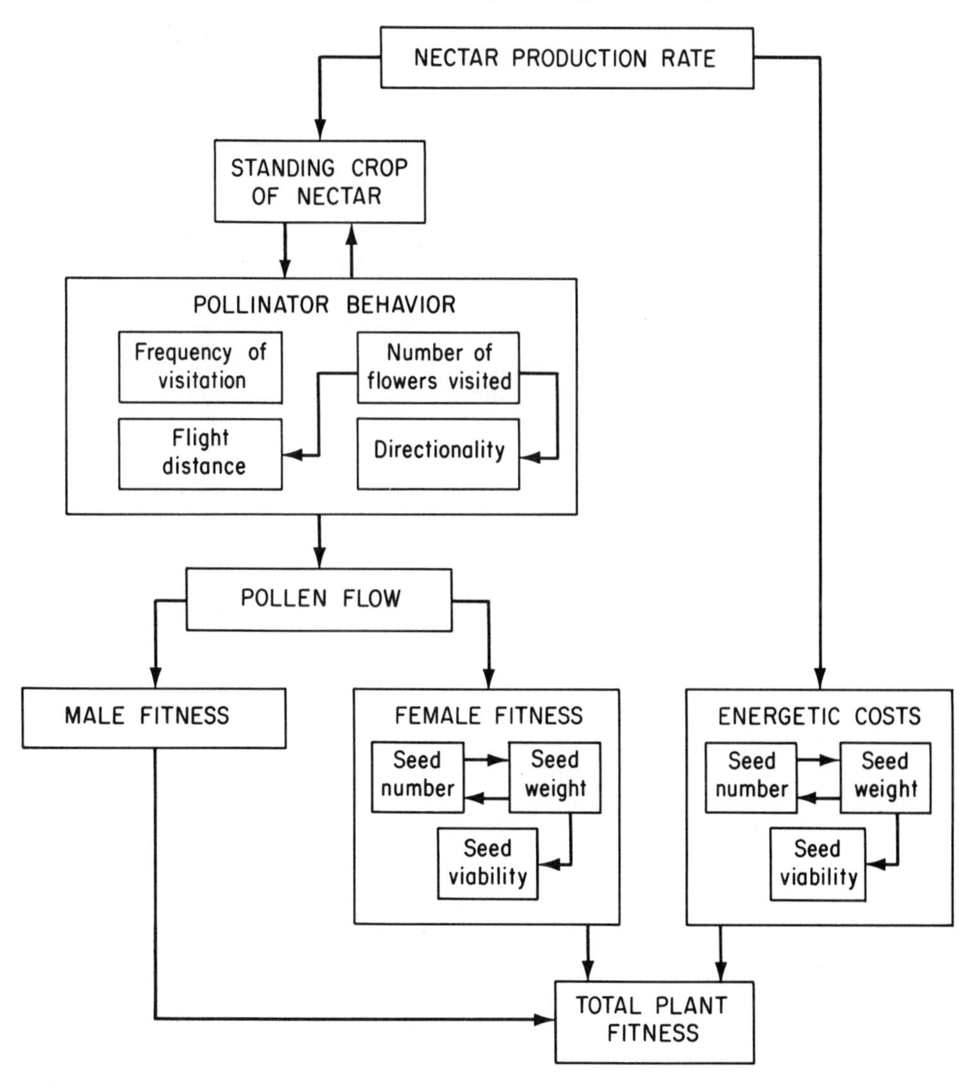

Fig. 8.1. A flow diagram detailing the ways in which floral nectar production interacts with pollinator foraging behavior to affect total plant fitness.

total seed output by the plant may, at a later time, be lowered either through reductions in seed set per flower or number of flowers produced.[88,174,216]

Polemonium foliosissimum provides an example of the confusion that can arise from hand-pollination data. When hand-pollinated flowers produced significantly more seeds than did control blossoms on control plants, Zimmerman[206,207] concluded that seed set was limited by pollen availability. Zimmerman and Pyke,[221] however, reached a different conclusion when they repeated the work incorporating two groups of control blossoms. One group consisted of flowers on the same plants as the experimental blossoms while the other was composed of flowers on control plants. As in earlier work,[206,207] Zimmerman and Pyke[221] found that hand-pollinated flowers set more seeds than did control blossoms. Control flowers on experimental plants, however, produced significantly fewer seeds per fruit than controls on control individuals. Over the season, the *Polemonium foliosissimum* plants whose flowers received extra pollen did not set significantly more seeds than untreated individuals. These results indicate that *Polemonium foliosissimum* individuals are able to redistribute resources among flowers within a plant, making a correct interpretation of hand-pollination data impossible without both types of controls. Most studies, however, have only one set of controls. Seed set values from hand-pollinated blossoms have been compared either to naturally pollinated flowers from the same plant[3,62,126,136,144,166] or from different plants.[6,7,12,61,83,113,142,187,200] Although such comparisons may reveal much about patterns of resource allocation among flowers within a plant (as in the *Polemonium foliosissimum* studies), they cannot demonstrate whether total seed set per plant is pollen limited.

Although the literature is replete with papers claiming that either pollen or other resources limit seed set, only Bierzychudek,[18,19] Motten,[115] and Stephenson[176] provided additional pollen to all flowers on experimental plants, while only Zimmerman and Pyke[221] pollinated a large fraction of flowers and established both sets of controls mentioned above. Of these studies, only Bierzychudek[18,19] found experimental individuals setting more seed than control plants. She also showed that enhanced seed set by experimental *Arisaema triphyllum* plants did not occur at the expense of reproduction the following year. Consequently, pollen limitation of seed set per plant has been adequately demonstrated just once. I do not mean to imply, however, that these data suggest that pollen is only rarely limiting under natural conditions. There are simply too few adequate studies for meaningful generalizations to be made. It is essential that properly designed experimentals be performed for each taxon under study.

When pollen is not limiting, increased pollinator visitation rates will not lead to enhanced seed set. Increased visitation rates may still be advantageous to the plant for two reasons. First, fitness is determined by both the quantity and quality of seeds produced by an individual. If the plant, once it had enough pollen for complete seed set, became selective with regard to which pollen it used to fertilize ovules,[13,86,116] then seed quality might improve as a function of visitation rate because of the increased variability of pollen brought to the plant. Second, total plant fitness is a result of both female function (seed production) and male function (pollen donation). Increased visitation rates increase the probability that a plant's pollen will be picked up and distributed by a pollinator. This area clearly needs further study for, as Janzen[87] says: "Pollination biology traditionally focuses on the incoming pollen . . . We know next to nothing of the biology of pollen donation or dispersal,[100] or how the male half of plant reproductive biology views the world."

Understanding and predicting how variability in a plant trait influences the reproductive success of the individual possessing that trait is dependent upon a clear picture of the factors limiting reproduction. For this reason, the first step toward determining if plants can manipulate pollinator behavior is to discover whether pollen is indeed limiting. Additionally, evaluation must be made of the extent to which total plant fitness is affected by enhanced pollen levels. The next step is to focus specifically on a plant trait that has a high probability of influencing pollinator behavior. Rate of nectar production is obviously one such trait.

RATE OF NECTAR PRODUCTION

An enormous amount of primarily descriptive work has been published on patterns of floral nectar production and standing crops. Studies have, for example, reported nectar production as a function of time of season,[11,48,93,128,145,222] time of day,[11,43,45,49,103,107,118,128,222] flower age,[30,48,63,93,128,138,158,168,222] flower location on a plant,[63,125,128,138,139,222] flower size,[23,67] plant size,[103,129,222] plant location,[23,50,101,153,222] and/or weather conditions.[15,38,40,121,128,132,172,193] Reports also indicate that nectar production may be affected by nectar removal from flowers,[17,20,46,48,52,107,123,128,132,150,170,173,222] by defoliation,[118] and by soil moisture.[128,215] Standing crops have been related to time of day,[11,130,210,222] time of season,[162,222] flower position,[16,41,53,79,143,190,222] nectar robbing,[8,107] and ambient temperature.[25,39,172] Some studies have also examined the distribution of nectar standing crops both within[209,222] and among plants.[130,196,209,212]

Although many of the above studies have discussed the patterns found in light of some aspect of the reproductive biology of the species involved, few have explicitly examined the adaptive significance of rate of nectar production.[128,129,140,196] Although the possible evolutionary advantages of some large-scale patterns such as the relationship between the energetic needs of pollinators and the average rate of production of visited plant species,[46,70,71,73,74,78,124] or the association between pollinator type and sugar concentration of nectar[5,21,26,77,149] have been discussed, only one article[140] has addressed the question of why an individual plant secretes the amount of nectar that it does. In attempting to answer this question, care must be taken to ensure that the selection pressures stemming from the interactions between plants and pollinators are properly defined. As Pyke and Waser[149] point out, many of the arguments advanced to explain the sugar concentration of floral nectar[5,21,26] can be interpreted to imply that plants are capable of evolving traits because those traits are advantageous to their pollinators. A more meaningful understanding of the selection pressures determining nectar production will come about when the evolutionary advantages of any particular trait are considered solely from the perspective of the individual plant.

The most productive approach is the one outlined by Pyke.[140] He claimed that, if the costs and benefits to an individual associated with a particular rate of nectar production could be determined, then the optimal rate of secretion could be calculated (Fig. 8.2). The optimal value would be the rate that constitutes an evolutionarily stable strategy.[105,106] In other words, individuals producing nectar at any rate other than this optimal value would experience reduced reproductive fitness and a population of plants producing nectar at this optimal rate could not be invaded by individuals with a different rate of secretion. It must be noted that even if such a theoretically optimal rate is found to exist, not all, or not even a majority, of the individuals in the population should necessarily be expected to exhibit the optimal rate.[47] If the optimal behavior is a positive function of the developmental state of the organisms, for example, we might expect only individuals above a certain age or size threshold to exhibit optimal phenotypes. Furthermore, the optimal strategy may actually be different for plants of different sizes. This problem notwithstanding, the optimal rate of nectar production should be the rate that yields the greatest difference between the cost and benefit curves of Fig. 8.2. The situation is complicated, however, by the fact that the shapes of the curves are a function of all of the variables in Fig. 8.1.

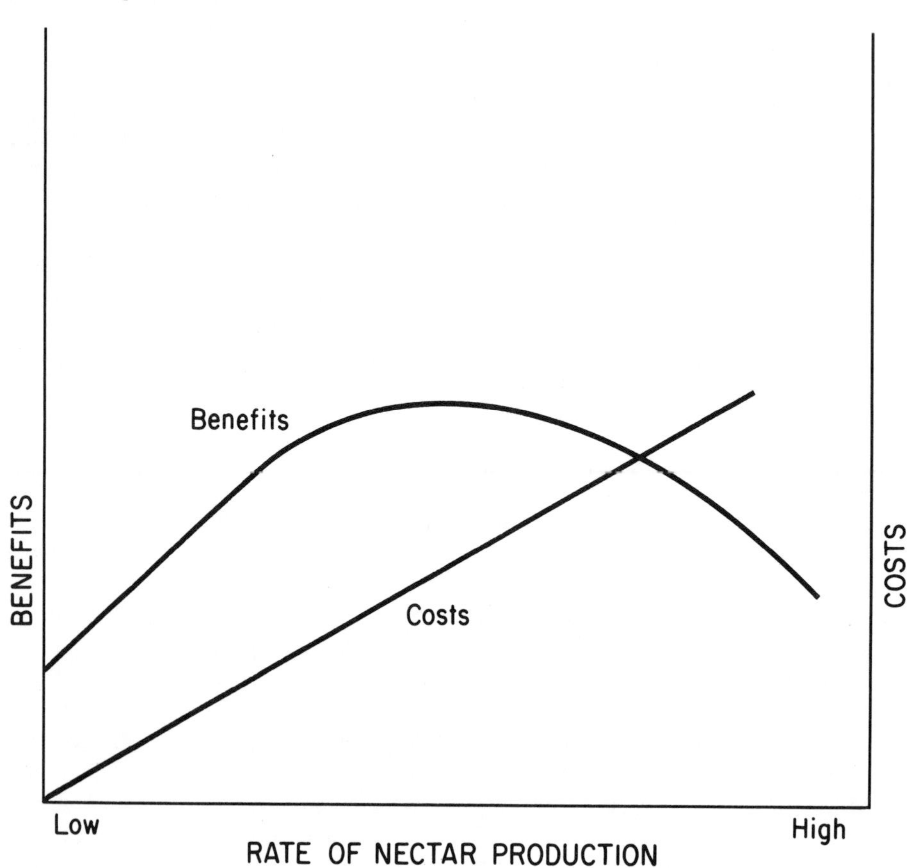

Fig. 8.2. Benefits and costs of reproduction associated with rates of nectar production. Note that no units have been placed on the benefit and cost axes. See text for a full discussion. (Modified from Pyke.[140])

Patterns and Heritability of Nectar Production

Rates of nectar production among plants in populations and among flowers within plants have been found to differ in variability. While Pleasants,[128] for example, reported high levels of interplant variability in *Ipomopsis agreggata,* Marden[103] found no evidence for phenotypic variation among *Impatiens capensis* plants. A number of studies have documented appreciable heterogeneity in rates of secretion among flowers within plants.[76,173,222] Therefore, the first step in any study attempting to examine the interplay between nectar production and pollinator behavior must be a full characterization, both temporal and spatial, of the patterns of nectar production for the population. Obviously variability (or lack of it) is of evolutionary interest only if nectar production has relatively high heritability, i.e., if it is susceptible to the effects of selection. This, in fact, has been found to be the case.[68,121,178–181] Hawkins[68] has, for example, shown that nectar production rates in clover increase significantly in just a few generations under artificial directional selection. Heritability has been assumed for natural populations[103,128,215] as well. While Marden[103] claimed that nectar production in *Impatiens capensis* is under strong stabilizing selection, Zimmerman[215] postulated that *Delphinium nelsonii* might be under directional pressure to increase its rate of production.

Standing Crop of Nectar

It is surprising that not a single study has dealt with the link between nectar production and standing crop of nectar (i.e., the second step in Fig. 8.1). A significant relationship between these parameters must exist if pollinators are to exert any selection pressure on rate of nectar production. Although it is difficult to conceive of circumstances in which a significant association does not exist, it is possible that the correlation might be so weak that pollinators could not exert any selection pressure. Such a situation might occur if high visitation rates per flower by pollinators were coupled with low nectar production rates. Under such conditions, standing crops in most blossoms would be small and pollinators might not be able to effectively differentiate among flowers offering slight rewards. In fact, standing crops of many flowers have been found to be quite small.[22,25,130,131,169,222]

The distribution of standing crops of nectar is a function of pollinator foraging behavior as well as of interplant variability in rate of production. Pleasants and Zimmerman[131] have demonstrated that the expected distribution of standing crops, assuming random foraging by pollinators and constant nectar production by plants, is one in which the mean nectar volume approximately equals the standard deviation of the distribution. Variability in nectar production tends to increase the variance relative to the mean while nonrandom foraging by pollinators (i.e., area-restricted foraging)[58,75,130,137,189,209,210,212,219] reduces the variance slightly. It is imperative, therefore, to characterize the foraging behavior of pollinators in order to make preliminary statements about pollen flow in the population as well as to assess the effect of foraging behavior on the standing crop distribution.

Pleasants[128] postulated that the amount of variability in nectar rewards offered by plants might influence pollinator behavior and, in turn, plant fitness. The effect, he claims, could be mediated through risk-aversive foraging.[29,48,152,192] Animals evidencing this behavior, when given a choice between a constant and a variable food reward of equal caloric value, prefer the low over the high variance type. Pleasants[128] therefore suggested that plants having flowers with a wide range of nectar production rates or having blossoms with highly variable standing crops should be less attractive to pollinators, and that fewer flowers would be visited consecutively on those individuals than on less variable plants. He thus predicted that variability in nectar reward might be an effective way for plants to encourage pollinators to leave and visit blossoms on other individuals after probing the optimal number of flowers. Zimmerman and Pyke[222] point out, however, that this explanation for variability in nectar levels is probably incorrect. The work on risk-aversive foraging is not applicable to a situation in which pollinators visit a single plant species. The confusion is due to the fact that different types of foraging decisions are being considered in the two instances. Pleasants is asking when pollinators decide to leave one plant for another, or, in other words, how departure decisions from food patches are influenced by reward variability. The risk-aversive foraging literature, on the other hand, is concerned with the way reward variability affects diet or patch choice. Throughout each risk-aversive foraging experiment the high variance reward was either in the same location (for sparrows)[29] or in experimental flowers of the same color (for bumblebees).[192] After sampling, animals began to show strong preference for the less variable reward. Preference could be demonstrated because foragers learned to associate high variability with a particular type or location of resource.

The situation is quite different, however, when pollinators are collecting nectar from flowers of a single plant species and where they can determine the amount of variability present in a plant only by actually sampling nectar in the blossoms of that individual. The literature on risk-aversive foraging, therefore, provides no insight into pollinator behavior on individual plants with variable nectar rewards. Such insight awaits future experimentation or computer simulation designed to examine the question directly. Although it is possible that, as Brink[22] claimed, species with particularly high variability in reward will be less preferred by pollinators employing risk-aversive foraging, such a response is a species-level phenomenon and should provide no selection pressure on nectar production at the individual level.

One final point needs to be made concerning the relationship between the standing crop of a population and that of individuals in the population. Ollason[119] presented a mathematical model of learning that demonstrated that, when pollinator travel time remains constant, departure decisions are independent of overall habitat quality. This means that pollinators will visit the same number of flowers per inflorescence even if the population significantly increases its standing crop of nectar. Ollason's model thus provides an actual mechanism for the behavior predicted by Charnov's[33] widely used marginal value theorem.[16,35,42,75,92,139,143,211] In experiments designed to test the learning model, Zimmerman and Cibula[218] found bumblebee behavior to be consistent with Ollason's prediction. The significance of this finding in the present context is twofold. First, if all individuals in a population increase their rate of nectar production, then the behavioral patterns of pollinators within inflorescences should not be altered. Selection pressure stemming from the departure decisions made by pollinators in response to alterations in nectar production rate will thus only come about as a result of changes in individual plant traits. Second, manipulations designed to examine the effect of nectar level on pollen flow must be carried out in such a way that the majority of the plant population remains unchanged.

Pollinator Foraging Behavior and Plant Fitness

Initial Approaches to Plants

Whether pollinators visit a plant has been assumed to be independent of the nectar production rate of that plant.[140] This assumption is based on the belief that pollinators are unable to remember the location of individual plants in a dense population. Although this point has not been examined directly it seems reasonable. It is possible that if nectar production is associated with some other phenotypic trait, then pollinators might selectively visit plants with high rates of production. Pollinators have been found, for example, to favor large plants over small ones.[55,65,139,155,202,206] In fact, however, examinations of nectar production as a function of number of flowers produced per plant or per inflorescence have yielded either no significant correlation or a negative one.[103,129,222] Although it is well documented that blossom color in many species is associated with reward level,[31,57,59,60,72,89,156,158] all species studied in this regard change petal color after pollination. The available data thus seem to indicate that even if pollinators can estimate the amount of nectar reward within flowers without actually entering them,[75,104] they are not able to assess the reward status of whole plants without actually visiting them. In general, then, the frequency with which pollinators visit an individual plant should be independent of that plant's nectar production rate.

Pollinator Behavior on Plants

There is a strong probability that the rate of nectar production will have significant effects on the behavior of pollinators already on the plant. Predictions from optimal foraging theory state that, as standing crop of nectar in flowers on a plant increases, the total number of blossoms visited by pollinators should also increase.[33,139,141] Although the number of flowers visited per plant is not independent of the number of blossoms available,[141,143] pollinators consistently visit more flowers on high rather than low reward quality inflorescences.[34,75,79,137,209] We can presume then that if a given plant produces nectar at a rate higher than that of all other individuals in the population, pollinators encountering that plant will visit more of its flowers. The effect on plant fitness (i.e., on the shape of the benefit curve in Fig. 8.2.) of such a simple behavioral change is not easy to predict. Among other factors, the effect will depend on whether the species is self-compatible, on the amount of available stigmatic space, on whether seed set is pollen limited, on whether the species can select among available pollen grains, and on the degree of pollen carryover. In general, it might be fair to assume that both female and male fitness increase as a function of the number of flowers visited per plant, although the rate of increase will differ among species. Pollen deposition on stigmas has actually been found to increase to an asymptotic value as a function of nectar level in a number of species.[122,133,184]

However, there are a number of special cases in which fitness might decrease if too many blossoms are visited at one time. Consider, for example, what happens when a pollinator visits large numbers of flowers on a self-incompatible individual. When an animal first encounters the plant, its pollen load consists entirely of pollen from other plants in the population. Some of this outcross pollen is deposited on the stigma of the first flower entered and is replaced by pollen shed by this blossom. As increasing numbers of flowers on the same plant are visited, the ratio of self to outcross pollen on the animal's body will shift until most, or all, of the grains available for deposition originated from that individual. At this extreme point only self-pollen will be deposited. If stigmatic surface is a limiting resource,[194] then continued visitation will have a negative impact on female fitness because of stigma clogging by self-incompatible pollen. Even if stigmatic space is only rarely limiting,[24,165] continued visitation might have a significant detrimental effect on fitness if the incompatibility mechanism is a postzygotic rather than a prezygotic one. If stigmatic space is limiting and/or if a postzygotic incompatibility mechanism is present, then reduced female fitness will occur well before the pollinator is completely covered with self-pollen. Female fitness might even decline with excessive flowers visited per individual on self-compatible species. Because the optimal balance of selfed versus outcrossed seeds will depend on a wide range of environmental conditions,[85] the additional number of selfed seeds arising from increased visitation may not be evolutionarily advantageous.

The exact shape of the female fitness curve will depend on the dynamics of pollen carryover on the body of pollinators. Although computer simulations of pollen carryover have been performed,[95] few studies have explored carryover in natural or experimental situations.[54,66,184,185,197,204] The results of these studies vary, but generally they suggest that pollen carryover may be longer and more variable than originally assumed. Thomson and Plowright,[184] for instance, noted pollen from a target *Erythronium americanum* blossom landing on the stigma of the fifty-fourth flower visited in succession. On the other hand, a number of studies presumed, for the sake of simplicity, that pollen was deposited on the next flower visited[9,97,101,160] while Cruden,[44] claiming that more than ~10 flowers visited per plant leads predominantly to geitonogamy, speculated that carryover was small. Galen and Plowright[54] actually found pollen carryover to be quite small in *Epilobium angustifolium*. The amount of outcross pollen deposited on stigmas of single inflorescences decreased exponentially; the stigma of the first flower visited averaged 50 outcross pollen grains while the fifth flower averaged only 5 grains. Similar results were found by Wolin et al.[204] for *Oenothera speciosa*. The decline in outcross pollen found in these two species is probably steeper than it would be for many others because both *Epilobium angustifolium* and *Oenothera speciosa* pollen grains are bound together by viscin threads. It is clear from the variability observed to date that general estimates of pollen carryover are not very useful and that actual pollen carryover patterns for the plant species in question must be determined.

Male fitness might also decline when too many blossoms are visited per plant. Pollen grains transported within a self-incompatible plant are lost reproductively, and as the number of flowers visited per plant increases, the number of self-pollen grains finding their way to stigmas must rise as well. In general, the total number of pollen grains picked up per plant by a pollinator will be a positive function of the number of flowers visited on that plant. The critical factor is the ultimate destination of those grains. The importance of outbreeding in many herbaceous plants can be seen by the frequency with which dichogamy is coupled with a gradient in the standing crop of nectar on plants with spiked inflorescences.[10,16,41,53,79,138,143,190] Such patterns encourage pollinators to visit (functionally) female flowers and deposit outcross pollen before picking up self-pollen from male blossoms. As with female function, the only way to assess the effect that a given number of flower visits has on reproductive fitness is to characterize the dynamics of pollen carryover. It should then become possible to make probability statements concerning the final location of the pollen grains produced by a plant and to predict how the addition or subtraction of a single flower visit per plant changes those probabilities. Although the work by Wolin et al.[204] and Galen and Plowright[53] is a step in this direction, such probability statements await far more detailed studies than have yet been completed.

An increase in the level of standing crop of nectar per flower will have a second major effect on foraging behavior while the pollinator is on the plant. As reward values increase, the time spent in each flower rises as well.[54,81,214] Since Thomson and Plowright[184] demonstrated that the number of pollen grains deposited on stigmas by bumblebees is a positive function of in-flower time, deposition should increase as a function of nectar level. Although the rate of pollen removal from anthers may vary in the same manner, no work yet addresses this phenomenon. Increased rates of pollen pick-up and deposition as a function of nectar reward suggest that plant fitness will also vary positively as a function of nectar level. The situation should be analogous in many respects to the case just discussed for length of visits to individual plants, i.e., fitness will probably rise to an asymptotic value and then decline when nectar levels are so high that pollinators remain in individual flowers for extended periods. The questions that must be answered are (1) How does a change in in-flower time affect pollen flow, or more specifically, the ultimate location of pollen grains? (2) What effect does the new pattern of pollen flow have on the quantity and quality of seeds produced? Again, answers to these questions await far more detailed studies on pollen carryover than have yet been conducted.

Pollinator Decisions upon Departing Plants

As a pollinator leaves a plant it must determine where to go next. The amount of nectar received can influence the two components of this decision: flight distance and degree of directionality. When pollinators employ area-restricted searching,[186] distance will be inversely and directionality positively correlated with reward received. Such a strategy is common.[54,58,75,137,212,219] By increasing their rate of turning and decreasing the interplant distance flown after encountering plants with large rewards, pollinators tend to stay in high-quality areas and quickly pass through poorer ones. On the few occasions when pollinators were found not to use area-restricted searching, they moved consistently between near neighbors.[69,80,212,213] The mean and range of flight distances is therefore much greater when area-restricted searching is used. At the population level, such behavior has a significant impact on gene flow and increases the genetic neighborhood size of the population in question.[213,219]

Plants with low standing crops will, on average, send their pollen grains farther than will plants with high reward values. Even ignoring the problems associated with our lack of knowledge of pollen carryover, the significance of this pattern for individual plants is not immediately clear. Optimal outcrossing distances have been found in the two species examined in detail *(Delphinium nelsonii* and *Ipomopsis aggregata),*[134,197] so any plant trait that encourages pollinators to make an interplant move approximating that distance will be favored. In general, it appears that the distances moved by pollinators are substantially shorter than those that are optimal from the plants' perspective.[134,197] Lower standing crops of nectar might therefore be advantageous.

A confounding point is that specific pairs of plants might be particularly effective at yielding high quality offspring.[13,201] As pointed out above, two questions about a change in nectar reward and the new pollen flow pattern must be addressed to assess fully the advantages associated with a particular rate of nectar production. Plants apparently make trade-offs between seed quantity and quality. (For further discussion of seed production and quality, see Lee[94] and Haig and Westoby,[64] this volume.) An assessment of the net effect of standing crop levels on total plant fitness thus demands determination of the relationships between seed number and both seed size and quality. The relationship between seed number and quality can best be examined in experimental gardens that allow various genotypes to experience common environmental conditions, while an analysis of the relationship between seed number and size requires hand-pollinations and an evaluation of the type of mechanisms (if any) allowing females to choose among pollen genotypes.[177]

Costs of Nectar Production

The final aspect that must be studied in order to achieve a complete picture of floral nectar production is the cost of the nectar produced. The principle of allocation implies that energy devoted to nectar production is unavailable to the plant for alternative use,[36] so when plants are energy limited[174,188,203,215] energy seems to be a reasonable currency in which to measure costs (but see Lovett Doust,[102]). However, plants may resorb nectar from aging flowers[38] and utilize its carbon in developing ovules.[91] Herbaceous perennials may, for example, devote between 8 and 15% of their annual net carbon gain to sexual reproduction.[112] Jurik[90] claims that most studies on perennials have actually underestimated the energetic contribution made to sexual reproduction while the percentage of available energy spent by annuals is slightly higher. Although many studies have estimated energy allocation to reproduction, most have failed to include the energy involved in the production of floral nectar.[1,2,32,37,154,182] The two studies that have explicitly examined this phenomenon concluded that plants partition a significant portion of their energy budget to nectar production.[147,171] Southwick,[171] for example, found that the nectar secreted by *Asclepias syriaca* accounts for between 4 and 37% of the daily photosynthate accumulated during the flowering period. He also found that the nectar produced by alfalfa contains almost twice the energy of its total seed crop. Apparently the cost of floral nectar production can be quite high.

For Fig. 8.2 to be meaningful, the cost and benefit curves must be expressed in the same currency.[140] The principle of allocation notwithstanding, it is not yet clear whether calories saved by producing slightly less nectar can be used to produce either more or larger seeds. Solving the currency problem is not going to be easy, either in the plant–pollinator system or in any other. As Fox[51] states: "Cost arguments are intuitive, convenient and pervasive in ecology. But for plant–herbivore interactions, as well as other areas of study, there are very few direct measurements of costs, or data suitable for testing . . . assumptions in a critical way."

Ultimately, when the above steps have been taken for a species, one should be able to determine the evolutionary advantages or disadvantages of situations such as the following: A plant increases its rate of nectar production slightly by devoting a slightly larger percentage of its energy budget to nectar instead of seeds. The change in nectar production promotes the dispersal of pollen, on average, to six instead of five subsequent plants but the six plants are somewhat closer than the five had been. Additionally, because of increased in-flower time, more outcross pollen is deposited on stigmas, but more self-pollen finds its way there as well.

FLOWERING PHENOLOGY

Studies on flowering phenology, like those concerning rate of nectar production, cover a wide range of topics. It has been suggested, for example, that flowering phenology is responsive to selection pressure from such variables as pollinator availability and interspecific competition for their services,[27,28,56,96,114,127,151,183,194,195] intraspecific competition for pollinators,[206,207,216] seed predation,[4,61,208,220] fruit dispersal,[167] environmental variables,[82,83,157] and mutualistic interactions among co-occurring species.[199] Numerous studies have shown that flowers opening at different times during the season vary in seed production per flower[7,30,111,157,159,194,205–208] while others have shown that plants with different flowering schedules differ in total seed set (i.e., in total female fitness).[4,61,135,163,195,220] Very few studies, however, have examined the interaction between flowering phenology and either pollen dispersal or outcrossing.[30,161,175,217]

The question to be considered in the present context is, Will a shift in time of flowering change the pollen flow dynamics in a population? The question can be examined either at the level of the population or at the level of the individual plant. At the population level, for example, Carpenter[30] hypothesized that the highest levels of outcrossing of mass-flowering species occur during the beginning and the end of the blooming season when the number of open blossoms per individual is smallest. Stephenson,[175] working with the tree species *Catalpa speciosa,* and Zimmerman,[217] studying the perennial herb *Polemonium foliosissimum,* indeed reported that most outbreeding took place late in the flowering season but found no evidence of the predicted early peak in outcrossing. Only one study[217] has focused on the flowering phenology of individual plants. Zimmerman[217] asked two related questions of *Polemonium foliosissimum:* (1) Does most outcrossing take place when individuals are at the peak of their bloom? (2) Do plants flowering either earlier or later than the population as a whole experience a different amount of outbreeding than those individuals whose peak of bloom coincides with the peak of the population? Results suggest that the flowering phenology of individuals is only pertinent when considered in the context of the population as a whole. At no particular time during the flowering schedules of individuals was outbreeding, as measured by either female or male function, enhanced. Individuals whose peak of bloom came after that of the population, however, were found to experience significantly higher levels of outcrossing than those plants whose bloom preceded or coincided with the population as a whole. In other words, *Polemonium foliosissimum* plants delaying their flowering were more likely to outcross.

Would such an increase in outbreeding be advantageous? Or, more to the point, would the possible benefits outweigh the costs of delayed blooming? This question is complicated because of the many factors that can influence both the costs and benefits. The optimal amount of outcrossing for any species will be determined by the mating dynamics of the population[134,197] and possibly by long-term environmental conditions,[85] as discussed above. The costs associated with delayed flowering could be considerable. Such factors as increased overlap of blooming time with sympatric species, less time for seeds to ripen, fewer conspecifics in bloom at the same time, few specialist pollinators about, etc. might reduce the number and/or quality of surviving seeds. In addition, a number of species have been found to produce less nectar late in the season[11,48,93,128,144,222] and thus all of the advantages and disadvantages associated with reward level must be considered as well. Any answer to this question for any species awaits the kind of comprehensive study outlined above.

The pollinator behavioral mechanism leading to increased outbreeding for late-blooming *Polemonium foliosissimum* individuals is not yet clear. All size classes of plants are represented in the blooming group throughout the season so pollinators are not responding exclusively to any one class at any particular point in time. The density of individuals in flower is significantly lower late in the season, but Zimmerman[217] has shown that the increased outcrossing is not simply due to the fact that the closest available plants are more distant than they were earlier in the season. Bumblebees apparently are deciding to fly to more distant plants late in the season. This was an unexpected result. Although pollinators' flight distances are usually,[98,99] but not always,[35] negatively correlated with plant density, bumblebees in the present situation are actively choosing to fly distances that are longer than necessary. More work clearly needs to be performed to understand the reasons for the flight patterns observed on populations of *Polemonium foliosissimum* and to determine how common this behavior actually is. Until pollinator behavior is better understood it is also impossible to estimate the frequency with which plant species can influence gene flow by altering their flowering phenology. Furthermore, the species studied to date in this light are all of mass-flowering taxa. The interaction between blooming schedule and pollinator behavior might be very different for species that produce few open blossoms at a time. Finally, for flowering phenology to be open to selection pressure stemming from altered patterns of gene flow, blooming time must be heritable. There is ample evidence for this,[108,109,117] with flowering time responding very rapidly to selection.[110,120]

CONCLUSIONS

As demonstrated above, pollinators respond to variability in nectar reward and flowering phenology. There is no reason to suspect that these two phenotypic characteristics would have any more or less of an effect on pollinator behavior than any of the other traits that could have been examined, such as inflorescence size and architecture, flower size, color, etc. The answer to the question, Can plants manipulate pollinator behavior? is thus yes. The more critical follow-up question, Can pollinators be manipulated to the plant's advantage? cannot yet be answered. As I have shown, an answer to this question involves a multistep process: (1) one must be able to predict pollinator responses to genotypic variability in the plant, (2) changes in pollinator behavior must be translated into new pollen flow patterns for the plant population, (3) the optimal pollen flow patterns from the plants' perspective must be determined, (4) the costs associated with a change in genotype must be estimated in a currency that is directly comparable to the number and/or quality of the seeds produced (maternal component of fitness) or the number and/or quality of the seeds to which pollen is contributed (paternal component of fitness) and (5) an evolutionarily stable strategy analysis must be performed to see what constellation of genotypes can coexist in a population. Gathering all of the information necessary to address the question for one plant trait in one species is a massive undertaking. Not a single study, or a collection of studies on a single taxon, comes close to having the necessary data.

Waddington[191] points out that a discrepancy between the optimal outcrossing distance[134,197] and the average flight distances of pollinators is not surprising. He suggests that the optimal outcrossing distance is not constant and that it actually increases as pollen dispersal distances increase. In other words, pollen flow will always be less than the optimal value. This fact, combined with the realization that the effect of many plant traits on pollinator behavior is frequency dependent, led Waddington to conclude that it is unlikely that plants might, in any significant evolutionary sense, influence pollen flow patterns. I feel that the available data indicate that Waddington's conclusion is premature. The proper question is not whether plants can achieve the ideal pollen flow patterns, but whether individuals can improve their relative fitness by altering where they send and from whom they receive pollen. As a result of frequency dependence, selection pressures might change over time. Additionally, Waser and Price[198] claim, but do not demonstrate, that increases in pollen flow distances will not have the simple effect of increasing the optimal outcrossing distance as Waddington claimed. Detailed studies of the sort outlined above should resolve the issue in no time.

ACKNOWLEDGMENTS

I thank David Hicks, John Pleasants, and Graham Pyke for numerous interesting discussions and many ideas that have found their way into the present chapter. I also thank Abby Frucht and David Hicks for reading and making needed improvements on an earlier version of the manuscript. I am also grateful for financial support from the National Science Foundation (BSR 82-19490) and the United States Department of Agriculture (83-CRCR-1-1242).

REFERENCES

1. Abrahamson, W. G., and Caswell, H., On the comparative allocation of biomass, energy, and nutrients in plants, *Ecology* **63,** 982–991 (1982).
2. Armstrong, R. A., A quantitative theory of reproductive effort in rhizomatous perennial plants, *Ecology* **63,** 679–686 (1982).
3. Arnold, R. M., Pollination, predation and seed set in *Linaria vulgaris* (Scrophulariaceae), *Am. Midl. Nat.* **107,** 360–369 (1982).
4. Augspurger, C. K., Reproductive synchrony of a tropical shrub: Experimental studies on effects of pollinators and seed predators on *Hypanthus prunifolius* (Violaceae), *Ecology* **62,** 775–788 (1981).
5. Baker, H. G., Sugar concentration in nectars from hummingbird flowers, *Biotropica* **7,** 37–41 (1975).
6. Barrett, S. C. H., The evolutionary breakdown of tristyly in *Eichhornia crassipes* (Mart.) Solms (Water Hyacinth), *Evolution* **33,** 499–510 (1979).
7. Barrett, S. C. H., Sexual reproduction in *Eichornia crassipes* (water hyacinth) II. Seed production in natural populations, *J. Appl. Ecol.* **17,** 113–124 (1980).
8. Barrows, E. M., Nectar robbing and pollination of *Lantana camara* (Verbenaceae), *Biotropica* **8,** 132–135 (1976).
9. Beattie, A. J., Plant dispersion, pollination, and gene flow in *Viola, Oecologia* **25,** 291–300 (1976).
10. Benham, B. R., Insect visitors to *Chamaenerion angustifolium* and their behavior in relation to pollination, *Entomologist* **102,** 221–228 (1969).
11. Bernhardt, P., and Calder, D. M., The floral ecology of sympatric populations of *Amyema pendulum* and *Amyema quandana* (Loranthaceae), *Bull. Torrey Bot. Club* **108,** 213–230 (1981).
12. Bertin, R. I., Floral biology, hummingbird pollination, and fruit production of trumpet creeper (*Campsis radicans,* Bignoniaceae), *Am. J. Bot.* **69,** 122–134 (1982).
13. Bertin, R. I., Paternity and fruit production in trumpet creeper *(Campsis radicans), Am. Nat.* **119,** 694–709 (1982).
14. Bertin, R. I., Paternity in plants, in *Plant Reproductive Ecology: Patterns and Strategies* (J. Lovett Doust and L. Lovett Doust, eds.), Chapter 2. Oxford Univ. Press, New York, 1988.
15. Bertsch, A., Nectar production of *Epilobium angustifolium* L. at different air humidities; nectar sugar in individual flowers and the optimal foraging theory, *Oecologia* **59,** 40–48 (1983).
16. Best, L. S., and Bierzychudek, P., Pollinator foraging on foxglove *(Digitalis purpurea):* A test of a new model, *Evolution* **36,** 70–79 (1981).
17. Beutler, R., Biologische-chemische Untersuchungen am Nektar von Immenblumen, *Z. Vergleich. Physiol.* **12,** 72–276 (1930).
18. Bierzychudek, P., Pollinator limitation of plant reproductive effort, *Am. Nat.* **117,** 838–840 (1981).
19. Bierzychudek, P., The demography of Jack-in-the-pulpit, a forest perennial that changes sex. *Ecol. Monogr.* **52,** 335–351 (1982).
20. Boetius, J., Uber den Verlauf der Nektarabsonderung einiger Blutenpflanzen, *Beih. Schweiz. Bienenztg.* **2,** 258–317 (1948).
21. Bolten, A. B., and Feinsinger, P., Why do hummingbird flowers secrete dilute nectar? *Biotropica* **10,** 307–309 (1978).
22. Brink, D. E., A bonanza-blank pollinator reward schedule in *Delphinium nelsonii* (Ranunculaceae), *Oecologia* **52,** 292–294 (1982).
23. Brink, D. E., and deWet, J. M., Interpopulation variation in nectar production in *Aconitum columbianum* (Ranunculaceae), *Oecologia* **47,** 160–163 (1980).
24. Brown, J. H., and Kodric-Brown, A., Convergence, competition and mimicry in a temperate community of hummingbird pollinated flowers, *Ecology* **60,** 1022–1035 (1979).

25. Brown, J. H., Kodric-Brown, A., Whitham, T. G., and Bond, H. W., Competition between hummingbirds and insects for the nectar of two species of shrubs, *Southwest. Nat.* **26,** 133–145 (1981).
26. Calder, W. A., On the temperature-dependency of optimal nectar concentration for birds, *J. Theor. Biol.* **78,** 185–196 (1979).
27. Campbell, D. R., Pollinator sharing and seed set of *Stellaria pubera:* Competition for pollination, *Ecology* **66,** 544–553 (1985).
28. Campbell, D. R., and Motten, A. F., The mechanism of competition for pollination between two forest herbs, *Ecology* **66,** 554–563 (1985).
29. Caraco, T., Martindale, S., and Whittam, T., An empirical demonstration of risk-sensitive foraging preferences, *Anim. Behav.* **28,** 820–830 (1980).
30. Carpenter, F. L., Plant-pollinator interactions in Hawaii: Pollination energetics of *Metrosideros collina* (Myrtaceae), *Ecology* **57,** 1125–1144 (1976).
31. Casper, B. B., and La Pine, T. R., Changes in corolla color and other floral characteristics in *Cryptantha humilis* (Boraginaceae): Cues to discourage pollinators? *Evolution* **38,** 128–141 (1984).
32. Chaplin, S. J., and Walker, J. L., Energetic constraints and adaptive significance of the floral display of forest milkweed, *Ecology* **63,** 1857–1870 (1982).
33. Charnov, E. L., Optimal foraging, the marginal value theorem, *Theor. Popul. Biol.* **9,** 129–136 (1976).
34. Charnov, E. L., Simultaneous hermaphroditism and sexual selection, *Proc. Natl. Acad. Sci. U.S.A.* **76,** 2480–2484 (1979).
35. Cibula, D. A., and Zimmerman, M., The effect of plant density on departure decisions: Testing the marginal value theorem using bumblebees and *Delphinium nelsonii, Oikos* **43,** 154–158 (1984).
36. Cody, M. L., A general theory of clutch size, *Evolution* **20,** 174–184 (1966).
37. Colosi, J. C., and Cavers, P. B., Pollination affects percent biomass allocated to reproduction in *Silene vulgaris* (Bladder Campion), *Am. Nat.* **124,** 299–306 (1984).
38. Corbet, S. A., Bee visits and the nectar of *Echium vulgare* L. and *Sinapsis alba* L., *Ecol. Entomol.* **3,** 25–37 (1978).
39. Corbet, S. A., Bees and the nectar of *Echium vulgare,* in *The Pollination of Flowers by Insects* (A. J. Richards, ed.), pp. 21–30. Academic Press, New York, 1978.
40. Corbet, S. A., Unwin, D. M., and Prys-Jones, O. E., Humidity, nectar, and insect visits to flowers, with special reference to *Crataegus, Tilia* and *Echium, Ecol. Entomol.* **4,** 9–22 (1979).
41. Corbet, S. A., Cuthill, J., Fallows, M., Harrison, T., and Hartley, G., Why do nectar-foraging bees and wasps work upwards on inflorescences? *Oecologia* **51,** 79–83 (1981).
42. Cowie, R. J., Optimal foraging in Great Tits *(Parus major), Nature (London)* **268,** 137–139 (1977).
43. Craig, J. L., and Douglas, M. E., Temporal partitioning of a nectar resource in relation to competitive asymmetries, *Anim. Behav.* **32,** 624–625 (1984).
44. Cruden, R. W., Intraspecific variation in pollen-ovule ratios and nectar secretion—preliminary evidence of ecotypic adaptation, *Ann. Missouri Bot. Gard.* **63,** 277–289 (1976).
45. Cruden, R. W., and Hermann-Parker, S. M., Butterfly pollination of *Caesalpinia pulcherrima,* with observations on a psychophilous syndrome, *J. Ecol.* **67,** 155–168 (1979).
46. Cruden, R. W., Hermann-Parker, S. M., and Peterson, S., Patterns of nectar production and plant-pollinator coevolution, in *Biology of Nectaries* (T. S. Elias and B. A. Bentley, eds.). Chapter 3. Columbia Univ. Press, New York, 1983.
47. Davis, M. A., The effect of pollinators, predators, and energy constraints on the floral ecology and evolution of *Trillium erectum, Oecologia* **48,** 400–406 (1981).
48. Feinsinger, P., Ecological interactions between plants and hummingbirds in a successional tropical community, *Ecol. Monogr.* **46,** 105–128 (1978).
49. Feinsinger, P., Linhart, Y. B., Swarm, L. A., and Wolfe, J. A., Aspects of the pollination biology of three *Erythrina* species on Trinidad and Tobago, *Ann. Missouri Bot. Gard.* **66,** 451–471 (1979).
50. Feinsinger, P., Wolfe, J. A., and Swarm, L. A., Island ecology: Reduced hummingbird diversity and the pollination biology of plants, Trinidad and Tobago, West Indies, *Ecology* **63,** 494–506 (1982).
51. Fox, L. R., Defense and dynamics in plant-herbivore systems, *Am. Zool.* **21,** 853–864 (1981).
52. Frost, S. K., and Frost, P. G. H., Sunbird pollination of *Strelitzia nicolai, Oecologia* **49,** 379–384 (1981).
53. Galen, C., and Plowright, R. C., Contrasting movement patterns of nectar-collecting and pollen-collecting bumble bees *(Bombus terricola)* on fireweed *(Chamaenerion angustifolium)* inflorescences, *Ecol. Entomol.* **10,** 9–17 (1985).
54. Galen, C., and Plowright, R. C., The effects of nectar level and flower development on pollen carry-over in inflorescences of fireweed *(Epilobium angustifolium)* (Onagraceae), *Can. J. Bot.* **63,** 488–491 (1985).

55. Geber, M. A., The relationship of plant size to self-pollination in *Mertensia ciliata, Ecology* **66,** 762–777 (1985).
56. Gentry, A. H., Flowering phenology and diversity in tropical Bignoniaceae, *Biotrópica* **6,** 64–68 (1974).
57. Gill, F. B., and Wolf, L. L., Foraging strategies and energetics of east African sunbirds at mistletoe flowers, *Am. Nat.* **109,** 491–510 (1975).
58. Gill, F. B., and Wolf, L. L., Nonrandom foraging by sunbirds in a patchy environment, *Ecology* **58,** 1284–1296 (1977).
59. Gori, D. F., Post-pollination phenomena and adaptive floral changes, in *Handbook of Experimental Pollination Biology* (C. E. Jones and R. J. Little, eds.), Chapter 2. Van Nostrand Reinhold, New York, 1983.
60. Gottsberger, G., Colour change of petals in *Malvaviscus arboreus* flowers, *Acta Bot. Neerl.* **20,** 381–388 (1971).
61. Gross, R. S., and Werner, P. A., Relationships among flowering phenology, insect visitors, and seed-set of individuals: Experimental studies on four co-occurring species of goldenrod *(Solidago:* Compositae), *Ecol. Monogr.* **53,** 95–117 (1983).
62. Haber, W. A., and Frankie, G. W., Pollination of *Luehea* (Tilliaceae) in Costa Rican deciduous forest, *Ecology* **63,** 1740–1750 (1982).
63. Haddock, R. C., and Chaplin, S. H., Pollination and seed production in two phenologically divergent prairie legumes *(Baptisia leucophaea* and *B. leucantha), Am. Midl. Nat.* **108,** 175–186 (1982).
64. Haig, D., and Westoby, M., Inclusive fitness, seed resources and maternal care, in *Plant Reproductive Ecology: Patterns and Strategies* (J. Lovett Doust and L. Lovett Doust, eds.), Chapter 3. Oxford Univ. Press, New York, 1988.
65. Hainsworth, F. R., Wolf, L. L., and Mercier, T., Pollination and pre-dispersal seed predation: net effects on reproduction and inflorescence characteristics in *Ipomopsis aggregata, Oecologia* **63,** 405–409 (1984).
66. Handel, S. N., Dynamics of gene flow in an experimental population of *Cucumis melo* (Curcurbitaceae), *Am. J. Bot.* **69,** 1538–1546 (1982).
67. Harder, L. D., Morphology as a predictor of flower choice by bumble bees, *Ecology* **66,** 198–210 (1985).
68. Hawkins, R. P., Selection for height of nectar in the corolla tube of English singlecut red clover, *J. Agric. Sci.* **77,** 348–350 (1971).
69. Haynes, J., and Mesler, M. R., Pollen foraging by bumblebees: Foraging patterns and efficiency on *Lupinus polyphyllus, Oecologia* **61,** 249–253 (1984).
70. Heinrich, B., The role of energetics in bumblebee-flower interrelationships, in *Coevolution of Animals and Plants* (L. E. Gilbert and P. H. Raven, eds.), pp. 141–158. Univ. of Texas Press, Austin, 1975.
71. Heinrich, B., Bee flowers: A hypothesis on flower variety and blooming times, *Evolution* **29,** 325–334 (1975).
72. Heinrich, B., Energetics of pollination, *Annu. Rev. Ecol. Syst.* **6,** 139–170 (1975).
73. Heinrich, B., Resource partitioning among eusocial insects: Bumblebees, *Ecology* **57,** 874–889 (1976).
74. Heinrich, B., The foraging specialization of individual bumblebees, *Ecol. Monogr.* **46,** 105–128 (1976).
75. Heinrich, B., Resource heterogeneity and patterns of movement in foraging bumblebees, *Oecologia* **40,** 235–245 (1979).
76. Herrera, C. M., and Soriguer, R. C., Inter- and intra-floral heterogeneity of nectar production in *Helleborus foetidus* L. (Ranunculaceae), *Biol. J. Linn. Soc.* **86,** 253–260 (1983).
77. Heyneman, A. J., Optimal sugar concentrations of floral nectar—dependence on sugar intake efficiency and foraging costs, *Oecologia* **60,** 198–213 (1983).
78. Hocking, B., Insect-flower associations in the high Arctic with special reference to nectar, *Oikos* **19,** 359–388 (1968).
79. Hodges, C. M., Bumble bee foraging: The threshold departure rules, *Ecology* **66,** 179–187 (1985).
80. Hodges, C. M., and Miller, R. B., Pollinator flight directionality and the assessment of pollen returns, *Oecologia* **50,** 376–379 (1981).
81. Hodges, C. M., and Wolf, L. L., Optimal foraging in bumblebees. Why is nectar left behind in flowers? *Behav. Ecol. Sociobiol.* **9,** 41–44 (1981).
82. Hodgkin, K. C., and Quinn, J. A., Environmental and genetic control of reproduction in *Danthonia caespitosa* populations, *Aust. J. Bot.* **26,** 351–364 (1978).
83. Hogan, K. P., The pollination biology and breeding system of *Aplectrum hyemale* (Orchidaceae), *Can. J. Bot.* 61, 1906–1910 (1983).
84. Jackson, M. T., Effects of microclimate on spring flowering phenology, *Ecology* **47,** 407–415 (1966).
85. Jain, S. K., The evolution of inbreeding in plants, *Annu. Rev. Ecol. Syst.* **7,** 469–495, (1976).

86. Janzen, D. H., A note on optimal mate selection in plants, *Am. Nat.* **111,** 365–371 (1977).
87. Janzen, D. H., Seed and pollen dispersal by animals: Convergence in the ecology of contamination and sloppy harvest, *Biol. J. Linn. Soc.* **20,** 103–113 (1983).
88. Janzen, D. H., De Vries, P., Gladstone, D. E., Higgins, M. L., and Lewinsohn, T. M., Self-pollination and cross-pollination of *Encyclia cordigera* (Orchidaceae) in Santa Rosa National Park, *Biotropica* **12,** 72–74 (1980).
89. Jones, C. E., and Buchmann, S. L., Ultraviolet floral patterns as functional orientation cues in Hymenopterous pollination systems, *Anim. Behav.* **22,** 481–485 (1974).
90. Jurik, T. W., Reproductive effort and CO_2 dynamics of wild strawberry populations, *Ecology* **64,** 1329–1342 (1983).
91. Kratashova, N., and Tsylenok, S., On the biologic role of nectaries and nectar. II. Studies on the resorption of nectar by flower parts, *12V, SIB OTD Akao Nauk. SSSR Ser. Biol. Med. Nauk.* **1,** 134–137 (1968).
92. Krebs, J. R., Ryan, J. C., and Charnov, E. L., Hunting by expectation or optimal foraging: A study of patch use by chickadees, *Anim. Behav.* **22,** 953–964 (1974).
93. Lack, A. J., Competition for pollinators in the ecology of *Centaurea scabiosa* L. and *Centaurea nigra* L. II. Observations on nectar production, *New Phytol.* **91,** 309–320 (1982).
94. Lee, T., Patterns of fruit and seed production, in *Plant Reproductive Ecology: Patterns and Strategies* (J. Lovett Doust and L. Lovett Doust, eds.), Chapter 9. Oxford Univ. Press, New York, 1988.
95. Lertzman, K. P., and Gass, C. L., Alternative models of pollen transfer, in *Handbook of Experimental Pollination Ecology* (C. E. Jones and R. J. Little, eds.), Chapter 24. Van Nostrand Reinhold, New York, 1983.
96. Levin, D. A., and Anderson, W. W., Competition for pollinators between simultaneously flowering species, *Am. Nat.* **104,** 455–467 (1970).
97. Levin, D. A., and Kerster, H. W., Local gene dispersal in *Phlox, Evolution* **23,** 560–571 (1968).
98. Levin, D. A., and Kerster, H. W., The dependence of bee-mediated pollen and gene dispersal upon plant density, *Evolution* **23,** 560–571 (1969).
99. Levin, D. A., and Kerster, H. W., Density-dependent gene dispersal in *Liatris, Am. Nat.* **103,** 61–74 (1969).
100. Levin, D. A., and Kerster, H. W., Gene flow in seed plants *Evol. Biol.* **7,** 138–220 (1974).
101. Linhart, Y. B., and Feinsinger, P., Plant-hummingbird interactions: Effects of island size and degree of specialization on pollination, *J. Ecol.* **68,** 745–760 (1980).
102. Lovett Doust, J. Experimental manipulation of patterns of resource allocation in the growth cycle and reproduction of *Smyrnium olusatrum* L. *Biol. J. Linn. Soc.* **13,** 155–166 (1980).
103. Marden, J. H., Intrapopulation variation in nectar secretion in *Impatiens capensis, Oecologia* **63,** 418–422 (1984).
104. Marden, J. H., Remote perception of floral nectar by bumblebees, *Oecologia* **64,** 232–240 (1984).
105. Maynard Smith, J., Evolution and the theory of games, *Am. Sci.* **64,** 41–45 (1976).
106. Maynard Smith, J., Optimization theory in evolution, *Annu. Rev. Ecol. Syst.* **9,** 31–56 (1978).
107. McDade, L. A., and Kinsman, S., The impact of floral parasitism in two neotropical hummingbird-pollinated plant species, *Evolution* **34,** 944–958 (1980).
108. McIntyre, G. I., and Best, K. F., Studies on the flowering of *Thlaspi arvense* L. IV. Genetic and ecological differences between early- and late-flowering strains, *Bot. Gaz.* **139,** 190–195 (1978).
109. McMillan, C., and Pagel, B. F., Phenological variation within a population of *Symphoricarpos occidentalis, Ecology* **39,** 766–770 (1958).
110. McNeilly, T., and Antonovics, J., Evolution in closely adjacent plant populations. IV. Barriers to gene flow, *Heredity* **23,** 205–218 (1968).
111. Melampy, M. N., and Hayworth, A. M., Seed production and pollen vectors in several nectarless plants, *Evolution* **34,** 1144–1154 (1980).
112. Mooney, H. A., The carbon balance of plants, *Annu. Rev. Ecol. Syst.* **3,** 315–346 (1972).
113. Morse, D. H., and Fritz, R. S., Contributions of diurnal and nocturnal insects to the pollination of common milkweed (*Asclepias syrica* L.) in a pollen-limited system, *Oecologia* **60,** 190–197 (1983).
114. Mosquin, T., Competition for pollinators as a stimulus for the evolution of flowering time, *Oikos* **22,** 398–402 (1971).
115. Motten, A. F., Reproduction of *Erythronium umbilicatum* (Liliaceae)—pollination success and pollinator effectiveness, *Oecologia* **59,** 351–359 (1983).
116. Mulcahy, D. L., Curtis, P. S., and Snow, A. A., Pollen competition in a natural population, in *Hand-*

book of Experimental Pollination Ecology (C. E. Jones and R. J. Little, eds.), Chapter 16. Van Nostrand Reinhold, New York, 1983.
117. Murfet, I. C., Environmental interaction and the genetics of flowering, *Annu. Rev. Plant Physiol.* **28,** 253–278 (1977).
118. Nuñez, J., Nectar flow by melliferous flowers and gathering flow by *Apis mellifera lingustica, J. Insect Physiol.* **23,** 265–275 (1977).
119. Ollason, J. G., Learning to forage—optimally? *Theor. Popul. Biol.* **18,** 44–56 (1980).
120. Paterniani, E., Selection for reproductive isolation between two populations of maize, *Zea mays* L., *Evolution* **23,** 534–547 (1969).
121. Pedersen, M. W., Environmental factors affecting nectar secretion and seed production in alfalfa, *Agron. J.* **45,** 359–361 (1953).
122. Pedersen, M. W., Seed production in alfalfa as related to nectar production and honeybee visitation, *Bot. Gaz.* **155,** 129–138 (1953).
123. Pedersen, M. W., and Bohart, G. E., Factors responsible for the attractiveness of various clones of alfalfa to pollen-collecting bumblebees, *Agron. J.* **45,** 548–551 (1953).
124. Percival, M. S., *Floral Biology.* Pergamon Press, London, 1965.
125. Percival, M. S., and Morgan, P., Observations on the floral biology of *Digitalis* species, *New Phytol.* **64,** 1–22 (1965).
126. Petersen, C., Brown, J. H., and Kodric-Brown, A., An experimental study of floral display and fruit set in *Chilopsis linearis* (Bignoniaceae), *Oecologia* **55,** 7–11 (1982).
127. Pleasants, J. M., Competition for bumblebee pollinators in Rocky Mountain plant communities, *Ecology* **61,** 1446–1459 (1980).
128. Pleasants, J. M., Nectar production patterns in *Ipomopsis aggregata* (Polemoniaceae), *Am. J. Bot.* **70,** 1468–1475 (1983).
129. Pleasants, J. M., and Chaplin, S. J., Nectar production rates in *Asclepias quadrifolia:* Causes and consequences of individual variation, *Oecologia* **59,** 232–238 (1983).
130. Pleasants, J. M., and Zimmerman, M., Patchiness in the dispersion of nectar resources: evidence for hot and cold spots, *Oecologia* **41,** 283–288 (1979).
131. Pleasants, J. M., and Zimmerman, M., The distribution of standing crop of nectar: What does it really tell us? *Oecologia* **57,** 412–414 (1983).
132. Plowright, R. C., Nectar production in the boreal forest lily *Clintonia borealis, Can. J. Bot.* **59,** 156–160 (1981).
133. Plowright, R. C., and Hartling, L. K., Red Clover pollination by bumble bees: A study of the dynamics of a plant-pollinator relationship, *J. Appl. Ecol.* **18,** 639–647 (1981).
134. Price, M. V., and Waser, N. M., Pollen dispersal and optimal out-crossing in *Delphinium nelsoni, Nature (London)* **277,** 294–296 (1979).
135. Primack, R. B., Variation in the phenology of natural populations of montane shrubs in New Zealand, *J. Ecol.* **68,** 849–862 (1980).
136. Primack, R. B., and Lloyd, D. G., Andromonoecy in the New Zealand montane shrub Manuka, *Leptospermum scoparium* (Myrtaceae), *Am. J. Bot.* **67,** 361–368 (1980).
137. Pyke, G. H., Optimal foraging: Movement patterns of bumblebees between inflorescences, *Theor. Popul. Biol.* **13,** 72–98 (1978).
138. Pyke, G. H., Optimal foraging in bumblebees and coevolution with their plants, *Oecologia* **36,** 281–293 (1978).
139. Pyke, G. H., Optimal foraging in hummingbirds: Testing the marginal value theorem, *Am. Zool.* **18,** 739–752 (1978).
140. Pyke, G. H., Optimal nectar production in a hummingbird pollinated plant, *Theor. Popul. Biol.* **20,** 326–343 (1981).
141. Pyke, G. H., Animal movements: An optimal foraging approach, in *The Ecology of Animal Movement* (I. R. Swingland and P. J. Greenwood, eds.), Chapter 2. Oxford Univ. Press, Oxford, 1981.
142. Pyke, G. H., Evolution of inflorescence size and height in the waratah *(Telopea speciosissima):* The difficulty of interpreting correlations between plant traits and fruit set, in *Pollination and Evolution* (J. A. Armstrong, J. M. Powell, and A. J. Richards, eds.), pp. 91–94. Royal Botanic Gardens, Sydney, 1982.
143. Pyke, G. H., Foraging in bumblebees: Rule of departure from an inflorescence, *Can. J. Zool.* **60,** 417–428 (1982).
144. Pyke, G. H., Fruit set in *Lambertia formosa* Sm. (Proteaceae), *Aust. J. Bot.* **30,** 39–45 (1982).

145. Pyke, G. H., Seasonal pattern of abundance of honeyeaters and their resources in healthland areas near Sydney, *Aust. J. Ecol.* **8,** 217–233 (1983).
146. Pyke, G. H., Optimal foraging theory: A critical review, *Annu. Rev. Ecol. Syst.* **15,** 523–575 (1984).
147. Pyke, G. H., and McNulty, I. B., The cost of floral nectar production in some subalpine herbaceous perennials, manuscript in preparation.
148. Pyke, G. H., Pulliam, H. R., and Charnov, E. L., Optimal foraging: A selective review of theory and tests, *Q. Rev. Biol.* **52,** 137–154 (1977).
149. Pyke, G. H., and Waser, N. M., The production of dilute nectars by hummingbird and honeyeater flowers, *Biotropica* **13,** 260–270 (1981).
150. Raw, G. R., The effect on nectar secretion of removing nectar from flowers, *Bee World* **34,** 23–25 (1953).
151. Reader, R. J., Competitive relationships of some bog ericads for major insect pollinators, *Can. J. Bot.* **53,** 1300–1305 (1975).
152. Real, L., Uncertainty and pollinator-plant interactions: The foraging behavior of bees and wasps on artificial flowers, *Ecology* **62,** 20–26 (1981).
153. Schaffer, W. M., Jensen, D. B., Hobbs, D. E., Gurevitch, J., Todd, J. R., and Schaffer, M. V., Competition, foraging energetics, and the cost of sociality in three species of bees, *Ecology* **60,** 976–987 (1979).
154. Schaffer, W. M., Inouye, R. S., and Whittam, T. S., Energy allocation by an annual plant when the effects of seasonality on growth and reproduction are decoupled, *Am. Nat.* **120,** 787–815 (1982).
155. Schaffer, W. M., and Schaffer, M. V., The adaptive significance of variations in reproductive habit in the Agavaceae II: Pollinator foraging behavior and selection for increased reproductive expenditure, *Ecology* **60,** 1051–1069 (1979).
156. Schemske, D. W., Pollinator specificity in *Lantana camara* and *L. trifolia* (Verbenaceae), *Biotropica* **8,** 260–264 (1976).
157. Schemske, D. W., Flowering phenology and seed set in *Claytonia virginica* (Portulacaceae), *Bull. Torrey Bot. Club* **104,** 254–263 (1977).
158. Schemske, D. W., Floral ecology and hummingbird pollination of *Combretum farinosium* in Costa Rica, *Biotropica* **12,** 169–181 (1980).
159. Schemske, D. W., Willson, M. F., Melampy, M. N., Miller, L. J., Verner, L., Schemske, K. M., and Best, L. B., Flowering ecology of some spring woodland herbs, *Ecology* **59,** 351–366 (1978).
160. Schmitt, J., Pollinator foraging behavior and gene dispersal in *Senecio* (Compositae), *Evolution* **34,** 934–943 (1980).
161. Schmitt, J., Density-dependent pollinator foraging, flower phenology, and temporal pollen dispersal patterns in *Linanthus bicolor, Evolution* **37,** 1247–1257 (1983).
162. Schmitt, J., Flowering plant density and pollinator visitation in *Senecio, Oecologia* **60,** 97–102 (1983).
163. Schmitt, J., Individual flowering phenology, plant size, and reproductive success in *Linanthus androsaceus,* a California annual, *Oecologia* **59,** 135–140 (1983).
164. Schoener, T. W., Theory of feeding strategies, *Annu. Rev. Ecol. Syst.* **2,** 369–404 (1971).
165. Silander, J. A., and Primack, R. B., Pollination intensity and seed set in the Evening Primrose *(Oenothera fruticosa), Am. Midl. Nat.* **100,** 213–216 (1978).
166. Snow, A. A., Pollination intensity and potential seed set in *Passiflora vitifolia, Oecologia* **55,** 231–237 (1982).
167. Snow, D. W., A possible selective factor in the evolution of fruiting seasons in tropical forest, *Oikos* **15,** 274–281 (1965).
168. Southwick, A. K., and Southwick, E. E., Aging effect on nectar production in two clones of *Asclepia syriaca, Oecologia* **56,** 121–125 (1983).
169. Southwick, E. E., Lucky hit nectar rewards and energetics of plant and pollinators, *Comp. Physiol. Ecol.* **7,** 51–55 (1982).
170. Southwick, E. E., Nectar biology and nectar feeders of common milkweed, *Asclepias syriaca* L., *Bull. Torrey Bot. Club.* **110,** 324–334 (1983).
171. Southwick, E. E., Photosynthate allocation to floral nectar: A neglected energy investment, *Ecology* **65,** 1775–1779 (1984).
172. Southwick, E. E., Loper, G. M., and Sadwick, S. E., Nectar production, composition, energetics and pollinator attractiveness in spring flowers of western New York, *Am. J. Bot.* **68,** 994–1002 (1981).
173. Steiner, K. E., Passerine pollination of *Erythrina megistophylla* Diels (Fabaceae), *Ann. Missouri Bot. Gard.* **66,** 490–502 (1979).

174. Stephenson, A. G., Flower and fruit abortion: Proximate causes and ultimate functions, *Annu. Rev. Ecol. Syst.* **12,** 253–279 (1981).
175. Stephenson, A. G., When does outcrossing occur in a mass-flowering plant? *Evolution* **36,** 762–767 (1982).
176. Stephenson, A. G., The regulation of maternal investments in an indeterminate flowering plant *(Lotus corniculatus), Ecology* **65,** 113–121 (1984).
177. Stephenson, A. G., and Bertin, R. I., Male competition, female choice, and sexual selection in plants, in *Pollination Biology* (L. Real, ed.), Chapter 6. Academic Press, New York, 1983.
178. Teuber, L. R., and Barnes, D. K., Breeding alfalfa for increased nectar production, *Proc. IVth Int. Symp. Pollination Maryland Agric. Exp. Stn. Spec. Misc. Publ.* **1,** 109–116 (1978).
179. Teuber, L. R., and Barnes, D. K., Environmental and genetic influences on alfalfa nectar, *Crop. Sci.* **19,** 874–878 (1979).
180. Teuber, L. R., Albertsen, M. C., Barnes, D. K., and Heichel, G. H., Structure of floral nectaries of alfalfa (*Medicago sativa* L.) in relation to nectar production, *Am. J. Bot.* **67,** 433–439 (1980).
181. Teuber, L. R., Barnes, D. K., and Rincker, C. M., Effectiveness of selection for nectar volume, receptacle diameter, and seed yield characteristics in alfalfa, *Crop Sci.* **23,** 283–289 (1983).
182. Thompson, K., and Stewart, A. J. A., The measurement and meaning of reproductive effort in plants, *Am. Nat.* **117,** 205–211 (1981).
183. Thomson, J. D., Skewed flowering distributions and pollinator attraction, *Ecology* **61,** 572–579 (1980).
184. Thomson, J. D., and Plowright, R. C., Pollen carryover, nectar rewards, and pollinator behavior with special reference to *Diervilla lonicera, Oecologia* **46,** 68–74 (1980).
185. Thomson, J. D., Maddison, W. P., and Plowright, R. C., Behavior of bumble bee pollinators of *Aralia hispida* Vent. (Araliaceae), *Oecologia* **54,** 326–336 (1982).
186. Tinbergen, N., Impekoven, M., and Franck, D., An experiment on spacing-out as a defense against predation, *Behaviour* **28,** 307–320 (1967).
187. Travis, J., Breeding system, pollination, and pollinator limitation in a perennial herb, *Amianthium muscaetoxicum* Liliaceae, *Am. J. Bot.* **71,** 941–947 (1984).
188. Udovic, D., and Aker, C. L., Fruit abortion and the regulation of fruit number in *Yucca whipplei, Oecologia* **49,** 245–248 (1981).
189. Waddington, K. D., Flight patterns of foraging honeybees in relation to artificial flower density and distribution of nectar, *Oecologia* **44,** 199–204 (1980).
190. Waddington, K. D., Factors influencing pollen flow in bumblebee-pollinated *Delphinium virescens, Oikos* **37,** 153–159 (1981).
191. Waddington, K. D., Pollen flow and optimal outcrossing distance, *Am. Nat.* **122,** 147–151 (1983).
192. Waddington, K. D., Allen, T., and Heinrich, B., Floral preferences of bumblebees *(Bombus edwardsii)* in relation to intermittent versus continuous rewards, *Anim. Behav.* **29,** 779–784 (1981).
193. Walker, A. K., Barnes, D. K., and Furgala, B., Genetic and environmental effects on quantity and quality of alfalfa nectar, *Crop Sci.* **14,** 235–238 (1974).
194. Waser, N. M., Competition for hummingbird pollination and sequential flowering in two Colorado wildflowers, *Ecology* **59,** 934–944 (1978).
195. Waser, N. M., Pollinator availability as a determinant of flowering time in ocotillo *(Fouquieria splendens), Oecologia* **39,** 107–121 (1979).
196. Waser, N. M., The adaptive nature of floral traits: Ideas and evidence, in *Pollination Biology* (L. Real, ed.), Chapter 10. Academic Press, New York, 1983.
197. Waser, N. M., and Price, M. V., Optimal and actual outcrossing in plants, and the nature of plant-pollinator interaction, in *Handbook of Experimental Pollination Biology* (C. E. Jones and R. J. Little, eds.), Chapter 17. Van Nostrand Reinhold, New York, 1983.
198. Waser, N. M., and Price, M. V., Experimental studies of pollen carryover: Effects of floral variability in *Ipomopsis aggregata, Oecologia* **62,** 262–268 (1984).
199. Waser, N. M., and Real, L. A., Effective mutualism between sequentially flowering plant species, *Nature (London)* **281,** 670–672 (1979).
200. Weller, S. G., Pollen flow and fecundity in populations of *Lithospermum caroliniense, Am. J. Bot.* **67,** 1334–1341 (1980).
201. Willson, M. F., and Burley, N., *Mate Choice in Plants: Tactics, Mechanisms, and Consequences.* Princeton Univ. Press, Princeton, New Jersey, 1983.
202. Willson, M. F., and Price, P. W., The evolution of inflorescence size in *Asclepias* (Asclepiadaceae), *Evolution* **31,** 495–511 (1977).

203. Willson, M. F., and Price, P. W., Resource limitation of fruit and seed production in some *Asclepias* species, *Can. J. Bot.* **58,** 2229–2233 (1980).
204. Wolin, C. L., Galen, C., and Watkins, L., The breeding system and aspects of pollination effectiveness in *Oenothera speciosa* (Onagraceae), *Southwest. Nat.* **29,** 15–20 (1984).
205. Woodell, S. R. J., Mattsson, O., and Philipp, M., A study in the seasonal reproductive and morphological variation in five Danish populations of *Armeria maritima, Bot. Tidsskr.* **72,** 15–30 (1977).
206. Zimmerman, M., An analysis of the reproductive strategies of *Polemonium* in Colorado. Ph.D. Dissertation, Washington University, St. Louis, 1979.
207. Zimmerman, M., Reproduction in *Polemonium:* Competition for pollinators, *Ecology* **61,** 497–501 (1980).
208. Zimmerman, M., Reproduction in *Polemonium:* Pre-dispersal seed predation, *Ecology* **61,** 502–506 (1980.)
209. Zimmerman, M., Nectar dispersion patterns in a population of *Impatiens capensis, Virginia J. Sci.* **32,** 150–152 (1981).
210. Zimmerman, M., Patchiness in the dispersion of nectar resources: Probable causes, *Oecologia* **49,** 154–157 (1981).
211. Zimmerman, M., Optimal foraging, plant density and the marginal value theorem, *Oecologia* **49,** 148–153 (1981).
212. Zimmerman, M., The effect of nectar production on neighborhood size, *Oecologia* **52,** 104–108 (1982).
213. Zimmerman, M., Optimal foraging: Random movement by pollen collecting bumblebees, *Oecologia* **53,** 394–398 (1982).
214. Zimmerman, M., Calculating nectar production rates: Residual nectar and optimal foraging, *Oecologia* **58,** 258–259 (1983).
215. Zimmerman, M., Plant reproduction and optimal foraging: Experimental nectar manipulations in *Delphinium nelsonii, Oikos* **41,** 57–63 (1983).
216. Zimmerman, M., Reproduction in *Polemonium:* A five year study of seed production and implications for competition for pollinator service, *Oikos* **42,** 225–228 (1984).
217. Zimmerman, M., Reproduction in *Polemonium:* Factors influencing outbreeding potential, *Oecologia* **72,** 624–632 (1987).
218. Zimmerman, M., and Cibula, D. A., Optimal foraging in bumblebees: The effect of memory on departure decisions, published as Cibula and Zimmerman, *Am. Midl. Nat.* **117,** 386–394 (1987).
219. Zimmerman, M., and Cook, S., Pollinator foraging, experimental nectar robbing and plant fitness in *Impatiens capensis, Am. Midl. Nat.* **113,** 84–91 (1985).
220. Zimmerman, M., and Gross, R. S., The relationship between flowering phenology and seed set in an herbaceous perennial plant, *Polemonium foliosissimum* Gray, *Am. Midl. Nat.* **111,** 185–191 (1984).
221. Zimmerman, M., and Pyke, G. H., Reproduction in *Polemonium:* Assessing the factors limiting seed set, *Am. Nat.,* in press.
222. Zimmerman, M., and Pyke, G. H., Reproduction in *Polemonium:* Patterns and implications of floral nectar production and standing crops, *Am. J. Bot.* **73,** 1405–1415 (1986).

9

Patterns of Fruit and Seed Production

THOMAS D. LEE

Ecological and evolutionary studies of plant reproduction have traditionally emphasized pollination and seed dispersal, while the development and maturation of fruits—and seeds within fruits—have received considerably less attention. In the past decade, however, a number of authors[63,88,89,153,187] have emphasized the importance of patterns of fruit and seed maturation and abortion. These patterns influence the size and quality of the seed crop and are thus intimately related to plant fitness.[153,155]

On any given plant, the number of ovules becoming seeds may be limited by (1) the number of ovules produced; (2) the quantity and quality of pollen transferred (pollen limitation); (3) the amount of nutrients and photosynthate available for allocation to fruits and seeds (resource limitation); (4) herbivores, predators, and disease; and (5) agents of the physical environment.[153] Clearly, this classification is oversimplified; distinguishing one limiting factor from another is often difficult. The distinction between pollen and resource limitation, for example, is often vague because these factors are interrelated in a complex way.[11,12,15,194] While all of the factors listed above probably limit fruit and seed production at certain times, it is clear that in many species from various habitats fruit and seed production are typically resource limited.[148,153,187]

This chapter focuses on plants in which fruit and seed production is resource limited. Assuming that such plants initiate a greater number of fruits and seeds than they can mature, this chapter addresses the question, What *proximate* factors determine which fruits and seeds mature and which do not? As reproductive abortion occurs at two levels, whole fruits and individual seeds within fruits, patterns of maturation and abortion at these two levels are described here in separate sections, fully recognizing that fruit and seed abortion are often related. Before describing the patterns, some physiological aspects of fruit and seed development are briefly reviewed and, in the final section of the chapter, some evolutionary implications of the patterns are discussed.

PHYSIOLOGY OF FRUIT AND SEED PRODUCTION

The physiological basis of fruit and seed maturation and abortion is not fully understood though the results of most studies conform to either of two general models. In the first model, fruits and seeds are "sinks" for resources, competing with one another as well as with vegetative sinks for limited photosynthate, nutrients, and water provided by "sources" (leaves, roots, etc.)[19,30,137] In the second model, phytohormones produced by a seed or fruit inhibit the growth and development of neighboring reproductive units.[36,59,172]

Central to the first or "source-sink" model is the concept of "sink strength," which is the ability of a fruit or seed to locally remove water and solutes from the phloem.[165] The strength of a sink appears to be determined by its metabolic activity, which in turn is related to the production of phytohormones, primarily (but not exclusively) by embryos and endosperms.[48,81] Developing seeds produce high levels of auxin, and removal of developing seeds from fruits frequently terminates fruit growth. Conversely, externally applied auxins can substitute for the presence of seeds in stimulating fruit development.[122] While the mode of action of auxin in developing fruits is still unknown, there is evidence that it causes significant increases in invertase, which hydrolyzes sucrose to hexoses,[128] thus promoting further sucrose import.[165] Auxins are not the only phytohormones associated with fruit growth. Gibberellins are probably produced by seeds and may be required for fruit development in some species.[126] Natural cytokinins, which are produced in the roots and translocated to shoots, are known to retard senescence and to stimulate sink strength.[121,158] Fruits have a strong affinity for cytokinins.[56] Exogenously applied gibberellins and synthetic cytokinins promote fruit initiation, growth, and even parthenocarpy.[30,87,126,158]

While phytohormone production (or uptake) apparently plays a significant role in determining the competitive ability of fruits and seeds, morphological constraints may also be important. Such constraints include the degree of development of vascular tissue supporting the fruit or seed[171] or the location of the fruit or seed in relation to other reproductive sinks and to the source of photosynthate and nutrients (see the next section of this chapter).

The source-sink model asserts that resources are limiting and fruits and seeds that are weaker sinks become "starved" for resources. In some studies, aborting fruits have been shown to contain lower levels of carbohydrates and have lower sink strength than nonaborting fruits, observations which are consistent with the model.[53,137]

The second model of fruit maturation and abortion, which involves chemical inhibition of aborting fruits and seeds by those that mature, is supported by several lines of evidence. First, studies on *Pyrus malus*[18] and soybean *(Glycine max)*[46] show no difference in soluble carbohydrates or protein between aborting and maturing fruits, suggesting that resource shortages may not directly cause abortion in these species. Second, extracts from young, developing fruits applied to inflorescences induce abortion of distal fruits in both *Lupinus luteus*[36,172] and *Glycine max.*[59] Data suggest that the inhibitor may be abscisic acid (ABA)[36] or auxin.[59] Third, in *Phaseolus vulgaris,* the presence of older, basal fruits in an inflorescence inhibits the growth of younger, distal fruits; the latter have high concentrations of ABA,[161] which promotes abscission.[3] Removal of older fruits results in a 50% reduction of the ABA content of younger fruits, suggesting direct inhibition of younger by older fruits.[161] However, it is possible that elevated ABA levels are induced by resource shortages.

In relating these two physiological models to what follows, two points should be kept in mind. First, our knowledge of the physiological basis of fruit maturation is incomplete; several of the patterns described later in this chapter have not been studied from a physiological perspective. Second, as Sage[137] has noted, the two models described above are not mutually exclusive; nutritional factors and inhibitory phytohormones may interact.

PATTERNS OF FRUIT PRODUCTION

Time of Initiation and Position

Perhaps the two most widely reported variables related to fruit maturation are time of fruit initiation and position of the fruit on the plant or in the inflorescence. In many species, early-formed fruits or those located closest to the sources of nutrients and photosynthate are more likely to mature than others.[153] As Stephenson[153] has observed, inflorescences often produce flowers acropetally, so that time of flowering (and presumably time of fruit initiation) is perfectly correlated with position in the inflorescence. It is thus difficult to determine which—position in the inflorescence or time of initiation—is more important.

The effect of timing or position may be seen at the level of the whole plant[52,57,75,77,91,118,142,162] as well as between and within inflorescences. Later-flowering inflorescences have lower fruit production in *Smyrnium olusatrum*[90] and, in *Asclepias tuberosa,* umbels closer to the main stem have a higher probability of setting fruit.[190,192] Within inflorescences, early-formed, proximal (basal) fruits have the highest probability of maturation[13,54,97,105,127,137,151,153,161,168,171] and, in several species, removal of early flowers or early-formed fruits demonstrates that later-formed, normally abortive fruits are viable and capable of maturing.[57,59,75,76,133,152,161,162,171,181] Inflorescences of *Arisaema triphyllum* show no decline in total fruit production toward the apex, although the proportion of barren fruits is greater there.[93]

In a few cases, the effects of time of fruit initiation and fruit position have been separated. Bookman[20] showed that in inflorescences of *Asclepias speciosa* early-formed fruits have a higher probability of maturation, apparently independent of fruit position. There is also evidence that position alone can play a role. By allowing only two whorls of flowers to form on inflorescences of *Lupinus luteus* and varying the location of the whorls, Van Steveninck[171] showed that distal whorls produce fruits with much slower growth rates than basal whorls. He noted that, prior to fruit development, vascular elements were almost completely absent at the top of the inflorescence, but well developed at the base of the inflorescence and suggested that resources may not have been adequately transported to distal flowers and fruits. Abortion of distal fruits in *Lupinus luteus* may thus be due to a morphogenetic and developmental contraint,[179] although it may be unwise to equate development of vascular tissue with rates of assimilate transport (T. L. Sage, personal communication). Such a constraint appears to be absent in *Catalpa speciosa,* where time of initiation and position in an inflorescence are unrelated to fruit production when only one flower per inflorescence is pollinated.[152]

While time of fruit initiation and position are often related to fruit production, the relationship is rarely perfect. A greater proportion of early-formed than late-formed fruits may mature, but rarely do *all* early-formed fruits mature and *all* late-formed fruits abort.[5,77,90,91,92,118,171,180] In some cases, the probability of maturation does not change with time of initiation or position on the plant or inflorescence.[28]

The lack of a perfect relationship between time of initiation (or position) and fruit maturation may be due, at least in part, to the tendency of fruits initiated synchronously to compete more intensely than those initiated at different times. Within umbels of *Asclepias,* for example, probability of fruit maturation increases with longer delays between pollinations, suggesting that interovary competition is less when pollination is asynchronous.[20,190,192] The same phenomenon may account for the tendency of *Cucurbita maxima* flowers that initiate fruits to alternate with flowers that abort.[26] These data, however, contrast with those from *Lupinus* spp., where the inhibition of later flowers by early ones becomes more marked as the differential in their stages of development increases.[129] Factors other than competition between synchronously formed fruits can cause nonsequential fruit maturation; these are discussed in the next three sections of this chapter.

What is the physiological basis for differential abortion of late-formed, distal fruits? There are at least four possible explanations. First, early-formed, proximal fruits may produce a growth inhibitor that induces abortion in late-formed, distal fruits[36,59,133,137,161,171] (see also the section Physiology of Fruit and Seed Production). Second, the temporal advantage of these fruits may allow them to generate more growth-stimulating phytohormones and to thus become more vigorous sinks, preempting resources that would otherwise be used by late-formed fruits[19,30,60,122,137] (see Physiology of Fruit and Seed Production). The third explanation is that proximal fruits are in a position to preempt resources, as photosynthate, nutrients, and water must pass them to reach distal fruits.[153,190] Fourth, distal fruits may abort due to a poorly developed vascular system in the distal portion of the inflorescence, as discussed above for *Lupinus luteus.*[171] The latter three explanations all suggest that late-formed distal fruits are "starved" for resources.

Another possible and as yet unexamined cause of differential abortion of distal fruits is pollen source. In a number of species, pollinators move upward on inflorescences[192,193] and thus basal fruits may be more likely than distal ones to receive cross-pollen. Differential success of cross-fruits (see below, Pollen Source) would thus lead to the greater retention of basal fruits. However, distal fruits are often aborted in self-pollinating species,[75,80,118] suggesting that pollen deposition patterns merely reinforce the influence of time of fruit initiation and fruit position.

Pollination Intensity and Seed Number

Two somewhat related factors often associated with fruit development and maturation are pollination intensity (the number of pollen grains deposited on the stigma) and the number of seeds per fruit. Both of these factors are known to vary among flowers or fruits in natural plant populations,[47,102,103,147,148,156] indicating at least a potential for them to influence fruit maturation patterns in nature.

Pollination and subsequent growth of pollen tubes can stimulate fruit growth and development. Swelling of ovaries between pollination and fertilization has been ascribed to pollination,[48,123] and in some agamospermous species pollination is required for fruit development even though fertilization of the egg does not occur.[150] It is the large quantity of auxin and gibberellin produced during pollen tube growth that initially stimulates fruit growth.[48,94,95,123]

Pollination intensity is usually positively correlated with the number of seeds per fruit.[4,77,102,146,147,173] This relationship may be exponential rather than linear as pollen germination and pollen tube growth may both be positively density dependent.[23,39,147] As pollination intensity and seed number are usually correlated, and as seeds are important sources of phytohormones,[48,123] a positive relationship between pollination intensity and fruit maturation may be the result of phytohormone production by either pollen tubes or seeds, or both. It is difficult to separate the effects of these two phytohormone sources and thus most of the patterns described below could be due to either or both. Apparently each can be important; while some fruits will develop and mature after pollination without fertilization,[48,81] it is also true that the removal or destruction of seeds halts fruit development in some species.[122,167]

In many species, a minimum number of pollen grains must be deposited on a stigma to produce a fruit and this may correspond to some minimum number of seeds formed.[4,12,102,103,147] Thresholds can be high—roughly 400 compatible grains are required for fruit production in *Campsis radicans*[12]—though in other species pollination with just a few grains can produce a fruit.[42] The threshold may change with environment; in *Cassia fasciculata* low densities of pollen initiate fruits under short day cycles but not under long day cycles.[79]

Fruits initiated with light pollen loads (and usually containing relatively few seeds) frequently have slower growth rates or lower probabilities of maturation than those pollinated with heavy loads.[4,12,41,66,77,79,103,157,173] The pollen density–fruit growth relationship in *Lycopersicon esculentum* holds whether intraspecific or interspecific pollen is used; with the latter (from *Lycopersicon peruvianum*) all initiated seeds abort early in development yet mature fruits are produced.[173] Self-incompatible species may show little or no relationship between pollen density on the stigma and probability of fruit maturation, as the proportion of self-pollen on the stigma may vary widely from flower to flower.

There are many reports of a positive relationship between seed number and fruit growth or probability of fruit maturation.[1,13,21,22,42,49,54,76,77,99,115,132,136,156,157] This relationship does not necessarily indicate an association of fruit maturation with pollination intensity as seed number may be determined by other factors, such as number of ovules per ovary, seed abortion due to lethal genes, seed abortion due to competition for maternal resources, or pollen source.

Experimental studies involving flower thinning provide strong evidence for differential abortion of fruits in which few viable seeds were formed. In *Pyrus malus* (apple) and the herbaceous legume *Lotus corniculatus,* random thinning of flowers results in mature fruits that contain fewer seeds than fruits from unthinned controls, indicating that few-seeded fruits are normally differentially aborted.[54,99,115,132,156] In these two species, few-seeded fruits are probably not the result of seed abortion induced by resource limitation. This statement can be made because flower thinning is performed before or just after fertilization; thus, if flowers had been well-pollinated with compatible pollen we would expect *more* seeds per mature fruit due to a reduction in competition-induced seed abortion[28,75,78] (see Patterns of Seed Production). Few-seeded fruits in apple and *Lotus* are apparently the result of self-pollination or inadequate pollination.[115,156]

Stephenson[153] has noted that the threshold seed number for fruit abortion can vary. In apples, few-seeded fruits mature in years when there is little competition among fruits, but abort when numerous many-seeded fruits are initiated.[132] In broad bean *(Vicia faba),* poor fertilization results in a high frequency of one-seeded fruits that are aborted only under stress.[136] Photoperiod appears to control the threshold in *Cassia fasciculata,* where few-seeded fruits, initiated with dilute pollen, grow more slowly and are differentially aborted under long days, while all fruits, regardless of seed number, grow rapidly and mature on plants grown under short day cycles.[79]

There is good evidence that pollen source influences both seed number per fruit and the probability of fruit maturation (see Pollen Source, and Bertin,[17] this volume). Fruits resulting from self-pollination often contain fewer seeds and have a lower probability of maturation than those initiated by crossing.[115,156] Bertin[13] showed that, in *Campsis radicans,* fruits from certain crosses are more likely to mature than others and that mature fruits from these crosses contain more seeds than those produced by less successful donors. In Bertin's study, seed number was rendered independent of pollination intensity as stigmas were hand-pollinated with abundant pollen.

Seed number per fruit may also be influenced by seed abortion during fruit development (see Patterns of Seed Production). In *Ribes nigrum,* aborted fruits have a lower average seed content than those that mature. However, lower seed number is not due to lack of fertilization, but to seed abortion that is more frequent in fruits toward the apex of the raceme, where the time of initiation is later and resources are assumed to be less available (Teaotia,[164] cited in Wright[189]). Similar patterns probably occur in other species (see Time of Initiation and Position). In these cases it is not clear which factor is mainly responsible for fruit abortion—seed number, time of initiation, or position—since all three are highly correlated. It is possible that reduced seed number is not the immediate cause of fruit abortion, and that seed abortion and eventually fruit abortion are *both* consequences of resource shortages associated with time of initiation and position in the inflorescence. Under resource limitation, some seeds within a fruit may abort early in development such that they are visually indistinguishable from unfertilized ovules while the remainder continue to grow until the whole fruit aborts, giving the impression that the fruit had, from the beginning, fewer developing seeds.

Comparison of seed numbers in aborting and mature fruits, then, may be misleading or simply uninformative unless various developmental stages of ovules and seeds can be precisely identified, a task which is often difficult.[117,120] Consequently, assertions about the importance of pollination intensity or pollen source in fruit development that are based on such comparisons[21,22,54,77] should be viewed with caution.

The relationship between seed number and fruit maturation may be complicated by interaction with other variables. In apple, for example, few-seeded fruits are typically aborted but on the same tree one can find some few-seeded fruits that matured and some many-seeded fruits that aborted.[54] Heinicke[54] argued that whether or not a particular fruit matures is influenced by the vigor of the "spur" (flowering branch) to which the fruit is attached. Vigorous spurs, which are heavier and have more leaves, tend to abort fewer fruits and retain some few-seeded fruits; weak spurs may abort all few-seeded fruits and some many-seeded fruits. Such intraplant variation in fruit abortion is a function of the physiological independence of reproductive modules (e.g., spurs in this case), a phenomenon whose significance has been emphasized by Watson and Casper.[179]

While a positive relationship between pollination intensity or seed number and fruit maturation seems to be widespread, it is apparently not universal. Research on *Vicia faba* by Rowland and Bond[135] indicates that few-seeded fruits are selectively aborted but Stoddard,[159] working with different varieties of the same species, reports no evidence for such a relationship.

Physiologically, the simplest explanation for the selective maturation of many-seeded fruits is that more seeds produce more growth substances and hence the fruit becomes a stronger sink for resources[77,153] or a stronger inhibitor of nearby fruits. However, when seed number is a function of pollination intensity, it is also possible that individual seeds in many-seeded fruits are stronger sinks because they were initiated under competition among male gametophytes. Male gametophyte competition has been shown to result in more vigorous offspring.[73,112] (See also Bertin,[17] this volume.)

Pollen Source

The source of pollen often determines whether or not a flower will produce fruit. The success of a particular pollen donor may be determined by prezygotic incompatibility mechanisms, postzygotic incompatibility mechanisms[44] (see Barrett,[8] this volume) or by differential abortion of fruits containing viable seeds[153] (referred to here as facultative abortion). A number of researchers have suggested or implied that these categories represent a continuum and are not discrete mechanisms.[13,113,155] Given the objective of this review, focus will be placed here on postzygotic incompatibility and facultative abortion.

In typical postzygotic incompatibility, fruit or seed mortality occurs with certainty, probably due to expression of lethal genes in embryo or endosperm.[37] In contrast, in facultative abortion all fruits contain genetically viable seeds and are capable of maturing but, due to interfruit competition, fruits initiated by pollen from particular sources tend to be weak sinks and abort. Such fruits do mature when interfruit competition is eliminated.

Postzygotic self-incompatibility has been reported for several species,[13,21,37,84,149] though the possibility of facultative abortion has rarely been experimentally excluded in these studies. Self-incompatibilities are sometimes only partial, with at least a small amount of fruit production occurring after selfing,[70,169] and in some species both self- and cross-fruits have relatively high probabilities of maturation but those for cross-fruits are higher.[5,54,64,114,175] Since fruit production is at least possible in these cases of partial incompatibility, some failure of self-fruits may be due to their inability to compete with cross-fruits. The role of such facultative abortion can be tested by manipulating the number of competing cross-fruits in selected inflorescences and following fruit maturation of self-fruits. In *Asclepias speciosa,* at least, failure of self-fruits is not due to facultative abortion as only 0.4% of self-fruit matured when initiation of competing fruits was prevented.[21]

Darwin[40] implied that facultative abortion of self-fruits occurs, and Stephenson[153] provided several examples.[26,51,169] It may also occur in apple, where self-fruits abort under normal soil conditions but set heavily when trees were provided with supplemental nitrogen.[55] There may be a similar explanation for the data in pear, where "the degree of self-sterility varies from year to year and in different trees of the same variety under different cultural treatments and in different localities" (Heinicke,[54] citing Waite[174]).

Some authors have suggested that there may be differential maturation of fruits initiated with pollen from different donors.[31,63,155,186] This occurs in *Pyrus malus,*[114] *Campsis radicans,*[13] *Asclepias speciosa,*[21] and *Raphanus sativus.*[96] In the *Campsis* study, Bertin showed that a pollen recipient differentially matures fruits initiated by certain donors, and that donors that are more successful in fathering fruits on one recipient are less so on others. When a *Campsis* donor is successful in producing fruits on a particular recipient the same is true when their sexual roles are reversed, suggesting that certain gene combinations result in more competitive fruits.[13] The situation is different in *Asclepias speciosa*[21] and *Raphanus sativus,*[96] where certain pollen donors are more successful than others over the whole range of recipients, indicating some strong paternal influences on fruit production.

Donor-related fruit abortion is at least partly facultative. An *Asclepias speciosa* donor's success in fathering fruit depends on whether or not its fruits are competing with fruits initiated by another donor.[21] Interfruit competition within inflorescences also seems to play a role in *Campsis radicans* fruit production; donors that initiate successful fruits in previously unpollinated inflorescences are less likely to do so when the inflorescences already contain developing fruits.[15] While the *Campsis* data strongly suggest facultative abortion, some fruit abortion in this species may be obligate, representing pre- or postzygotic incompatibility.[13,15]

Not all species tested have shown differential maturation based on pollen source. In several species, crosses with near and far neighbors or with different varieties of the same crop species have no effect on the probability of fruit production.[11,77,118]

Janzen[63] suggested that natural selection should favor maternal plants that obtain some optimal number of donors to contribute to their seed crop and there is some evidence for this. In inflorescences of the orchid *Encyclia cordigera,* fruit production is greater when each flower is crossed with a different donor than when all flowers are crossed with the same donor or are selfed.[64] (Only the differences between selfed and multiple donor inflorescences are statistically significant.) Studies on four species, *Tabebuia rosea,*[11] *Caesalpinia eriostachys,*[11] *Campsis radicans,*[16] and *Cassia fasciculata,*[77] reveal no differences in the probability of fruit maturation when individual stigmas receive either pollen from one donor or a mixture of pollen from several donors. In *Raphanus sativus,* mixed pollinations result in more seed weight per fruit, but have no significant effect on the probability of fruit maturation.[96]

While number of donors does not seem to affect fruit production to any great extent, the identity of the donors in the pollen mixture does. In *Campsis radicans* the probability of fruit maturation for a multiple donor pollination is lower than when the most successful donor in the mixture is used alone.[16]

The physiological explanation for the effect of pollen donor on fruit production is unknown but there are several hypotheses. First, under self-pollination or pollination by "unfavored" donors, fewer pollen grains may germinate and fewer pollen tubes may reach ovules, resulting in fewer seeds being formed. Fewer pollen tubes and seeds would mean less induction of fruit growth (see above, Pollination Intensity and Seed Number) and poorer competitive ability on the part of the fruits. Second, differences in seed numbers among fruits with different parentage may result from differential seed abortion. Bertin[13] suggested that seeds from unfavored donors may have more homozygous recessive lethal or deleterious alleles, resulting in more embryo abortion,[37,83] fewer developing seeds per fruit, and ultimately more fruit abortion. Lower numbers of seeds are sometimes found in self-fruits[115,156] or fruits initiated by unfavored donors[13] and these observations lend credence to these first two hypotheses.

Third, even if all initiated fruits have similar numbers of developing seeds, the seeds in fruits initiated by unfavored donors may be less competitive due to the presence of deleterious alleles or allele combinations, or overall inferiority of paternal genes. Vigor of certain seeds and fruits may be due to heterosis and again, the *Campsis* data are consistent with this hypothesis. Bookman,[21] however, noted that heterosis is not a likely explanation for differential fruit production in *Asclepias speciosa,* as there is no correlation between relative pod production in reciprocal crosses. In *Asclepias speciosa,* pollen donor success and the number of seeds per fruit are not significantly correlated, but Bookman did find a strong correlation between donor success and pollen tube growth on artificial medium, indicating paternal effects independent of seed number per fruit.[21]

Most authors assume a genetic basis for donor-related, differential fruit production, but it is possible that pollen quality is conditioned by the environment of the donor individual. A genetic influence is very likely in *Campsis radicans,* however, as donor success is consistent from year to year.[15]

Damage

Stephenson[153] thoroughly reviewed the literature and found that damage to fruits caused by temperature extremes, seed predation by insects, or wounding of the pericarp by insects can induce fruit abortion. Fruit abortion in *Sesbania spp.* increases as the number of experimentally damaged ovules per fruit increases,[97] and even slight damage can induce abortion of *Asclepias exaltata* fruits.[131] There is evidence that fungal infection of fruits can also induce fruit abortion.[67]

PATTERNS OF SEED PRODUCTION

Resource Limitation

While there is good evidence that fruit abortion often results from limited resources, the role of resources in within-fruit seed abortion has received less study. One probable reason for the lack of research in this area is the difficulty involved in assessing and quantifying abortion.[27,78,120] Evidence for seed abortion due to limited resources comes from studies of yield component compensation, in which a reduction in one component of yield, say fruit number, results in an increase in seed number per fruit.[2] One explanation for this pattern is that when fewer fruits are initiated there is less competition among reproductive sinks and thus there is less seed abortion per fruit. An alternative explanation is that ovule number in new flowers increases when the number of developing fruits is reduced.[97] Experimental manipulations provide more direct evidence. Reduction of the number of fruits in a developing fruit crop sometimes reduces seed abortion in the remaining fruits,[75,78] and Casper has shown that reduction of competition among the four ovules of *Cryptantha flava* flowers (achieved through experimental ovule destruction) results in increased probability of maturation for the remaining ovules.[28] Supplementation of nutrients and water to greenhouse-grown individuals of *Cryptantha flava* also reduces seed abortion compared to field-grown plants.[28] Experimentally applied drought and defoliation treatments may induce seed abortion.[62,78,98,134] Reduction of competition between vegetative and reproductive sinks, caused by application of chemical inhibitors to the shoot apex, results in less seed abortion per pod in *Vicia faba.*[30] Successful culture of normally abortive embryos[117,119,166] also suggests that abortion can result from resource limitation. Frequently, however, resource addition or fruit thinning do not increase seed number per fruit.[28,91,97,101,182] These latter studies do not necessarily suggest that aborted ovules are inviable, as other components of yield (e.g., fruit production) may have increased in response to greater resource availability.

Time of Initiation and Position

The time of initiation of a seed, or its location in a fruit or on an inflorescence or plant, may influence its probability of abortion. As is the case with whole fruits, time of initiation and spatial location of seeds are often positively correlated (though there are some notable exceptions in within-fruit patterns); thus, separating these two factors can be difficult.

At the level of the whole inflorescence, seed production is often greatest in early-initiated fruits or those closest to the source of nutrients and photosynthate.[9,32,43,93,141,145,164,170,189,191] It is usually uncertain, however, whether this pattern is due to greater fertilization success in the proximal portion of the inflorescence or to less seed abortion there[93] (see Zimmerman,[194] this volume). It is also quite possible that seed production per fruit is limited by the number of ovules per ovary and that this number is lower in the distal portion of an inflorescence, perhaps due to lower resource levels there. Delph[43] demonstrated experimentally that reduced seed production in late-formed fruits of *Lesquerella gordonii* is due to resource limitation and not inadequate pollination. However, the roles of seed abortion and ovule number were not distinguished in her study. In *Cryptantha flava,* where fruits may contain one or sometimes two seeds, two-seeded fruits tend to be found closer to the main stem where resources are perhaps more available.[28] All *Cryptantha flava* flowers contain four ovules and typically almost all are fertilized[27,28,29] so the relationship between seed production and position is clearly a result of differential abortion. Pechan and Morgan[125] suggest that greater seed abortion in distal fruits of *Brassica napus* is due to reduced assimilate supplies there or to an inhibitory concentration of growth substances produced in and transported from the older pods. Consistent with their hypothesis is the observation that removal of proximal fruits in *Brassica napus* sometimes results in distal fruits containing more seeds.[163] Some species show no relationship between seed abortion and position in an inflorescence; in *Lupinus luteus* there is no clear trend toward a greater proportion of aborted seeds in the distal portion of an inflorescence[171] and in *Phaseolus vulgaris,* fruits from early and late cohorts of flowers had similar numbers of seeds.[91,118]

Within fruits there is a diversity of patterns of seed abortion. In many of the Leguminosae, seeds at the fruit base (proximal to the pedicel) have a higher probability of abortion than do those at the distal end;[10,11,35,69,78,117,119,159,180] however, other species show the reverse pattern. In five Neotropical legumes, Wyatt[191] found a high frequency of seed maturation at the bases of fruits, and stylar seeds aborted more frequently in two crucifers.[106,125] Still other species have highest maturation at positions toward the middle of the fruit. Horowitz et al.[58] report highest seed set at the second ovule position from the base in *Lupinus nanus* and one cultivar of *Medicago sativa,* with set declining toward the distal and basal ends. In *Vicia faba,* the second position from the distal end has lower seed set[159] and, in *Pisum sativum,* seeds at both ends of the fruit abort more frequently than those in the center.[85]

Superimposed on the pattern of within-fruit seed abortion is the tendency for alternate seeds to abort. In *Medicago sativa,* seed production is higher in even-numbered ovule positions (numbered from fruit base); these ovules are connected to the vascular bundle in the upper pod valve.[58] Similar patterns occur in other *Medicago* species,[58] and Bawa and Buckley[10] report an alternation of very high and very low abortion probabilities within fruits of *Bauhinia ungulata, Cassia biflora,* and *Cassia emarginata.*

Several hypotheses have been advanced to explain these within-fruit position effects. Watson and Casper[179] suggest that within-fruit patterns of seed abortion may be interpreted in light of vasculature and translocation of resources. The greater abortion of stylar seeds in certain species[106,125,191] may result from preemption of resources by basal seeds that are closer to the source of nutrients and assimilates.[78,117,119] Vasculature and assimilate distribution may be involved in abortion of alternate seeds in legume fruits, as alternate ovules in legumes are associated with the vascular system of one of the two fruit valves.[110] It has also been suggested that alternate abortion is due to greater interference among neighboring seeds,[10] but, given that alternate seeds are associated with different vascular bundles, it is unlikely that such interference is caused by resource competition. Little is known of assimilate distribution and fruit vasculature in most species and further research in this area should be rewarding.

A number of authors have suggested that time of fertilization and rate of pollen tube growth influence within-fruit abortion patterns.[11,73,78,117,119,180] Ovules fertilized first may have a temporal advantage in competing for resources[69,109,160,176,185] and may even prevent the fertilization of other ovules.[109] In many species fertilization is basipetal, i.e., from stylar to basal end of the ovary.[7,35,68,78,136] Thus, seeds in the stylar portion of the fruit would have a temporal advantage in growth and hence would be better competitors for resources. An alternative explanation involves gametophyte competition, with the assertion being that rapid pollen tube growth results from either a superior pollen genotype[112] or a more vigorous combination of pollen and maternal genomes,[113] and that seeds initiated by fast-growing pollen tubes are thus genetically superior competitors for maternal resources[11,73] (see Pollination Intensity and Seed Number). Consequently, as faster-growing pollen tubes tend to fertilize stylar ovules, abortion of developing seeds in stylar positions should be less frequent than at the fruit base.

Based on the timing and gametophyte competition hypotheses, we would expect abortion patterns to reflect the pattern of fertilization, even when it is not basipetal. This is apparently the case in *Lupinus nanus,* where high rates of abortion at the stylar end correlate with the later occurrence of fertilization there.[58,65] However, in *Medicago sativa* abortion is frequent at the stylar end[58] but fertilization is apparently basipetal.[35]

As time of fertilization and pollen tube growth rate are likely to be highly correlated, it is difficult to separate the effects of these two factors; either or both would account for the differential abortion of basal seeds that occurs in so many legumes. However, some data indicate a role for male gametophyte competition. When mixtures of pollen from different donors are used, paternal genes occur with different frequencies at different ovule positions in the ovary,[58,96] indicating some genetic "sorting" in relation to ovule position and the potential for position-related abortion to be genetically selective. Meinke[106,107] attributes greater seed abortion in stylar halves of *Arabidopsis thaliana* fruits to greater expression of embryo-lethal genes there and he suggests that this position effect results from gene expression during pollen tube growth. Pollen bearing the lethal genes apparently grow at a different rate than normal pollen. In *Phaseolus vulgaris,* the disappearance of a position effect as crosses become more outbred suggests a genetic influence on abortion, rather than just a temporal one.[117,119]

The differential maturation of seeds toward the middle of fruits as in *Pisum sativum,*[85] *Vicia faba,*[159] and other legumes may be due to an interaction between time of fertilization and proximity to resources.[72] Seeds in the middle of the fruit may be initiated early enough to be better sinks for resources than basal seeds but may also have greater access to resources than distal seeds; hence they are least likely to abort. It is also possible that the mechanical constraints of pod taper prevent the necessary enlargement of developing seeds at the ends of the fruit. This is suggested by the lower amount of proximal and distal seed abortion in blunt-podded varieties of *Pisum sativum.*[124]

Within-fruit patterns of seed abortion may change over the course of fruit development. In four tropical legumes seed abortion was classified as either early or late (it was assumed that most ovules were fertilized), and in each species early abortion shows a different pattern from late abortion.[10,11] In *Bauhinia ungulata,* for example, early abortion is highest at the proximal end and lowest in the middle of fruits; late abortion is highest at the distal end.[11,180] Bawa and Buckley[10] postulate that differences in the pattern of early and late abortion occur because genotypes most successful during pollen tube growth or in early embryo development may be less successful than other genotypes in drawing maternal resources later in development.

Within-fruit patterns of abortion do not appear to be fixed for a particular species. In *Phaseolus vulgaris* the tendency to abort basal seeds is reduced when cross-pollinations are more outbred.[119] In *Vicia faba,* irrigation of field-grown plants causes an increase in abortion at the stylar seed position and a dramatic decrease in basal positions.[159]

Pollen Source

The source of pollen is known to influence the number of seeds produced per fruit. Self-pollination often results in fewer mature seeds per fruit than cross-pollination,[45,82,115,142,143,156] though it is rarely known with certainty whether this reduction arises from lack of fertilization or postzygotic abortion (see also Barrett,[8] this volume). As pollen tube growth can be inhibited in self styles,[44,113,140] it is likely that at least some ovule abortion is due to failure of fertilization. However, as will be shown below, abortion of fertilized ovules does occur, and it is likely that both factors are involved in determining seed number per fruit.[35,140]

Greater embryo abortion occurs following selfing in both conifers[138,139,160] and angiosperms.[34,37,82,83,140] The effect of selfing, however, varies greatly among species. In *Phlox drummondii,* seeds initiated via selfing are only 17% less likely to mature,[83] though in *Medicago sativa* seeds from selfing are five times less likely to do so.[24] Levin[83] suggests that the relatively low inbreeding depression in *Phlox drummondii* may be related to the low levels of heterozygosity in this species, an argument that has been made for annual plants in general by Wiens.[184]

Cross-pollination with different individuals may also result in different amounts of abortion. As the spatial distance (and perhaps genetic dissimilarity[82,155,178,187]) between pollen donor and recipient increases, the probability of seed abortion after hand-pollination often decreases,[33,82,83] a pattern that may result from the tendency of near neighbors to be closely related and hence to demonstrate inbreeding depression when crossed.[33,130] In other species, however, there is an optimal distance for seed production, below *and* above which seed production declines.[6,82,130,178] In *Delphinium nelsoni* and *Ipomopsis aggregata,* experiments involving hand-pollinations show that seed production per fruit is maximal when parents are separated by 1–100 m[130,178] and, in two species of *Stylidium,* seed set per capsule is highest for interparental distances of 20–40 km.[6] In neither study was seed abortion distinguished from lack of fertilization, but in both cases hand-pollination likely resulted in more than adequate pollen transfer. In addition, aborted *Stylidium* ovules exhibited some development before failing[6] suggesting that they had been fertilized. It is important to note that not all studies of pollen donor effects show significant differences among donors in average seed production per fruit.[100,111]

The relationship between pollen source and seed production has led several researchers to suggest that abortion is related to overall genetic similarity of maternal and paternal genomes. Greater seed abortion resulting from mating of genetically very similar or very dissimilar individuals may be due to the expression of deleterious gene combinations[82,155,178,187] or to the presence of one or a few alleles in the genome of the pollen donor.[118] These alternatives await testing.

Wiens[184] observed that seed abortion is related to plant life history and probably mating system. In a survey of 191 species, he found that the ratio of *mature seeds per fruit* to *ovules per fruit* was significantly greater in annuals (85%) than for herbaceous perennials (57%) and that woody perennials had the lowest seed/ovule ratios (33%). Wiens argued that annuals, which are typically inbreeders, have low heterozygosity and thus little opportunity for interembryo selection. Perennials, especially woody perennials, tend to be outcrossers and have higher heterozygosity. Thus, ovule abortion in these plants might be a consequence of selection against homozygotes or lethal or semilethal allele combinations. Data from a sample of 14 tropical legumes are consistent with Wiens' hypothesis; inbreeders have higher seed/ovule ratios than selfers.[10] Breeding system is not only correlated with the amount of ovule abortion in a fruit but with the position of abortion as well. In perennial, outcrossing *Cryptantha* species "the location of successful ovules within the fruit is random, while in annual, autogamous species the position is always the same."[179]

While the discussion so far has emphasized the effects of single pollen donors, it is also possible that the number of seeds per fruit may be influenced by the number of pollen donors. Based on kin selection theory, Kress[71] has suggested that more abortion should occur when seeds in a fruit are half-sibs as opposed to full sibs. However, while there is slight evidence for this prediction in *Vaccinium corymbosum,*[170] mixtures of pollen from different donors do not affect seed number per fruit over single crosses in *Costus allenii,*[144] *Campsis radicans,*[16] or *Cryptantha* spp.[27]

Seed abortion in association with pollen source is often attributed to lethal alleles or allele combinations,[37,38,82,83,184] but it is usually unclear whether such alleles are inherently lethal or whether differential mortality occurs only when fruits and seeds are competing for limited parental resources.[78] Several authors have suggested that, in outbreeding species, seeds initiated by self-fertilization are simply inferior competitors for resources[25,86,138,139,160] and Bawa and co-workers have argued that in general seed maturation within fruits is "selective" and based on the quality of the pollen or embryo genotypes.[10,11] When seed abortion occurs in the absence of competition among seeds[108] it is likely due to lethal genes, but abortion that occurs when seeds and fruits are competing with each other or with vegetative sinks may be due to resource limitation. As noted earlier, ovule abortion is sometimes facultative[28,75,77] but whether or not seed abortion related specifically to pollen source is facultative awaits further investigation. Even if seed abortion *is* facultative, determination of *which* seeds abort may have a strong genetic basis. High levels of homozygosity due to inbreeding, or deleterious but not necessarily lethal gene combinations resulting from inbreeding or excessive (long-distance) outbreeding, respectively, may result in embryos or endosperms that are weak sinks for resources.[13,155] In *Raphanus sativus,* seeds resulting from different pollen donors differed in size, suggesting a link between paternity and vigor of seeds during development.[96]

EVOLUTIONARY ASPECTS

Many of the fruit and seed maturation patterns described herein enhance the fitness of the maternal plant, its offspring, or both. Consequently, these patterns may have arisen, or are at least maintained, by natural selection. Fitness effects involve (1) resource efficiency, (2) offspring quality, and (3) efficiency of seed dispersal.

Several of the patterns may enhance fitness because they result in efficient use of resources by the maternal plant. Nakamura[118] observed that "a maternal plant conserves resources if it invests in those reproductive organs that require the least amount of future parental investment to complete development." The widely reported differential abortion of late-formed fruits thus conserves resources and is functionally analogous to "brood reduction" in animals, in which the youngest juveniles in a brood have the lowest survivorship.[118,153] No one, however, has precisely quantified the resource savings resulting from differential abortion of late- versus early-formed fruits or seeds. The gain is likely to be minute in cases of within-fruit seed abortion; a seed initiated a few minutes or hours later than its siblings will certainly not differ appreciably in resource content from others in the same ovary. It is thus possible that differential abortion of late-initiated fruits and seeds is an *exaptation*[50] (a trait whose evolutionary origin was not due to the selective forces that presently maintain it) rather than an *adaptation* (a trait currently maintained by the same selective forces that caused it to evolve initially[50]). The pattern may have originated as a consequence of the fundamental laws of plant growth, i.e., stronger sinks outcompete weaker sinks and early-formed fruits tend to be stronger sinks.

Differential abortion of fruits and seeds based on position alone may also be advantageous in terms of resource efficiency. Selective abortion of distal fruits in an infructescence or distal seeds within a fruit may reduce the amount of support tissue that must be thickened and strengthened to bear the mature fruits and seeds.[14,191] In addition, translocation costs may be reduced. Neither of these potential benefits has been quantified.

Another example that may involve resource efficiency is differential abortion of few-seeded fruits. Such fruits cost more in terms of packaging material (pericarp) *per seed* than do many-seeded fruits.[21,77,173,188] A seed in a one-seeded fruit of *Cassia fasciculata,* for example, is associated with three times more pericarp than a seed in a 12-seeded fruit (T. D. Lee, unpublished data).

The average fitness of offspring may also be enhanced by differential abortion. Many authors suggest that selection will favor plants that differentially mature offspring of high quality (= greater fitness).[13,31,63,153,155,186,187] Indeed, many have suggested that overinitiation of offspring, including polyembryony in conifers, provides an opportunity for differential abortion of low-quality offspring,[31,138,139,153,160,187] but data are sparse. In species that typically cross-pollinate, differential abortion of fruits or seeds resulting from self-pollination may well enhance maternal fitness, as inbred progeny are usually inferior to outbred progeny.[33,40,82,143] Data from several species indicate that differential abortion of fruits initiated by different pollen donors also enhances offspring quality. When *Asclepias speciosa* fruits fathered by different donors were in competition for maternal resources, fruits fathered by successful donors produced heavier seeds with better germination and which gave rise to larger seedlings than did seeds from fruits grown in the absence of interdonor competition.[21] Working with *Campsis radicans,* Bertin[13] found that donors that were successful in producing fruit on recipient plants tended to yield fruits containing more and better-quality seeds. Marshall and Ellstrand[96] found in *Raphanus sativus* that the pollen donor resulting in the greatest percent fruit production and number of seeds per fruit also fathered the largest seeds.

Offspring quality may also be affected by differential maturation of fruits and seeds initiated under different levels of gametophyte competition. Fruits resulting from heavy pollination may contain higher-quality seeds due either to the fertilization of ovules by more vigorous pollen tubes[112] or to the fast growth of pollen tubes whose genomes combine favorably with that of the recipient plant.[113] There is considerable evidence that gametophyte competition results in faster growing, more competitive offspring.[104,112] In at least two species that abort lightly pollinated flowers, *Cassia fasciculata*[79] and *Cucurbita pepo* (zucchini),[157] seeds from fruits initiated under heavy pollination produce more vigorous plants than do those from light pollinations. In *Lotus corniculatus,* which differentially matures many-seeded fruits, seeds from few-seeded fruits produce plants with lower growth rates and lower fecundity than seeds from many-seeded fruits.[156] In this species, however, few seeded fruits may result from either light pollen loads *or* self-pollination, and thus differential abortion may be adaptive in that it eliminates the inferior products of inbreeding as well as progeny fathered by slow pollen tubes.[156] Within fruits of legumes, the differential abortion of basal seeds should enhance seed quality because, in many species, basal seeds are initiated by slower growing pollen tubes.[11,73,78] Data from *Phasaeolus vulgaris* are not consistent with this prediction as mature seeds from basal and distal ends of pods do not differ in percentage germination or rate of seedling growth.[119] However, in this study, abortion of some basal ovules may have already eliminated most inferior genotypes, so the lack of a difference may not be surprising. Tests of effects of position on seed quality will require methods that allow all or almost all fertilized ovules to mature as seeds.

An additional fitness advantage of differential fruit maturation may involve seed dispersal. For example, differential maturation of large (many-seeded) fruits may enhance seed dispersal if animal dispersers selectively devour large fruits. In ballistically dispersed *Cassia fasciculata,* many-seeded fruits, which are differentially matured, may produce a broader seed shadow than few-seeded fruits.[74] If a broad seed shadow allows escape from predators or colonization of microsites that are more favorable for establishment, fitness may be enhanced by differential fruit maturation. On the other hand, Casper and Wiens argue that if whole fruits are dispersed as units, individual plants that differentially mature *few-seeded* fruits may have greater fitness.

Such a pattern would reduce sib-competition and, if fruits are wind dispersed, would lighten dispersal units and allow them to travel farther.[29]

Nakamura has suggested that inclusive fitness may be affected by fruit and seed abortion. Thus, under certain circumstances, selection may favor maturation of embryos that are closely related to the maternal plant at the expense of other progeny.[116]

Clearly, many of the patterns described herein affect plant fitness and, therefore, may have arisen by natural selection. It is curious, however, that although the patterns are diverse and influence fitness in several different ways, they may have a similar physiological basis. The physiological literature (see Physiology of Fruit and Seed Production) suggests that whether a fruit matures or aborts is highly dependent on either its sink strength or its ability to chemically inhibit neighboring fruits. While mechanisms are not fully understood and while debate continues on the roles of sink strength and production of inhibitors, there seems to be a consensus that the success of a particular reproductive unit is influenced by its own metabolic activity and consequent production of phytohormones.

Since much of vegetative plant growth is a function of hormonally mediated competition among sinks, it could be argued that patterns of fruit and seed maturation are consequences of a physiological law of plant growth and development—weak sinks ultimately fail.[179] Under this interpretation, fruit and seed production patterns that enhance fitness are exaptations;[50] their origin is a consequence of the laws of plant growth, even though they currently enhance fitness and may be maintained by selection. On the other hand, fitness effects of reproductive patterns may be partially responsible for the evolution of regulatory mechanisms based on sink strength and hormone production; thus the patterns could be considered adaptations. Resolution of this dichotomy may come from detailed studies of the roles of phytohormones and source-sink relations in the reproduction of evolutionarily less advanced plant groups.

A trait essential to the patterns described in this chapter is the initiation of more fruits and seeds than can be matured with available resources. Such "over-initiation" of fruits and seeds causes competition to occur among them, with subsequent maturation or abortion based on sink strength. Initiation of "excess" fruits and seeds may be favored by several environmental factors, including unpredictability of the physical environment, unpredictability of pollen quantity and quality, and fruit and seed predation.[11,77,153] In an unpredictable physical environment plants initiating a few extra fruits have a fecundity advantage in years when favorable conditions allow such fruits to mature.[153] When pollination regime is uncertain, overinitiation allows differential survival of fruits and seeds initiated with large amounts of or high-quality pollen,[11,40,153] though it is likely that such differential maturation is a passive rather than an active process.[21,177] An individual that overinitiates fruits may also satiate predators, allowing at least some fruits to mature.[61,153] For these reasons, fruit and seed overinitiation may be an adaptation.

Of course, it is possible that overinitiation is not adaptive. Plants may overinitiate fruits simply because selection has favored the overinitiation of flowers. Excess flowers may be favored due to pollinator attraction or pollen donation[153,155] and the fact that a greater number of flowers are pollinated than matured to fruit may have no adaptive significance. However, we would expect selection to favor relatively early abortion of such excess fruits, and, as many plants retain fruits and invest additional resources in them before abortion,[77,89,153,154] it is likely that excess fruit initiation in these species is not merely a consequence of flower overinitiation.

Recent authors have used the term "female choice" to describe differential abortion of fruits and seeds.[153,187] If the physiological basis of differential abortion is the ability of fruits and seeds to compete for resources, then the choice involved is not an active process but a passive one.[21] Such choice primarily involves overinitiation of fruits and seeds and subsequent competition among them, with the "chosen" being those that are most vigorous in acquiring resources or inhibiting others from doing so. Lack of active choice does not, however, mean lack of maternal control over the developing seed crop; Westoby and Rice[183] have argued that, while success of developing seeds is a function of their sink strength, the maternal plant limits the extent of uncontrolled "scramble competition" by inducing early abortion of less competitive sinks and setting an upper limit to the size of more competitive seeds. Westoby and Rice further suggest that such control may be achieved through the integuments, maternal tissues that surround the offspring during development and possibly act as a "sealing layer."

CONCLUSIONS

In plants in which reproduction is resource limited, the pattern of fruit and seed abortion is decidedly nonrandom. At the whole-fruit level, abortion is related to time of fruit initiation, proximity of the fruit to resources, pollination intensity, number of developing seeds in the fruit, source of pollen, number of pollen donors, and damage to fruits and seeds. At the level of the individual seed, abortion is related to the time of seed initiation, position within the fruit, time of fruit initiation, position of the fruit in the inflorescence, and pollen source.

Many of the factors related to fruit and seed maturation patterns are intercorrelated with one another. Pollen source, position in the inflorescence, and pollination intensity are frequently positively related to seed number. Time of fertilization within an ovary is usually related to speed of pollen tube growth. Such correlations make it difficult to establish the proximate causes of abortion patterns. Even when variables associated with abortion are not intercorrelated, their relative importances are not known. For example, if pollen source, pollination intensity, and damage to fruits are significant, which one is responsible for the greatest share of fruit abortion? A first step has been taken by Nakamura[118] who modeled the interaction between time of initiation and pollen source, but experimental data are needed.

The patterns of fruit and seed maturation described herein can be interpreted proximally in terms of hormonally mediated source-sink relationships. Fruits or seeds with greater metabolic activity and greater hormone production or uptake are more likely to mature than are weaker sinks.

Patterns of fruit and seed production may positively affect fitness of maternal plants or their progeny by increasing efficiency of resource use, improving average quality of offspring, and increasing the effectiveness of seed dispersal. While such traits are thus probably maintained and perhaps enhanced by natural selection, they may have originated as a consequence of source-sink relationships which evolved due to other selective forces. Conversely, source-sink relationships may in part be an evolutionary result of the fitness value of various patterns of maturation of reproductive structures.

ACKNOWLEDGMENTS

I thank V. J. Apsit, R. I. Bertin, A. P. Hartgerink, J. W. McClure, and R. R. Nakamura for critical review of the manuscript. Support during preparation of the manuscript was provided by the New Hampshire Agricultural Experiment Station (H-273) and the U.S. Department of Agriculture (Grant No. 85 CRCR-1-1643).

REFERENCES

1. Aalders, L. E., and Hall, I. V., Pollen incompatibility and fruit set in lowbush blueberries, *Can. J. Genet. Cytol* **3,** 300–307 (1961).
2. Adams, M. W., Basis of yield component compensation in crop plants with special reference to the field bean, *Phaseolus vulgaris, Crop Sci.* **7,** 505–510 (1967).
3. Addicott, F. T., *Abscission,* pp. 126–130. Univ. California Press, Berkeley, 1982.
4. Akamine, E. K., and Girolami, G., Pollination and fruit set in the yellow passion fruit, *Hawaii Agric. Exp. St. Tech. Bull. No. 39,* 1959.
5. Aker, C. K., and Udovic, D., Oviposition and pollination behavior of the yucca moth, *Tegeticula maculata* (Lepidoptera: Prodoxidae), and its relationship to the reproductive biology of *Yucca whipplei* (Agavaceae), *Oecologia* **49,** 96–101 (1981).
6. Banyard, B. J., and James, S. H., Biosystematic studies in the *Stylidium crassifolium* species complex (Stylidiaceae), *Aust. J. Bot.* **27,** 27–37 (1979).
7. Barnes, D. K., and Cleveland, R. W., Genetic evidence for nonrandom fertilization in alfalfa as influenced by differential pollen tube growth, *Crop Sci.* **3,** 295–297 (1963).
8. Barrett, S. C. H., The evolution, maintenance, and loss of self-incompatibility systems, in *Plant Reproductive Ecology: Patterns and Strategies* (J. Lovett Doust and L. Lovett Doust, eds.), Chapter 5. Oxford Univ. Press, New York, 1988.
9. Baumann, T. E., and Eaton, G. W., Competition among berries on the cranberry upright, *J. Am. Soc. Hortic. Sci.* **111,** 869–872 (1986).
10. Bawa, K. S., and Buckley, D., Seed/ovule ratios and mating systems in the Leguminosae, manuscript in preparation.
11. Bawa, K. S., and Webb, C. J., Flower, fruit, and seed abortion in tropical forest trees: Implications for the evolution of paternal and maternal reproductive patterns, *Am. J. Bot.* **71,** 736–751 (1984).
12. Bertin, R. I., Floral biology, hummingbird pollination and fruit production of trumpet creeper (*Campsis radicans,* Bignoniaceae), *Am. J. Bot.* **69,** 122–134 (1982).
13. Bertin, R. I., Paternity and fruit production in trumpet creeper *(Campsis radicans), Am. Nat.* **119,** 694–709 (1982).
14. Bertin, R. I., The ecology of sex expression in red buckeye, *Ecology* **63,** 445–456 (1982).
15. Bertin, R. I., Non-random fruit production in *Campsis radicans:* Between-year consistency and effects of prior pollination, *Am. Nat.* **126,** 750–759 (1985).
16. Bertin, R. I., Consequences of multiple paternity in *Campsis radicans, Oecologia* **70,** 1–5 (1986).
17. Bertin, R. I., Paternity in plants, in *Plant Reproductive Ecology: Patterns and Strategies* (J. Lovett Doust and L. Lovett Doust, eds.), Chapter 2. Oxford Univ. Press, New York, 1988.
18. Beruter, J., Sugar accumulation and changes in the activities of related enzymes during development of the apple fruit. *J. Plant Physiol.* **121,** 331–341 (1985).
19. Binnie, R. C., and P. E. Clifford, Flower and pod production in *Phaseolus vulgaris, J. Agric. Sci. Cambridge* **97,** 397–402 (1981).
20. Bookman, S. S., Effects of pollination timing on fruiting in *Asclepias speciosa* Torr. (Asclepiadaceae), *Am. J. Bot.* **70,** 897–905 (1983).
21. Bookman, S. S., Evidence for selective fruit production in *Asclepias, Evolution* **38,** 72–86 (1984).
22. Brewer, J. W., and Dobson, R. C., Seed count and berry size in relation to pollinator level and harvest date for the highbush blueberry *Vaccinium corymbosum, J. Econ. Entomol.* **62,** 1353–1356 (1969).
23. Brink, R. A., The physiology of pollen IV. Chemotropism, the effects on growth of grouping grains, formation and function of callose plugs, summary and conclusions, *Am. J. Bot.* **11,** 351–416 (1924).
24. Brink, R. A., and Cooper, D. C., Somatoplastic sterility in *Medicago sativa, Science* **90,** 545–546 (1939).

25. Buccholz, J. T., Suspensor and early embryo of *Pinus, Bot. Gaz.* **66,** 185–228 (1918).
26. Bushnell, J. W., The fertility and fruiting habit in *Cucurbita, Proc. Am. Soc. Hortic. Sci.* **17,** 47–51 (1920).
27. Casper, B. B., The efficiency of pollen transfer and rates of embryo initiation in *Cryptantha* (Boraginaceae), *Oecologia* **59,** 262–268 (1983).
28. Casper, B. B., On the evolution of embryo abortion in the herbaceous perennial *Cryptantha flava, Evolution* **38,** 1332–1349 (1984).
29. Casper, B. B., and Wiens, D., Fixed rates of ovule abortion in *Cryptantha flava* (Boraginaceae) and its possible relation to seed dispersal, *Ecology* **62,** 866–869 (1981).
30. Chapman, G. P., and Sadjadi, A. S., Exogenous growth substances and internal competition in *Vicia faba* L., *Z. Pflanzenphysiol.* **104,** 265–273 (1981).
31. Charnov, E. L., Simultaneous hermaphroditism and sexual selection, *Proc. Natl. Acad. Sci. U.S.A.* **5,** 2480–2484 (1979).
32. Chauhan, Y. S., and Bhargava, S. C., Comparison of the terminal raceme of rapeseed (*Brassica campestris* var. Yellow Sarson) and mustard *(Brassica juncea)* cultivars, *J. Agric. Sci. Cambridge* **107,** 469–473 (1986).
33. Coles, J. F., and Fowler, D. P., Inbreeding in neighboring trees in two white spruce populations, *Silvae Genet.* **25,** 29–34 (1976).
34. Cooper, D. C., and Brink, R. A., Partial self-incompatibility and the collapse of fertile ovules as factors affecting seed formation in alfalfa, *J. Agric. Res.* **60,** 453–472 (1940).
35. Cooper, D. C., Brink, R. A., and Albrecht, H. R., Embryo mortality in relation to seed formation in alfalfa *(Medicago sativa), Am. J. Bot.* **24,** 203–213 (1937).
36. Cornforth, J. W., Milborrow, B. V., Ryback, G., Rothwell, K., and Wain, R. L., Identification of the yellow lupin growth inhibitor as (+)− Abscisin II ((+)− Dormin), *Nature (London)* **211,** 742–743 (1966).
37. Crowe, L. K., The polygenic control of outbreeding in *Borago officinalis, Heredity* **27,** 111–118 (1971).
38. Crumpacker, D. W., Genetic loads in maize (*Zea mays* L.) and other cross-fertilized plants and animals, *Evol. Biol.* **1,** 306–424 (1967).
39. Cruzan, M. B., Pollen tube distribution in *Nicotiana glauca:* Evidence for density-dependent growth, *Am. J. Bot.* **73,** 902–907 (1986).
40. Darwin, C., *The Effects of Cross- and Self-fertilization in the Vegetable Kingdom,* pp. 399–400. D. Appleton and Co., New York, 1892.
41. Davis, R. M., Jr., Smith, P. G., Schweers, V. H., and Scheuerman, R. W., Independence of floral fertility and fruit-set in the tomato, *Proc. Am. Soc. Hortic. Sci.* **86,** 552–556 (1965).
42. Dempsey, W. H., and Boynton, J. E., Effect of seed number on tomato fruit size and maturity, *Proc. Am. Soc. Hortic. Sci.* **86,** 575–581 (1965).
43. Delph, L. F., Factors regulating fruit and seed production in the desert annual *Lesquerella gordonii, Oecologia* **69,** 471–476 (1986).
44. de Nettancourt, D., *Incompatibility in Angiosperms.* Springer-Verlag, Berlin, 1977.
45. Dobrofsky, S., and Grant, W. F., An investigation into the mechanism for reduced seed yield in *Lotus corniculatus, Theor. Appl. Genet.* **57,** 157–160 (1980).
46. Dybing, C. D., Ghiasi, H., and Paech, C., Biochemical characteristics of soybean ovary growth from anthesis to abscission of aborting ovaries, *Plant Physiol.* **81,** 1069–1074 (1986).
47. Garwood, N. C., and Horvitz, C. C., Factors limiting fruit and seed production of a temperate shrub, *Staphylea trifolia* L. (Staphyleaceae), *Am. J. Bot.* **72,** 453–466 (1985).
48. Goodwin, P. B., Phytohormones and fruit growth, in *Phytohormones and Related Compounds—A Comprehensive Treatise, Volume II* (D. S. Lethan, P. B. Goodwin, and T. J. V. Higgins, eds.), Chapter 3. Elsevier/North Holland Biomedical Press, Amsterdam, 1978.
49. Gorchov, D. L., Fruit ripening asynchrony is related to variable seed number in *Amelanchier* and *Vaccinium, Am. J. Bot.* **72,** 1939–1943 (1985).
50. Gould, S. J., and Vrba, E. S., Exaptation—a missing term in the science of form, *Paleobiology* **8,** 4–15 (1982).
51. Haber, E. S., Inbreeding the Table Queen (Des Moines) squash, *Proc. Am. Soc. Hortic. Sci.* **25,** 111–114 (1928).
52. Handel, S. N., and Mishkin, J. L., Temporal shifts in gene flow and seed set: evidence from an experimental population of *Cucumis sativus, Evolution* **38,** 1350–1357 (1984).
53. Hanft, J. M., and Jones, R. J., Kernal abortion in maize. I. Carbohydrate concentration patterns and acid invertase activity of maize kernals induced to abort *in vitro, Plant Physiol.* **81,** 503–510 (1986).

54. Heinicke, A. J., Factors influencing the abscission of flowers and partially developed fruits of the apple (*Pyrus malus* L.), *Cornell Univ. Agric. Exp. Stn. Bull. No. 393,* 1917.
55. Hill-Cottingham, D. G., and Williams, R. R., Effect of time of application of fertilizer nitrogen on the growth, flower development and fruit set of maiden apple trees, var. Lord Lambourne, and on distribution of total nitrogen within the trees, *J. Hortic. Sci.* **42,** 319–338 (1967).
56. Hoad, G. V., Loveys, B. R., and Skene, K. G. M., The effect of fruit removal on cytokinin and gibberellin-like substances in grape leaves, *Planta* **136,** 25–30 (1977).
57. Holtsford, T. P., Nonfruiting hermaphroditic flowers of *Calochortus leichtlinii* (Liliaceae): potential reproductive functions, *Am. J. Bot.* **72,** 1687–1694 (1985).
58. Horovitz, A., Meiri, L., and Beiles, A., Effects of ovule positions in fabaceous flowers on seed set and outcrossing rates, *Bot. Gaz.* **137,** 250–254 (1976).
59. Huff, A., and Dybing, C. D., Factors affecting shedding of flowers in soybean (*Glycine max* (L.)Merrill), *J. Exp. Bot.* **31,** 751–762 (1980).
60. Hurd, R. G., Gay, A. P., and Mountfield, A. C., The effect of partial flower removal on the relation between root, shoot, and fruit growth in the indeterminate tomato, *Ann. Appl. Biol.* **93,** 77–89 (1979).
61. Janzen, D. H., Seed predation by animals, *Annu. Rev. Ecol. Syst.* **2,** 465–492 (1971).
62. Janzen, D. H., Effect of defoliation on fruit-bearing branches of the Kentucky coffee tree, *Gymnocladus dioicus* (Leguminosae), *Am. Midl. Nat.* **95,** 474–478 (1976).
63. Janzen, D. H., A note on optimal mate selection by plants, *Am. Nat.* **111,** 365–371 (1977).
64. Janzen, D. H., DeVries, P., Gladstone, D. E., Higgins, M. L., and Lewisohn, T. M., Self- and cross-pollination of *Encyclia cordigera* (Orchidaceae) in Santa Rosa National Park, Costa Rica, *Biotropica* **12,** 72–74 (1980).
65. Jaranowski, J., Fertilization and embryo development in cases of autogamy, *Genet. Pol.* **3,** 209–242 (1962).
66. Jennings, D. L., and Topham, P. B., Some consequences of raspberry pollen dilution for its germination and for fruit development, *New Phytol.* **70,** 371–380 (1971).
67. Jones, A. L., *Diseases of Tree Fruits.* Cooperative Extension Service of the Northeastern States, New Brunswick, New Jersey, 1976.
68. Jones, D. F., *Selective Fertilization,* pp. 10–12, 68. Univ. of Chicago Press, Chicago, 1928.
69. Kambal, A. E., Flower drop and fruit set in field beans, *Vicia faba* L., *J. Agric. Res. Cambridge* **72,** 131–138.
70. Kephart, S. R., Breeding systems in *Asclepias incarnata* L., *A. syriaca* L. and *A. verticillata* L., *Am. J. Bot.* **68,** 226–232 (1981).
71. Kress, W. J., Sibling competition and evolution of pollen unit, ovule number, and pollen vector in angiosperms, *Syst. Bot.* **6,** 101–112 (1981).
72. Labeyrie, V., and Hossaert, M., Ambiguous relations between *Bruchus affinis* and the *Lathyrus* group, *Oikos* **44,** 107–113 (1985).
73. Lee, T. D., Patterns of fruit maturation: A gametophyte competition hypothesis, *Am. Nat.* **123,** 427–432 (1984).
74. Lee, T. D., Effects of seed number per fruit on seed dispersal in *Cassia fasciculata* (Caesalpiniacea), *Bot. Gaz.* **145,** 136–139 (1984).
75. Lee, T. D., Patterns of fruit and seed production in a Vermont population of *Cassia nictitans* L. (Caesalpiniaceae), manuscript in preparation.
76. Lee, T. D., and Bazzaz, F. A., Regulation of fruit and seed production in an annual legume, *Cassia fasciculata, Ecology* **63,** 1363–1373 (1982).
77. Lee, T. D., and Bazzaz, F. A., Regulation of the fruit maturation pattern in an annual legume, *Cassis fasciculata, Ecology* **63,** 1374–1388 (1982).
78. Lee, T. D., and Bazzaz, F. A., Maternal regulation of fecundity: Non-random ovule abortion in *Cassia fasciculata* Michx., *Oecologia* **68,** 459–465 (1986).
79. Lee, T. D., and Hartgerink, A. P., Pollination intensity, fruit maturation pattern, and offspring quality in *Cassia fasciculata* (Leguminosae), in *Biotechnology and Ecology of Pollen* (D. L. Mulcahy et al., eds.), pp. 417–422. Springer-Verlag, New York, 1986.
80. Lee, T. D., and Willson, M. F., Reproductive ecology of five herbs common in central Illinois, *Mich. Bot.* **23,** 23–28 (1983).
81. Leopold, A. C., and Kriedemann, P. E., *Plant Growth and Development,* Second Ed. Chapter 13. McGraw-Hill, New York, 1975.
82. Levin, D. A., Plant parentage: an alternative view of the breeding structure of populations, in *Population*

Biology, Retrospect and Prospect (C. E. King and P. S. Dawson, eds.), Chapter 7. Columbia Univ. Press, New York, 1983.

83. Levin, D. A., Density and proximity-dependent crossing success in *Phlox drummondii, Evolution* **38,** 116–127 (1984).
84. Lewis, D., *Sexual Incompatibility in Plants,* p. 24. Univ. Park Press, Baltimore, 1979.
85. Linck, A. J., The morphological development of the fruit of *Pisum sativum* var. *alaska, Phytomorphology* **11,** 79–84 (1961).
86. Lindgren, D., The relationship between self-fertilization, empty seeds, and seeds originating from selfing as a consequence of polyembryony, *Stud. For. Suec.* **126** (1975).
87. Lis, E. K., and Antoszewski, R., Do growth substances regulate the phloem as well as the xylem transport of nutrients to the strawberry receptacle? *Planta* **156,** 492–495 (1982).
88. Lloyd, D. G., Parental strategies of angiosperms, N.Z. *J. Bot.* **17,** 595–606 (1979).
89. Lloyd, D. G., Sexual strategies in plants I. An hypothesis of serial adjustment of maternal investment during one reproductive session, *New Phytol.* **86,** 69–79 (1980).
90. Lovett Doust, J., Floral sex ratios in andromonoecious Umbelliferae, *New Phytol.* **85,** 265–273 (1980).
91. Lovett Doust, J., and Eaton, G. W., Demographic aspects of flower and fruit production in bean plants, *Phaseolus vulgaris* L., *Am. J. Bot.* **69,** 1156–1164 (1982).
92. Lovett Doust, J., and Harper, J. L., The resource costs of gender and maternal support in an andromonoecious umbellifer, *Smyrnium olusatrum* L., *New Phytol.* **85,** 251–264 (1980).
93. Lovett Doust, L., Lovett Doust, J., and Turi, K., Fecundity and size relationships in jack in the pulpit, *Arisaema triphyllum* (Araceae), *Am. J. Bot.* **73,** 489–494 (1986).
94. Lund, H. A., Growth hormones in the styles and ovaries of tobacco responsible for fruit development, *Am. J. Bot.* **43,** 562–568 (1956).
95. Lund, H. A., Growth hormones in the styles and ovaries of tobacco responsible for fruit development, *Am. J. Bot.* **43,** 562–568 (1956).
96. Marshall, D. L., and Ellstrand, N. C., Sexual selection in *Raphanus sativus:* experimental data on nonrandom fertilization, maternal choice, and consequences of multiple paternity, *Am. Nat.* **127,** 415–445 (1986).
97. Marshall, D. L., Levin, D. A., and Fowler, N. L., Plasticity in yield components in response to fruit predation and date of fruit initiation in three species of *Sesbania* (Leguminosae), *J. Ecol.* **73,** 71–80 (1985).
98. Marshall, D. L., Levin, D. A., and Fowler, N. L., Plasticity of yield components in response to stress in *Sesbania macrocarpa* and *Sesbania vesicaria* (Leguminosae), *Am. Nat.* **127,** 508–521 (1986).
99. Martin, D., Lewis, T. L., and Cerny, J., Jonathan Spot—three factors related to incidence: Fruit size, breakdown, and seed numbers, *Aust. J. Agric. Res.* **12,** 1039–1049 (1961).
100. Mazer, S. J., Snow, A. A., and Stanton, M. L., Fertilization dynamics and parental effects upon fruit development in *Raphanus raphanistrum:* Consequences for seed size variation, *Am. J. Bot.* **73,** 500–511 (1986).
101. McAlister, D. F., and Krober, O. A., Response of soybeans to leaf and pod removal, *Agron. J.* **50,** 674–677 (1958).
102. McDade, L. A., Pollination intensity and seed set in *Trichanthera gigantea* (Acanthaceae), *Biotropica* **15,** 122–124 (1983).
103. McDade, L. A., and Davidar, P., Determinants of fruit and seed set in *Pavonia dasypetala* (Malvaceae), *Oecologia* **64,** 61–67 (1984).
104. McKenna, M. A., and Mulcahy, D. L., Gametophytic competition in *Dianthus chinensis:* Effect on sporophyte competitive ability, in *Pollen: Biology and Applications in Plant Breeding* (D. L. Mulcahy, ed.), pp. 419–434. Elsevier, Amsterdam, 1983.
105. McKone, M. J., Reproductive biology of several brome grasses *(Bromus):* breeding system, pattern of fruit maturation, and seed set, *Am. J. Bot.* **72,** 1334–1339 (1985).
106. Meinke, D. W., Embryo-lethal mutants of *Arabidopsis thaliana:* Evidence for gametophytic expression of mutant genes, *Theor. Appl. Genet.* **63,** 381–386 (1982).
107. Meinke, D. W., Embryo lethal mutants of *Arabidopsis thaliana:* Analysis of mutants with a wide range of lethal phases, *Theor. Appl. Genet.* **69,** 543–552 (1985).
108. Mikkola, L., Observations on interspecific sterility in *Picea, Ann. Bot. Fenn.* **6,** 285–339 (1969).
109. Mogensen, H. L., Ovule abortion in *Quercus* (Fagaceae), *Am. J. Bot.* **62,** 160–165 (1975).
110. Moore, J. A., The vascular anatomy of the flower in the Papillionaceous Legumes, *Am. J. Bot.* **23,** 279–290 (1936).

111. Motten, A. F., Reproduction of *Erythronium umbilicatum* (Liliaceae): Pollination success and pollinator effectiveness, *Oecologia* **59,** 351–359 (1983).
112. Mulcahy, D. L., The rise of the angiosperms: A genecological factor, *Science* **206,** 20–23 (1979).
113. Mulcahy, D. L., and Mulcahy, G. B., Gametophytic self-incompatibility re-examined, *Science* **220,** 1247–1251 (1983).
114. Murneek, A. E., The nature of shedding of immature apples, *Missouri Agric. Exp. Stn. Res. Bull. No. 201,* 1933.
115. Murneek, A. E., The embryo and endosperm in relation to fruit development with special reference to the apple, *Malus sylvestris, Proc. Am. Soc. Hortic. Sci.* **64,** 573–582 (1954).
116. Nakamura, R. R., Plant kin selection, *Evol. Theor.* **5,** 113–117 (1980).
117. Nakamura, R. R., Reproductive capacity and kinship in *Phaseolus vulgaris* L. Ph.D. thesis, Yale University, New Haven, Connecticut, 1983.
118. Nakamura, R. R., Maternal investment and fruit abortion in *Phaseolus vulgaris* L., *Am. J. Bot.* **73,** 1049–1057 (1986).
119. Nakamura, R. R., The ecology of bean embryos, manuscript in preparation.
120. Nakamura, R. R., and Stanton, M. L., Cryptic seed abortion and the estimation of seed set, *Can. J. Bot.*, in press.
121. Nesling, F. A. V., and Morris, D. A., Cytokinin levels and embryo abortion in interspecific *Phaseolus* crosses, *Z. Pflanzenphysiol.* **91,** 345–358 (1979).
122. Nitsch, J. P., Growth and morphogenesis of the strawberry as related to auxins, *Am. J. Bot.* **37,** 211–215 (1950).
123. Nitsch, J. P., Plant hormones and the development of fruits, *Q. Rev. Biol.* **27,** 33–57 (1952).
124. Pate, J. S., and Flinn, A. M., Fruit and seed development, in *The Physiology of the Garden Pea* (J. F. Sutcliffe and J. S. Pate, eds.), Chapter 15. Academic Press, London, 1977.
125. Pechan, P. A., and Morgan, D. G., Defoliation and its effects on pod and seed development in oil seed rape (*Brassica napus* L.) *J. Exp. Bot.* **36,** 458–468 (1985).
126. Pharis, R. P., and King, R. W., Gibberellins and reproductive development in seed plants, *Annu. Rev. Plant Physiol.* **36,** 517–568 (1985).
127. Picken, A. J. F., A review of pollination and fruit set in the tomato (*Lycopersicon esculentum* Mill.), *J. Hortic. Sci.* **59,** 1–13 (1984).
128. Poovaiah, B. W., and Veluthambi, K., Auxin-regulated invertase activity in strawberry fruits, *J. Am. Soc. Hortic. Sci.* **110:** 258–261 (1985).
129. Porter, N. G., Interaction between lateral branch growth and pod set in primary inflorescences of lupin, *Aust. J. Agric. Res.* **33,** 957–965 (1982).
130. Price, M. V., and Waser, N. M., Pollen dispersal and optimal outcrossing distance in *Dephinium nelsoni, Nature (London)* **277,** 294–296 (1979).
131. Queller, D. C., Proximate and ultimate causes of low fruit production in *Asclepias exaltata, Oikos* **44,** 373–381 (1985).
132. Quinlan, J. D., and Preston, A. P., Effects of thinning blossoms and fruitlets on growth and cropping on sunset apple, *J. Hortic. Sci.* **43,** 373–381 (1968).
133. Rawson, H. M., and Evans, L. T., The pattern of grain growth within the ear of wheat. *Aust. J. Biol. Sci.* **23,** 753–764 (1970).
134. Robins, J. S., and Domingo, C. E., Moisture deficits in relation to the growth and development of dry beans, *Agron. J.* **48,** 67–70 (1956).
135. Rowland, G. G., and Bond, D. A., The relationship between number of seeds and the frequency of ovule fertilization in field beans (*Vicia faba* L.), *J. Agric. Sci. Cambridge* **100,** 35–41 (1983).
136. Rowland, G. G., Bond, D. A., and Parker, M. L., Estimates of the frequency of fertilization in field beans *(Vicia faba)* L., *J. Agric. Sci. Cambridge* **100,** 25–33 (1983).
137. Sage, T., Relationship of carbon availability to abscission in reproductive tissue of *Phaseolus vulgaris,* in *Proc. 13th Annu. Meet. Plant Growth Regulator Soc.* (A. R. Cooke, ed.), pp. 268–276. Plant Growth Regulator Society, 1986.
138. Sarvas, R., Investigations on the flowering and seed crop of *Pinus sylvestris, Commun. Inst. For. Fenn. No. 53,* 1962.
139. Sarvas, R., Investigations on the flowering and seed crop of *Picea abies, Commun. Inst. For. Fenn. No. 67,* 1968.
140. Sayers, E. R., and Murphy, R. P., Seed set in alfalfa as related to pollen tube growth, fertilization frequency and post-fertilization ovule abortion, *Crop Sci.* **6,** 365–368 (1966).

141. Schemske, D. W., Flowering phenology and seed set in *Claytonia virginica* (Portulacaceae), *Bull, Torrey Bot. Club* **104,** 254-263 (1977).
142. Schemske, D. W., Floral convergence and pollinator sharing in two bee-pollinated tropical herbs, *Ecology* **62,** 946–954 (1981).
143. Schemske, D. W., Breeding system and habitat effects on fitness components in three neotropical *Costus* (Zingiberaceae), *Evolution* **37,** 523–539 (1983).
144. Schemske, D. W., and Pautler, L. P., The effects of pollen composition on fitness components in a neotropical herb, *Oecologia* **62,** 31–36 (1984).
145. Sheldrake, A. R., and Saxena, N. P., Comparisons of earlier- and later-formed pods of chickpeas (*Cicer arietinum* L.), *Ann. Bot. (London)* **43,** 467–473 (1979).
146. Silander, J. A., and Primack, R. B., Pollination intensity and seed set in the evening primrose *(Oenothera fruiticosa), Am. Midl. Nat.* **109,** 213–216 (1978).
147. Snow, A. A., Pollination intensity and potential seed set in *Passiflora vitifolia, Oecologia* **55,** 231–237 (1982).
148. Snow, A. A., Pollination dynamics in *Epilobium canum* (Onagraceae): Consequences for gametophytic selection, *Am. J. Bot.* **73,** 139–157 (1986).
149. Sparrow, F. K., and Pearson, N. L., Pollen incompatibility in *Asclepias syriaca, J. Agric. Res.* **77,** 187–199 (1948).
150. Stebbins, G. L., *Variation and Evolution in Plants,* p. 385. Columbia Univ. Press, New York, 1950.
151. Stephenson, A. G., An evolutionary examination of the fruiting display of *Catalpa speciosa* (Bignoniaceae), *Evolution* **33,** 1200–1209 (1979).
152. Stephenson, A. G., Fruit set, herbivory, fruit reduction, and the fruiting strategy of *Catalpa speciosa* (Bignoniaceae), *Ecology* **61,** 57–64 (1980).
153. Stephenson, A. G., Flower and fruit abortion: Proximate causes and ultimate functions, *Annu. Rev. Ecol. Syst.* **12,** 253–279 (1981).
154. Stephenson, A. G., The cost of over-initiating fruit, *Am. Midl. Nat.* **112,** 379–386 (1984).
155. Stephenson, A. G., and Bertin, R. I., Male competition, female choice, and sexual selection in plants, in *Pollination Biology* (L. Real, ed.), Chapter 6. Academic Press, Orlando, Florida, 1983.
156. Stephenson, A. G., and Winsor, J. A., *Lotus corniculatus* regulates offspring quality through selective fruit abortion, *Evolution* **40,** 453–458 (1986).
157. Stephenson, A. G., Winsor, J. A., and Davis, L. E., Effects of pollen load size on fruit maturation and sporophyte quality in zucchini, in *Biotechnology and Ecology of Pollen* (D. L. Mulcahy et al., eds.), pp. 429–434. Springer-Verlag, New York, 1986.
158. Stevens, G. A., and Westwood, M. N., Fruit set and cytokinin-like activity in the xylem sap of sweet cherry *(Prunus avium)* as affected by rootstock, *Physiol. Plant.* **61,** 464–468 (1984).
159. Stoddard, F. L., Effects of irrigation, plant density and genotype on pollination, fertilization and seed development in spring field beans (*Vicia faba* L.), *J. Agric. Sci., Cambridge* **107,** 347–355 (1986).
160. Sweet, G. B., Shedding of reproductive structures in forest trees, in *Shedding of Plant Parts* (T. T. Kozlowski, ed.), Chapter 9. Academic Press, New York, 1973.
161. Tamas, I. A., Wallace, D. H., Ludford, P. M., and Ozbun, J. L., Effect of older fruits on abortion and abscisic acid concentration of younger fruits in *Phaseolus vulgaria* L., *Plant Physiol.* **64,** 620–622 (1979).
162. Tayo, T. O., Flower and pod production at various nodes of *Phaseolus vulgaris* L., *J. Agric. Sci., Cambridge* **107,** 29–36 (1986).
163. Tayo, T. O., and Morgan, D. G., Factors influencing flower and pod development in oil-seed rape (*Brassica napus* L.), *J. Agric. Sci., Cambridge* **92,** 363–373 (1979).
164. Teaotia, S. S., Studies on fruit development and fruit drop in black currants. Ph.D. thesis, Univ. of Bristol, Bristol, UK, 1953.
165. Thorne, J. H., Phloem unloading of C and N assimilates in developing seeds, *Annu. Rev. Plant Physiol.* **36,** 317–343 (1985).
166. Tukey, H. B., Artificial culture of sweet cherry embryos, *J. Hered.* **24,** 7–12 (1933).
167. Tukey, H. B., Development of cherry and peach fruits as affected by destruction of the embryo, *Bot. Gaz.* **98,** 1–24 (1936).
168. Udovic, D., and Aker, C., Fruit abortion and the regulation of fruit number in *Yucca whipplei, Oecologia* **49,** 245–248 (1981).
169. Urata, U., Pollination requirements of macadamia, *Hawaii Agric. Exp. Stn. Tech. Bull. No. 22,* 1954.

170. Vander Kloet, S. P., and Tosh, D., Effects of pollen donors on seed production, seed weight, germination, and seedling vigor in *Vaccinium corymbosum* L., *Am. Midl. Nat.* **112,** 392–396 (1984).
171. Van Steveninck, R. F. M., Factors affecting the abscission of reproductive organs in yellow lupins (*Lupinus luteus* L.). I. The effect of different patterns of flower removal, *J. Exp. Bot.* **8,** 373–381 (1957).
172. Van Steveninck, R. F. M., Factors affecting the abscission of reproductive organs in yellow lupins. (*Lupinus luteus* L.). II. Endogenous growth substances in virus-infected and healthy plants and their effect on abscission, *J. Exp. Bot.* **10,** 367–376 (1959).
173. Verkerk, K., The pollination of tomatoes, *Neth. J. Agric. Sci.* **5,** 37–54 (1957).
174. Waite, M. B., The pollination of pear flowers, *U.S. Div. Veg. Pathol. Bull.* **5,** 1–86 (1984).
175. Waller, D., The relative costs of self- and cross-fertilized seeds in *Impatiens capensis* (Balsaminaceae), *Am. J. Bot.* **66,** 313–320 (1979).
176. Ward, H. M., *The Oak.* D. Appleton Co., New York, 1892.
177. Wareing, P. F., and Phillips, I. D. H., *The Control of Growth and Development,* Second Ed., pp. 123–128. Pergamon Press, Oxford, 1978.
178. Waser, N. M., and Price, M. V., Optimal and actual outcrossing in plants, and the nature of plant-pollinator interaction, in *Handbook of Experimental Pollination Biology* (C. E. Jones and R. J. Little, eds.), Chapter 17. Von Nostrand Reinhold Co., New York, 1983.
179. Watson, M. A., and Casper, B. B., Morphogenetic constraints on patterns of carbon distribution in plants, *Annu Rev. Evol. Syst.* **15,** 233–258 (1984).
180. Webb, C. J., and Bawa, K. S., Patterns of fruit and seed production in *Bauhinia ungulata* (Leguminosae), *Plant Syst. Evol.* **151,** 55–65 (1985).
181. Webb, R. A., Terblanche, J. H., Purves, J. V., and Beech, M. G., Size factors in strawberry fruit, *Sci. Hortic. (Amsterdam)* **9,** 347–356 (1978).
182. Weller, S. G., Pollen flow and fecundity in populations of *Lithospermum carolinense, Am. J. Bot.* **67,** 1334–1341 (1980).
183. Westoby, M., and Rice, B., Evolution of the seed plants and inclusive fitness of plant tissues, *Evolution* **36,** 713–724 (1982).
184. Wiens, D., Ovule survivorship, brood size, life history, breeding system, and reproductive success in plants, *Oecologia* **64,** 47–53 (1984).
185. Williams, D. D. F., Influence of soil moisture level on flower abscission, ovule abortion, and seed development in the snap bean, *Diss. Abstr.* **22,** 2933 (1962).
186. Willson, M. F., Sexual selection in plants, *Am. Nat.* **113,** 777–790 (1979).
187. Willson, M. F., and Burley, N., *Mate Choice in Plants: Tactics, Mechanisms, and Consequences.* Princeton University Press, Princeton, New Jersey, 1983.
188. Willson, M. F., and D. W. Schemske, Pollinator limitation, fruit production, and floral display in pawpaw *(Asimina triloba), Bull. Torrey Bot. Club* **107,** 401–408 (1980).
189. Wright, S. T. C., Studies of fruit development in relation to plant hormones III. Auxins in relation to fruit morphogenesis and fruit drop in the black currant, *Ribes nigrum, J. Hortic. Sci.* **31,** 196–211, (1956).
190. Wyatt, R., The reproductive biology of *Asclepias tuberosa:* I. Flower number, arrangement, and fruit-set, *New Phytol.* **85,** 119–131 (1980).
191. Wyatt, R., Components of reproductive output in five tropical legumes, *Bull. Torrey Bot. Club* **108,** 67–75 (1981).
192. Wyatt, R., Inflorescence and architecture: How flower number, arrangement, and phenology affect pollination and fruit-set, *Am. J. Bot.* **69,** 585–594 (1982).
193. Wyatt, R., Pollinator-plant interactions and the evolution of breeding systems, in *Pollination Biology* (L. Real, ed.), Chapter 4. Academic Press, Orlando, Florida, 1983.
194. Zimmerman, M., Nectar production, flowering phenology, and strategies for pollination, in *Plant Reproductive Ecology: Patterns and Strategies* (J. Lovett Doust and L. Lovett Doust eds.), Chapter 8. Oxford Univ. Press, New York, 1988.

10

Plant Morphology and Reproduction

DONALD M. WALLER

A plant's reproductive behavior depends in both obvious and subtle ways on its size, growth form, and longevity. Some of these associations are the unavoidable results of the great plasticity found in plants, such as the general correlation between plant size and total flower and fruit production. Others probably reflect adaptation to particular growing conditions, such as the association noted by Salisbury[114] between trees growing in shaded habitats and large seeds. This chapter explores how plant size, longevity, and patterns of construction influence the adoption of particular flowering and fruiting habits. To do so, I view plant form as the physical embodiment of its life history, i.e., in terms of its effects on reproductive fitness.

Consistent associations between certain types of reproduction and certain plant forms immediately raise the question of whether these associations reflect adaptive "strategies" evolved in response to particular selective forces, or are instead incidental consequences of how plants grow and develop. They can be both. The ways that branches form, stems elongate, and crowns develop obviously set the stage by favoring certain flowering and fruiting patterns, possibly by favoring certain pollinators or dispersal agents. Conversely, we have often been quick to infer that a given association represents an adaptive response before exploring how it might have arisen from some mechanical or physiological constraint (e.g., the apparent increases in reproductive effort with plant size; see Weiner,[140] this volume). How these physiological and developmental details constrain flowering and fruiting remains unclear in most species. We may be safe in inferring an adaptive response when phylogenetically unrelated species facing similar selective regimes display a common suite of morphological and reproductive characters, but such convergence has rarely been documented. Ultimately, we need studies of the quantitative genetics of morphological and life history characters to distinguish truly adaptive responses from correlated responses to selection (see below, The Genetic Basis for Morphological Variation).

In this chapter, I will first review the manner in which morphological traits may be incorporated into life history analyses to assess their contribution to reproductive success. Then the evolutionary response of plants to selection, which depends on the genetic basis of their morphology, is discussed. This is followed by a survey of some of the many associations noted between patterns of flowering and fruiting and particular plant forms. These associations could arise for a variety of reasons, requiring us to carefully assess the relative roles of various possible causal factors. Finally, I discuss some of the mechanisms by which plant form might affect flowering and fruiting, emphasizing both the plasticity and constraints imposed by modular construction.

Plant morphology (the size, shape, and positioning of plant parts) affects reproductive performance on several scales: whole plant, branch or ramet, inflorescence, flower, and seed. This chapter considers how the forms of plants and their parts affect patterns of flowering and fruiting. Treatments of several related topics have already appeared, including the details of how flowers are arranged into inflorescences,[33,144] associations between particular plant forms and the timing of reproductive events,[110,146] the ecological role of particular types of flowers, fruits and seeds, the scaling of reproductive allocation (see Weiner,[140] this volume), and the qualitative and quantitative descriptions of branching and plant architecture.[12,135] This chapter aims to provide an overview of a rather broad territory rather than a detailed and systematic survey.

A great diversity of both *intrinsic* and *extrinsic* factors affect the reproductive behavior of plants. The extrinsic factors include both abiotic growing conditions (e.g., temperature, light, and wind) and biotic factors, like competitors, pollinators, and dispersers. Some of these extrinsic factors are discussed elsewhere in this book (Chapters 2, 4, 5, 8, 11, and 12) and will only be considered here insofar as they mediate plant reproductive responses through changing plant morphology. Intrinsic factors, like the network of vascular supply tissue, carbon and nutrient balances within the plant, and the modular nature of plants, are just as important, and have rarely been considered in this context.[138,139]

THE CONTEXT: PLANT LIFE HISTORIES

The ultimate criterion for judging a plant's success, or the contribution of any character to that success, is the number of descendants it leaves. Thus, the adaptive significance of any aspect of plant form rests on its effect on that plant's reproduction. Although life history theory has traditionally assessed how growth and reproduction at each *age* contribute to reproductive success (e.g., Ref. 104), these models generalize easily to represent size classes, or *stages,* that are more appropriate than age classes for most plants.[19,20,22,78]

What are the reproductive consequences of differentiating many branches instead of just a few? Of accelerating or retarding development? Of producing a flower instead of a branch or vice versa at some particular position? Answers to these questions could be pursued experimentally using plant hormones or some other treatment to divert a plant's development from its normal course (given appropriate controls, of course, to ensure that the treatment does not otherwise disrupt the plant). Results from agronomy and horticulture suggest that such manipulations have dramatic effects. If realized patterns of growth are optimal, however, no manipulation should increase fitness, even in slowly growing plants. Their slow growth may represent their adjustment to slow rates of nutrient supply, whereas rapid growth could bring high transpiration and nutrient demands later that could not be met by the plant.[24,58] Similarly, trees that often appear to be "sink-limited" should not benefit from a more rapid proliferation of branches (although the reproductive accounting would have to be continued for its entire life span). Rapid crown growth could outstrip the ability of a tree's roots and trunk to support further growth, and any temporary flush of growth enjoyed by a rapidly spreading crown might come at the expense of stature and ability to compete later for light. Thus, a plant's morphology should affect how we interpret its physiology: the existence of sink-limited growth does not indicate that plants are so rich in carbohydrate that carbon allocation is not an important factor, but only that growth is being restricted to certain axes.

Vegetative Growth versus Reproduction

To assess which life history is theoretically optimal, we need to determine that schedule of somatic growth and sexual reproduction that maximizes total reproduction (or, more properly, the rate of increase of a genome). Growth and reproduction often compete, and may do so on a structural level as when meristems differentiate into specialized vegetative or reproductive axes. In such cases, buds or meristems that develop flowers or inflorescences are unavailable for vegetative growth and vice versa (although some of the nutrients contained in vegetative structures can be, and often are, reused for reproduction). This may not appear to be much of a constraint, in that many plants, especially trees, seem to have an abundant supply of such buds or are capable of differentiating more such buds quickly, as with epicormic and root sprouts. However, plants growing in close competition (or herbaceous plants growing in the shade) may be limited by how many meristems they can develop.[137]

Vegetative and reproductive axes in some plants are produced in characteristic, predictable positions, but in other cases, a particular meristem may give rise to either a vegetative shoot or a flower or inflorescence. In such taxa we can study the proximate and ultimate factors that favor each mode of growth, or even experimentally trick plants into differentiating the "wrong" type of axis. (This, of course, is a common horticultural practice as when photoperiod is adjusted to induce flowering.) Plant physiologists have elucidated much about the change from a vegetative to a reproductive axis, but have generally been more concerned with proximate cause than morphological and reproductive consequences. It may also be possible to reciprocally transplant populations or related taxa that display different types of axes in the same position in different environments to see how each type of structure affects performance in each environment. This would also illuminate the genetic basis for such a response.

Water hyacinth is a plant in which meristems may develop into either new vegetative shoots or into an inflorescence. By adjusting photoperiod, Watson[137] induced flowering in one population while monitoring the vegetative growth of both it and a vegetative control. Even before the flowering axis was large enough to constitute an appreciable material investment, it significantly decreased the vegetative growth rate of its axis. This clearly demonstrates the "opportunity" cost of diverting meristems to reproduction.

Form and Reproduction in Annuals

In annuals, reproductive success is summed up by the number of descendants they produce during their brief life. This reduces their life history strategy to two problems: (1) when to initiate flowering and (2) whether to mix vegetative growth and reproduction. Simple models that assume a predictable end to the growing season[75,102,103] predict that annuals should invest exclusively in vegetative growth for a set period, then switch to invest all further resources in reproduction (as well as retranslocating available stored material[27]). If the end of the season is unpredictable, this "bang–bang" strategy may be replaced by a strategy of mixed growth and reproduction.[29,30] If continued growth brings high rewards, as when plants compete strongly with their neighbors for light, then a slight edge in height could allow a plant to dominate its neighbors. Clearly, the best strategy depends on what neighboring plants do,[118] necessitating the use of the evolutionary stable strategy (ESS) theory.[48,92] This approach developed from game theory, and assumes that the "payoff" associated with any particular pattern of growth depends on the frequency of it and other strategies in the local population. Under competitive conditions, plants might be selected to play an evolutionary game of "chicken," postponing reproduction until the last possible moment. Alternatively, if survival is uncertain, or if the rates of flower and fruit development are slowed under competitive conditions, selection could favor earlier reproduction. In either case, annuals should display traits that allow them to ripen seeds with speed and certainty (see below, Flower and Seed Polymorphisms).

What are the reproductive consequences of producing a flower instead of a branch at a particular position? Instead of the manipulative experiments proposed above, we might do "Gedanken" experiments to estimate the fitness of alternative patterns of growth and reproduction that might conceivably occur even when they do not in nature. This requires that we be able to estimate accurately the growth rates of alternative forms together with their eventual payoff (seed set). Using extensive demographic data on rates of branch growth, survival, and reproductive potential in *Floerkea proserpinocoides* from several environments, Smith[123] estimated the expected returns from producing a branch or a flower at each of several nodes. In almost every population, plants switch from vegetative growth to reproduction at the "correct" time to maximize their total seed set in that particular environment. This technique is a potentially powerful tool for investigating whether observed morphologies and patterns of growth indeed increase reproductive success.

Growth and Reproduction in Perennials

Givnish[48] used an ESS model to ask how tall herbs should grow as a function of the density of competing plants and the allometrically increasing cost of supporting higher leaves. If light and other resources allow a continuous cover of herbs to form, taller, more competitive herbs are favored. Because the cost of constructing stems is high (scaling with the fourth power of height,[87] tall plants cannot invest as heavily (in a proportionate sense) in reproduction as shorter herbs. This simple trade-off is probably the basis for the high yields obtained with dwarf strains of rice and wheat in the "Green Revolution" (where the density and genetic identities of neighbors are, of course, artificially managed to reduce interplant competition). Some agronomists have recognized this trade-off and emphasized the importance of selecting crops for architectural traits.[38] Where growth continues for several years, however, selection favors perennials that reduce or postpone reproduction, accumulating resources instead that allow them to overtop and dominate their neighbors (or subdue them in some other way, e.g., allelopathy).

Perennial forest herbs, living in a shady environment that cannot be improved by growing taller, often adopt various forms of clonal growth. They also tend to restrict their flowering, except in patches of higher light.[114] It would be interesting to explore (perhaps with experiments) whether the costs of producing seeds in the shade would reduce the ability of these herbs to persist in these environments. Close clonal growth (termed "phalanx" by Lovett Doust[84]), also acts to "cover the flank" of central stems, permitting them to grow relatively short yet still compete with neighbors.[135] The savings in reduced stem tissue could then be invested in flowers and fruit (or further organs of clonal spread). The clonal herb *Laportea canadensis* appears in some situations to be shorter than Givnish's theory would predict,[95] providing a possible example. Closely packed clonal shrubs and trees may also be practicing such defensive tactics.

Grime[58,59] argues that plants have evolved three primary strategies (ruderal, stress-tolerant, and competitive) in response to disturbance, stress, and the consequent levels of competition they permit. Like the idea of *r* and *K* selection proposed by MacArthur and Wilson,[85] Grime's scheme heuristically illustrates the trade-offs that constrain a plant's success to one or a few habitats. Grime[59] also attempts to quantitatively categorize all plants with respect to these characteristics according to various morphological and physiological measures, an effort that appears less successful as different species can respond to the same environmental perturbation (e.g., a flood) as either a stress (grasses, sedges, and trees) or a disturbance (tall herbs and shrubs), depending on their morphology.[96]

If the assumptions of life history theory are correct, there should exist trade-offs between vegetative performance and later growth and reproduction. These trade-offs are conspicuous in many trees and shrubs where they often take the form of a reduction in growth during or following a season of heavy investment in reproduction.[39,113,128] Foresters and fruit tree growers have long been aware of such negative interactions between growth and reproduction, particularly in trees that mast-fruit, i.e., produce an abundant fruit crop in some years but not others. Trade-offs between growth and reproduction are assumed to stem from access by the individual to a single pool of resources. However, large modular plants may contain many, relatively autonomous subunits,[138,139] suggesting that trade-offs might occur locally within these modules. Trade-offs between growth and reproduction occur within both stems and branches of the dioecious, clonal shrub, smooth sumac (D. Waller, unpublished data).

Comparing trade-offs among plants is more difficult, however, because their plasticity leads to positive phenotypic correlations between growth and reproduction. Such trade-offs should take the form of negative genetic correlations among fitness components. This makes it necessary either to conduct the breeding experiments typical of quantitative genetics, or, at a minimum, to control statistically for the effects of plant size (and possibly age).

Experimental work has revealed negative genetic correlations between viability and fertility in barley *(Hordeum vulgare)*,[28] among various yield components in *Plantago lanceolata*,[106] between early and late stage of development in *Papaver dubium*,[44] and between juvenile size and eventual fecundity in the winter annual, *Geranium carolinianum*[112] (see also Ref. 83). One exception is *Impatiens capensis*,[97] where negative correlations may have been precluded by its inbred population structure.

Lande[76] developed a general theory based on quantitative genetics of how genetic correlations among life history traits should constrain and channel a population's evolution. Such theory has potential for predicting how populations will respond to selection on various life history and morphology characters, at least in the short term. However, it is likely that these genetic correlations vary from population to population and through time, making it hard to determine how they might channel long-term evolutionary responses.[98] In addition, the short-term response to selection that is implied by the correlation structure may not reflect the ultimate constraints imposed by genetics or form. Perhaps the genetic correlations ultimately constraining plant form could be identified by calculating genetic correlation structures from a variety of populations and/or related species and looking for common elements. Alternatively, a detailed knowledge of the developmental and physiological basis for growth and reproduction in a species might allow one to predict, on mechanistic grounds, the nature of these ultimate genetic constraints. Neither of these research programs has been attempted, perhaps because of the size of the tasks, or perhaps because of the gaps that have traditionally separated physiology, morphology, and quantitative genetics.

Opportunities for Clonal Growth

Life history models may be generalized to explicitly include the consequences of vegetative "reproduction."[21] As with simpler models, it is the contribution of each activity (local growth, vegetative propagation, or reproduction) at each age or size to total seed production, together with the trade-offs among these activities, that determines the optimal pattern of allocation. As opportunities for gaining fitness through stem and clonal growth decrease, sexual reproduction is increasingly favored. Evidence for a trade-off between clonal growth and allocation to female (but not male) sexual reproduction now exists for at least one species, *Laportea canadensis.*[95]

By making particular assumptions regarding the mode(s) of clonal growth and trade-offs between growth and reproduction, it is possible to make theoretical predictions regarding the relative allocation to clonal growth versus sexual reproduction. Armstrong[4] developed such a model, assuming that (1) total seed production is proportional to the area covered, (2) radial growth rate and seed production per unit area depend linearly on allocation to each activity, and (3) that each ramet's reproductive behavior is independent of its age and position within the clone. It predicts a fixed ratio of one-third allocation to seeds and two-thirds to clonal growth (ramet growth is not a variable in this model). He recently extended this model[5] to incorporate leaf cover values that vary in response to investment in clonal growth and that bring different energetic returns. Under conditions where clonal allocation increased cover more than it increased radial growth (as it would in central ramets when only peripheral ramets support the expansion of new clonal ramets), no simple prediction could be made about the expected ratio of allocations. Similarly, when clonal allocation increased radial growth much more than local cover, the optimal allocation depended on the efficiency with which allocation to "rhizomes" increased radial growth. Specifically, the optimal ratio of clonal versus sexual reproduction was $2/(B - 2)$ where B is the exponent to which rhizome length is raised to calculate the energetic cost per unit rhizome. Thus, clonal plants are not all alike, and quantitative predictions of allocation will require knowledge of the internal physiological and biomechanical constraints, the patterns of growth, and the nature of the external environment that such plants face.

The Genetic Basis for Morphological Variation

Most differences in organ shape, developmental order, and the relative placement of leaves and flowers have, at least partly, a genetic basis. Surveys have revealed significant genetic components to morphological variation in barley,[100] *Spartina patens,*[121] *Borrichia fructescens,*[2] *Salix repens,*[42] and *Dryas octopetala.*[86]

The genetic basis for morphological characters is probably some mixture of many genes of minor effect, producing quantitative variation, and one or a few genes of major effect, segregating in classic Mendelian fashion. In a recent review, Gottlieb[53] concluded that although differences in plant size usually reflect the combined action of many genes of individually small effect, qualitative morphological differences within and especially between species could often be attributed to one or a few genes of major effect. Gottlieb argued that the relatively open development of plants via the indeterminate production of repeated parts might accommodate genetic change via such major genes better than the intricate and tightly integrated development of animals. Coyne and Lande[34] challenged this interpretation, arguing that deleterious pleiotropic effects resulting from major gene substitutions could occur as often as in animals, and that the resemblance between species differences and those induced by major gene substitutions did not necessarily imply that such genes cause the interspecific differences. Gottlieb[54] argued back that deleterious pleiotropic effects need not be as pervasive in plants, and that his hypothesis that initial major gene substitutions cause differentiation between lines does not preclude the later selection of multiple modifier loci of minor effect. Both sides agree that more information is needed to determine the nature and number of genes causing morphological differences between taxa.

ASSOCIATIONS AMONG GROWTH FORM, LIFE HISTORY, AND REPRODUCTIVE BEHAVIOR

Having discussed the way in which a plant's form determines much of its life history, and how this provides a context for interpreting plant morphology, let us turn to the patterns associating particular forms with various types of reproductive behavior. This section presents some of these associations, together with suggestions about why they might exist.

Plant forms can be categorized in many ways,[135] by, among other things, developmental morphology,[62] physiognomy,[15,136] and function.[72] Hallé et al. describe in detail how vegetative and reproductive characteristics combine to produce 23 morphologically distinct architectural models. In the simpler models, architecture is intimately tied up with the pattern of reproduction. A monocaulous (single-stemmed) tree, for example, may end in an inflorescence (Holttum's model), making it necessarily monocarpic. Species in several palm genera display this architecture, including *Corypha, Arenga,* and *Metroxylon.*[32] If several monoaxial trunks derive from one plant (Tomlinson's model), the individual shoots may be monocarpic (as in the banana *Musa*), as can the entire genet (as in several bamboos), but this is not necessarily the case. Thus, ecological conditions that favor these architectures could indirectly favor monocarpy (or vice versa). Monocarpy has, however, also evolved in other architectures where it presumably requires some hormonal cue.

Associations between reproductive and morphological characters should be analyzed very carefully, since many factors may be confounded. For example, the relative abundances of various plant forms like those of Raunkiaer[108] change predictably in response to climatic gradients, but so do the abundances of possible pollinators and dispersal agents. Changes in reproductive characteristics along a mountain gradient, for example, could reflect all of these influences.[6] This will make it difficult to identify which factor(s) cause a particular association. Where possible, we should obtain data on plant form, breeding system, pollination and dispersal syndrome, and relevant environmental variables and analyze the data together, adjusting for the effects of extraneous variables.

Breeding System

What associations exist between particular breeding systems and growth forms? This section will first address the broad-scale patterns that have been noted in the literature, then consider several subsidiary questions of how particular aspects of plant morphology may have influenced the likelihood of adopting certain breeding systems and how breeding systems may affect selection on morphology.

There is a rather dramatic association between plant stature and breeding system (Table 10.1), with monoecy and dioecy usually associated with woodiness and tree size. Maynard Smith[91] interprets this to be the result of increased selection for outcrossing in large, long-lived plants. Some have argued, however, that a genetic incompatibility mechanism would be sufficient to enforce outcrossing, and that selection of another kind must be operating (see Barrett,[10] Cox,[33] and Schlessman,[118] this volume). How woodiness, pollinator type, and the presence of fleshy fruit simultaneously affect the likelihood of adopting dioecy awaits detailed analyses.[43]

Dioecy allows possibilities for sexual dimorphism, limited, of course, by the genetic correlations between forms expressed in each sex. This is true whether gender is determined genetically or environmentally if whatever cue that determines gender can also influence form (see Meagher,[93a] this volume). These species therefore provide a valuable context for examining how selection on reproductive characters can modify vegetative characters. Differences in growth rate or form between male and female plants have been noted in several dioecious species (e.g., *Silene alba*[60]). In some clonal species, males have more extensive clonal propagation (e.g., *Rumex* spp.[107]; *Thalictrum dioicum*[94]), possibly because female investment in ripening seeds and fruits exhausts more resources.[25] In other clonal species, however, females show enhanced vegetative growth, as in *Populus tremuloides*[55] and *Rhus glabra* (unpublished data). This could occur either because males in these species invest more than females in

Table 10.1. Dependence of the Breeding System on Plant Growth Form[a,b]

Layer	Hermaphrodite	Monoecious	Dioecious
Tree	50–57	14–30	20–33
Shrub	60–89	0–38	0–23
Field	0–27	0–9	0–9

[a]After Maynard Smith.[91]

[b]Figures show the percentages of plants in four types of deciduous woodland in Britain that are in each layer.

reproduction (presumably in response to differing selective pressures acting between the sexes) or, perhaps more parsimoniously, because of the earlier timing of male investment. Such investment may compete directly with growth and may have to be supplied from stored and retranslocated photosynthate, incurring these additional costs.

Monoecy provides for some of the same functional isolation between the sexes as found in dioecy, but the isolation is within rather than between plants. While morphology may be less labile than in dioecious plants, the positioning and number of male and female flowers may provide clues as to how gender function is affected by local differences in plant form. One advantage of monoecy may be that such plants can arrange the flowers of each sex to match particular microenvironments.[71] It may also allow plants to shift easily their relative investment in male or female flowers in response to local environmental conditions or their own size or vigor.[143]

The placement of male and female flowers in different parts of the plant or plant module automatically affects the relative allocation to gender when variable environmental conditions affect plant size and form. In *Ambrosia trifida,* male inflorescences are terminal and female ones are axillary. Thus, the fact that short, shaded plants are mostly female, while the taller plants also have male inflorescences[1] reflects the effects of architecture. Similarly, more-branched *Xanthium strumarium* plants produced a preponderance of staminate inflorescences.[77] In pines, male cones occur in clusters at the base of the current year's growth and are abundant on lower branches, while female cones cluster along upper branches[109] where they may have better access to photosynthate.

Grasses and sedges, primarily wind pollinated like the conifers, vary greatly in their placement of male and female flowers, which, in turn, can affect the type and degree of herkogamy they express. With a hierarchy of reproductive structures (floret, spikelet, spike, and inflorescence), gender differentiation can occur on any level.[33] There is remarkable diversity even within the single large genus *Carex.* Radford et al.[109] list 45 species where male and female flowers co-occur in the same inflorescence, 20 where the male flowers develop first ("androgynous"), and 25 where the female flowers do ("gynecandrous"). Further herkogamy may result from the placement of flowers into distinct male and female spikes. Male spikes are usually terminal, female spikes lateral, favoring protandry in determinate inflorescences. However, the terminal spikes may themselves be gynecandrous, leading to a female/male/female alternation. Some Cyperaceae follow the reverse sequence. A comprehensive survey of the ecological correlates of the positioning of male and female flowers in the grasses and sedges (controlling for phylogenetic relationship) might generate some interesting hypotheses about the function these differences serve. Examining the patterns of development and vascular anatomy might, alternatively, reveal them to be the necessary consequence of design constraints.

The transition from the wild grass teosinte (*Zea mays* ssp. *mexicana*) with its tiny ears and rock-hard seed cases to modern maize (*Zea mays* ssp. *mays*) has long been an enigma.[88] Teosinte has a central stem with long lateral branches, both flanked with lateral female spikes and terminated by a tassel (the male inflorescence). In maize, the position of the branch tassel is occupied by an ear, and the lateral branches are greatly telescoped. Iltis[73] has offered the startling hypothesis that the ear is, by homology, the "feminized" central spike of the male tassel (Fig. 10.1). He suggests that the lateral branches were initially shortened by some disease or mutation, bringing the terminal

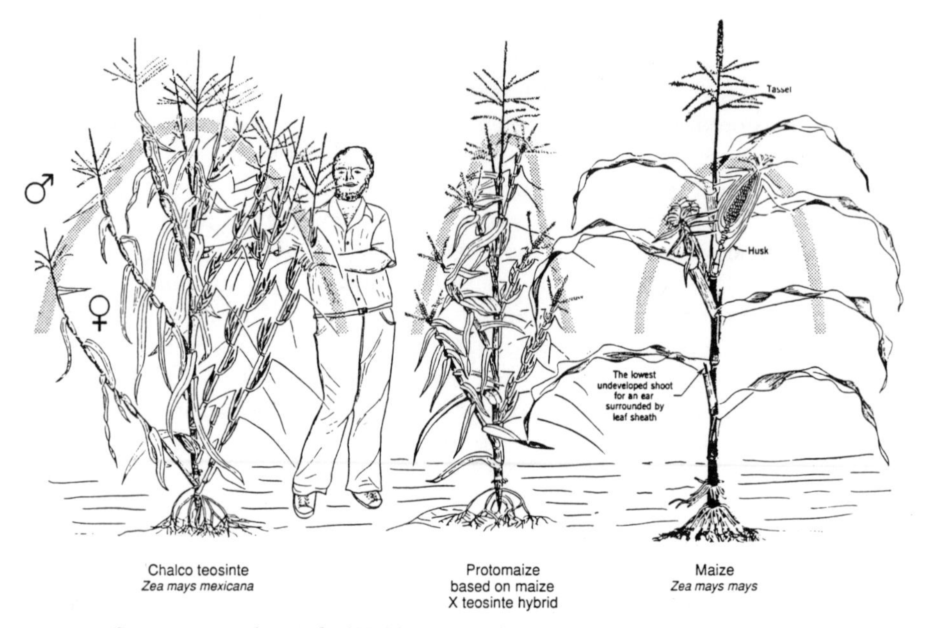

Fig. 10.1. Gross comparison of wild Chalco teosinte, the presumed ancestor of maize, with a teosinte–maize hybrid and maize showing habit (left sides) and internal diagrams (right sides). Iltis's[73] "catastrophic sexual transmutation" theory holds that some, perhaps intially environmental, factor contracted the teosinte branch internodes to the point where their terminal male tassels were brought through a transition zone (shaded) into a zone of female expression. Subsequent apical dominance presumably supressed the lower lateral female inflorescences. (Modified from Iltis.[73])

male spikes into a physiological zone of "feminization" where strong apical dominance caused further condensation and suppression of lateral inflorescences. Humans presumably seized on these developmental accidents and selected strongly to fix this trait. While the genetic and physiological bases for this postulated "catastrophic sexual transmutation" are still unclear, the morphological evidence strongly supports it. This evidence includes the frequent presence of staminate "tails" on maize ears and the curiously thickened central male spikelet in corn but not teosinte (presumably a correlated response to selection on the ear). If true, such a shift well illustrates the intimate connection between morphology and reproductive behavior in monoecious species.

In both monoecious plants and those with hermaphroditic flowers, natural selection will tend to favor a roughly equal level of investment in male and female function (assuming outcrossing and the absense of various other special circumstances[25]). Artificial selection, however, may modify this original pattern of allocation, both by changing the breeding system (generally more toward self-fertilization, which favors female function) and by allowing male and female traits to be selected somewhat independently. Indeed, the independence of male and female structures in monoecious species like cucurbits and corn may have preadapted them to artificial selection for increased seed output in that such selection could act on the hormonal cues that determine flower sex. Plants that diverted resources away from male flowers to the production of more ovules and seeds would be favored by plant breeders (up to the limit where fruiting was pollen limited). It would therefore be interesting to compare the relative allocation to male and female structures in monoecious crops to that in their wild relatives to see whether such reallocation has accompanied domestication.

In plants with hermaphroditic flowers, the relative allocation to male and female function may also shift with flower position or overall plant form. In the Compositae, flowers usually develop centripetally within heads, so protandry produces a spatial arrangement of peripheral female and central male flowers.[127] This may also occur within the umbels of the Umbelliferae or among their primary and secondary inflorescences.[35,80]

Pollination

Plant form, in conjunction with habitat and phenology, affects the probability of adopting particular pollination syndromes. Here some associations are suggested, without the statistical analysis that will be necessary to unequivocally demonstrate their existence and untangle related factors.

Tall herbs that grow densely in open habitats (e.g., *Ambrosia, Artemisia,* and many grasses) are frequently wind pollinated, especially when insect pollinators are scarce.[142] Trees that occupy low diversity stands tend also to be wind pollinated, and probably for the same reason: they have access to winds away from the boundary layer at ground level (especially before leaf-out in deciduous canopies). These associations seem natural, since taller plants are more exposed to the wind and release their pollen at a greater height. Because most wind-dispersed pollen is distributed in the immediate vicinity of its parent, this mode of pollination is ineffective for plants distributed singly at some distance from each other. Self-incompatible clonal plants might have the opposite problem, with self-pollen saturating the stigmas within the clone (see Bertin,[13] this volume). This is not a problem, of course, in dioecious (or self-compatible) species and may, in part, account for the association of dioecy with wind pollination.[47] Although wind pollination appears to some to be "inefficient" and primitive, it is a derived condition in many angiosperms and therefore represents active selection away from animal to wind pollination.

Wind-pollinated flowers in trees are usually placed on the periphery of the crown or in long pendant catkins that provide greater access to the wind. The spikes of grasses and the long, open racemes of other wind-pollinated herbs do the same. The branches and inflorescence structures within plants may also serve to deflect breezes to facilitate pollen dispersal or capture. Niklas[101] claims that the structures of female conifer cones produce aerodynamic vortices that preferentially precipitate pollen of their own species over that of other species.

Woodland herbs often have solitary, relatively large flowers placed terminally or in axils and pollinated by specialized insects. Such sparsely distributed plants lack clear access to the wind, a situation that probably favors pollinators capable of ensuring accurate pollen delivery over relatively long distances.

The timing of sexual phases within and between flowers also has an obvious morphological component (see also Cox,[33] this volume). Within a flower, the normal sequence of centripetal development tends to produce protandry, the maturation of anthers before styles and stigmas. This morphological/developmental foundation may provide a proximate explanation for why protandry occurs more frequently than protogyny:[145] 37% versus 8% in 235 hermaphroditic New Zealand plants,[51] even though protogyny is a more effective way to prevent self-pollination in that it gives foreign pollen a head start over self-pollen.[40] In monoecious plants, however, protogyny (defined, obviously, in relation to the different flowers on a plant) occurs more frequently than protandry, often reflecting morphology, as discussed above.

Finally, a species' pattern of flowering may also depend directly on plant form. In *Linanthus androsaceus,* stabilizing selection has apparently been effective in favoring a particular date of first flowering, but larger plants continue flowering longer, producing a positively skewed distribution of population flowering in the population. It had been argued that such a population distribution represents an adaptive pattern which enhances pollinator visitation,[119] yet in this species it appears instead to be the natural consequence of a population's size structure.

Flower and Seed Polymorphisms

Variation within plants like that just described has given rise in many species to distinct floral and seed polymorphisms where the effect of plant form on reproduction is often pronounced. A familiar dimorphism is the presence of both ray and disk florets in the heads of most of the Compositae and the less extreme differentiation present in the clumped flowers of other families (e.g., the asymmetrically large-petalled marginal flowers and central dark flower in the umbellifer *Daucus carota*). Darwin[36] discussed many examples of floral polymorphism, including male versus female flowers, heterostyly, and cleistogamy, which is the production of small flowers specialized for self-fertilization. These polymorphisms generally reflect some specialization of function, as with the gender specializations discussed earlier. Within inflorescences, the specializations are usually for attraction versus pollen or seed production.

Cleistogamous flowers are a conspicuous floral polymorphism that responds to plant form in a characteristic way. In many of the grasses, cleistogamous flowers are produced in set locations (e.g., at the base of a spike within the leaf sheath).[18] This tends to produce a relatively constant fraction of cleistogamous flowers on all plants. In most other cases, however, the proportion of cleistogamous flowers and fruit is highly variable, responding to local environmental conditions, a situation termed "environmental cleistogamy."[129] In fact, there is usually a hierarchy of investment such that some cleistogamous flowers are produced by all plants, but only larger, more vigorous plants produce outcrossed (chasmogamous) flowers.[120,132]

The conditions that favor cleistogamous flowers are invariably "adverse," e.g., cold, drought, and shade. Darwin[36] suggested that this correlation might reflect the vulnerability of such plants to abandonment by their specialized pollinators under conditions inappropriate for their activity. He noted that a high proportion of cleistogamous species have bilaterally symmetric flowers, suggesting their dependence on specialized, and perhaps fickle, pollinators. However, field studies have found no particular dearth of pollinators in these areas, and such variable pollinator levels would not, in any case, account for why large, vigorous plants also produce cleistogamous flowers. It seems more likely that these patterns reflect internal constraints: bilaterally symmetric flowers attractive to specialized pollinators tend to be large, full of nectar, and hence energetically expensive. Plants (or localized parts of plants) with few resources may not be able to "afford" their higher cost. The patterns of producing cleistogamous flowers within plants reinforce this hypothesis: they are produced on lower and inner branches in *Impatiens capensis*[132] and during the summer in woodland violets when the plants are shaded. These observations suggest that reduced (local) growth rates may directly favor cleistogamous flowers, due to either their energetic efficiency[117,131] or accelerated development.[81]

Such floral polymorphisms predispose certain taxa to the evolution of seed polymorphisms. Probably the most general type is dormancy polymorphism, most conspicuous in the seeds of many weeds and desert annuals.[31,66] In most cases, these polymorphisms are imposed by the seed coat, a maternal tissue (as expected since progeny are under less selection to disperse[63]). Such polymorphisms could allow seed type to react to position within the mother plant or external environmental conditions. Thus, their presence could allow selection to adjust the relative number of flower or seed types among plants of different size or growing under different conditions, as with cleistogamy.

Differences in dormancy may be accompanied by physical seed dimorphisms, especially in the Compositae with their distinct disk and ray florets. In *Hypochaeris glabra,* plants growing at high density produce mostly short, rough achenes adapted for epizoochory, while plants at low density produce more long achenes with a pappus adapted for dispersal via wind.[9] Similarly, in *Heterotheca latifolia,* the ray achenes have delayed germination, responsive to growing conditions, while the disk achenes have a pappus for wind dispersal and display little dormancy.[130] Larger plants produce more plumed disk achenes, but late in the season the heads are smaller and produce more ray achenes. Although such flexibility could allow a plant to adjust the relative production of each seed type in an apparently adaptive way, we should also consider how structural and developmental constraints might provide a more parsimonious explanation. In *Xanthium strumarium,* fruits each contain two seeds, differing in size and dormancy, but this dimorphism does not appear to depend on plant form or growing conditions.[65]

Seed polymorphisms also result from *amphicarpy,* the production of different sizes of fruit. This usually results from the production of underground cleistogamous flowers that produce much larger seeds (probably reflecting the importance of seed predation). Amphicarpic plants usually display a hierarchy of investment similar to other cleistogamous plants: even small plants ripen one or a few subterranean cleistogamous seeds, but only larger plants growing under favorable conditions invest in aerial flowers and fruit.[26] This hierarchy is particularly pronounced in *Amphicarpaea bracteata,* where a tripartite hierarchy exists: a plant first produces one to a few subterranean cleistogamous seeds, then several aerial cleistogamous flowers, and finally, aerial open flowers and seeds.[120] Open flowers therefore only appear on the largest plants.

Annual plants display a disproportionate share of these floral and seed polymorphisms, perhaps because the uncertainty they face has severe consequences. By germinating over several years via dormancy polymorphisms, they clearly hedge bets against catastrophic mortality in any one year. These plants are also strongly constrained by the length of their growing season and the necessity to produce some seed under all circumstances.

Fruit Dispersal and Plant Form

A plant's size and the positions of its fruits probably affect both the rate and quality of dispersal. Such effects are most obvious with wind-dispersed seeds, e.g., spinning samaras where dispersal distance and dispersion are both directly related to height of release.[56] Therefore it is not surprising that tall trees and herbs of open habitats are frequently wind dispersed, while understory trees and shrubs often rely on fleshy fruits.[126] As with pollen, a variety of structures serve to present fruit to the wind, including "salt shakers" (e.g., *Papaver*), dehiscent pods *(Asclepias),* and the scapes of many of the Compositae (e.g., *Hieracium, Taraxacum*). Augspurger[7] has experimentally modified the size, weight, and aerodynamic loading of artificial fruits to assess their influence on the dispersal patterns produced. The importance of these patterns for reaching "safe sites" is suggested by the result in a wind-dispersed tropical tree adapted to colonizing gaps that showed an increased incidence of fungal disease among seedlings that were clumped, shaded, or did not disperse as far.[8] Later successional trees, frequently with larger seeds, may have seeds better defended against pathogens. It would be natural to extend these efforts to explore the effects of other aspects of morphology, such as leaf and branch obstacles and the aggregations of some winged fruits, on subsequent dispersal and establishment.

Associations between growth forms and the modes of seed dispersal adopted by species have not been systematically investigated. Salisbury[114] described the "reproductive capacities" (seed set and seed size) of a great many British plants to argue against the notion that reproductive outputs were in some way adjusted to the characteristic level of mortality faced by each species. In doing so, he documented systematic variation in seed weights obtained from plants of various stature and occupying various habitats (Fig. 10.2). Mean seed weight increases with the shadiness and often the dryness of the seedling habitat, and, to some extent, with the stature of the plant producing it. This appears to be an adaptive response, since larger seeds provide more reserves for penetrating the leaf litter layer and enduring shade and drought. Salisbury also noted a tendency for vegetative propagation to occur under shaded conditions (probably for the same reason), and for obligately parasitic plants to have minute seeds, which would increase their number and dispersal distance while not compromising their growth as much as it would in autotrophic plants.

Positive correlations between seed size and the shadiness of seedling habitats and overall plant stature are not restricted to Britain. In a comprehensive survey of seed sizes in tropical trees with animal-dispersed fruit, Foster and Janson[41] found that seeds requiring large light gaps for establishment were smaller than those from other trees. They also found that larger seeds are associated with taller trees, mammal- instead of bird-dispersed fruit, and, to a lesser extent, trees rather than liana growth forms. In each comparison, they correctly controlled for the statistical effects of other variables, ensuring that these associations are real.

Turning to deciduous forest herbs, Givnish et al.[49] noted that spring ephemerals that mature their seeds early in the summer rely mostly on ant dispersal, while tall herbs that retain their foliage through the summer much more often rely on epizoochory (Table 10.2). Herbs ripening seeds late in the season depend mostly on birds (abundant and hungry during migration season) and mammals or the wind (perhaps because these herbs tend to be the tallest). It would appear natural that the plants producing epizoochorous fruit would be of intermediate stature—tall herbs or short shrubs, both because these plants tend to occupy the edge habitats frequented by mammals and because their stature puts them into proximity with these animals.

It might seem that fruit placement and form generally would be less important for the dispersal of animal-ingested fruit, yet structural details can also have dramatic

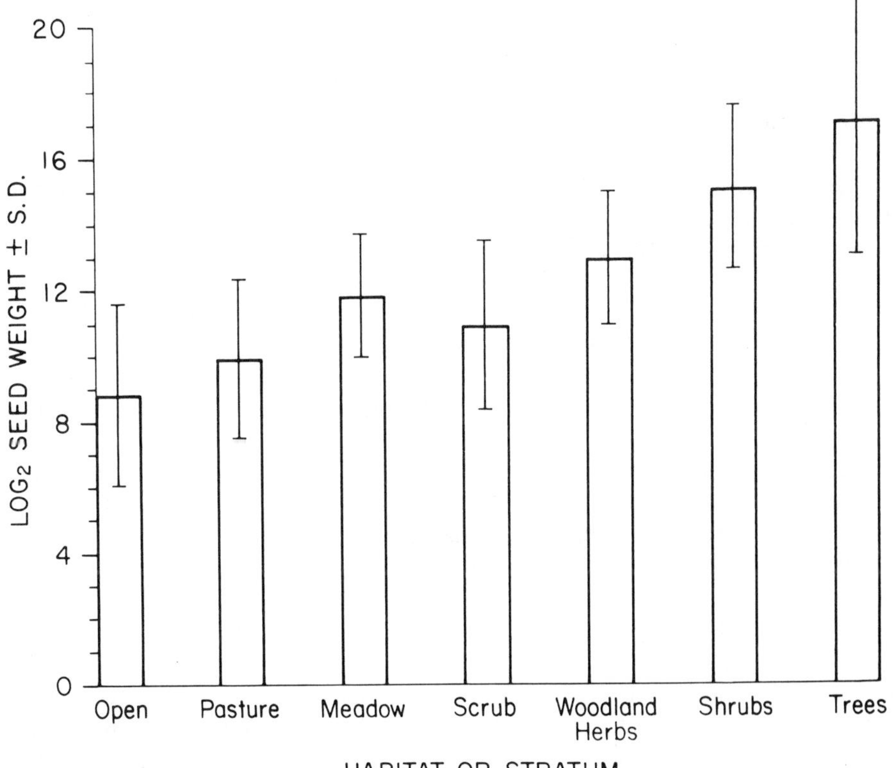

Fig. 10.2. Variation in (fresh) seed weight among several British habitats and plant strata. Note $\log_2$ scale and variability within groups. (Data from Salisbury.[114])

Table 10.2. How Mode of Seed Dispersal Depends on Season of Dispersal in a Virginia Herb Community[a,b]

	Date of fruit ripening		
	Before 6/15	6/15–8/15	After 8/15
Barochory	7	14	1
Ant	6	5	
Bird or mammal		2	8
Epizoochory	1	14	
Ballistic		1	2
Wind		7	6
	—	—	—
Total	14	43	17

[a]From Givnish et al.[49]

[b]These phenological effects reflect, in part, differences in plant form. Numbers indicate number of species displaying that mode of seed dispersal.

importance in these species. Stiles[124] hypothesized that some bird-dispersed deciduous plants cease photosynthesis early and degrade their chlorophyll, producing brightly colored "flag" leaves that attract dispersers. Denslow and Moermond[37] tied normally terminal fruit clusters into both terminal and axillary locations and discovered that terminal infructescences are harder for bird dispersers to reach, but suffer far less predation by noctural mammals than axillary infructescences. Birds also discriminate between fruit clusters, based on both their attractiveness (dependent on their nutritional content) and the level of difficulty in obtaining them, allowing the researchers and perhaps the plants to "titrate" attractiveness against proximity.[79]

INTERACTIONS BETWEEN PLANT FORM AND REPRODUCTION

Given the patterns just described, it behooves us to explore the nature of the connections between plant form and reproduction in more detail. What are the mechanisms by which morphology constrains reproduction? The two aspects of plant development, the unfolding of vegetative form and the production of flowers and fruits, are not related as cause and effect. Rather, each influences the other and both are components of the plant's overall life history.

Size and Fitness

We expect reproductive success to correlate with the total number of flowers and fruit on a plant, which, in turn, usually parallels plant size. However, plant size usually must exceed some threshold to produce any successful fruit (or even flowers).[140] What this threshold is undoubtedly depends on the characteristic morphology (and associated physiology) of the plant, and deserves some investigation.

While the association between plant size and fecundity may commonly hold, it needs to be tested far more often than it has been so far. Larger plants may not always produce more flowers than smaller plants, fruit set may not parallel flower production, and negative correlations between seed size and seed number may reduce the assumed association between reproductive output and fitness. Even where it is possible to demonstrate that success as a female (through seed) depends on plant size, it may be difficult or impossible to follow the success of pollen. In cases where this has been done, male reproductive success does not always parallel female success.[118]

Perhaps the most precise method that exists for measuring the ultimate reproductive success of individuals is to determine genetically the maternity and paternity of surviving seedlings in the field. This has now been done for the lily *Chamaelirium luteum* (Ref. 93b and Meagher, personal communication), where there does not appear to be a positive relation between plant size and reproductive success. In fact, larger males frequently had *lower* fitness than small males in this population, a warning against automatically equating size with reproductive success.

The correlations observed between overall size and various fitness components, when they do exist, are still phenotypic rather than genotypic. Thus, they usually reflect the highly variable circumstances that plants invariably experience during growth rather than genetically based differences. To untangle the relative contributions of genotype and environment to plant performance requires common garden, reciprocal transplant, or some more elaborate breeding experiment. A study of a natural population of *Stephanomeria exigua* found that differences in size were not genetic.[52] Other studies have detected significant, although variable, genetic components of performance in *Plantago lanceolata,*[106] *Phlox*[11] (also Levin, personal communication), and *Impatiens capensis.*[134] Growth in conifers, or its variability, also appears to be correlated with heterozygosity.[61,99]

Plasticity of Plant Form

Much has been written on the degree to which plants can respond to varying conditions: this is commonly termed plasticity.[14] Variation in size mostly reflects the variable number of subunits produced by a plant rather than their individual size or shape,[66,141] but plant form may also vary, even within subunits. Familiar examples include thicker sun versus shade leaves, and dissected submerged versus entire emergent leaves in aquatic plants.[122] The degree of plasticity parallels other aspects of plant growth: ruderal and competitive plants (sensu Grime[58]) often express considerable plasticity, while stress-tolerant plants generally respond less to temporary flushes of resources, either because they occupy habitats chronically short of such resources or because opportunistic responses could overextend the plant's physiological ability to tolerate subsequent stresses.[24,57]

Morphological plasticity is favored when plants occupy different environments during their ontogeny. Some lianas display considerable vegetative plasticity as a result of growing in first dark, then very bright environments. Studies are few, due to their size and location, but in a species of *Monstera,* Ray[111] documented an extreme degree of morphological plasticity and the associated effects on its flowering and fruiting.

Such vegetative plasticity is often matched by reproductive plasticity, generally in the number of parts, but sometimes in the placement or type of reproductive structures. Many deterministically flowering species in the Compositae and Umbelliferae differentiate secondary and tertiary heads or umbels when conditions permit continued growth and flowering. Such subsequent inflorescences are usually smaller, and may also differ in their relative gender and seed sets (see Hendrix,[69] this volume, and Refs. 70 and 82).

Even without dimorphic flowers or inflorescences, seed weight tends to vary with position, timing of development,[23] or physiological condition. Such nonrandom intraplant variation has been noted in *Rumex crispus,*[90] *Pastinaca,*[70] *Lomatium prayi,*[125] *Impatiens capensis,*[133] and *Sesbania.*[89] Again, it is tempting to interpret these patterns of variation as adaptive adjustments to changing selective forces, but it seems likely that at least some of this variation is the unavoidable consequence of local source/sink relations. For example, increased competition for nutrients may occur when many ovules develop in close proximity, reducing seed size. In a review of assimilate partitioning, physiologists[46] concluded that "it is the ability of sinks to take up sugars . . . which determines partitioning, not the relative adequacy of vascular connections between sources and sinks, nor the relative activity of various sources." Ecologists[138,139] reviewing much of the same literature, however, argue that vascular connections and/or local hormonal controls can divide plants into semiautonomous "independent physiological units." Of course, species differ considerably in their vascular architecture, and also in the relative advantages to be gained by integration versus local autonomy, making generalizations dangerous. In *Impatiens,* seed weight increased toward the top of the plant, as would be expected if seed size responded developmentally to local levels of photosynthate.[133] Similarly, in three species of *Sesbania,* plant size and the local environment were unrelated to seed size, while position within the plant affected seed mass in two species.[89] An experiment on the Kentucky coffee tree *(Gymnocladus dioicus)* showed that defoliating certain branches, although an insignificant fraction of the total crown, caused all the fruits on those branches to abort.[74] Perhaps a quantitative measure like the "repeatability" of quantitative genetics could be used to assess the degree of within-plant control on morphological and reproductive characters.

Clonal Growth: Opportunities and Constraints

Many plant species spread clonally via stolons, rhizomes, bulbs, bulbils, and root sprouts, presumably to exploit spatially distributed resources.[67] Incorporating these various modes of vegetative propagation into life history models,[4,5,21] makes their relative contribution to fitness explicit. Hierarchical models can also be constructed to represent both multiplication of ramets and their eventual seed production.

Associations occur between certain types of clonal propagation and habitat features. Most herbs inhabiting shaded forest understories have some means of clonal growth. In such habitats, large reserves may be necessary to initiate a new plant, and organs of clonal spread allow a larger level of investment than seeds.[114] Such organs could also allow such plants to grow preferentially in patches of high resource availability and/or cut off their subsidy to ramets experiencing poor growth. They may allow flexibility for responding to locally variable resource abundances. Some clonal species of *Viola* (R. Cook, personal communication), *Aster,*[105] and *Ambrosia*[115] appear to explore their habitat in this way, "choosing" sites with better growing conditions. Such flexibility mirrors the flexible branching and limb growth in many angiosperm trees that allows them to respond tactically to light gaps in the canopy,[135] reminding us of the essential similarity of these two forms of growth.

Clonal growth is also frequently associated with early successional or recently disturbed habitats, where it may serve as a means for quickly exploring and claiming space. L. Lovett-Doust[84] coined the terms "guerilla" and "phalanx" to represent the alternative extremes of rapid, linear growth of runners versus the slow and steady advance of a clumped array of contiguous clonal ramets. This dichotomy was, in fact, anticipated by Salisbury,[114] who distinguished between "organs of migration" (e.g., thin stolons) and structures that permit "continuity of occupation and preclusion of colonization by other species" (e.g., tightly packed rhizomes and bulbs). Such patterns of clonal growth clearly impinge on their reproductive biology. In phalanx species, the inner ramets usually flower and fruit, while peripheral ramets both conquer adjacent space and "cover the flank" of interior ramets. Inner, older ramets in clones of smooth sumac resprout repeatedly when cut or injured whereas peripheral young ones do not (D. Waller, unpublished data). This suggests that the new, subsidized ramets represent "venture capital" for testing the suitability of nearby habitat while interior ramets become organs for persisting at a site and producing seeds. A continuity of clonal growth forms exist, from those designed for exploring and quickly exploiting space, to those effective at continuously dominating space.

The initiation of flowering is often synchronous within a plant, and is apparently the result of some hormonal trigger.[45] This integration contrasts with the relative autonomy of modules within plants in terms of their quantitative level of flowering and fruiting. The degree of autonomy varies among individual ramets and even branches within ramets, with modules often responding to very local growing conditions. For example, Pitelka and Ashmun[105] contrast the relative autonomy or ramets in the opportunistic herb *Aster acuminatus* to the more integrated and long-lived ramets in *Clintonia borealis.* In *Solidago canadensis,* new ramets at the edge of a clone are dependent on imported resources, and ramets apparently reintegrate in response to the temporary shading of some ramets.[68] Within clones of both *Rhus glabra* (D. Waller, unpublished data) and *Rhus typhina* (J. and L. Lovett Doust, personal communication), the output of an inflorescence or infructescence depends largely on the size or leaf area of the associated branch.

Life history theory predicts certain consequences of continued clonal growth. First, by allowing indefinite survival of the genet, it favors reduced flowering and fruiting. Many clonal plants, shade-tolerant herbs in particular, persist for long periods with purely vegetative growth. As growth continues and clones come to occupy large areas, opportunities for seedling establishment are reduced. Indeed, many researchers report finding few, if any, successful seedlings at all in the vicinity of long-lived established clones.[66] This, of course, favors further reductions in sexual reproduction, as well as increased dispersal and/or dormancy. Such growth also reduces the likelihood of successful cross-pollination while increasing the frequency of intraclone pollen transfer. This favors both the evolution of mechanisms to ensure effective outcrossing (as discussed earlier) and possibly further reductions in sexual reproduction generally. As a result, many clonal species (as well as large trees) are either dioecious or monoecious with marked dichogamy.[25,91] Even with mechanisms that increase outcrossing, clonal growth can limit the number of genets in an area and so reduce effective population size.[3,64] This promotes random genetic drift and consequent population differentiation, yet at least one clonal species *(Trifolium repens)* maintains considerable local genotypic diversity,[16,17] possibly in response to the stature and form of neighboring plants.

Clonal growth, by contributing to longevity, also slows rates of genetic change in populations while augmenting the possibility of intense and protracted competition with other plants and interactions with pathogens and herbivores. These could select, in turn, for increased sexual recombination. Finally, by contributing to lateral spread, clonal growth can expand the genetic neighborhood of a population, just as enhanced movement of pollen and seeds does.[50]

CONCLUSIONS

Theoretical and empirical research into the dynamics of plant reproductive behavior can no longer assume that reproductive traits are free to evolve without regard to plant form. The details of a plant's structural organization both constrain its reproductive behavior and provide opportunities for subsequent selection by extrinsic pollination and dispersal agents. While some of these connections are obvious, others are subtle. Besides the patterns already evident, there are probably other associations between particular flowering or fruiting syndromes and plant growth forms that remain to be discovered. Further surveys of the distributions of reproductive traits across species with contrasting form are needed, but they should be sensitive to statistical difficulties.

Detailed morphological, ecological, and genetic analyses of the interactions occurring between form and reproductive performance in particular species are also needed to identify the mechanisms behind these associations.

A plant's form is as integral a part of its life history as its schedule of seed production, and may, in fact, determine it. It may also provide a physical representation of past developmental choices. In addition, by modifying their reproductive behavior, plant forms influence the maintenance of genetic variation and hence the evolutionary prospects of plant populations.

ACKNOWLEDGMENTS

I thank A. Bell, J. Harper, T. Jayasingam, and J. and L. Lovett Doust for commenting on a draft of this paper.

REFERENCES

1. Abul-Fatih, H. A., and Bazzaz, F. A., The biology of *Ambrosia trifida* L. III. Growth and biomass allocation, *New Phytol.* **83,** 829–838 (1979).
2. Antlfinger, A. E., The genetic basis of microdifferentiation in natural and experimental populations of *Borrichia frutescens* in relation to salinity, *Evolution* **35,** 1056–1086 (1981).
3. Antonovics, J., and Levin, D. A., The ecological and genetic consequences of density dependent regulation in plants, *Annu. Rev. Ecol. Syst.* **11,** 411–452 (1980).
4. Armstrong, R. A., A quantitative theory of resource partitioning in rhizomatous perennial plants, *Ecology* **63,** 679–686 (1982).
5. Armstrong, R. A., On the quantitative theory of resource partitioning in rhizomatous perennial plants: The influences of canopy structure, rhizome branching pattern, and self-thinning, *Ecology* **64,** 703–709 (1983).
6. Arroyo, M. T. K., Primack, R., and Armesto, J., Community studies in pollination ecology in the high temperate Andes of central Chile I. Pollination mechanisms and altitudinal variation, *Am. J. Bot.* **69,** 82–97 (1982).
7. Augspurger, C. K., Morphology and dispersal potential of wind-dispersed diaspores of neotropical trees, *Am. J. Bot.* **73,** 353–363 (1986).
8. Augspurger, C. K., and Kelley, C. K., Pathogen mortality of tropical tree seedlings: Experimental studies of the effects of dispersal distance, seedling density, and light conditions, *Oecologia* **61,** 211–217 (1984).
9. Baker, G. A., and O'Dowd, D. J., Effects of parent plant density on the production of achene types in the annual *Hypochaeris glabra, J. Ecol.* **70,** 201–216 (1982).
10. Barrett, S. C., The evolution, maintenance and loss of self-incompatibility systems, in *Plant Reproductive Ecology: Patterns and Strategies* (J. Lovett Doust and L. Lovett Doust, eds.), Chapter 5. Oxford Univ. Press, New York, 1988.
11. Bazzaz, F. A., Levin, D. A., and Schmierbach, M. R., Differential survival of genetic variants in crowded populations of *Phlox, J. Appl. Ecol.19,* 8911–9000 (1982).
12. Bell, A., Dynamic morphology: A contribution to plant population ecology, in *Perspectives in Plant Population Ecology* (R. Dirzo and J. Sarukhan, eds.), Chapter 2. Sinauer, Sunderland, Massachusetts, 1984.
13. Bertin, R., Paternity in plants, in *Plant Reproductive Ecology: Patterns and Strategies* (J. Lovett Doust and L. Lovett Doust, eds.), Chapter 2. Oxford Univ. Press, New York, 1988.
14. Bradshaw, A. D., Evolutionary significance of phenotypic plasticity in plants, *Adv. Genet.* **13,** 115–155 (1965).
15. Brunig, E. F., Tree forms in relation to environmental conditions: An ecological viewpoint, in *Tree Physiology and Yield Improvement* (M. G. R. Cannell and F. T. Last, eds.), pp. 139–156. Academic Press, London, 1976.
16. Burdon, J. J., Intra-specific diversity in a natural population of *Trifolium repens* L., *J. Ecol.* **68,** 717–735 (1980).
17. Cahn, M. G., and Harper, J. L., The biology of the leaf mark polymorphism in *Trifolium repens* L. 1. Distribution of phenotypes at a local scale, *Heredity* **37,** 309–325 (1976).

18. Campbell, C. S., Quinn, J. A., Cheplick, G. P., and Bell, T. J., Cleistogamy in grasses, *Annu. Rev. Ecol. Syst.* **14,** 411–441 (1983).
19. Caswell, H., Optimal life histories and the maximization of reproductive value: a general theorem for complex life cycles, *Ecology* **63,** 1218–1222 (1982).
20. Caswell, H., Stable population structure and reproductive value for populations with complex life cycles, *Ecology* **63,** 1223–1231 (1982).
21. Caswell, H., Evolutionary demographic analysis of clonally reproducing organisms, in *The Population Biology and Evolution of Clonal Organisms,* (J. B. C. Jackson, L. W. Buss, and R. E. Cook, eds.), Chapter 6. Yale University Press, New Haven, Connecticut, 1986.
22. Caswell, H., and Werner, P. A., Transient behavior and life history analysis of teasel (*Dipsacus sylvestris* Huds.), *Ecology* **59,** 53–66 (1977).
23. Cavers, P. B., and Steel, M. G., Patterns of change in seed weight over time on individual plants, *Am. Nat.* **124,** 324–335 (1984).
24. Chapin, F. S., The mineral nutrition of wild plants, *Annu. Rev. Ecol. Syst.* **11,** 233–260 (1980).
25. Charnov, E. L., *The Theory of Sex Allocation.* Princeton Univ. Press, Princeton, New Jersey, 1982.
26. Cheplick, G. P., and Quinn, J. A., *Amphicarpum purshii* and the "pessimistic strategy" in amphicarpic annuals with subterranean fruits, *Oecologia* **52,** 327–332 (1982).
27. Chiarello, N., and Roughgarden, J., Storage allocation in seasonal races of an annual plant: Optimal vs. actual allocation, *Ecology* **65,** 1290–1301 (1984).
28. Clegg, M. T., Kahler, A. L., and Allard, R. W., Estimation of life cycle components of selection in an experimental plant population. *Genetics* **89,** 765–792 (1978).
29. Cohen, D., Maximizing final yield when growth is limited by time or limiting resources, *J. Theor. Biol.* **33,** 299–301 (1971).
30. Cohen, D., The optimal timing of reproduction, *Am. Nat.* **110,** 801–807 (1976).
31. Cook, R., The biology of seeds in the soil, in *Demography and Evolution in Plant Populations* (O. T. Solbrig, ed.), Chapter 6. Blackwells, New York, 1980.
32. Corner, E. J. H., *The Natural History of Palms.* Weidenfeld and Nicholson, London, 1968.
33. Cox, P. A., Monomorphic and dimorphic sexual strategies: a modular approach, in *Plant Reproductive Ecology: Patterns and Strategies* (J. Lovett Doust and L. Lovett Doust, eds.), Chapter 4. Oxford Univ. Press, New York, 1988.
34. Coyne, J. A., and Lande, R., The genetic basis of species differences in plants, *Am. Nat.* **126,** 141–145 (1985).
35. Cruden, R., and Hermann-Parker, S. M., Temporal dioecism: An alternative to dioecism? *Evolution* **26,** 373–389 (1977).
36. Darwin, C., *Different Forms of Flowers on Plants of the Same Species.* Murray & Co., London, 1877.
37. Denslow, J. S., and Moermond, T. C., The effect of accessibility on rates of fruit removal from tropical shrubs: An experimental study, *Oecologia* **54,** 170–176 (1982).
38. Donald, C. M., and Hamblin, J., The biological yield and harvest index of cereals as agronomic and plant breeding criteria, *Adv. Agron.* **28,** 361–405 (1976).
39. Eis, S., Garman, E. H., and Ebell, L. F., Relation between cone production and diameter increment of Douglas Fir *(Pseudotsuga menziessi),* Grand Fir *(Abies grandis),* and Western White Pine *(Pinus monticola), Can. J. Bot.* **43,** 1553–1559 (1965).
40. Faegri, K., and van der Pijl, L., *The Principles of Pollination Ecology.* Pergamon Press, New York, 1966.
41. Foster, S. A., and Janson, C. H., The relationship between seed size and establishment conditions in tropical woody plants, *Ecology* **66,** 773–780 (1985).
42. Fowler, N., Zasada, J., and Harper, J. L., Genetic components of morphological variation in *Salix repens, New Phytol.* **95,** 121–131 (1983).
43. Fox, J., Incidence of dioecy in relation to growth form, *Oecologia* **67,** 244–249 (1985).
44. Gale, J. S., and Eaves, L. J., Variation in wild populations of *Papaver dubium.* V. The application of factor analysis to the study of variation, *Heredity* **29,** 135–149 (1972).
45. Galston, A. W., and Davies, P. J., *Control Mechanisms in Plant Development.* Prentice Hall, Englewood Cliffs, New Jersey, 1970.
46. Gifford, R. M., and Evans, L. T., Photosynthesis, carbon partitioning, and yield, *Annu. Rev. Plant Physiol.* **32,** 485–509 (1981).
47. Givnish, T. J., Ecological constraints on the evolution of breeding systems in seed plants: dioecy and dispersal in gymnosperms, *Evolution* **34,** 959–972 (1980).
48. Givnish, T. J., On the adaptive significance of leaf height in forest herbs, *Am. Nat.* **120,** 353–381 (1982).

49. Givnish, T. J., Terborgh, J. W., and Waller, D. M., Plant form, temporal community structure, and species richness in forest herbs of the Virginia piedmont, unpublished manuscript.
50. Gliddon, C., Behhassen, E., and Gouyon, P. H., Genetic neighbourhoods in plants with diverse systems of mating and different patterns of growth, *Heredity* **59,** 29–32 (1987).
51. Godley, E. J., Flower biology in New Zealand, *N.Z. J. Bot.* **17,** 441–466 (1979).
52. Gottlieb, L. D., Genotypic similarity of large and small individuals in a natural population of the annual plant *Stephanomeria exigua* ssp. *coronaria* (Compositae), *J. Ecol* **65,** 127–134 (1977).
53. Gottlieb, L. D., Genetics and morphological evolution in plants, *Am. Nat.* **123,** 681–709 (1984).
54. Gottlieb, L. D., Reply to Coyne and Lande, *Am. Nat.* **126,** 146–150 (1985).
55. Grant, M. C., and Mitton, J. B., Elevational gradients in adult sex ratios and sexual differentiation in vegetative growth rates of *Populus tremuloides* Michx., *Evolution* **33,** 914–918 (1979).
56. Green, D. S., The terminal velocity and dispersal of spinning samaras, *Am. J. Bot.* **67,** 1218–1224 (1980).
57. Grime, J. P., Shade avoidance and shade tolerance in flowering plants, in *Light as an Ecological Factor* (G. C. Evans, R. Bainbridge, and O. Rackhamn, eds.), pp. 281–301. Blackwells, London, 1966.
58. Grime, J. P., Evidence for the existence of three primary strategies in plants and its relevance to ecological and evolutionary theory, *Am. Nat.* **111,** 1169–1194 (1977).
59. Grime, J. P., *Plant Strategies and Vegetation Processes.* John Wiley & Sons, New York, 1979.
60. Gross, K. L., and Soule, J. D., Differences in biomass allocation to reproductive and vegetative structures of male and female plants of a dioecious, perennial herb, *Silene alba* (Miller) Krause, *Am. J. Bot.* **68,** 801–807 (1981).
61. Guries, R. P., and Ledig, F. T., Genetic diversity and population structure in pitch pine (*Pinus rigida* Mill.), *Evolution* **36,** 387–402 (1982).
62. Hallé, F., Oldemann, R. A. A., and Tomlinson, P. B., *Tropical Trees and Forests: An Architectural Analysis.* Springer-Verlag, Berlin, 1978.
63. Hamilton, W. D., and May, R. M., Dispersal in stable habitats, *Nature (London)* **269,** 578–581 (1977).
64. Handel, S. N., The intrusion of clonal growth patterns on plant breeding systems, *Am. Nat.* **125,** 367–384 (1985).
65. Hare, J. D., Variation in fruit size and susceptibility to weed predation among and within populations of the cocklebur, *Xanthium strumarium, Oecologia* **46,** 217–222 (1980).
66. Harper, J. L., *Population Biology of Plants.* Academic Press, New York, 1977.
67. Harper, J. L., Modules, branches, and the capture of resources, in *The Population Biology and Evolution of Clonal Organisms* (J. B. C. Jackson, L. W. Buss, and R. E. Cook, eds.), Chapter 1. Yale Univ. Press, New Haven, Connecticut, 1986.
68. Hartnett, D. C., and Bazzaz, F. A., Physiological integration among intraclonal ramets in *Solidago canadensis, Ecology* **64,** 779–788 (1983).
69. Hendrix, S. D., Herbivory and its impact on reproduction, in *Plant Reproductive Ecology: Patterns and Strategies* (J. Lovett Doust and L. Lovett Doust, eds.), Chapter 12. Oxford Univ. Press, New York, 1988.
70. Hendrix, S. D., and Trapp, E. J., Plant-herbivore interactions: Insect induced changes in host plant sex expression and fecundity, *Oecologia* **49,** 119–122 (1981).
71. Heslop Harrison, J., Sexuality of angiosperms, in *Plant Physiology: A Treatise* (F. C. Steward, ed.), Vol. 6C. Academic Press, New York, 1972.
72. Horn, H. S., *The Adaptive Geometry of Trees.* Princeton Univ. Press, Princeton, New Jersey, 1971.
73. Iltis, H. H., From teosinte to Maize: The catastrophic sexual transmutation, *Science* **222,** 886–894 (1983).
74. Janzen, D. H., Effect of defoliation on fruit bearing branches of the Kentucky Coffee Tree, *Gymnocladus dioicus* (Leguminosae), *Am. Midl. Nat.* **95,** 474–478 (1976).
75. King, D., and Roughgarden, J., Energy allocation patterns of the California grassland annuals *Plantago erecta* and *Clarkia rubicunda, Ecology* **64,** 16–24 (1982).
76. Lande, R., A quantitative genetic theory of life history evolution, *Ecology* **63,** 607–615 (1982).
77. Lechowicz, M. J., The effects of individual variation in physiological and morphological traits on the reproductive capacity of the common cocklebur *Xanthium strumarium* L., *Evolution* **38,** 833–844 (1984).
78. Lefkovitch, L. P., The study of population growth in organisms grouped by stages, *Biometrics* **21,** 1–18 (1965).
79. Levey, D. J., Moermond, T. C., and Denslow, J. S., Fruit choice in neotropical birds: The effect of distance between fruits on preference patterns, *Ecology* **65,** 844–850 (1984).

80. Lindsey, A. H., Floral phenology patterns and breeding systems in *Thaspium* and *Zizia* (Apiaceae), *Syst. Bot.* **7,** 1–12 (1982).
81. Lord, E. M., Cleistogamy: A tool for the study of floral morphogenesis, function, and evolution, *Bot. Rev* **47,** 421–449 (1981).
82. Lovett Doust, J., Experimental manipulation of patterns of resource allocation in the growth cycle and reproduction of *Smyrnium olusatrum, Biol. J. Linn. Soc.* **13,** 155–166 (1980).
83. Lovett Doust, J., Lovett Doust, L., and Eaton, W., Sequential yield component analysis and models of growth in bush bean (*Phaseolus vulgaris* L.) *Am. J. Bot.* **70,** 1063–1070 (1983).
84. Lovett Doust, L., Population dynamics and local specialization in a clonal perennial *(Ranunculus repens)* I. The dynamics of ramets in contrasting habitats, *J. Ecol.* **69,** 743–755 (1981).
85. MacArthur, R. H., and Wilson, E. O., *Island Biogeography.* Princeton Univ. Press, Princeton, New Jersey, 1967.
86. McGraw, J. B., and Antonovics, J., Experimental ecology of *Dryas octopetala* ecotypes. I. Ecotypic differentiation and life cycle stages of selection, *J. Ecol.* **71,** 879–897 (1983).
87. MacMahon, T. A., and Kronauer, R. E., Tree structures: Deducing the principle of mechanical design, *J. Theor. Biol.* **59,** 443–466 (1976).
88. Mangelsdorf, P. C., *Corn: Its Origin, Evolution, and Improvement.* Harvard Univ. Press, Cambridge, Massachusetts, 1974.
89. Marshall, D. L., Fowler, N. C., and Levin, D. A., Plasticity in yield components in natural populations of three species of *Sesbania, Ecology* **66,** 753–761 (1985).
90. Maun, M. A., and Cavers, P. B., Seed production and dormancy in *Rumex crispus.* II. The effects of removal of various proportions of flowers at anthesis, *Can. J. Bot.* **49,** 1841–1848 (1971).
91. Maynard Smith, J., *The Evolution of Sex.* Cambridge Univ. Press, Cambridge, 1978.
92. Maynard Smith, J., and Price, G. R., The logic of animal conflict, *Nature (London)* **246,** 15–18 (1973).
93a. Meagher, T., Sex determination in plants, in *Plant Reproductive Ecology: Patterns and Strategies* (J. Lovett Doust and L. Lovett Doust, eds.), Chapter 6. Oxford Univ. Press, New York, 1988.
93b. Meagher, T. R., Analysis of paternity within a natural population of *Chamaelirium luteum.* 1. Identification of most-likely male parents, *Am. Nat.* **128,** 199–215 (1986).
94. Melampy, M. N., Sex-linked niche differentiation in two species of *Thalictrum, Am. Midl. Nat.* **106,** 325–334 (1981).
95. Menges, E. S., Biomass allocation and geometry of the clonal forest herb, *Laportea canadensis:* Adaptive responses or allometric constraints? *Am. J. Bot.* **74,** 551–563 (1987).
96. Menges, E. S., and Waller, D. M., Plant strategies in relation to elevation and light in floodplain herbs, *Am. Nat.* **122,** 454–473 (1983).
97. Mitchell-Olds, S. T., Quantitative genetics of survival and growth in *Impatiens capensis, Evolution,* **40,** 107–116 (1986).
98. Mitchell-Olds, S. T., and Rutledge, J. J., Quantitative genetics in natural plant populations: A review of the theory, *Am. Nat.* **127,** 379–402 (1986).
99. Mitton, J. B., and Grant, M. C., Observations on the ecology and evolution of quaking aspen, *Populus tremuloides,* in the Colorado Front Range. *Am. J. Bot.* **67,** 202–209 (1980).
100. Nevo, E., Zohary, D., Brown, A. H. D., and Haber, M., Genetic diversity and environmental associations of wild barley *Hordeum spontaneum* in Israel, *Evolution* **33,** 815–833 (1979).
101. Niklas, K. J., The motion of windborne pollen grains around conifer ovulate cones: implication on wind pollination, *Am. J. Bot.* **71,** 356–374 (1984).
102. Paltridge, G. W., and Denholm, J. V., Plant yield and the switch from vegetative to reproductive growth, *J. Theor. Biol.* **44,** 23–34 (1974).
103. Paltridge, G. W., Denholm, J. V., and Conner, D. J., Determinism, senescence, and the yield of plants, *J. Theor. Biol.* **110,** 383–398 (1984).
104. Pianka, E. R., and Parker, W. S., Age specific reproductive tactics, *Am. Nat.* **109,** 453–464 (1975).
105. Pitelka, L. F., and Ashmun, J. W., The physiology and ecology of connections between ramets in clonal plants, in *The Population Biology and Evolution of Clonal Organisms* (J. B. C. Jackson, L. W. Buss, and R. E. Cook, eds.), Chapter 11. Yale Univ. Press, New Haven, Connecticut, 1986.
106. Primack, R. B., and Antonovics, J., Experimental ecological genetics in *Plantago.* V. Components of seed yield in the ribwort plantain, *Plantago lanceolata* L., *Evolution* **35,** 1069–1079 (1981).
107. Putwain, P. D., and Harper, J. L., Studies on the dynamics of plant populations, V: Mechanisms governing the sex ratio in *Rumex acetosa* and *R. acetosella, J. Ecol.* **60,** 113–128 (1972).

108. Raunkaier, C., *The Life Forms of Plants and Statistical Plant Geography,* the collected papers of C. Raunkaier. Oxford Univ. Press, Oxford, 1934.
109. Radford, A. E., Ahles, H. E., and Bell, C. R., *Flora of the Carolinas.* Univ. of North Carolina Press, Chapel Hill, 1951.
110. Rathcke, B., and Lacey, E. P., Phenological patterns of terrestrial plants, *Annu. Rev. Ecol. Syst.* **16,** 179–214 (1985).
111. Ray, T., *Monstera tenuis* (Chirravaca, Mano de Tigre, Monstera), in *Costa Rican Natural History* (D. Janzen, ed.), pp. 278–280. Univ. of Chicago Press, Chicago, 1983.
112. Roach, D., Life history variation in *Geranium carolinianum* I. Covariation between characters at different stages of the life cycle, *Evolution,* in press.
113. Rohmeder, E., Beziehungen zwishen Fruch-bzw. Samenerzeugung und Holzerzeugung der Waldbaume, *Allg. Forst Z.* **22,** 33–39 (1967).
114. Salisbury, E. J., *The Reproductive Capacity of Plants.* Bell and Sons, London, 1942.
115. Saltzman, A., Habitat selection in a clonal plant, *Science* **228,** 603–604 (1985).
116. Schaffer, W. M., Some observations on the evolution of reproductive rate and competitive ability in flowering plants, *Theor. Popul. Biol.* **11,** 90–104 (1977).
117. Schemske, D. W., Evolution of reproductive characteristics in *Impatiens* (Balsaminaceae): The significance of cleistogamy and chasmogamy, *Ecology* **59,** 596–613 (1978).
118. Schlessman, M., Gender diphasy ("Sex choice"), in *Plant Reproductive Ecology: Patterns and Strategies* (J. Lovett Doust and L. Lovett Doust, eds.), Chapter 7. Oxford Univ. Press, New York, 1988.
119. Schmitt, J., Individual flowering phenology, plant size, and reproductive success in *Linanthus androsaceus,* a California annual, *Oecologia* **59,** 135–140 (1983).
120. Schnee, B. K., and Waller, D. M., Reproductive behavior of *Amphicarpaea bracteata* (Leguminosae), an amphicarpic annual, *Am. J. Bot.* **73,** 376–386 (1986).
121. Silander, J. A., and Antonovics, J., The genetic basis of the ecological amplitude of *Spartina patens* I. Morphometric and physiological traits, *Evolution* **33,** 1114–1127 (1979).
122. Sculthorpe, C. D., *The Biology of Aquatic Vascular Plants.* Edward Arnold, London, 1967.
123. Smith, B. H., The optimal design of a herbaceous body, *Am. Nat.* **123,** 197–211 (1984).
124. Stiles, E. W., Fruit flags: Two hypotheses, *Am. Nat.* **120,** 500–509 (1982).
125. Thompson, J. N., Variation among individual seed masses in *Lomatium prayi* (Umbelliferae) under controlled conditions: Magnitude and partitioning of the variance, *Ecology* **65,** 626–631 (1984).
126. Thompson, J. N., and Willson, M. F., Evolution of temperate fruit/bird interactions: Phenological strategies, *Evolution* **33,** 973–983 (1979).
127. Thomson, J. D., and Plowright, R. C., Pollen carryover, nectar rewards, and pollinator behavior with special reference to *Diervilla lonicera, Oecologia* **46,** 68–74 (1980).
128. Tuomi, J., Niemela, P., and Mannila, R., Resource allocation on dwarf shoots of birch *(Betula pendula):* Reproduction and leaf growth, *New Phytol.* **91,** 483–487 (1982).
129. Uphof, J. C. T., Cleistogamic flowers, *Bot. Rev.* **4,** 21–49 (1938).
130. Venable, D. L., and Levin, D. A., Ecology of achene polymorphism in *Heterotheca latifolia* I. Achene structure, germination and dispersal, *J. Ecol.* **73,** 133–145 (1985).
131. Waller, D. M., The relative costs of self- and cross-fertilized seeds in *Impatiens capensis* (Balsaminaceae), *Am. J. Bot.* **66,** 313–320 (1979).
132. Waller, D. M., Environmental determinants of outcrossing in *Impatiens capensis* (Balsaminaceae), *Evolution* **34,** 747–761 (1980).
133. Waller, D. M., Factors influencing seed weight in *Impatiens capensis* (Balsaminaceae), *Am. J. Bot.* **69,** 1470–1475 (1982).
134. Waller, D. M., The genesis of size hierarchies in seedling populations of *Impatiens capensis* Meerb., *New Phytol.* **100,** 243–260 (1985).
135. Waller, D. M., and Steingraeber, D., Branching and modular growth: Theoretical models and empirical patterns, in *The Population Biology and Evolution of Clonal Organisms* (J. B. C. Jackson, L. W. Buss, and R. E. Cook, eds.), Chapter 7. Yale Univ. Press, New Haven, Connecticut, 1986.
136. Warming, E., *Oecology of Plants: An Introduction to the Study of Plant Communities.* Oxford Univ. Press, Oxford, 1909.
137. Watson, M. A., Developmental constraints: Effect on population growth and patterns of resource allocation in a clonal plant, *Am. Nat.* **123,** 411–426 (1984).
138. Watson, M. A., Integrated physiological units in plants, *Trends Ecol. Evol.* **1,** 119–123 (1986).

139. Watson, M. A., and Casper, B. B., Morphogenetic constraints on patterns of carbon distribution in plants, *Annu. Rev. Ecol. Syst.* **15,** 233–258 (1984).
140. Weiner, J., The influence of competition on plant reproduction, in *Plant Reproductive Ecology: Patterns and Strategies* (J. Lovett Doust and L. Lovett Doust, eds.), Chapter 11. Oxford Univ. Press, New York, 1988.
141. White, J., Plant metamerism, in *Perspectives in Plant Population Ecology* (R. Dirzo and J. Sarukhan, eds.), Chapter 1. Sinauer, Sunderland, Massachusetts, 1984.
142. Whitehead, D. R., Wind pollination in the angiosperms. Evolutionary and environmental considerations, *Evolution* **23,** 28–35 (1969).
143. Willson, M. F., *Plant Reproductive Ecology.* Wiley Interscience, New York, 1984.
144. Wyatt, R., Inflorescence architecture: How flower number, arrangement, and phenology affect pollination and fruit set, *Am. J. Bot.* **69,** 585–594 (1982).
145. Wyatt, R., Pollinator-plant interactions and the evolution of breeding systems, in *Pollination Ecology* (L. Real, ed.), Chapter 4. Academic Press, New York, 1983.
146. Zimmerman, M., Nectar production, flowering phenology, and strategies for pollination, in *Plant Reproductive Ecology: Patterns and Strategies* (J. Lovett Doust and L. Lovett Doust, eds.), Chapter 8. Oxford Univ. Press, New York, 1988.

11

The Influence of Competition on Plant Reproduction

JACOB WEINER

Competition or interference is ubiquitous in its influence on plants. It is rare to find a plant which has not been affected negatively by neighboring plants. For the plant population ecologist, competition or interference is best defined as any negative effect due to the proximity of neighbors.[16] Thus, interference is something that occurs between individuals, reducing their growth and/or increasing their probability of death. In this chapter, I use the terms competition and interference interchangeably, although the latter term may be a better one to encompass all negative neighbor effects. I do not discuss competition for the services of pollinators, as this is a separate topic that has recently been reviewed elsewhere.[49]

For the purpose of this chapter, I will interpret the concept of reproduction broadly to include both sexual and asexual means of propagation. Several authors (e.g., Harper[17] and Abrahamson[3]) have pointed out the similarity between the growth of plants and what has been called "vegetative reproduction." Plant growth is usually modular, and if some of the modules are capable of independent existence (such as bramble shoots that have layered or strawberry ramets that have become separate) the result has been referred to as vegetative reproduction. Because of this similairty, Harper[18] has suggested that we consider such forms of propagation as clonal growth, while the term reproduction should be reserved for processes that produce offspring sexually. In fact, since asexually produced offspring are genetically identical to the parent, while sexually produced offspring are genetically different from both parents, it would perhaps be more logical to refer to asexual propagation as "reproduction" (*re*—Latin: again) while sexual forms of propagation be called "neoproduction" (*neo*—Latin: new). Since such a change in terminology appears unlikely, the traditional terminology is used here and both sexual reproduction and clonal growth are considered as different modes of reproduction.

In discussing the influence of a factor such as interference there can be confusion between ecological and evolutionary effects. There is no reason to assume that an individual plant's plastic response to a factor is similar to the effect of natural selection in response to that factor. Some debates on the influence of competition on reproduction (e.g., the exchange between Abrahamson[2] and Law et al.[33]) appear to be based on the confusion that occurs when some of the same terminology is used to refer to ecological (e.g., plastic) and evolutionary (genetic) responses. In this chapter, I address first the ecological question: How does competition affect the reproductive behavior of individuals and, therefore, populations? I do not attempt to review the literature on life history strategies in plants, as this has been done elsewhere (Willson[75]; other chapters in this volume, e.g., 2, 9, 13–15), although I refer to this literature when it bears directly on individual plants' responses to interference. Herein I briefly review the literature on the effects of competition on the reproductive behavior of plants, and compare alternative evolutionary hypotheses to explain these observations.

There are several possible effects of competition on the reproductive behavior of plants. Interference may

1. Reduce the probability that an individual will reproduce or reduce the amount of reproduction (number or size of seeds or ramets produced)[32];
2. Change the plant's reproductive allocation (proportion of resources in reproductive tissues)[11];
3. Change the timing of reproduction (onset and duration of reproductive activities);
4. Change the mode of reproduction (e.g., the proportion of ramet versus genet production, or sexual versus apomictic seed production);
5. Change mating behavior (e.g., gender allocation, proportion of cleistogamous versus chasmogamous flowers) (see Zimmerman,[78] this volume);

Also, competition may change the frequency and spatial distribution of these behaviors within a population (e.g., the amount of individual variation).

Each of these effects is discussed below.

DENSITY-YIELD RELATIONSHIPS AND REPRODUCTIVE ALLOCATION

When plants are grown at a range of densities for a given period of time, yield per unit area increases with increases in density. At higher densities, progressively smaller increases in total yield are observed. Above a certain density, yield no longer increases with density. This "asymptotic" relationship between density and yield ("law of constant final yield") has been documented for many plant species[17,56] (Fig. 11.1a,b). This effect is due to density-dependent mortality and the plastic reduction in size of individual plants. The asymptote reflects the limit on total biomass that is determined by the total amount of available resources. Numerous field and experimental studies have demonstrated that density can reduce almost every aspect of individual yield. Several mathematical relationships between yield and density have been described.[17,29,64,73]

Although the response of total yield to density is usually asymptotic, different components of yield can respond quite differently. Reproductive yield, as measured by fruit or seed production, may follow a pattern similar to that of total yield,[41,66] but fruit and seed production usually increases and then decreases as plant density is increased (Fig. 11.1c,d). This "parabolic (sic) relationship" between density and reproductive components of yield has been observed in many crop[73] and weed[58] species. Thus, as density is increased a greater proportion of the population's biomass is in vegetative structures as opposed to reproductive tissues. I know of only one case in which increased density resulted in increased sexual reproductive allocation: Assemat and

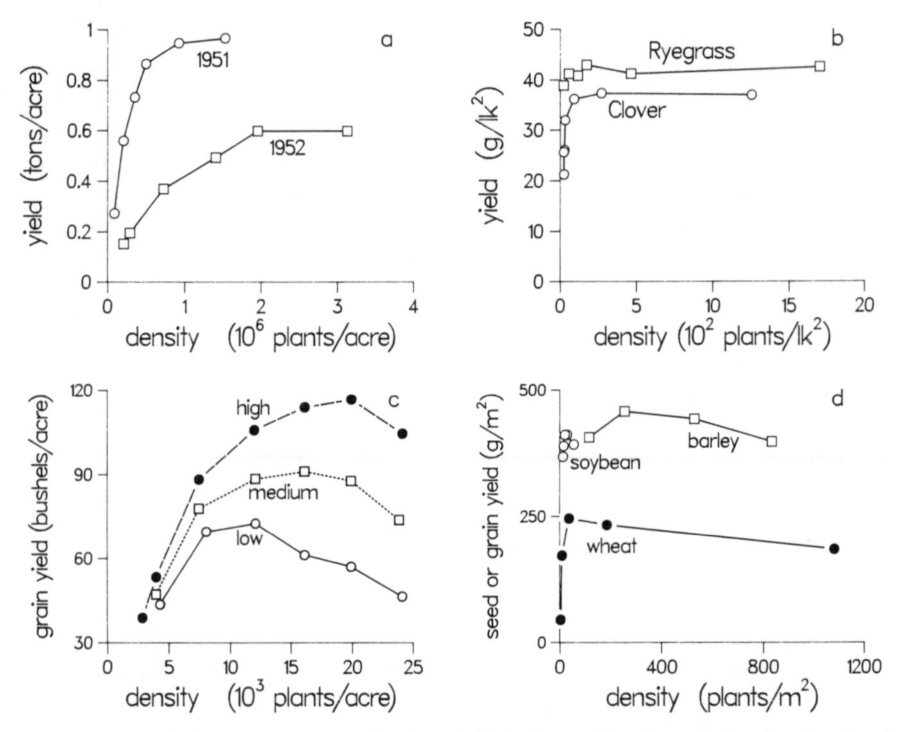

Fig. 11.1. Total yield (a,b) and reproductive yield (c,d) as a function of planting density for selected crops. (a) Total dry matter of Essex Giant rape; (b) total dry matter of Wimmera ryegrass and subterranean clover, units are plants/link^2; (c) mean grain yield of maize grown at low, medium, and high levels of nitrogen; (d) reproductive yield for soybean, barley, and wheat. (a–c after Willey and Heath;[73] d after Willey,[72] from various sources.)

Oka[5] report higher reproductive allocation in the weed *Echinochloa crus-galli* at high density than at low density.

Since most of the data on yield versus density are obtained at the population level (i.e., total yield versus density),[17,64,72] it is not clear in most cases whether the reduction in reproductive yield at high density is due to a change in reproductive allocation (harvest index) of most individuals or a change in the reproductive allocation of the smaller plants only. Decreasing reproductive allocation in individuals grown at higher densities has been observed in *Plantago major,*[24] *Plantago coronopus,*[62] *Chamaesyce hirta,*[58] *Coix mayuen,*[31] *Bellis perennis,* and *Prunella vulgaris.*[53]

Recently there have been attempts to quantify the effect of competition on individuals. In this approach, the growth, size, or reproductive output of a plant is considered to be a function of its neighborhood conditions, such as the number, distance, size, and angular dispersion of neighbors.[12,36,55,63,67,68] In two of these studies,[55,67] neighborhood formulations have been shown to account for a large fraction of the variation in individual reproductive output.

We conclude that competition reduced reproductive output and, in most cases, reproductive allocation as well. The relationship between density and reproductive yield is related to the effects of density on plant size and timing of reproduction (see below, Competition and the Timing of Reproduction, and Competition and Reproductive Allocation: A Question of Size and Allometry?)

While the reproductive yield of individuals decreases at higher densities, one component of reproductive yield, mean weight of seeds produced by an individual, appears to be affected much less than other yield components.[17] While data have been accumulating that show the mean seed size is plastic (e.g., Marshall et al.,[37] Pitelka et al.,[46] Cavers and Steel[10]), the observation that mean seed weight is one of the least plastic characters of plant growth[20,27,48] still stands as a valid generalization. Although mean seed weight may differ among populations of the same species or even among co-occurring individuals within the same population[75] it appears to be relatively non-plastic for a given genotype. Plants reduce their seed output primarily by producing fewer, and, secondarily, by producing smaller, seeds.

COMPETITION AND THE TIMING OF REPRODUCTION

Often, interference delays the onset of reproductive activity. High density has been shown to delay heading in cereal crops.[9] Removal of neighbors advanced the flowering date in *Senecio viscosus* but had no effect on flower initiation in *Serecio sylvaticus.*[41] In some cases, annuals reproduce later in the season when grown at higher densities. There are several examples of monocarpic plants that normally behave as biennials reproducing in their third, or even later years at higher densities.[15,61,71] Some "biennials" may even flower in their first year of growth under very favorable conditions, including low density. A wide range of reproductive timing in response to density has been shown for *Plantago coronopus.*[30] Individuals with few and distant neighbors reproduced sooner and sometimes showed repeated reproduction, whereas plants with many and near neighbors showed delayed reproduction and, in some cases, died without producing.

The influence of competition on the timing of reproduction and on reproductive allocation is often confounded. If interference delays reproduction, individuals grown at low density will show higher allocation to reproductive structures than individuals grown at higher densities after a given period of growth. It is possible that plants suffering from competition may achieve the same reproductive allocation as plants grown at lower density if the former were given a longer period to develop, even if they do not achieve the same size as low-density individuals. Very few studies have taken this possibility into account. A notable exception is the study by Waite and Hutchings,[62] in which energy allocation patterns were studied in *Plantago coronopus.* They found that plants at high density in the field and greenhouse still show lower reproductive allocation when allowed to complete their life cycle.

Interference seems either to delay the time of reproduction or have no effect on it. In mineral nutrition experiments, nitrogen deficiency can produce precocious flowering, so one might expect competition for nitrogen to have a similar effect. I have found only one report of such a response in the published literature: Palmblad[42] reported that flowers appeared earlier in stands of increasing density in *Silene anglica* and *Bromus tectorum,* although he presents no data. Although it might seem adaptive for some plants to respond to a deteriorating environment (perhaps caused by increasing interference from neighbors) by reproducing sooner, I know of no data which show earlier reproduction due to competition. This suggests that there are qualitative constraints on strategies that can evolve in response to competition.

COMPETITION AND THE MODE OF REPRODUCTION

Interference not only reduces reproductive output, it can also influence the mode of reproduction. Many plants are able to propagate themselves vegetatively via stolons, rhizomes, roots, tubers, bulbs, etc., as well as performing sexual reproduction resulting in seed production. (In some species seeds may be produced asexually.) There appear to be trade-offs between these two modes of reproduction. Plants that utilize both methods usually do so at different times during the growing season,[75] although there are cases in which plants practice both forms of reproduction at the same time.[8]

Abrahamson[3] has studied vegetative and seed reproduction in *Rubus* and *Fragaria* and has developed a model that predicts changes in the mode of reproduction in response to density. His model is based upon life history strategies and the costs and benefits of sexual versus asexual reproduction (clonal growth), and predicts that a strategy with flexible allocation to the two different modes of reproduction should be found in plants that grow in "stable" habitats. When density is low in such habitats, he argues that vegetative spread should be favored as an efficient way to colonize available area and propagate a genotype that is successful in that habitat. At high density, it may be advantageous to emphasize sexual reproduction through seeds as a way to produce highly dispersible novel genotypes that have the best chance of reaching and colonizing new sites. Thus, his model predicts that at low density more resources should be allocated to vegetative propagation, whereas at high density there should be a shift to sexual seed reproduction. He found data consistent with such a switch in field populations of *Rubus hispidus* and *Rubus trivialis* growing in different successional stages.[1] He hypothesized that plants were changing their reproductive behavior in response to nutrient conditions, but his data do not exclude the alternative hypothesis that genotypes in field and forest were different.[33] Holler and Abrahamson[28] tested the model with *Fragaria virginiana.* They found a decrease in percent allocation to clonal growth at high density but no change in allocation to seed production with density. Similar results were reported in density experiments on *Hieracium floribundom.*[3] In density experiments on *Tussilago farfara,* Ogden[38] observed a decrease in percent allocation to vegetative reproduction but no change in allocation to seeds at high density. These results are similar to those for *Fragaria.* Other studies have not supported Abrahamson's model. The herbaceous forest perennial *Aster acuminatus* showed no change in allocation to sexual reproduction or clonal growth over several densities.[45] Removal of weeds did not affect the ratio of sexual reproduction to clonal growth in *Fragaria virginiana.*[57] Percent allocation to tubers as opposed to inflorescences increased considerably at high density in the weed *Cyperus rotundus,*[74] a result opposite to the model's prediction. This is the only case of which I am aware in which increases in density resulted in increases in vegetative reproductive allocation. The limited data available suggest that interference usually decreases allocation to clonal growth more than allocation to seed reproduction, but perhaps such a generalization is premature.

Interference can also serve to change the proportion of propagule types in species that produce polymorphic seeds or fruits. The annual *Hypochoeris glabra* produces two distinctly different achene types, beaked and unbeaked. Due to changes in receptacle size with density, a greater proportion of the beaked achene type is produced at low density than at high density.[6]

COMPETITION AND GENDER

Interference can influence the gender allocation of a plant or the sex ratio in a population of plants. In dioecious plants, male and female plants may show niche differention that can result in differential responses to interference. Male spinach *(Spinacia oleracea)* plants show significantly higher survivorship at high density than do females, resulting in a population that is predominantly male.[39] At low density there was no difference in survivorship between males and females. Populations of *Silene alba* are typically female biased, and this bias increased with increasing density in field and glasshouse populations.[35]

Interference may affect gender distribution through mechanisms other than survivorship. Competition seems to reduce sexual expression in females more than in males in the cycad *Macrozamia riedlei.*[40] Gender has been shown to be associated with size in *Arisaema triphyllum,* an herbaceous forest perennial that can change sex.[7,34,47] Larger individuals tend to be female, smaller individuals are more likely to be male. While the effect of interference has not been studied directly, we might expect it to increase the proportion of male plants in the population by decreasing mean plant size. *Acer saccharinum* trees can be male, female, or bisexual, and individuals may switch from female to bisexual.[51] Bisexual trees tend to be larger than male or female trees. Since mean size is smaller at higher densities, higher density populations might have a higher proportion of unisexual trees.

COMPETITION AND VARIATION IN REPRODUCTIVE BEHAVIOR

Considerable data are accumulating in support of the generalization that competition among plants increases variability in many aspects of plant performance, including individual size and reproductive output.[69,70] At higher densities a larger proportion of the population has no offspring at all. In experimental populations of the annuals *Trifolium incarnatum* and *Lolium multiflorum,* I found that variability in size always increased at higher densities, and that variability in reproductive output was always greater than variability in size (e.g., Fig. 11.2). Increased variability in reproductive behavior due to competition may provide phenotypic differences within populations that may not be expressed at low density. Thus, competition may increase the range of material on which natural selection can act.

The effect of competition on variability in size and reproductive output seems to be a function of the mechanism of competition: competition for light appears to increase variability, but competition for nutrients may tend to decrease such variability.[64] On a sandy beach where the competition was for water and nutrients and the canopy was not closed, higher-density patches of *Xanthium strumarium* showed lower variability in fruit production.[77] This situation may be the exception. A review of density experiments shows increased variability in size at higher densities in 14 out of 16 cases.[70]

Interference seems to increase variability in phenology, as well as size and reproductive output. Density was correlated with increased variability in the duration of flowering and therefore reduced flowering synchrony in *Linathus androsaceus.*[54]

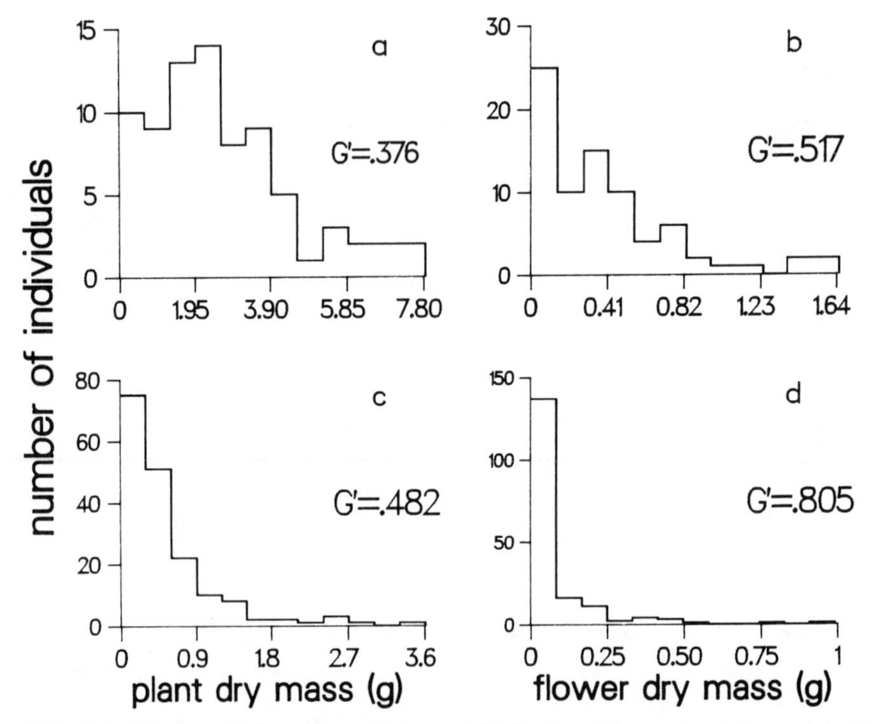

Fig. 11.2. Distribution of above-ground dry weight (a,c) and flower head dry weight (b,d) for experimental populations of *Trifolium incarnatum* grown at low (a,b) and high (c,d) densities with fertilizer. (Corrected from Weiner,[69] Figs. 4 and 7.) G′, unbiased Gini coefficient, a measure of variability or inequality.

ARE PLANTS' REPRODUCTIVE RESPONSES TO COMPETITION PART OF A "REPRODUCTIVE STRATEGY"?

A plant's response to competition may be the result of physiological or developmental constraints and may not represent a genetically based "strategy." While many would agree, in principle, that we should not assume every attribute or behavior of an organism is adaptive,[13,19] the goal of many a research program is to find the adaptive significance of specific traits or behaviors. Lack of empirical support for a hypothesized adaptive significance often leads to a search for alternative adaptive hypotheses, rather than a questioning of the assumption of adaptation. This very prevalent attitude is usually implicit but it is sometimes stated explicitly, as in an important recent paper[62]: "Since a weight-related plastic reproductive allocation has been demonstrated under both laboratory and field conditions it can be considered an evolved adaptive trait which may be explained in terms of individual plant fitness." While the conclusion may be correct, it does not follow from the premise. If we substitute "response to a new herbicide" for "plastic reproductive allocation" in the sentence, the assumptions are illuminated. Although plants may show a size-related response to a new herbicide, we would not consider this response to be an adaptation. That a behavior is adaptive should be an hypothesis, not an assumption, and alternative hypotheses should be sought if the adaptive hypothesis is to obtain strong support.

The assumption that behaviors are adaptive becomes even more problematic when one is referring to specific instances of individual variation within a population. An organism may have a fixed behavior pattern or a flexible behavior. In either case we can attempt to test the hypothesis that the observed pattern of behavior is adaptive (i.e., confers an increase in fitness). However, the hypothesis that any particular manifestation of plastic behavior is adaptive is much more risky, and requires more assumptions than the hypothesis that the behavioral plasticity itself is adaptive. Even if a plastic behavior pattern is adaptive, it does not follow that every manifestation of this plasticity confers fitness. Thus, the life history of a species is the product of natural selection, but changes in this life history pattern due to stresses such as competition may not be adaptive. Students of plant physiology do not assume that the behaviors of plants under experimental conditions of specific nutrient deficiencies are part of an evolved strategy to maximize fitness under such conditions, yet this may be exactly what ecologists are doing when they assume that responses of organisms to some of the difficult conditions nature often presents are strategic. Even if ultimate biological constraints do not limit the possibilities for adaptive responses, the behavior of organisms under extreme conditions (sugh as high density) may be the result of selection to optimize fitness under conditions that are more typical, conditions that usually have a greater effect on fitness, or conditions for which possible changes in the genome will have a greater effect on fitness.

The advantage of the adaptationist or "strategy" approach is that it readily leads to testable hypotheses, whereas alternative nonadaptationist perspectives may not generate hypotheses as readily. The ability to produce testable hypotheses is a necessary attribute of a scientific theory, and there is great need to develop alternative hypotheses that are not based on adaptation.

I propose three alternative interpretations of a plant's response to interference from neighbors. (1) It is possible that the plant's response is not adaptive in any sense. The plant may not have the ability to deal with the degree of interference it is experiencing. Even if an organism is adapted to tolerate a particular stress, or even maximize its fitness in the face of this stress, there will be levels of the stress above which the adaptive response breaks down. Such situations are beyond the range of adaptive tolerance, even if the organism is able to continue to live. Many of the behaviors observed in organisms at high density may fall into this category. In many cases a high degree of stress may result in increased variation in individual behavior within a population. For example, *Senecio vulgaris* individuals growing in very small pots showed much more variation in reproductive behavior than individuals growing in larger pots.[21] Since organisms cannot do everything, what may be selected in such a case may not be a plastic strategy that maximizes fitness under conditions of either small or large root volumes, but an ability to tolerate the extreme conditions. There may be trade-offs between the ability to survive and grow under extreme conditions and the ability to grow well under better conditions.[14] Such trade-offs may provide the basis for alternative hypotheses relating plant behavior to natural selection.

(2) Plants may have evolved fixed responses to the environment that are independent of competition per se. For example, plants may grow as best they can until they reach a threshold size for reproduction. If and when they attain this size, a certain proportion of resources above this level is then allocated to reproduction (Fig. 11.3). In this model competition is no different from any other resource limitation, and responses are fixed. This can be expected to be true for some plants; simple strategies for

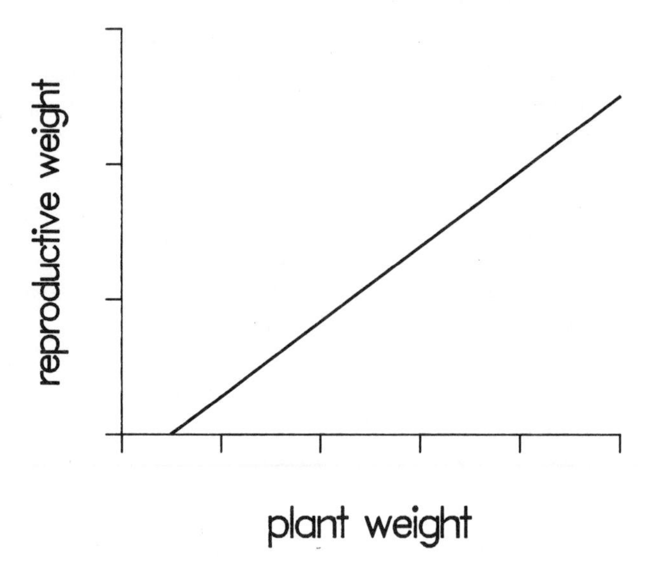

Fig. 11.3. Theoretical relationship between weight of a plant and weight of its reproductive structures (e.g., flowers, fruits, seeds).

dealing with particular problems are usually found in some organisms whenever more complex strategies for addressing the same problem are found in other organisms. Simple solutions to problems will be favored unless the selection pressure for a more complex solution outweighs the additional costs involved. Also, simple solutions are more likely to be available within the population's genetic variation. There is evidence (see below, Competition and Reproductive Allocation: A Question of Size and Allometry?) that the reproductive behavior of many plants is determined primarily by the size they achieve.

(3) The ecological responses of a genotype to different competitive regimes may indeed represent an evolved, flexible strategy that is the result of natural selection (i.e., individuals that had this capacity generally showed higher fitness than those that did not). Plants are neither responding to conditions outside their range of adaptive response, nor are they following simple fixed behavior patterns as limited by their environment. Rather, plants have evolved a suite of possible responses or tactics, each of which will tend to increase fitness in particular environments. This is the approach taken by Waite and Hutchings[62] in interpreting the results for *Plantago coronopus.* Presumably, there are costs involved in having such an ability, as there are for any developmental flexibility. Strong support for an hypothesis of adaptive developmental flexibility requires data that are not consistent with hypotheses that do not invoke flexibility because (1) Occam's razor obliges us to accept the simpler of two explanations if they are equal in other ways, and (2) because natural selection will favor a simpler (and therefore cheaper) solution when it is otherwise equal to another, more complex solution. The argument here is similar to those concerning the evolution of intelligence in animals. The benefits of developmental flexibility must exceed the increased costs over fixed behavior patterns if such flexibility is to be selected.

Thus, the hypothesis that a plant's plastic response to interference is a reflection of an evolved, flexible strategy rests upon a specific set of assumptions. The plant's ancestors must have evolved in environments with varying degrees of interference, and selection must have favored different behaviors under different competitive regimes. Individuals that showed the plastic response pattern must have had higher fitness than those that showed simpler, fixed responses.

In summary, organisms may do things not only because they have been selected to do them, but because they cannot help doing them. This may be because of direct constraints, overload and breakdown of their adaptive mechanisms, or because they have the ability to do something else that solves a more important (or more soluble) problem.

COMPETITION AND REPRODUCTIVE ALLOCATION: A QUESTION OF SIZE AND ALLOMETRY?

In many cases, changes in time of flowering and seed production in plants grown at different densities can be explained in terms of plastic growth and size-dependent reproduction of individuals. Many plants must achieve a minimum size if they are to produce flowers and fruits.[43,59,71,76] Above this threshold size there may be a relatively simple relationship between size and seed production (Fig. 11.3). This could account for the "parabolic" response of reproductive yield to density and the decrease in reproductive allocation of individuals grown at higher densities. At higher densities individual plants are smaller and, because the x-intercept of the relationship between reproductive biomass and vegetative biomass is positive, they have a smaller proportion of their biomass in reproductive tissues. As examples of an allometric view of reproductive allocation, I have reanalyzed the data on reproductive allocation in field populations of *Verbascum thapsus* presented by Reinartz,[50] and that of experimental populations of *Plantago major* presented by Hawthorn and Cavers.[23,24] Reinartz's data on 12 field populations of *Verbascum thapsus* (Fig. 11.4a) show a trend of increasing fraction of plant biomass in seeds with increasing mean plant size. When replotted, the data show a simple linear relationship between vegetative biomass and seed biomass: there is a minimum vegetative biomass required for reproduction and above that minimum a constant proportion of the additional biomass is in seeds. A linear regression of seed weight on vegetative weight accounts for 96% of the variation in weight of seeds produced (Fig. 11.4b), and the residuals are consistent with the assumptions of regression analysis. If there is a minimum size for reproduction and the relationship between reproductive weight and vegetative weight is linear, the relationship between percent reproductive allocation and total or vegetative weight will not be linear: it will shown an ever-decreasing slope approaching a constant percent allocation to reproduction.

Similar results are obtained from reanalysis of data on reproductive allocation in density experiments on *Plantago major*[22,23,24] (Fig. 11.5). Data that would be curvilinear if percent allocation were plotted against weight appear to be linear when reproductive weight is plotted against plant weight (or, more appropriately, against vegetative weight). Ninety-eight percent of the variation in mean spike weight produced by individuals with different neighbor conditions[23] (Fig. 11.5a) and 72% of the variation in seed weight produced by *Plantago major* individuals that did produce seeds in monocultures grown at three densities,[22,24] (Fig. 11.5b) can be accounted for by this simple model. A similar relationship accounts for almost all the variation in repro-

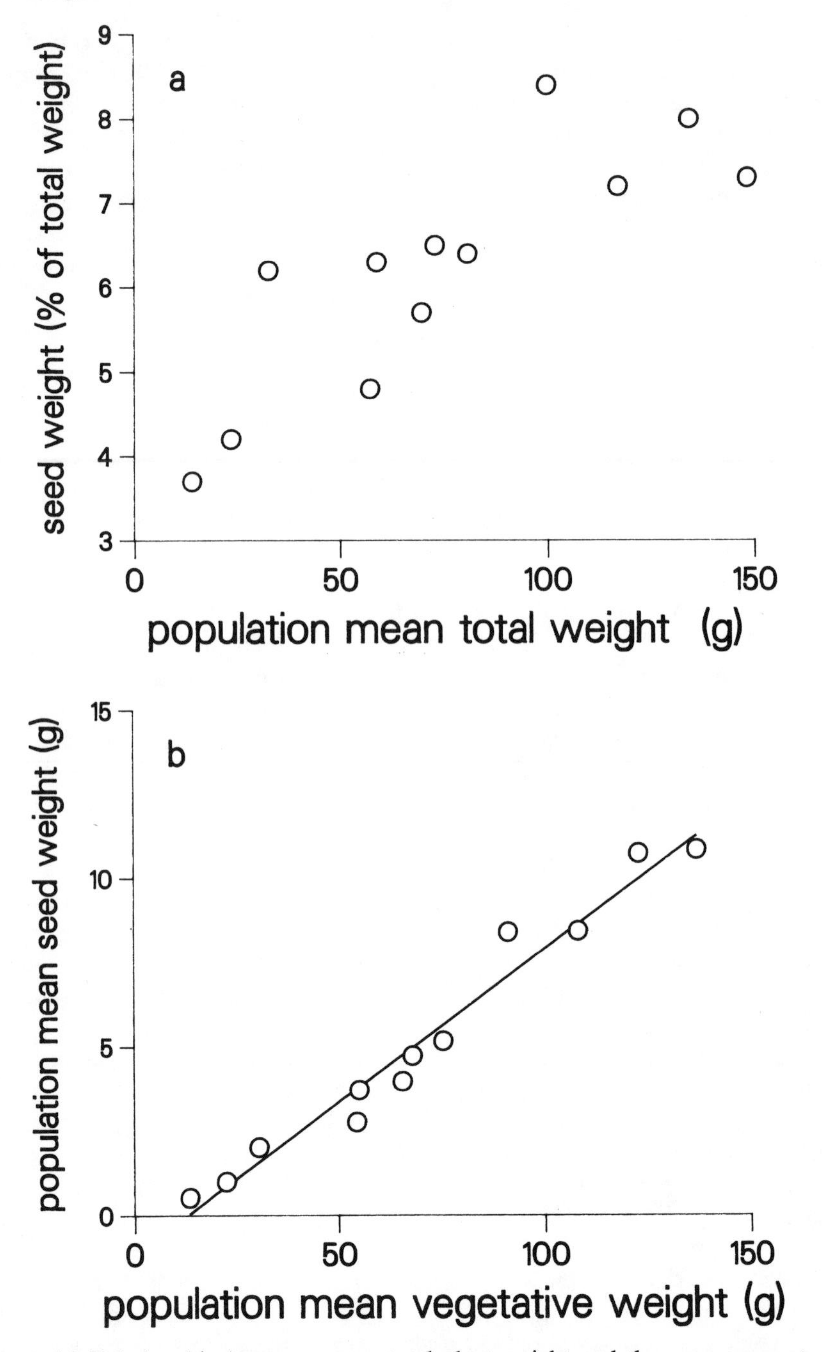

Fig. 11.4. (a) Relationship between mean total plant weight and the mean percentage of dry matter in seeds in 12 field populations of *Verbascum thapsus* from North America (after Reinartz[50]). (b) Same data replotted as mean vegetative plant weight versus mean seed weight (r^2 = 0.96).

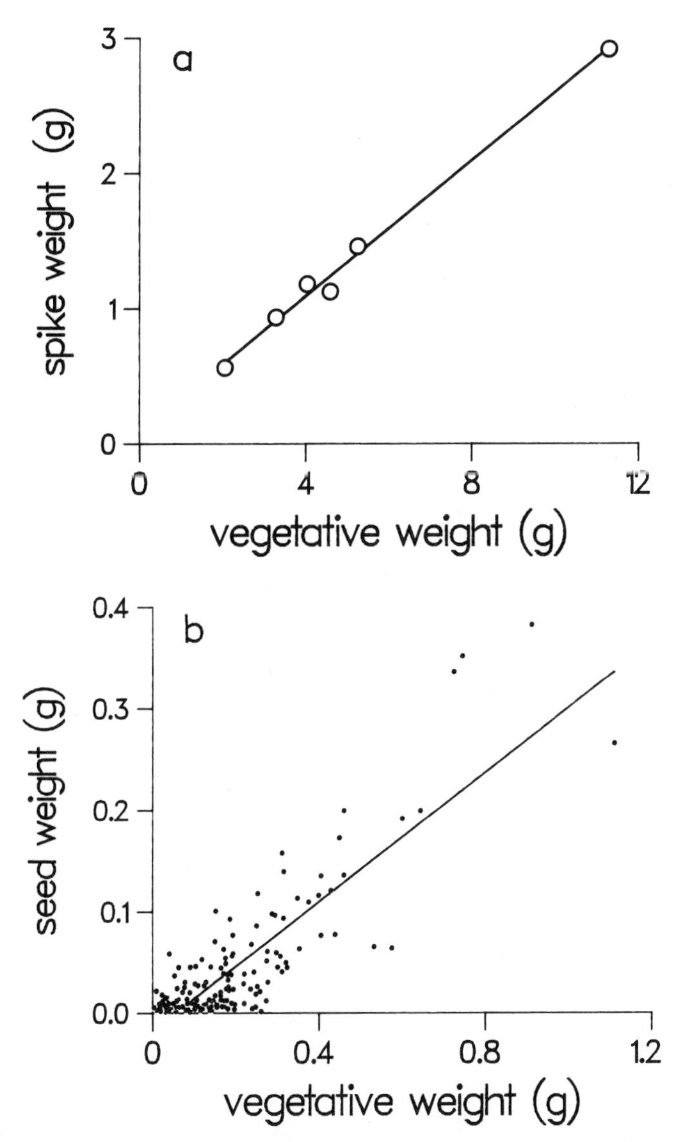

Fig. 11.5. (a) Relationship between mean spike weight and mean vegetative weight for *Plantago major* plants grown with different numbers and kinds of neighbors (after Hawthorn and Cavers[23]). (b) Relationship between seed weight and vegetative weight for *P. major* individuals grown in single plantain stands without grass at three sowing densities (Hawthorn and Cavers[24] and unpublished data from W. Hawthorn[22]).

ductive allocation in *Plantago coronopus*[61] and has been observed in several other studies.[25,63,65] Linear relationships between reproductive output and size have been studies by Samson and Werk.[52] They found evidence for such relationships in 18 species of desert winter annuals, and in published studies on 8 annual and 4 perennial species.

Changes in reproductive allocation with size may be seen in terms of simple allometric relationships. Since there is a minimum size for reproduction, percent reproductive allocation increases with size. Similarly, the concept of maximum reproductive allocation under ideal conditions[60] may be nothing more than a reflection of optimal growth.

A linear model of size-dependent reproduction is consistent with (although not dependent upon) the notion that plant growth is modular. If all modules are reproductive, a simple proportion of biomass would be expected to be in reproductive tissues, as in *Vulpia.*[65] If some nonreproductive modules are produced first, and all (or a constant proportion of all) subsequent modules are reproductive, the linear relationship seen in *Plantago* and *Verbascum* would be expected. Because of the modular nature of plant growth and reproduction, the reproductive behavior of plants may often be reducible to simple rules for plant growth, rather than complex, flexible reproductive strategies.

A linear relationship between reproductive weight and vegetative weight may be a reflection of inherent constraints operating on plant growth and reproduction. The biological plant may be analogous to an industrial plant, a factory in which offspring (seeds) are the products being manufactured. To build such a factory, a significant amount of capital (resources) must be invested to build machinery, etc. before the first unit can be produced. Thereafter the cost of producing each additional unit is not nearly as great, and may be relatively fixed, resulting in a linear relationship. Reproductive versus vegetative weight curves are analogous to marginal cost curves in microeconomics. Mathematically and biologically, it may be more useful in most cases to look at a plant's reproductive output as a function of size or other variables, than to look at its percent reproductive allocation. Nonlinear relationships between size and reproductive output may also exist. For woody perennials, biomechanical constraints may require that the weight of vegetative support tissues increase more than the weight of other organs at larger sizes. Thus, we might expect the slope of the relationship between reproductive weight and vegetative weight to decrease as plant size gets very large.

Competition may affect reproductive output through mechanisms in addition to simple size-dependent reproduction. While vegetative weight alone accounts for 72% of the variation in seed weight produced by *Plantago major* individuals in monocultures grown at three densities (Fig. 11.5b),[22] plants that achieve a given vegetative weight while growing at higher densities do tend to have slightly lower seed weight than plants of the same vegetative weight that grew at lower density. Density accounts for an additional 3.7% of the variation in seed weight produced by individuals ($p <$ .01). This may be a reflection of a flexible strategy or constraints on growth through which plants at higher densities have more of their weight in vegetative tissues. It appears that size-independent differences due to competition are small compared to those that can be attributed to size-dependent reproduction.

Size differences caused by some environmental factors may not always fit the linear size-dependent reproduction model as well as size differences caused by interference. Annual knotweeds (*Polygonum* spp.) show differences in reproductive output in different local environments, but, unlike *Verbascum thapsus,* these differences in reproductive output do not appear to be correlated with size.[26] When size differences were generated by environmental manipulations, only 56% of the variation in reproductive weight of *Diplacus auranthiacus* stems could be accounted for by differences in size. Shaded plants had greater vegetative weight than plants that were watered or fertilized, but the former group showed lower reproductive weight.[4] This could be because the levels and ratios of mineral nutrient, light, and water have direct physiological effects on growth and development (e.g., nitrogen surplus will increase growth and suppress flowering). Density does not change the relative concentrations of these resources available to the population, rather, it determines the number of individuals among which these nutrients will be (equally or unequally) divided. Pitelka et al.[44] have argued that it is not size but the accumulation of some critical store of resources (or the attainment of a physiological state indicating that adequate resources will be available) that determines if *Aster acumunatus* plants flower. Pitelka et al. used height as a measure of size and evaluated reproduction in percent allocation. But a regression of mean reproductive weight versus mean vegetative weight for the different environmental treatments accounts for over 70% of the variation in reproductive weight (and it accounts for 80% if the one case in which reproductive allocation was not significantly different from zero is excluded). The correlation between number of flower heads per ramet and ramet weight is 0.96. If the resource being stored before a plant will reproduce is photosynthate, the level of accumulation will be reflected in plant weight, although not necessarily plant height. In plants, size may be a measure of stored resources, and an indication of future available resources.

If reproductive output is a function of size, the allometric relationship between vegetative weight and reproductive weight may be considered as an essential part of the plant's reproductive strategy. Different life histories may be a reflection of different reproductive allometries, which are the products of natural selection. This may be true whether or not the allometric relationship between size and reproductive output is as simple as it appears to be for *Plantago coronopus* or *Verbascum thapsus*. Alternatively, different allometries may be the result of constraints imposed by different growth forms, and may be merely satisfactory, rather than optimal.

As discussed above, the effects of density on reproduction may be closely related to the time and rate of growth and development. As plants grow, at any given point they will have a certain size and a certain reproductive yield (Fig. 11.3). At higher densities plants may grow more slowly and they will have a smaller size, and therefore be at a point of lower reproductive allocation than plants grown at lower density for the same period of time. There is considerable evidence[15,56,71] that fecundity and other demographic parameters are better viewed as functions of size rather than age. Thus, within limits, if plants growing at higher densities are given a longer period to develop so that they could achieve the same size, they might show reproductive behavior similar to plants grown at low density for a shorter period. Ideally, one would like to compare the effects of density at equal mean plant weights, as well as equal time for development. At high densities, however, high mean plant weights cannot be achieved because of the constraints imposed by density-dependent mortality. Also, even if high- and low-density populations show similar size-dependent reproductive allocation, density may change the population size structure and therefore the distribution of reproduction among individuals within the population.

Hypotheses based upon flexible strategies may yield the same predictions as an allometric hypothesis. For example, Hickman[26] has suggested that competition may result in more assimilate being allocated to structures that increase competitive ability, leaving less assimulate for reproduction. While the allometric approach has the advantage of simplicity, if weight-independent changes in energy allocation in response to density can be demonstrated, this would provide strong support for an hypothesis of flexible reproductive strategy. Hutchings and Waite[30] see the effects of competition on the timing of reproduction in *Plantago coronopus* as independent of the effects of competition on size. They looked at life history in *Plantago coronopus* as a function of the proximity of neighbors, as measured by nearest neighbor distance, mean distance to several neighbors, and area of Thiessen polygons based upon neighbor locations. They found that individuals that had fewer and distant neighbors tended to reproduce earlier than those with more and close neighbors. Plants that failed to reproduce had the smallest territories. They report that these differences in phenology could not be attributed to differences in plant size as measured by rosette diameter, although they present very little data on size differences and rosette diameter is probably not a good measure of size. Even if a plant's response to neighbors is a reflection of an evolved, flexible reproductive strategy, plant size may be the mechanism through which variation in environmental conditions (including neighbor proximity) is detected and translated into variation in reproductive behavior.

Future research into the effects of competition (or other factors) on plant reproduction should (1) include data on individuals of different sizes, (2) emphasize reproductive output, rather than allocation, and (3) look at size-dependent effects. Such data may help us to distinguish between nonadaptive responses, simple size-dependent effects, and flexible reproductive strategies.

ACKNOWLEDGMENTS

This work was supported by a Smithsonian Postdoctoral Fellowship while the author was in residence at the Smithsonian Environmental Research Center, Edgewater, Maryland. I thank W. Hawthorn for providing me with raw data on *Plantago major,* S. C . Thomas for help with statistical analyses, P. E. Paston for administrative assistance, and S. Kinsman, D. Samson, M. Hutchings, L. Wagner, and J. and L. Lovett Doust for helpful comments on an earlier draft of this chapter.

REFERENCES

1. Abrahamson, W. G., Reproductive strategies in dewberries, *Ecology* **56,** 721–726 (1975).
2. Abrahamson, W. G., A comment on vegetative and seed reproduction in plants, *Evolution* **33,** 517–519 (1979).
3. Abrahamson, W. G., Demography and vegetative reproduction, in *Demography and Evolution in Plant Populations* (O. T. Solbrig, ed.), Chapter 5. Blackwell, Oxford, 1980.
4. Alpert, P., Newell, E. A., Chu, C., Glyphis, J., Gulmon, S., Hollinger, D. Y., Johnson, N. D., Mooney, H. A., and Puttick, G., Allocation to reproduction in the chaparral shrub, *Diplacus aurantiacus, Oecologia* **66,** 309–316 (1985).
5. Assemat, L., and Oka, H. I., Neighbor effects between rice (*Oryza sativa* L.) and barnyard grass (*Echinocholoa crus-galli* Beauv.) strains, *Oecologia Plantarum* **1,** 371–393 (1980).
6. Baker, G. A., and O'Dowd, D. J., Effects of parent plant density on the production of achene types in the annual, *Hypochoeris glabra, J. Ecol.* **70,** 201–215 (1982).
7. Bierzychudek, P., Determinants of gender in jack-in-the-pulpit: The influence of plant size and reproductive history, *Oecologia* **65,** 14–18 (1984).
8. Bishop, G. F., and Davy, A. J., Density and the commitment of apical meristems to clonal growth and reproduction in *Hieracium pilosella, Oecologia* **66,** 417–422 (1985).
9. Busch, R. H., and Luizzi, D., Effects of intergenotypic competition on plant height, days to heading, and grain yield of F2 through F5 bulks of spring wheat, *Crop. Sci.* **19,** 815–819 (1979).
10. Cavers, P. B., and Steel, M. G., Patterns of change in seed weight over time on individual plants, *Am. Nat.* **124,** 324–335 (1984).

11. Evenson, W. E., Experimental studies of reproductive energy allocation in plants, in *Handbook of Experimental Pollination Biology* (C. E. Jones, and R. J. Little, eds.)., Chapter 12. Scientific and Academic Editions, New York, 1983.
12. Goldberg, D. E., Neighborhood competition in an old-field plant community, *Ecology* **68,** 1211–1223 (1987).
13. Gould, S. J., and Lewontin, R. C., The spandrels of San marco and the Panglossian paradigm: A critique of the adaptationist programme, *Proc. R. Soc. London B* **205,** 581–598 (1979).
14. Grime, J. P., *Plant Strategies and Vegetation Processes.* Wiley, Chichester, U.K., 1979.
15. Gross, K. L., Predictions of fate from rosette size in four"biennial" species: *Verbascum thapsus, Oenothera biennis, Daucus carota,* and *Tragopogon dubius, Oecologia* **48,** 290–213 (1981).
16. Harper, J. L., The nature and consequences of interference amongst plants, in *Genetics Today, Proc. XI Int. Cong. Genetics,* Vol. 2, pp. 465–482. Hague, 1965.
17. Harper, J. L., *Population Biology of Plants.* Academic, London, 1977.
18. Harper, J. L., The concept of population in modular organisms, in *Theoretical Ecology: Principles and Applications* (R. M. May, ed.), 2nd Ed., pp. 55–77. Blackwell, Oxford, 1981.
19. Harper, J. L., After description, in *The Plant Community as a Working Mechanism* (E. I. Newman, ed.), pp. 11–25. Blackwell, Oxford, 1982.
20. Harper, J. L., Lovell, P. H., and Moore, K. G., The shapes and sizes of seeds, *Annu. Rev. Ecol. Syst.* **1,** 327–356 (1970).
21. Harper, J. L., and Ogden, J., The reproductive strategy of higher plants. I. The concept of strategy with special reference to *Senecio vulgaris* L., *J. Ecol.* **58,** 682–698 (1970).
22. Hawthorn, W. R., unpublished data listing.
23. Hawthorn, W. R., and Cavers, P. B., Resource allocation in young plants of two perennial species of *Plantago, Can. J. Bot.* **56,** 2533–2537 (1978).
24. Hawthorn, W. R., and Cavers, P. B., Dry weight and resource allocation patterns among individuals in populations of *Plantago major* and *P. rugelii, Can. J. Bot.* **60,** 2424–2439 (1982).
25. Hedley, C. L., and Ambrose, M. J., Designing "leafless" plants for improving yields of the dried pea crop, *Adv. Argon.* **34,** 255–277 (1981).
26. Hickman, J. C., Environmental unpredictability and plastic energy allocation strategies in the annual *Polygonum cascadense* (Polygonaceae), *J. Ecol.* **63,** 689–701 (1975).
27. Hodgson, G. L., and Blackman, G. E., An analysis of the influence of plant density on the growth of *Vicia faba.* Part I. The influence of density on the pattern of development, *J. Exp. Bot.* **7,** 147–165 (1957).
28. Holler, C. C., and Abrahamson, W. G., Seed and vegetative reproduction in relation to density in *Fragaria virginiana* (Rosaceae), *Am. J. Bot.* **64,** 1003–1007 (1977).
29. Holliday, R. J., Plant population and crop yield, *Nature (London)* **186,** 22–24 (1960).
30. Hutchings, M. J., and Waite, S., Cohort behaviour and environmental determination of life histories within a natural population of *Plantago coronopus* L., in *Structure and Functioning of Plant Populations 2. Phenotypic and Genotypic Variation in Plant Populations* (J. Haeck and J. W. Woldendorp, eds.), pp. 171–184. North-Holland, Amsterdam, 1985.
31. Kawano, S., and Hayashi, S., Plasticity in growth and reproductive energy allocation of *Coix ma-yuen* Roman. cultivated at varying densities and nitrogen levels, *J. Coll. Lib. Arts, Toyama Univ. Jpn. (Nat. Sci.)* **10,** 1–32 (1977).
32. Krebs, C. J., *Ecology: The Experimental Analysis of Distribution and Abundance,* 3rd Ed. Harper and Row, New York, 1985.
33. Law, R., Bradshaw, A. D., and Putwain, P. D., Reply to W. G. Abrahamson, *Evolution* **33,** 519–520 (1979).
34. Lovett Doust, J., and Cavers, P. B., Sex and gender dynamics in jack-in-the-pulpit, *Arisaema triphyllum* (Araceae), *Ecology* **63,** 797–808 (1982).
35. Lovett Doust, J., O'Brien, G., and Lovett Doust, L., Effect of density on secondary sex characteristics and sex ratio in *Silene alba* (Caryophyllaceae), *Am. J. Bot.* **74,** 40–46 (1987).
36. Mack, R. N., and Harper, J. L., Interference in dune annuals: Spatial pattern and neighbourhood effects, *J. Ecol.* **65,** 345–363 (1977).
37. Marshall, D. L., Fowler, N. L., and Levin, D. A., Plasticity in yield components in natural populations of three species of *Sesbania, Ecology* **66,** 753–761 (1985).
38. Ogden, J., The reproductive strategy of higher plants. II. The reproductive strategy of *Tussilago farfara* L., *J. Ecol.* **62,** 291–324 (1974).

39. Onyekwelu, S. S., and Harper, J. L., Sex ratio and niche differentiation in spinach (*spinacia oleracea* L.), *Nature (London)* **282,** 609–611 (1979).
40. Ornduff, R., Male-biased sex ratios in the cycad *Macrozamia riedlei* (Zamiaceae), *Bull. Torrey Bot. Club* **112,** 393–397 (1986).
41. Palmblad, I. G., Comparative response of closely related species to reduced competition. I. *Senecio sylvaticus* and *S. viscosus, Can. J. Bot.* **46,** 225–228 (1968).
42. Palmblad, I. G., Competition in experimental populations of weeds with emphasis on the regulation of population size, *Ecology* **49,** 26–34 (1968).
43. Peterson, D. L., and Bazzaz, F. A., Life cycle characteristics of *Aster pilosus* in early successional habitats, *Ecology* **59,** 1005–1013 (1978).
44. Pitelka, L. F., Ashmun, J. W., and Brown, R. L., The relationships between seasonal variation in light intensity, ramet size, and sexual reproduction in natural and experimental populations of *Aster acuminatus* (Compositae), *Am. J. Bot.* **72,** 311–319 (1985).
45. Pitelka, L. F., Stanton, D. S., and Peckenham, M. O., Effects of light and density on resource allocation patterns in a forest herb, *Aster acuminatus* (Compositae), *Am. J. Bot.* **67,** 942–948 (1980).
46. Pitelka, L. F., Thayer, M. E., and Hansen, S. B., Variation in achene weight in *Aster acuminatus, Can. J. Bot.* **61,** 1414–1420 (1983).
47. Policansky, D., Sex choice and the size advantage model in jack-in-the-pulpit *(Arisaema triphyllum), Proc. Natl. Acad. Sci. U.S.A.* **78,** 1306–1308 (1981).
48. Puckeridge, D. W., and Donald, C. M., Competition among wheat plants sown at a wide range of densities, *Aust. J. Agric. Res.* **18,** 193–211 (1967).
49. Rathcke, B., Competition and facilitation among plants for pollination, in *Pollination Biology* (L. Real, ed.), Chapter 12. Academic Press, New York, 1983.
50. Reinartz, J. A., Life history variation of common mullein *(Verbascum thapsus)* II. Plant size, biomass partitioning and morphology, *J. Ecol,* **72,** 913–925 (1984).
51. Sakai, A. K., and Oden, N. L., Spatial pattern of sex expression in silver maple (*Acer saccharinum* L.): Morisita's index spatial autocorrelation, *Am. Nat.* **122,** 489–508 (1983).
52. Samson, D. A., and Werk, K. S., Size-dependent effects in the analysis of reproductive effort in plants, *Am. Nat.* **127,** 667–680 (1986).
53. Schmid, B., and Harper, J. L., Clonal growth in grassland perennials. I. Density and pattern dependent competition between plants with different growth forms, *J. Ecol.* **73,** 793–808 (1985).
54. Schmitt, J., Individual flowering phenology, plant size, and reproductive success in *Linanthus androsaceus,* a California annual, *Oecologia* **59,** 135–140 (1983).
55. Silander, J. A., and Pacala, S. W., Neighborhood predictors of plant performance, *Oecologia* **66,** 256–263 (1985).
56. Silvertown, J. W., *Introduction to Plant Population Biology,* 2nd Ed. Longman, London, 1987.
57. Smith, C. C., The distribution of energy into sexual and asexual reproduction in wild strawberries *(Fragaria virginiana),* in *Third Midwest Prairie Conf. Proc, Manhattan, Kansas,* pp. 55–60, 1972.
58. Snell, T. W., and Burch, D. G., The effects of density on resource partitioning in *Chamaesyce hirta* (Euphorbiaceae), *Ecology* **56,** 742–746 (1975).
59. Thompson, D. A., and Beattie, A. J., Density-mediated seed and stolon production in *Viola* (Violaceae), *Am. J. Bot.* **68,** 383–388 (1981).
60. Thompson, K., and Stewart, A. J. A., The measurement and meaning of reproductive effort in plants, *Am. Nat.* **77,** 205–211 (1981).
61. Van der Meijden, E., and Van der Waals-Kooi, R. E., The population ecology of *Senecio jacobaea* in a sand dune system, *J. Ecol.* **67,** 131–153 (1979).
62. Waite, S., and Hutchings, M. J., Plastic energy allocation patterns in *Plantago coronopus, Oikos* **38,** 333–342 (1982).
63. Waller, D. M., Neighborhood competition in several violet populations, *Oecologia* **52,** 116–122 (1981).
64. Watkinson, A. R., Density-dependence in single-species populations of plants, *J. Theor. Biol.* **83,** 345–357 (1980).
65. Watkinson, A. R., Factors affecting the density response of *Vulpia fasciculata, J. Ecol.* **70,** 149–161 (1982).
66. Watkinson, A. R., and Harper, J. L., The demography of a sand dune annual: *Vulpia fasciculata.* I. The natural regulation of populations, *J. Ecol.* **66,** 15–33 (1978).
67. Weiner, J., A neighborhood model of annual-plant interference, *Ecology* **63,** 1237–1241 (1982).
68. Weiner, J., Neighbourhood interference amongst *Pinus rigida* individuals, *J. Ecol.* **72,** 183–195 (1984).

69. Weiner, J., Size hierarchies in experimental populations of annual plants, *Ecology* **66,** 743–752 (1985).
70. Weiner, J., and Thomas, S. C., Size variability and competition in plant monocultures, *Oikos* **47,** 211–222 (1986).
71. Werner, P. A., Predictions of fate from rosette size in teasel (*Dipsacus fullonum* L.), *Oecologia* **20,** 197–201 (1975).
72. Willey, R. W., Plant population and crop yield, in *Handbook of Agricultural Productivity,* Vol. 1, pp. 201–207. CRC Press, Boca Raton, Florida, 1982.
73. Willey, R. W., and Heath, S. B., The quantitative relationships between plant population and crop yield, *Adv. Agron.* **21,** 281–321 (1969).
74. Williams, R. D., Quimby, P. C., and Frick, K. F., Intraspecific competition of purple nutsedge *(Cyperus rotundus)* under greenhouse conditions, *Weed Sci.* **25,** 477–481 (1977).
75. Willson, M. F., *Plant Reproductive Ecology.* Wiley, New York, 1985.
76. Wolfe, L. M., The effect of size on reproductive characteristics in *Erythronium, Can. J. Bot.* **61,** 3489–3493 (1983).
77. Zimmerman, J. K., and Weis, M., Factors affecting survivorship, growth, and fruit production in a beach population of *Xanthium strumarium, Can. J. Bot.* **62,** 2122–2117 (1984).
78. Zimmerman, M., Nectar production, flowering phenology, and strategies for pollination, in *Plant Reproductive Ecology: Patterns and Strategies* (J. Lovett Doust and L. Lovett Doust, eds.), Chapter 8. Oxford Univ. Press, New York, 1988.

12

Herbivory and Its Impact on Plant Reproduction

STEPHEN D. HENDRIX

The currency of fitness for plants as well as animals is the number of descendants that an individual produces. In plants, adaptations leading to maximization of lifetime reproductive output are diverse (Cousens,[23] DeWreede and Klinger,[35] and Mishler,[105] this volume). This diversity is due in part to differences in life histories and the nature and intensity of selection pressures on plants of different taxonomic affinities, morphologies (Waller,[141] this volume), and habitats.

One important selection pressure faced by all or nearly all plants results from the removal of vegetative and reproductive structures by herbivores. Herbivores have a long evolutionary association with plants[57,126] and, although today the number of orders of animals utilizing plants either partially or completely as a food source is relatively small,[130] the diversity of herbivores within certain of these orders is enormous. Most striking is that approximately 50% of all insect species and 65% of mammal species are partially or completely phytophagous (Table 12.1). Furthermore, no plant tissue is immune to attack by herbivores. Both mammals and insects with chewing mouthparts consume directly leaves, stems, roots, flowers, and seeds or fruits, while insects with piercing mouthparts feed on plant sap or cell contents.

Given the ubiquitous presence of herbivores and their ability to utilize all the different plant tissues available, it is surprising that herbivory has not always been recognized as having a significant impact on the reproductive ecology of plants.[46] Nevertheless, the accumulated evidence from numerous studies of effects of herbivory on plants in both agricultural and natural settings leads to the inescapable conclusion that herbivory is a major factor influencing the reproductive success of plants. Herbivory also affects both the structure and dynamics of plant populations and communities[12,31,47] as well as the evolution of chemical defenses,[22,123] but these aspects of plant–herbivore interactions are beyond the scope of this chapter. This chapter focuses primarily on the direct and indirect effects of herbivory on mortality, growth, and reproduction of individual plants.

METHODS OF ANALYZING THE EFFECTS OF HERBIVORY ON GROWTH AND REPRODUCTION

Before discussing the effects of herbivore damage to vegetative parts on plant reproduction, it is necessary to review the methods used to describe such effects because different methods have different biases and different inherent problems. The most common method utilized is artificial removal of vegetative tissue, usually by clipping. While artificial defoliation has the advantage of permitting randomization of treatments and the removal of exact amounts of tissue,[80] it rarely is done in a manner that adequately simulates natural defoliation. With respect to grazing mammals, clipping often removes more tissue than the herbivore would and this exaggerates the severity of the effects.[72] Furthermore, tissues may be removed indiscriminately when clipped, whereas grazing herbivores are often selective.[86] Because different types and ages of

Table 12.1. Numerical Relationships Among Phytophagous and Nonphytophagous Species of Mammals and Insects[a]

Order	Number of species	Number phytophagous species	Percentage phytophagous	Percentage phytophagous of all phytophagous
		MAMMALS[b]		
Marsupialia	242	117	48.3	4.6
Chiroptera	853	280	32.8	11.1
Primates	166	166	100.0	6.6
Lagomorpha	63	63	100.0	2.5
Rodentia	1,685	1,685	100.0	66.8
Perissodactyla	16	16	100.0	0.6
Artiodactyla	170	170	100.0	6.7
Nonphytophagous[c]	700	0	0.0	—
Others[d]	41	25	61.0	1.0
Total	3,936	2,522	64.1[e]	99.9
		INSECTS[a,f]		
Collembola	2,000	50	2.5	0.01
Orthoptera	23,000	20,700	90.0	5.8
Isoptera	2,100	2,100	100.0	0.6
Thysanoptera	4,700	4,650	99.0	1.3
Hemiptera	23,500	11,750	50.0	3.3
Homoptera	33,000	33,000	100.0	9.2
Coleoptera	300,000	141,000	47.0	39.5
Lepidoptera	113,000	111,740	98.0	31.0
Diptera	90,000	22,500	25.0	6.3
Hymenoptera	108,000	10,800	10.0	3.0
Nonphytophagous[g]	26,334	0	0.0	—
Total	725,634	358,290	49.2[e]	100.0

[a]From Hendrix.[57]

[b]Includes omnivorous mammals.

[c]Includes the orders Carnivora, Cetacea, Insectivora, Monotremata, Pholidota, and Pinnipedia.

[d]Includes Dermoptera, Edentata, Hydracoidea, Proboscidea, and Sirenia.

[e]Total number of phytophagous species/total number of species.

[f]Certain small orders with a few phytophagous species are not included.

[g]Includes Ephemeroptera, Mallophaga, Neuroptera, Odonata, Plecoptera, Siphonaptera, and Tricoptera.

tissues contribute differentially to growth and reproduction, the effects of clipping can differ from those of natural herbivory. For example, clipping may remove apices where regrowth could occur whereas herbivore damage may be confined to leaves, allowing axillary buds to develop.

Artificial tissue removal and natural insect damage may differ in other subtle but important ways that can increase the impact of artificial herbivory on plant reproduction. Chewing insects leave midribs, petioles, or some of the leaf blade[80] but when clipping is used whole leaves are often removed.[55,75,82] Also, when chewing insects are in the growth phase they gradually remove ever-increasing amounts of tissue over a number of days whereas clipping is often done over a shorter period of time.[55,82,116] Another problem is that the addition of saliva by feeding herbivores is hypothesized to stimulate tissue regrowth,[15,32] although in some studies such effects have not been found.[33]

Considering the problems associated with artificial herbivory, the method must be applied with care. Natural levels of herbivory on individual plants in their normal habitat should be determined first and then the temporal and spatial patterns found should be mimicked. Unfortunately, only a few studies utilizing artificial herbivory have been designed in the above manner.[94,134] As a consequence, in most studies the treatments and resulting effects on reproduction are often extreme and rarely reflect what actually occurs in nature. However, they are useful in demonstrating extreme effects of herbivory and potential response characteristics of host plants (see below, Compensatory Responses of Plants to Herbivory).

A second technique used to study effects of herbivory involves comparisons of naturally damaged plants with plants protected from herbivores. Exclosures are used to protect plants from mammalian herbivores,[47,172] while pesticides may be used to protect plants from insects.[77,91,106,131,142] In some instances, levels of damage are partially controlled by caging known numbers of herbivores on the plant.[6,16] These methods avoid the problems of artificial herbivory outlined above, although care must be taken in the experimental design to ensure adequate replication for statistical analyses.[69]

A third method of analyzing effects of herbivory involves no manipulation; instead, natural levels of damage to vegetative parts are measured or estimated and then correlated with growth and/or reproductive characteristics.[80,132] While this method avoids the problems of the alternative methods outlined above, the results must be interpreted with care because many aspects of the experimental design are uncontrolled. Differences in growth and reproductive performance of individuals may reflect important but subtle microhabitat differences rather than, or in addition to, differences in levels of herbivory. Also, herbivores may preferentially attack larger[137] or more vigorous individuals[47,108] which could, in turn, affect the results if control plants are not of equivalent size and physiological condition. Despite the difficulties in interpreting results, this method is likely to be used more frequently because of the recent invention of portable leaf area meters. This equipment is also valuable because of the relative ease with which one can determine temporal and spatial patterns of damage for simulated herbivory experiments or assess levels of herbivory in exclusion experiments. Equally importantly, it allows sequential, nondestructive monitoring of foliar damage.

Utilization of any of these methods with crop plants presents additional problems when results are extrapolated to all plants. Crop plants are often genetically altered and the experiments are usually conducted under row crop conditions where plants are not stressed by competition. Furthermore, crop plants in a given field are often genetically uniform. Thus, the effects of herbivory may be substantially different from those that occur in nature. Nevertheless, studies of crop plants and their herbivores can provide useful insights into certain aspects of plant–herbivore interactions. For example, Via[139] found that when individuals of the pest species *Liriomyza sativae* were reared on different crops, performance as measured by pupal weight and developmental time differed, suggesting the potential for evolution of genetically based host plant specialization.

It is clear from the above discussion that no one method is necessarily better than any other. All have specific advantages and disadvantages and the particular methods any researcher uses will depend on the characteristics of the system under study and the hypotheses to be tested. In most instances, a combination of methods is most useful and such a unified approach is likely to be used more often in the future.

EFFECT OF REMOVAL OF VEGETATIVE PLANT PARTS ON MORTALITY, GROWTH, AND REPRODUCTION

Herbivore-Mediated Plant Mortality

The most obvious way in which herbivores can reduce plant reproduction is either by preventing plants from entering the reproductive stage or by shortening the life of the individual. Seedlings are particularly vulnerable to herbivore-induced mortality. Vertebrates can cause extensive damage either by directly consuming all or nearly all above ground tissue or, in the case of large vertebrates, by trampling.[1] For example, in California chaparral, small mammals cause seedling mortality levels of 44 and 19%, respectively, in the shrubs *Ceanothus greggii* and *Ademonostoma fasciculatum.*[103] As a consequence, high seedling mortality due to vertebrates can have the long-term effect of decreasing plant population recruitment.[90,114]

In contrast, seedling mortality due to insects may be less common when smaller amounts of tissue are removed. Natural insect herbivory on tree seedlings of *Shorea leprosula* and *Shorea maxwelliana* in the Malaysian lowland rain forest resulted in losses of 13 and 5% of total leaf area, respectively, and did not affect survival of 2-year-old seedlings.[4] Similarly, insects feeding on seedlings of the chaparral shrubs mentioned above[103] as well as those of *Colliguaya odorifera* did not increase mortality.[104] However, grasshoppers caused high mortality of *Gutierrezia microcephala* seedlings[113] and nymphs of the shield-backed katydid *(Atlanticus testaceous)* may cause high mortality of *Melampyrum lineare* seedlings.[14]

The differences in the results of these studies on the effects of herbivory on seedlings suggests that seedlings of some species are resilient to herbivory while others are not. The number and frequency of periods of defoliation will be important in determining regrowth potential following herbivory, although other factors such as the amount of stored food reserves, availability of axillary buds, and microhabitat conditions should also play significant roles in determining the fate of seedlings suffering herbivory.

The death of adult plants resulting from vertebrate feeding is often associated with high points in herbivore population cycles. Large populations of elephants,[81] deer,[78] and hares and rodents[19,117] can cause extensive plant mortality. At such times host species may actually be eliminated from plant communities.[47] Even when vertebrate population levels are not excessive, bark-ringing by animals such as goats, sheep, rabbits, and squirrels can kill woody plants through the destruction of phloem and vascular cambium.[24] In contrast to woody plants grazed by vertebrates, grassland vegetation is resilient to the effects of even high levels of herbivory.[100,102]

Mortality in adult plants resulting directly from insects usually occurs following repeated complete or nearly complete defoliation of host plants. In temperate forests of the northern hemisphere outbreaks of many introduced insects cause extensive host mortality,[80,133] although some trees are more susceptible than others.[80] For example, when oak forests in New Jersey were defoliated by the gypsy moth *(Porthetria dispar)* in three successive years, mortality was 31% in *Quercus rubra* but only 0–4% in three other *Quercus* species.[132]

Plant mortality may result also when levels of insect damage are below 100%. In east Texas, larvae of the pipevine swallowtail butterfly, *Battus philenor,* normally consume about 45% of the annual leaf crop of *Aristolochia reticulata,* resulting in mortality rates 4.5 times greater than among protected plants.[115] Plant mortality due to insect herbivory in the shrub *Sarothamnus scoparius* was twice as high in an unsprayed plot as in a pesticide-sprayed plot after 9 years,[142] although differential mortality in the two plots was not apparent for the first 5 years because whole bushes rarely die in one season. In other ecosystems, routinely high levels of herbivory may result in plant mortality. For example, in Australia, insect herbivores of *Eucalyptus* commonly consume 20–50% of the foliage[106] and, in the tropics, leaf-cutter ants (*Atta* spp.) and flea beetles (*Oedionychus* spp.) can cause high levels of damage.[121,122] Even when mean levels of damage to a species are low, as is most often the case,[9,83] some individuals may sustain high levels of damage that may eventually affect their longevity.[21,94]

Although herbivores can directly kill their host plants if feeding pressure is sufficiently intense, the final killing agent is often something other than the primary herbivore. Initial herbivore damage may weaken an individual and increase its susceptibility to drought,[68] other insects, vertebrates, or fungi,[26,80,132] or decrease its competitive abilities.[5,47]

Reductions in Growth and Reproduction Following Herbivory

While plant mortality is the most dramatic means by which herbivores affect plant reproduction, decreases in growth and subsequent reproduction rather than outright death are the more usual effects of herbivore damage to vegetative parts. In the tropical shrub, *Piper arieianum,* both shoot length and seed production were reduced following simulated herbivory[94] (Fig. 12.1), with both the amounts of tissue removed and plant size affecting subsequent growth and reproduction. Likewise, experimental defoliations in *Catalpa speciosa* resulting in increased abortion of fruits.[134] Louda[91] found that removal (using pesticides) of the leaf-feeding flea beetles from *Cardamine cordifolia* resulted in significant increases in plant height and reproduction, and Myers[107] found that natural defoliation of *Rosa nutkana* by tent caterpillars reduced rosehip production 10-fold.

Reductions in growth following herbivory are usually proportional to the amount of tissue lost,[79,80,82,94] although for many species there is a threshold level below which growth and/or reproduction is not significantly reduced (Refs. 8, 138, but see Ref. 25). For example, in *Piper arieianum* defoliation that was less than 30% did not significantly affect growth[94] (Fig. 12.1). Similarly, continuous defoliation of two-thirds of the

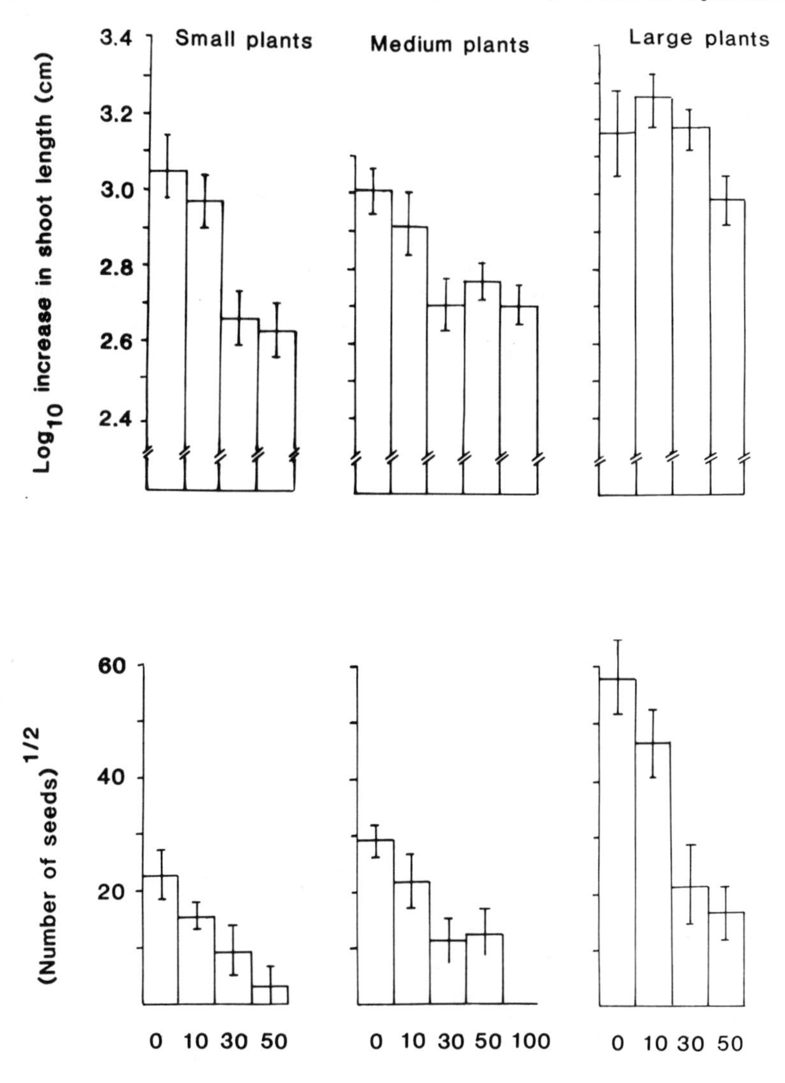

Fig. 12.1. The effects of varying levels of simulated herbivory on growth and reproduction for three size classes of *Piper arieianum. (Upper)* Growth as measured by the total length of new shoots produced over the entire plant during the first year after a single defoliation (±1 SE). *(Lower)* Seed production during the first year after defoliation (±1 SE). Small plants: 20–33 leaves; medium plants: 34–55 leaves; large plants: 56–120 leaves. (From Marquis,[94] copyright 1984 by the AAAS.)

leaves of *Ambrosia artemisiifolia* and *Arctium minus* did not significantly reduce growth as measured by height of plants that survived.[116] In crop species such as snap-beans[45] and soybeans[136] up to 50% defoliation may not affect yield. The existence of threshold levels of damage before reductions in growth and/or reproduction occur suggest that plants have the ability to compensate at least partially for herbivore damage (see below Compensatory Responses of Plants to Herbivory).

A variety of factors besides levels of damage interact to determine the impact of herbivore damage on growth and reproduction. Loss of older vegetative tissue is usually less detrimental than loss of newer tissue because the former contributes less to reproduction,[80,124] although in white pine removal of 1-year-old needles was more harmful than removal of either current-year or 2-year-old foliage.[87] In evergreen trees, loss of new tissue has the additional effect of reducing production of hormones necessary for utilization of stored photosynthate.[80]

Timing of defoliation is important also in determining the effects of tissue loss on growth and reproduction. In trees and shrubs that depend on stored photosynthate for growth, early defoliation after food reserves are depleted is more deleterious than late defoliation when reserves have built up again.[18,80,132] However, late defoliations may result in winter damage to twigs because of decreased lignification.[80] In contrast, mid- and late-season foliar damage markedly affects growth in species that utilize current photosynthate for continuous or recurrent foliage production.[79] With respect to reproduction, removal of vegetative tissues at the time fruits are filling depresses yields most strongly.[3,49] For example, in soybeans, removal of up to two-thirds of vegetative tissue before flowering started did not decrease yields, but removal of one-sixth of the leaves at the time pods were filling did.[20]

Other factors can mediate the effects of vegetative tissue removal on growth and reproduction. Small plants are generally less tolerant of herbivore damage than large plants[3,94,115] (Fig. 12.1) and plants in low-competition situations are less affected by tissue loss than are plants in high-competition situations.[82] Furthermore, extreme weather conditions can accentuate the negative effects of herbivory on plant growth and reproduction, although in *Dactylis glomerata* intense grazing has resulted in selection for a form with short, prostrate leaves that is also tolerant of waterlogging.[39]

Indirect Effects of Vegetative Herbivory on Plant Characteristics

In addition to the direct effects of vegetative tissue removal on growth and reproduction, a variety of indirect effects of herbivory may alter the reproductive capacity of host plants. Removal of above-ground tissue results in a retardation of new root growth with a decrease in the life span of existing roots.[5,29,61,72,120] This may, in turn, restrict water and nutrient uptake[72] with the result that reproductive output is decreased. The degree to which root growth is negatively affected is determined in part by both the amount of above-ground damage[29,72,101] and the degree to which resources from roots are allocated for regrowth of above-ground tissues.[120]

Premature leaf shedding and alteration of leaf characteristics can occur following herbivory. When aphids fed on *Tilla × vulgaris* or *Acer pseudoplatanus,* leaves were shed earlier, were heavier per unit area, and contained more nitrogen than leaves on plants that were not grazed.[36,37] Aphid attack one year resulted in smaller leaves the following year in *Tilia* but not *Acer.* Likewise, leaf-mining insects on *Betula tortuosa* reduced leaf size by 20%.[52] In both oaks[40] and *Ilex aquifolium*[112] leaf miners induce early leaf abscission. Chewing insects can also reduce leaf longevity and initiation of new leaves.[91] Such changes in leaf longevity and leaf characteristics should result in reduced growth and ultimately, reproduction, but quantification of their impact on reproductive success is lacking.

Herbivory may affect seed quality as well as seed number when vegetative tissues are removed. In *Piper arieianum* 50% defoliation reduced seed viability by 25%.[94]

Grazing by the beetle *Gastrophysa viridula* on *Rumex obtusifolius* and *Rumex crispus* reduced seed weight.[6] In another study, although overall germination in *Rumex crispus* was not affected by seed weight, smaller seeds germinated more rapidly and under a wider range of conditions.[97] Also, stem gall insects of *Solidago canadensis* reduced propagule weight[50] as well as clonal growth.[135]

EFFECTS OF REMOVAL OF REPRODUCTIVE STRUCTURES ON PLANT FITNESS

Herbivore destruction of flowers, seeds, and fruits by both invertebrates and vertebrates can substantially reduce the reproductive output of plants.[48,74] For example, some individuals in 15 out of 23 species of Central American Leguminosae had at least 50% of their seed crop destroyed by bruchid beetles,[73] and similar levels of bruchid attack on *Astragalus* species in Utah have also been reported.[44] In desert ecosystems ants and/or rodents may consume as much as 75% of all seeds produced,[11] and in grasslands seed consumption by mice varied from 37% of total production for *Bromus diandrus* to 75% for *Avena fatua.*[7] Likewise, birds, rodents, and deer can destroy large proportions of the reproductive output of trees and shrubs in deciduous and coniferous forests either by immediately eating the nuts and seeds or, in the case of birds and rodents, caching them for later use.[27,74,129] Flower-feeding herbivores can also significantly reduce seed production. For instance, caterpillars of the lycaenid butterfly, *Glaucopsyche lygdamus,* feeding on the flowers of *Lupinus amplus* routinely reduce seed set by 50% in some populations.[10] A similar reduction in seed set occurred in small individuals of *Pastinaca sativa* following flower feeding by *Depressaria pastinacella.*[56]

In contrast to the extensive literature detailing reductions in reproductive output due to seed predation, far fewer studies document whether or not such reductions translate into decreases in fitness as measured by differential recruitment from damaged and undamaged plants. Because patterns of recruitment in plant populations can be regulated by factors such as safe sites for seed germination[48] or the presence of a seed bank[113] rather than by seed production, it is erroneous to assume that reductions in seed production necessarily result in proportional reductions in recruitment. For example, the nearly complete herbivore destruction of the seed crop of the arid grassland shrub *Gutierrezia microcephala* did not reduce population recruitment the next year, possibly due to the existence of a large seed bank.[113] However, if high levels of seed destruction continued, a decrease in population recruitment would result when the seed bank became depleted.

In one well-documented case seed predation by insects has been directly tied to decreases in population recruitment. Louda[88,89] found that, in the sage scrub community of southern California, removal of flower and seed eaters of the native shrub, *Haplopappus squarrosus,* with a pesticide increased recruitment (Fig. 12.2). Furthermore, recruitment was proportional to the number of viable seeds produced and juvenile recruitment was proportional to seedling establishment. In both desert[28,71] and grassland ecosystems[7] exclusion of seed predators resulted in changes in plant numbers of certain species, implying that recruitment in these species was altered by seed predation either directly through changes in seed number or indirectly through changes in competitive hierarchies.

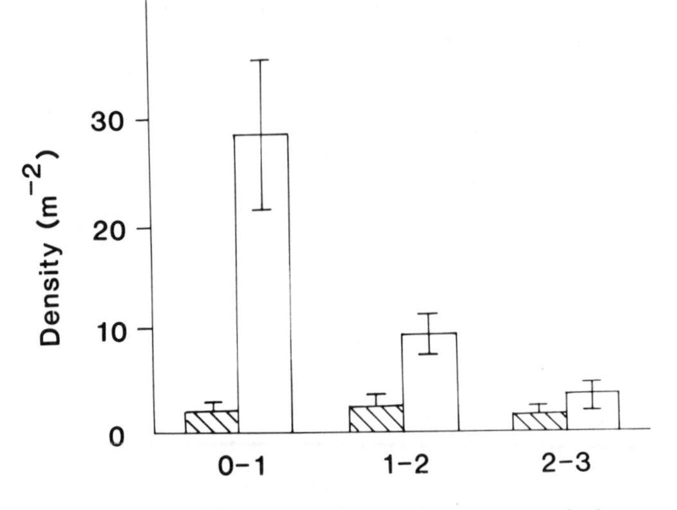

Fig. 12.2. Variation in mean seedling density (±1 SE) over distance around five experimental *Happlopappus squarrosus* plants after insecticide-sprayed (open columns) and water-sprayed (hatched columns) treatments. (From Louda,[88] copyright 1982 by Blackwell Scientific Publications Ltd.)

Analysis of the potential effect of seed predators on seed production and subsequent population recruitment is considerably complicated by the fact that some of these animals also function as seed dispersers. Despite the fact that seed and fruit eaters such as monkeys, squirrels, birds, rodents, and ants can destroy large proportions of a plant's reproductive output,[11,27,63,65,74,129] their role in population recruitment can be critical. For example, in central Oregon germination of *Pinus ponderosa* and *Purshia tridentata* seeds in unrecovered rodent caches accounted for 15 and 90%, respectively, of all seedlings in these two species.[144]

Although it is difficult to follow the fate of seeds dispersed by animals, destruction of large proportions of the seed crop by disperser-predators may be balanced by subtle advantages hypothesized to accrue to seeds handled by these organisms.[66] Removal of seeds from the vicinity of the parent plant may increase the probability of recruitment among these more widely dispersed propagules, if seed and seedling predators concentrate their activities under the parent plant.[65] Another potential advantage is that seeds may be dropped or stored in specific locations favoring seed germination or seedling survivorship.[74,76] In other instances, disperser-predators may broadcast seeds over a wide area, which could increase the probability of seeds hitting safe sites if those near the parent plant are saturated.

In addition to the ecological effects of seed predation on plant reproduction, evolutionary changes in fruit and seed characteristics can result from long-term plant–seed predator interactions. With respect to bruchid beetles, Janzen[73] found legume species that were attacked had significantly smaller seeds, more seeds per m^3 of canopy, and more grams of seeds per m^3 of canopy. In contrast, vertebrate seed predation on conifers has resulted in selection for increased amounts of protective tissue in the form of cone scales and decreased seed number, but seed weight remains constant.[38,128]

The differences in the evolutionary effects of vertebrate and invertebrate seed predators may be due to ecological differences between the two types of predators as well as ecological differences among the plants attacked.[38] For example, invertebrate seed predators take a great length of time to destroy a seed crop and they generally have narrow host plant specificity. The dispersal system of the plant may also be important.[38] In species with wind-dispersed seeds, such as conifers, changes in morphological characteristics of support structures (size and shape of cones) are unlikely to affect dispersal whereas similar changes in species with animal-dispersed seeds might negatively affect the attractiveness of the reproductive unit to potential dispersers.[38,67]

Seed predators can also affect temporal patterns of reproduction in long-lived species. Janzen[74] hypothesized that intense predispersal seed predation selected for high levels of plant reproduction some years followed by a number of years of low or no reproduction ("masting"). During years of low reproduction, population levels of seed predators are also low. When high levels of reproduction occur during mast years, seed predator numbers are insufficient to utilize fully the increased food resources and some seeds escape attack. The intervening years of low reproduction prevent seed predator numbers from increasing in response to mast years, although as a defense mechanism, masting appears to be more effective against insects than against vertebrates.[111] Silvertown's[127] analysis of the relationship between seed crop size and predispersal seed predation in 25 tree species indicated that in 16 species masting reduced seed predation. In the species in which masting did not function as a defense strategy against seed predators as well as in species in which it did, it is likely that the interval between mast years is due to depletion of the tree's food reserves in years of high reproduction followed by a number of years of low reproduction during which food reserves again accumulate. Also, masting is rare in shrubs that may be too-short-lived to effectively use masting as a defense mechanism against seed predators. It is also rare in shrubs and trees with fleshy dispersal units.[127] In the latter instance, dependence on frugivores for dispersal may necessitate relatively consistent fruit production from year to year.[64]

COMPENSATORY RESPONSES OF PLANTS TO HERBIVORY

Traditionally, herbivory is considered to have a negative effect on plant reproduction and certainly the majority of available evidence supports this view. As a result of this selection pressure, it is hypothesized that plants have evolved chemical and mechanical defenses.[41,119] A comprehensive review of the interactions of herbivores and secondary plant compounds is provided by Rosenthal and Janzen.[123]

In addition to evolutionary responses to herbivory, plants show a variety of other rapid responses or reactions that result from disruption of the complex physiological processes governing patterns of growth and development. Such responses differ in two important respects from evolutionary responses to herbivory. First, unlike evolutionary responses that only occur via differential selection over one or more generations, "induced" responses occur during the life span of the individual that is attacked. Second, unlike evolutionary responses that result in essentially permanent changes in plant characteristics, present even when herbivory does not occur, induced responses result in changes in plant characteristics that are usually not permanent and that only occur in response to herbivore attack.

A wide variety of rapid, inducible responses to herbivory have now been identified. Particularly well studied is the induction of chemical defenses following herbivory.[2,17,118] Evidence for induction of chemical defenses has been found in 14 families[2,13,17,118,145] representing annuals, monocarpic perennials, herbaceous perennials, and both deciduous and evergreen woody perennials. Furthermore, attack on one plant may induce chemical defenses in neighboring plants,[2] although the likelihood that this phenomenon occurs has been challenged.[42] In terms of plant reproduction, induction of chemical defenses may be important because, if levels of defenses remain high for a sufficiently long period of time before dropping to preattack levels, future herbivore attack may decrease.[53,125]

Herbivory can also induce changes in resource allocation. When actively growing regions of plants are damaged, rates and/or amounts of resources translocated to various regions of the plant are altered. For example, in sugarcane, removal of young leaves with no damage to the growing tip increased translocation to that area from remaining leaves.[51] Other grasses show similar responses to tissue removal.[43,95] Also, removal or destruction of flowers and immature fruits can alter the balance between vegetative growth and reproduction,[70,93] increase the size of remaining fuits,[93,96,98] or increase the number of flowers or fruits produced later in the growing season.[56,92]

A third induced response to herbivory is a delay in the decrease in net photosynthetic rates normally associated with leaf maturation.[34,143] Prevention of the usual decline in photosynthetic rates may be due to increased rates of assimilate transportation from leaves to damaged regions,[109] increased carbon dioxide uptake via alteration of stomatal or mesophyll resistance to gas exchange,[62] or increased levels of carboxylating enzymes.[110] Maintenance of higher-than-normal photosynthetic rates in tissues remaining after herbivory may at least partially offset the decreases in growth and subsequent reproduction normally associated with herbivory (see above, Reductions in Growth and Reproduction following Herbivory).

Delays in whole-plant senescence (often called "rejuvenation") are also associated with removal of reproductive tissues.[30,84] This delay may result from the prevention of assimilate mobilization into fruits, although shifts in hormonal balances responsible for the senescence process are also involved.[84] An important side effect of flower removal and the associated delay of senescence is that vegetative growth can increase, and this can result in increased reproduction later in the growing season or increased life span of the individual.[85]

Last, in some andromonoecious species (i.e., plants with male and hermaphroditic flowers on the same individual), herbivore destruction of flowers early in the reproductive phase can alter sex expression in inflorescences developing later in the season. For example, in both *Pastinaca sativa*[59] and *Heracleum lanatum,*[58] herbivore destruction of the first inflorescence produced (the predominantly hermaphroditic primary umbel) resulted in increased hermaphroditic flower production in the later orders of umbels that developed. In both these species, these later umbels produced higher seed set following floral herbivory on the primary umbel. This increase at least partially offsets the loss of seeds that would have been produced on the primary umbel.[56,58]

One net effect of these compensatory responses is that productivity is often unaffected or even increases following herbivory. For example, in the Serengeti Plains, grazing vertebrates reduced plant height but increased biomass concentration (mg/10 cm^3).[99] Also, in many crop plants yields are not reduced by herbivory.[3,49] In all cases the ability of plants to offset decreases in productivity is strongly dependent on the amount of damage, with compensation occurring at low to moderate levels of herbivory.[3,101,140]

Despite the beneficial nature of the interaction at the community level, it is important to note that increased vegetative productivity should not be equated with the Darwinian fitness of individuals and that, in most cases, it is unlikely that damage is offset competely. With grasses, increases in vegetative productivity are associated with decreases in sexual reproduction.[101] Likewise, induced chemical defenses may prevent high levels of future damage but these chemical require utilization of energy and nutrients that could have been used in future growth or reproduction. For example, in leaves of *Betula pubescens* the percentage of nonphenolic foliage nitrogen is inversely related to the percentage of foliage phenolics,[54] suggesting a trade-off between defenses and growth. Changes in resource allocation and photosynthetic rates may allow more reproduction than occurs in damaged plants that do not respond similarly. However, reproduction is still likely to be less than in undamaged plants, if current growth and/or reproduction reduces future growth or reproduction. Even in *Pastinaca sativa,* where population recruitment from plants whose primary umbel was destroyed by herbivory was equal to that from undamaged plants in 1 of 2 years, full compensation was made in terms of maternal but probably not paternal fitness.[60] In other cases of compensatory responses to herbivory, the exact degree to which these responses offset damage await studies that analyze quantitatively the effects of these reactions on reproductive output and population recruitment.

CONCLUSIONS

Thirty years ago the vast majority of research on the impact of herbivory on plants was conducted by agricultural scientists interested in reducing herbivore damage to crop plants. Since then, basic research by biologists interested in wild plants growing in their natural habitats has greatly expanded our understanding of both the evolutionary and ecological effects of herbivory on plant growth and reproduction. Nevertheless, our knowledge in this area is far from complete. For example, we know little about the factors influencing whether or not seedlings survive herbivory or, if they do, how this affects future growth and development. Likewise, other than for a few commercially grown trees, we know very little about the long-term effects of herbivory on growth and development of perennial plants. Also, our understanding of compensatory reactions and their role in offsetting herbivore damage is rudimentary. In addition, advances made in other areas such as plant demography and plant quantitative genetics are only now being applied to the study of plant–herbivore interactions. Clearly, many more years of fruitful research on the impact of herbivory on plants lie ahead.

ACKNOWLEDGMENTS

I thank R. W. Cruden, J. Howard, H. Howe, and the editors for helpful comments on early drafts of this chapter.

REFERENCES

1. Adams, S. N., Sheep and cattle grazing in forests: A review, *J. Appl. Ecol.* **12,** 143–152 (1975).
2. Baldwin, I. T., and Schultz, J. C., Rapid changes in tree leaf chemistry induced by damage: Evidence for communication between plants, *Science* **221,** 277–279 (1983).
3. Bardner, R., and Fletcher, K. E., Insect infestations and their effects on growth and yield of field crops: A review, *Bull. Entomol. Res.* **64,** 141–160 (1974).
4. Becker, P., Effects of insect defoliation and artificial defoliaiton on survival of *Shorea* seedlings, in *Tropical Rain Forest: Ecology and Management* (S. L. Sutton, T. C. Whitmore, and A. C. Chadwick, eds.), pp. 241–252. Blackwell Scientific Publ., London, 1983.
5. Bentley, S., and Whittaker, J. B., Effects of grazing by a chrysomelid beetle, *Gastrophysa viridula.* on competition between *Rumex obtusifolius* and *Rumex crispus, J. Ecol.* **67,** 79–90 (1979).
6. Bentley, S., Whittaker, J. B., and Malloch, A. J. C., Field experiments on the effects of grazing by a chrysomelid beetle *(Gastrophysa viridula)* on seed production and quality in *Rumex obtusifolius* and *Rumex crispus, J. Ecol.* **68,** 671–674 (1980).
7. Borchert, M. I., and Jain, S. K., The effect of rodent seed predation on four species of California annual grasses, *Oecologia* **33,** 101–113 (1978).
8. Boscher, J., Modified reproduction strategy of leek *Allium porrum* in response to a phytophagous insect, *Acrolepiopsis assectella, Oikos* **33,** 451–456 (1979).
9. Bray, J. R., Primary consumption in three forest canopies, *Ecology* **45,** 165–167 (1964).
10. Breedlove, D. E., and Ehrlich, P. R., Plant-herbivore coevolution: lupines and lycaenids, *Science* **162,** 671–672 (1968).
11. Brown, J. H., Reichman, O. J., and Davidson, D. W., Granivory in desert ecosystems, *Annu. Rev. Ecol. Syst.* **10,** 201–227 (1979).
12. Brown, V. K., Secondary succession: Insect-plant relationships, *Bioscience* **34,** 710–716 (1984).
13. Bryant, J. P., Phytochemical deterrence of snowshoe hare browsing by adventitious shoots of four Alasksan trees, *Science* **213,** 889–890 (1981).
14. Cantlon, J. E., The stability of natural populations and their sensitivity to technology. *Brookhaven Symp. Biol.* **22,** 197–205 (1969).
15. Capinera, J. L., and Roltsch, W. J., Response of wheat seedlings to actual and simulated migratory grasshopper defoliation, *J. Econ. Entomol.* **73,** 258–261 (1980).
16. Cardona, C., Gonzalez, R., and Schoonhoven, A. V., Evaluation of damage to common beans by larvae and adults of *Diabrotica balteata* and *Cerotoma facialis, J. Econ. Entomol.* **75,** 324–327 (1982).
17. Carroll, C. R., and Hoffman, C. A., Chemical feeding deterrent mobilized in response to insect herbivory and counteradaptation by *Epilachna tredecimnotata, Science* **209,** 414–416 (1980).
18. Chapin, F. S., III, Nutrient allocation and responses to defoliation in tundra plants, *Arct. Alp. Res.* **12,** 553–563 (1980).
19. Chew, R. M., The impact of small mammals on ecosystem structure and function, in *Populations of Small Mammals under Natural Conditions: A Symposium,* Pymatuning Laboratory of Ecology, Special Publ. No. 5, pp. 167–180, 1978.
20. Coggin, D. L., and Dively, G. P., Effects of depodding and defoliation on yield and quality of lima beans, *J. Econ. Entomol.* **73,** 609–614 (1980).
21. Coley, P. D., Intraspecific variation in herbivory on two tropical tree species, *Ecology* **64,** 426–433 (1983).
22. Coley, P. D., Bryant, J. P., and Chapin, F. S., III, Resource availability and plant antiherbivore defense, *Science* **230,** 895–899 (1985).
23. Cousens, M., Reproductive strategies of pteridophytes, in *Plant Reproductive Ecology: Patterns and Strategies* (J. Lovett Doust and L. Lovett Doust, eds.), Chapter 15, Oxford Univ. Press, New York, 1988.
24. Crawley, M. J., *Herbivory: The Dynamics of Animal-Plant Interactions,* p. 32. Univ. of California Press, Berkeley, 1983.
25. Crawley, M. J., Reduction of oak fecundity by low density herbivore populations, *Nature (London)* **314,** 163–164 (1985).
26. Danell, K., and Huss-Danell, K., Feeding by insects and hares on birches earlier affected by moose browsing, *Oikos* **44,** 75–81 (1985).
27. Darley-Hill, S., and Johnson, W. C., Acorn dispersal by the blue jay *(Cyanocitta cristata), Oecologia* **50,** 231–232 (1981).

28. Davidson, D. W., Samson, D. A., and Inouye, R. S., Granivory in the Chihuahuan Desert: Interactions within and between trophic levels, *Ecology* **66,** 486–502 (1985).
29. Davidson, J. L., and Milthorpe, F. L., The effect of defoliation on the carbon balance in *Dactylis glomerata, Ann. Bot. (London)* **118,** 185–198 (1966).
30. Derman, B. D., Rupp, D. C., and Nooden, L. D., Mineral distribution and monocarpic senescence in anoka soybeans, *Am. J. Bot.* **65,** 205–213 (1978).
31. DeSteven, D., Reproductive consequences of insect seed predation in *Hamamelis virginiana, Ecology* **64,** 89–98 (1983).
32. Detling, J. K., and Dyer, M. I., Evidence for potential plant growth regulators in grasshoppers, *Ecology* **62,** 485–488 (1981).
33. Detling, J. K., Dyer, M. I., Procter-Gregg, C., and Winn, D. T., Plant-herbivore interactions: Examination of potential effects of bison saliva on regrowth of *Bouteloua gracilis* (H.B.K.) Lag., *Oecologia* **45,** 26–31 (1980).
34. Detling, J. K., Dyer, M. I., and Winn, D. T., Net photosynthesis, root respiration, and regrowth of *Bouteloua gracilis* following simulated grazing, *Oecologia* **41,** 127–134 (1979).
35. DeWreede, R., and Klinger, T., Reproductive strategies in algae, in *Plant Reproductive Ecology: Patterns and Strategies* (J. Lovett Doust and L. Lovett Doust, eds.), Chapter 13. Oxford Univ. Press, New York, 1988.
36. Dixon, A. F. G., The role of aphids in wood formation. I. The effect of the sycamore aphid, *Drepanosiphum platanoides* (Schr.) (Aphididae), on the growth of sycamore, *Acer pseudoplatanus* (L.), *J. Appl. Biol.* **8,** 165–179 (1971).
37. Dixon, A. F. G., The role of aphids in wood formation. II. The effect of the lime aphid, *Eucallipterus tiliae* L. (Aphididae) on the growth of lime *Tilia* × *vulgaris* Hayne, *J. Appl. Ecol.* **8,** 393–399 (1971).
38. Elliot, P. F., Evolutionary responses of plants to seed-eaters: Pine squirrel predation on lodegepole pine, *Evolution* **28,** 221–231 (1974).
39. Etherington, J. R., Relationship between morphological adaptation to grazing, carbon balance and waterlogging tolerance in clones of *Dactylis glomerata* L., *New Phytol.* **98,** 647–658 (1984).
40. Faeth, S. H., Connor, E. F., and Simberloff, D., Early leaf abscission: a neglected source of mortality for folivores, *Am. Nat.* **117,** 409–415 (1981).
41. Feeny, P. P., Plant apparency and chemical defense, in *Biochemical Interactions between Plants and Insects* (P. Wallace and R. Mansell, eds.), *Recent Adv. Phytochem.* **10,** 1–40 (1976).
42. Fowler, S. V., and Lawton, J. H., Rapidly induced defenses and talking trees: the devil's advocate position, *Am. Nat.* **126,** 181–195 (1985).
43. Gifford, R. M., and Marshall, C., Photosynthesis and assimilate distribution in *Lolium multiflorum* Lam. following differential tiller defoliation, *Aust. J. Biol. Sci.* **26,** 517–526 (1973).
44. Green, T. W., and Palmblad, I. G., Effects of insect seed predators on *Astragalus cibarius* and *Astragalus utahensis* (Leguminosae), *Ecology* **56,** 1435–1440 (1975).
45. Guene, G. L., and Minnick, D. R., Snap bean yields following simulated insect defoliation, *Proc. Fla. State Hortic. Soc.* **80,** 132–134 (1967).
46. Hairston, N. G., Smith, F. E., and Slobodkin, L. B., Community structure, population control, and competition, *Am. Nat.* **94,** 421–425 (1960).
47. Harper, J. L., The role of predation in vegetational diversity, in *Diversity and Stability in Ecological Systems* (G. M. Woodwell and H. H. Smith, eds.), pp. 48–62. Symp. Brookhaven Nat. Lab., Upton, New York, 1969.
48. Harper, J. L., *Population Biology of Plants,* Chapters 5, and 15. Academic Press, New York, 1977.
49. Harris, P., A possible explantion of plant yield increases following insect damage, *Agro-Ecosystems* **1,** 219–225 (1974).
50. Hartnett, D. C., and Abrahamson, W. G., The effects of stem gall insects on life history patterns in *Solidago canadensis, Ecology* **60,** 910–917 (1979).
51. Hartt, C. E., Kortschak, H. P., and Burr, G. O., Effects of defoliation, deradication, and darkening the blade upon translocation of C^{14} in sugarcane, *Plant Physiol.* **39,** 15–22 (1964).
52. Haukioja, E., Measuring consumption in *Eriocrania* (Eriocraniidae, Lep.) miners with reference to interaction between leaf and miner, *Rep. Kevo Subarct. Res. Stn.* **11,** 16–21 (1974).
53. Haukioja, E., On the role of plant defences in the fluctuation of herbivore populations, *Oikos* **35,** 202–213 (1980).
54. Haukioja, E., Niemela, P., and Siren, S., Foliage phenols and nitrogen in relation to growth, insect dam-

age, and ability to recover after defoliation, in the mountain birch *Betula pubescens* ssp. *tortuosa, Oecologia* **65,** 214–222 (1985).
55. Heichel, G. H., and Turner, N. C., Branch growth and leaf numbers of red maple (*Acer rubrum* L.) and red oak (*Quercus rubra* L.): Response to defoliation, *Oecologia* **62,** 1–6 (1984).
56. Hendrix, S. D., Compensatory reproduction in a biennial herb following insect defloration, *Oecologia* **42,** 107–118 (1979).
57. Hendrix, S. D., An evolutionary and ecological perspective of the insect fauna of ferns, *Am. Nat.* **115,** 171–196 (1980).
58. Hendrix, S. D., Reactions of *Heracleum lanatum* to floral herbivory by *Depressaria pastinacella, Ecology* **65,** 191–197 (1984).
59. Hendrix, S. D., and Trapp, E. J., Plant-herbivore interactions: Insect induced changes in host plant sex expression and fecundity, *Oecologia* **49,** 119–122 (1981).
60. Hendrix, S. D., and Trapp, E. J., Floral herbivory in *Pastinaca sativa*: The role of compensatory responses in offsetting reductions in fitness, in preparation.
61. Hodgkinson, K. C., and Baas Becking, H. G., Effect of defoliation on root growth of some arid zone perennial plants, *Aust. J. Agric. Res.* **29,** 31–42 (1977).
62. Hodgkinson, K. C., Smith, N. G., and Miles, G. E., The photosynthetic capacity of stubble leaves and their contribution to growth of the lucerne plant after high level cutting, *Aust. J. Agric. Sci.* **23,** 225–228 (1972).
63. Howe, H. F., Monkey dispersal and waste of a neotropical fruit, *Ecology* **61,** 944–959 (1980).
64. Howe, H. F., and Estabrook, G. F., On intraspecific competition for avian dispersers in tropical trees, *Am. Nat.* **111,** 817–832 (1977).
65. Howe, H. F., Schupp, E. W., and Westley, L. C., Early consequences of seed dispersal for a neotropical tree *(Virola surinamensis), Ecology* **66,** 781–791 (1985).
66. Howe, H. F., and Smallwood, J., Ecology of seed dispersal, *Annu. Rev. Ecol. Syst.* **13,** 201–228 (1982).
67. Howe, H. F., and Vande Kerckhove, G. A., Nutmeg dispersal by tropical birds, *Science* **210,** 925–926 (1980).
68. Huffaker, C. B., Fundamentals of biological control of weeds, *Hilgardia* **27,** 101–157 (1957).
69. Hulbert, S. H., Pseudoreplication and the design of ecological field experiments, *Ecol. Monogr.* **54,** 187–211 (1984).
70. Hurd, R. G., Gay, A. P., and Mountifield, A. C., The effect of partial flower removal on the relation between root, shoot, and fruit growth in the indeterminate tomato, *Ann. Appl. Biol.* **93,** 77–89 (1979).
71. Inouye, R. S., Byers, G. S., and Brown, J. H. Effects of predation and competition on survivorship, fecundity, and community structure of desert annuals, *Ecology* **61,** 1344–1351 (1980).
72. Jameson, D. A., Responses of individual plants to harvesting, *Bot. Rev.* **29,** 532–594 (1963).
73. Janzen, D. H., Seed-eaters versus seed size, number, toxicity and dispersal, *Evolution* **23,** 1–27 (1969).
74. Janzen, D. H., Seed predation by animals, *Annu. Rev. Ecol. Syst.* **2,** 465–492 (1971).
75. Janzen, D. H., Effect of defoliation on fruit-bearing branches of the Kentucky coffee tree, *Gymnocladus dioicus* (Leguminosae), *Am. Midl. Nat.* **95,** 474–478 (1976).
76. Jensen, T. S., Seed-seed predator interactions of European beech, *Fagus silvatica* and forest rodents, *Clethrionomys glareolus* and *Apodemus flavicollis, Oikos* **44,** 149–156 (1985).
77. Kinsman, S., and Platt, W. J., The impact of a herbivore upon *Mirabilis hirsuta,* a fugitive prairie plant, *Oecologia* **65,** 2–6 (1984).
78. Klein, D. R., Food selection by North American deer and their response to over-utilization of preferred plant species, in *Animal Populations in Relation to their Food Sources* (W. Adam, ed.), pp. 25–46. Blackwell Scientific Publ. Oxford, 1970.
79. Kozlowski, T. T., Tree physiology and forest pests, *J. For.* **67,** 118–123 (1969).
80. Kulman, H. M., Effects of insect defoliation on growth and mortality of trees, *Annu. Rev. Entomol.* **16,** 289–324 (1971).
81. Laws, R. M., Parker, I. S. C., and Johnstone, R. C. B., *Elephants and Their Habitat,* Chapter 11. Clarendon Press, Oxford, 1975.
82. Lee, T. D., and Bazzaz, F. A., Effects of defoliation and competition on growth and reproduction in the annual plant *Abutilon theophrasti, J. Ecol.* **68,** 813–821 (1980).
83. Leigh, E. G., Jr., and Smythe, N., Leaf production, leaf consumption, and the regulation of folivory on Barro Colorado Island, in *The Ecology of Arboral Folivores* (G. G. Montgomery, ed.), pp. 33–50. Smithsonian Institute Press, Washington, D.C., 1978.
84. Leopold, A. C., Senescence in plant development, *Science* **134,** 1727–1732 (1961).

85. Leopold, A. C., Niedergang-Kamien, E., and Janick, J., Experimental modification of plant senescence, *Plant Physiol.* **34,** 570–573 (1959).
86. Lindroth, R. L., Diet optimization by generalist mammalian herbivores, *The Biologist* **61,** 41–58 (1979).
87. Linzon, S. N., The effect of artificial defoliation of various ages of leaves upon white pine growth, *For. Chron.* **34,** 50–56 (1958).
88. Louda, S. M., Distribution ecology: Variation in plant recruitment over a gradient in relation to seed predation, *Ecol. Monogr.* **52,** 25–41 (1982).
89. Louda, S. M., Limitation of the recruitment of the shrub *Haplopappus squarrosus* (Asteraceae) by flower and seed-feeding insects, *J. Ecol.* **70,** 43–53 (1982).
90. Louda, S. M., Seed predation and seedling mortality in the recruitment of a shrub, *Happlopappus venetus* Blake (Asteraceae) along a climatic gradient, *Ecology* **64,** 511–521 (1983).
91. Louda, S. M., Herbivore effect on stature, fruiting, and leaf dynamics of a native crucifer, *Ecology* **65,** 1379–1386 (1984).
92. Lovett Doust, J., and Eaton, G. W., Demographic aspects of flower and fruit production in bean plants, *Phaseolus vulgaris* L., *Am. J. Bot.* **69,** 1156–1164 (1982).
93. Maggs, D. H., The reduction in growth of apple trees brought about by fruiting, *J. Hortic. Sci.* **38,** 119–128 (1963).
94. Marquis, R. J., Leaf herbivores decrease fitness of a tropical plant, *Science* **226,** 537–539 (1984).
95. Marshall, C., and Sagar, G. R., The distribution of assimilates in *Lolium multiflorum* Lam. following differential defoliation, *Ann. Bot. (London)* **32,** 715–719 (1968).
96. Marshall, D. L., Levin, D. A., and Fowler, N. L., Plasticity in yield components in response to fruit predation and date of fruit initiation in three species of *Sesbania* (Leguminosae), *J. Ecol.* **73,** 71–81 (1985).
97. Maun, M. A., and Cavers, P. B., Seed production and dormancy in *Rumex crispus.* I. The effects of removal of cauline leaves at anthesis, *Can. J. Bot.* **49,** 1123 (1971).
98. McAlister, D. F., and Krober, O. A., Response of soybeans to leaf and pod removal, *Agron. J.* **50,** 674 (1958).
99. McNaughton, S. J., Serengeti migratory wildebeest: Facilitation of energy flow by grazing, *Science* **191,** 92–94 (1976).
100. McNaughton, S. J., Grassland-herbivore dynamics, in *Dynamics of an Ecosystem* (A. R. E. Sinclair and M. Norton-Griffiths, eds.), Chapter 3. Univ. of Chicago Press, Chicago, 1979.
101. McNaughton, S. J., Grazing as an optimization process: Grass-ungulate relationships in the Serengeti, *Am. Nat.* **113,** 691–703 (1979).
102. McNaughton, S. J., Wallace, L. L., and Coughenour, M. B., Plant adaptation in an ecosystem context: Effects of defoliation, nitrogen, and water on growth of an African C_4 sedge, *Ecology* **64,** 307–318 (1983).
103. Mills, J. N., Herbivory and seedling establishment in postfire southern California chaparral, *Oecologia* **60,** 267–270 (1983).
104. Mills, J. N., Effects of feeding by mealybugs (*Planococcus citri,* Homoptera: Pseudococcidae) on the growth of *Colliguaya odorifera* seedlings, *Oecologia* **64,** 142–144 (1984).
105. Mishler, B., Reproductive ecology of bryophytes, in *Plant Reproductive Ecology: Patterns and Strategies* (J. Lovett Doust and L. Lovett Doust, eds.), chapter 14. Oxford Univ. Press, New York, 1988.
106. Morrow, P. A., and LaMarche, V. C., Jr., Tree ring evidence for chronic insect suppression of productivity in subalpine *Eucalyptus, Science* **201,** 1244–1246 (1978).
107. Myers, J. H., Interactions between western tent caterpillars and wild rose: A test of some general plant herbivore hypotheses, *J. Anim. Ecol.* **50,** 11–25 (1981).
108. Myers, J. H., Effect of physiological condition of the host plant on the ovipositional choice of the cabbage white butterfly, *Pieris rapae, J. Anim. Ecol.* **54,** 193–204 (1985).
109. Neales, T. F., and Incoll, L. D., The control of leaf photosynthesis rate by the level of assimilate concentration in the leaf: a review of the hypothesis, *Bot. Rev.* **34,** 107–125 (1968).
110. Neales, T. F., Treharne, K. J., and Wareing, P. F., A relationship between net photosynthesis, diffusive resistance, and carboxylating enzyme activity in bean leaves, in *Photosynthesis and Photorespiration* (M. D. Hatch, C. B. Osmond, and R. E. Slayter, eds.), pp. 89–96. Wiley-Interscience, New York, 1971.
111. Nilsson, S. G., Ecological and evolutionary interactions between reproduction of beech *Fagus silvatica* and seed eating animals, *Oikos* **44,** 157–164 (1985).

112. Owen, D. F., The effect of a consumer, *Phytomyza ilicis,* on seasonal leaf-fall in the holly, *Ilex aquifolium, Oikos* **31,** 268–271 (1978).
113. Parker, M. A., Size-dependent herbivore attack and the demography of an arid grassland shrub, *Ecology* **66,** 850–860 (1985).
114. Peterken, G. F., Mortality of holly *(Ilex aquifolium)* seedlings in relation to natural regeneration in the New Forest, *J. Ecol.* **54,** 259–269 (1966).
115. Rausher, M. D., and Feeny, P., Herbivory, plant density, and plant reproductive success: The effect of *Battus philenor* on *Aristolochia reticulata, Ecology* **61,** 905–917 (1980).
116. Reed, F. C., and Stephenson, S. N., The effects of simulated herbivory on *Ambrosia artemisiifolia* L. and *Arctium minus* Schk., *Mich. Acad.* **4,** 359–364 (1972).
117. Reichardt, P. B., Bryant, J. P., Clausen, T. P., and Wieland, G. D., Defense of winter-dormant Alaska paper birch against snowshoe hares, *Oecologia* **65,** 58–69 (1984).
118. Rhoades, D., Evolution of plant chemical defenses against herbivores, in *Herbivores: Their Interaction with Secondary Plant Metabolites* (G. A. Rosenthal and D. H. Janzen, eds.), pp. 3–54. Academic Press, New York, 1979.
119. Rhoades, D., and Cates, R., Towards a general theory of plant anti-herbivore chemistry, in *Biochemical Interactions between Plants and Insects* (J. Wallace and R. Mansell, eds.), *Recent Adv. Phytochem.* **10,** 168–213 (1976).
120. Richards, J. H., Root growth response to defoliation in two *Agropyron* bunchgrasses: Field observations with an improved root periscope, *Oecologia* **64,** 21–25 (1984).
121. Rockwood, L. L., The effect of defoliation on seed production of six Costa Rican tree species, *Ecology* **54,** 1363–1369 (1973).
122. Rockwood, L. L., Seasonal changes in the susceptibility of *Crescentia alata* leaves to the flea beetle, *Oedionychus* sp. *Ecology* **55,** 142–148 (1974).
123. Rosenthal, G. A., and Janzen, D. H., eds., *Herbivores: Their Interaction with Secondary Plant Metabolites.* Academic Press, New York, 1979.
124. Sackston, W. E., Effects of artificial defoliation on sunflowers, *Can. J. Plant Sci.* **39,** 108–118 (1959).
125. Schultz, J. C., and Baldwin, I. T., Oak leaf quality declines in response to defoliation by gypsy moth larvae, *Science* **217,** 149–151 (1982).
126. Scott, P. D., and Taylor, T. N., Plant/animal interactions during the upper carboniferous, *Bot. Rev.* **49,** 259–307 (1983).
127. Silvertown, J. W., The evolutionary ecology of mast seeding in trees, *Biol. J. Linn. Soc.* **14,** 235–250 (1980).
128. Smith, C. C., The coevolution of pine squirrels *(Tamiasciurus)* and conifers, *Ecol. Monogr.* **40,** 349–371 (1970).
129. Sork, V. L., Examination of seed dispersal and survival in red oak, Quercus rubra (Fagaceae), using metal-tagged acorns, *Ecology* **65,** 1020–1022 (1984).
130. Southwood, T. R. E., Interaction of plants and animals: Patterns and processes, *Oikos* **44,** 5–11 (1985).
131. Stamp, N. E., Effect of defoliation by checkerspot caterpillars *(Euphydryas phaeton)* and sawfly larvae (*Macrophya nigra* and *Tenthredo grandis*) on their host plants (*Chelone* spp.), *Oecologia* **63,** 275–280 (1984).
132. Stalter, R., and Serrao, J., The impact of defoliation by gypsy moths on the oak forest at Greenbrook Sanctuary, New Jersey, *Bull. Torrey Bot. Club* **110,** 526–529 (1983).
133. Stephens, G. R., The relation of insect defoliation to mortality in Connecticut forests, *Conn. Agric. Exp. Stn. Bull.* **723,** 1–16 (1971).
134. Stephenson, A. G., Fruit set, herbivory, fruit reduction, and the fruiting strategy of *Catalpa speciosa* (Bignoniaceae), *Ecology* **61,** 57–64 (1980).
135. Stinner, B. R., and Abrahamson, W. G., Energetics of the *Solidago canadensis*-stem gall insect-parasitoid guild interaction, *Ecology* **60,** 918–926 (1979).
136. Thomas, G. D., Ignoffo, C. M., Biever, K. D., and Smith, D. B., Influence of defoliation and depodding on yield of soybeans, *J. Econ. Entomol.* **67,** 683–685 (1974).
137. Thompson, J. N., Within-patch structure and dynamics in *Pastinaca sativa* and resource availability to a specialized herbivore, *Ecology* **59,** 443–448 (1978).
138. Torres, J. C., Guitierrez, J. R., and Fuentes, E. R., Vegetative responses to defoliation of two Chilean matorral shrubs, *Oecologia* **46,** 161–163 (1980).
139. Via, S., The quantitative genetics of polyphagy in an insect herbivore. I. Genotype-environment interaction in larval performance on different host plant species, *Evolution* **38,** 881–895 (1984).

140. Vickery, P. J., Grazing and net primary production of a temperate grassland, *J. Appl. Ecol.* **9,** 307–314 (1972).
141. Waller, D., Plant morphology and reproduction, in *Plant Reproductive Ecology: Patterns and Strategies* (J. Lovett Doust, and L. Lovett Doust, eds.), Chapter 10. Oxford Univ. Press, New York, 1988.
142. Waloff, N., and Richards, O. W., The effect of insect fauna on growth, mortality and natality of broom, *Sarothamnus scoparius, J. Appl. Ecol.* **14,** 787–798 (1977).
143. Wareing, P. F., Khalifa, M. M., and Trehare, K. J., Rate-limiting processes in photosynthesis at saturating light intensities, *Nature (London)* **220,** 453–457 (1968).
144. West, N. E., Rodent-influenced establishment of ponderosa pine and bitterbush seedlings in central Oregon, *Ecology* **49,** 1009–1011 (1968).
145. White, J., Flagging: Host defences versus oviposition strategies in periodical cicadas (*Magicicada* spp., Cicadidae, Homoptera), *Can. Entomol.* **113,** 727–738 (1981).

III

Reproductive Strategies of Non-Angiosperms

13

Reproductive Strategies in Algae

R. E. DE WREEDE and T. KLINGER

Data on algal reproductive strategies are few and scattered. Indeed, in his treatment of the theory of sex allocation, Charnov[16] notes that "... algae ... are not even mentioned." This chapter represents a synthesis of current information, but many gaps in knowledge and understanding remain. The various reproductive pathways in algae are explored largely in the light of theory proposed for other groups of organisms, both higher plants and animals. Because the reader must be familiar with the complex life histories of algae, these will be briefly reviewed here.

The algae exhibit an immense range of size, structural complexity, and morphology. Size varies from micrometers (e.g., the unicell *Chlorella*) to meters (e.g., the giant kelp *Macrocystis*). Structural complexity varies from undifferentiated single cells to highly differentiated multicellular kelps, which may even have cells specialized for translocation of photosynthates. Morphology varies from unicells and colonies to finely divided filaments, to blades with stipes and holdfasts. It is impossible to examine reproductive behavior for all these types in a single chapter, so we have chosen to limit the following discussion to the macroscopic marine algae, the red, brown, and green seaweeds.

Few rigorous experiments have been performed to elucidate algal reproductive strategies. As a result, only correlations can be drawn between environmental patterns and trends in reproduction. Stearns[89] argues forcefully that such correlations and any conclusions drawn from them are "... less than convincing. ..." We can only agree, and urge caution in their interpretation. Data reviewed below are best used to suggest the experiments that clearly remain to be done.

ALGAL LIFE HISTORIES

Three fundamental algal life histories are generally recognized. The monophasic life history maintains a single dominant free-living stage, which may be haploid or diploid. Among taxa in which the diploid phase dominates, meiosis occurs in the gametangium (hence, meiosis is gametic). Two such taxa are *Fucus* (the common brown rockweed) and *Codium* (a green seaweed). In contrast, where the haploid phase is dominant, meiosis occurs in the zygote, which constitutes the sole diploid phase in the life history (hence meiosis is described as zygotic). An example of a zygotic monophasic taxon is *Spirogyra*, a green alga commonly found in ponds.

The biphasic life history maintains two more or less codominant stages (one diploid, the other haploid). These phases may be either iso- or heteromorphic, and usually alternate with each other. In the biphasic life history, the diploid sporophyte generates haploid spores meiotically (hence, meiosis is sporic); spores give rise to haploid gametophytes, which in turn produce gametes mitotically. Biphasic life histories can be either isomorphic (sporophyte and gametophyte look much the same), as in the sea lettuce *Ulva,* or heteromorphic, as in the kelps. Among heteromorphic life histories, either the sporophyte or the gametophyte can be the larger, more conspicuous, stage.

The triphasic life history is restricted to the red algae (Rhodophyta). In this case, an additional diploid spore-producing phase (the carposporophyte) is superimposed on a fundamentally biphasic life history consisting of a diploid tetrasporophyte and a haploid gametophyte. Diploid, mitotically produced spores liberated by the carposporophytes give rise to the tetrasporophytes by mitosis. Tetrasporophytes in turn produce haploid spores in tetrasporangia (hence, meiosis is sporic).

More than one of the characteristic life histories may occur in the same species. For example, the green alga *Blidingia minima* exhibited four different life history patterns among individuals collected at different sites and in different seasons, and parthenogenesis was common among the isogametes.[91] Life history variations among the red algae[99] and brown algae[17,71,76,100] are reviewed elsewhere.

Clearly the algae show wide variation in life history, and in the response of life history to environmental conditions. An understanding of such variation may enhance our understanding of life history evolution in these taxa.

RESOURCE ALLOCATION AND REPRODUCTION

The theoretical consequences of resource allocation have not been explored specifically for the algae, but ideas proposed for other organisms may be appropriate, and are considered here.

It is generally held that reproduction imposes a cost on an organism, either in terms of slower growth or an increased chance of mortality.[18,34,41,56,57,96] Consequently, models of the various life history and reproductive alternatives generally assume such a trade-off.

We predict the following from the assumption of resource trade-off: (1) an organism must attain a certain size in order to begin reproduction; (2) because resources must be partitioned, a reduction or cessation of growth is likely at the onset of reproduction; (3) an organism is more likely to die after producing and releasing reproductive structures or propagules; (4) for a perennial species, large reproductive output in one year should be inversely correlated with growth in that year, or with survival or growth in the following year; (5) if more gametes or spores are produced from a given quantity of resource, fewer offspring or decreased growth or survival of these offspring may occur due to less resource having been allocated to each gamete or spore.

According to Bell,[6] the existence of reproductive cost is a necessary condition for the optimization of life histories. However, in laboratory studies, the same author[6,7] generally found a positive or zero correlation between present and future reproduction or survival among several taxa of asexual freshwater invertebrates. Giesel[41] has discussed some consequences of cost-free reproduction. Also, Watson and Casper[99] cite several studies in which the green reproductive structures of terrestrial plants possibly repayed a portion of the cost of their formation. Resource allocation and reproduction-associated costs may not be strictly comparable between heterotrophic animals and autotrophic plants.

The algae deserve close attention in any evaluation of the cost hypothesis because algae (with the exception of the Charophyta) produce no ancillary reproductive structures, and oogonia and sporangia are normally pigmented and potentially capable of photosynthesis prior to liberation.[51,58,69] Therefore the existence of cost-free or cost-reduced reproduction may very likely be found among the algae because they are photoautotrophic and lacking in ancillary reproductive structures.

There is some evidence for algae that suggests the existence of a resource trade-off. In agreement with predictions 1 and 2 above, Schiel[84] found that, for some species of *Landsburgia* and *Carpophyllum,* only the larger plants reproduced in any given year. In three species of *Sargassum,* allocation to rhizome (holdfast) production and sexual reproduction (as receptacle production) was shown to represent alternative strategies, and the existence of an energetic trade-off between growth and reproduction was suggested.[68] In *Dumontia incrassata* the female plants initiate reproductive branches only after these plants reach their maximum size, i.e., after growth has stopped.[29] In agreement with prediction 3, the rare occurrence of sexual reproduction in *Halimeda* is inevitably followed by the death of the parent thallus, and this may be pervasive among the order Caulerpales.[47]

Other data may not be consistent with the resource trade-off hypothesis. Maximum reproduction and maximum growth occur simultaneously in *Gigartina stellata,*[10] and *Codium fragile.*[38] These findings conflict with predictions 1 and 2 above. In natural populations of *Iridaea cordata* blade growth continues after the blade becomes reproductive,[43] although in culture this same species stops growing at the onset of reproduction.[97] Similarly, *Gracilaria verrucosa* continues to grow as it becomes reproductive.[49] For *Sargassum sinclairii* Schiel[84] found that plants that reproduced in one year were the ones most likely to do so again the following year. Those individuals that did not reproduce in a given year frequently failed to do so in the following year and suffered disproportionately large mortality[84] as well. This is contrary to prediction 4 above.

Schiel[84] has argued that the cost of reproduction necessarily includes the cost of producing the structures that bear the sporangia or gametangia. Dawkins[22] has made a similar argument, though not explicitly for the algae. However, for organisms such as seaweeds, in which the blades constitute the primary photosynthetic organ and only occasionally (seasonally) bear reproductive structures, the cost of blade production is arguably uncoupled from the cost of reproduction. Indeed, for some algal species such as *Pterygophora californica* even the lateral blades (sporophylls) function solely as photosynthetic structures for most of the year, and are reproductive for only 3–4 months.[27] In this species, sporophylls constitute the primary photosynthetic organ, by weight and by surface area. For many other species (e.g., *Laminaria setchellii*) the reproductive structures are formed directly on the vegetative blade.[53] In such cases, the additional cost of reproduction lies solely in the cost of production of oogonia, antheridia, and sporangia.

The available evidence does not clearly indicate the existence of resource trade-off or reproduction-associated cost among the algae. Further, it has not been demonstrated that algal growth and reproduction are limited by the same resources. Among the algae, reproduction may conceivably be cost-free or, more probably, cost-reduced.

This condition differs markedly from the situation in higher plants, which may bear extensive ancillary reproductive structures, and from that in heterotrophic animals, which may experience a very finite energy supply due to constraints on feeding and nutrient intake.

AGE-SPECIFIC REPRODUCTION AND REPRODUCTIVE EFFORT

Both semelparous and iteroparous species exist among the algae. Alternative reproductive pathways are shown in Fig. 13.1, as are some of the conditions that may select for them. In theory, semelparity is favored by conditions under which the probability of reproductive success at a given time is high, and iteroparity is favored when the chance of reproductive success is low.[19] Thus, the semelparous organism may be at an advantage under conditions of low juvenile mortality and high adult mortality; these same conditions could also favor an earlier age of reproduction among iteroparous organisms. The iteroparous organism may be at an advantage under the alternative conditions of high juvenile mortality and low adult mortality; these conditons may also favor delayed reproduction among iteroparous individuals.

Does theory allow any predictions with regard to age of first reproduction? If limited resources must be allocated, and if increased size enhances survival, then reproduction should occur relatively earlier in semelparous than in iteroparous organisms. This is so because the same selection pressures that give rise to semelparity should favor precocious reproduction. One might further expect a semelparous organism to devote its resources to a single reproductive event and subsequently die. Given these assumptions, early reproduction among iteroparous organisms is delayed because available resources are instead allocated to growth and maintenance. In contrast, one could argue that if resources are not limiting, iteroparous as well as semelparous organisms should exhibit early reproduction. The occurrence of a consistently earlier age of reproduction among semelparous organisms is therefore consistent with a resource-partitioning model.

Reproductive effort was defined by Pianka[77] as " . . . the effects of various current levels of reproduction upon future reproductive success. . . . " If a resource trade-off does exist between reproduction and growth and survival (but see above, Resource Allocation and Reproduction) then reproductive effort in any given reproductive period should theoretically be greater in the semelparous than in the iteroparous organisms. However, this expectation pertains to strictly comparable organisms; comparisons between taxa may be confounded by other factors, and must be interpreted with caution.

There is little agreement on the appropriate measurement of reproductive effort,[45,57,77,93] but researchers have previously used size and number of gametes or spores, and ratios of reproductive versus nonreproductive tissues.

Survivorship and Semelparity versus Iteroparity

Relative survival of juveniles versus adults is an important selective agent for semelparity versus iteroparity. High juvenile and low adult mortality is described by an inverse logarithmic (type III) curve.[75] Survival of the iteroparous sporophyte stages of

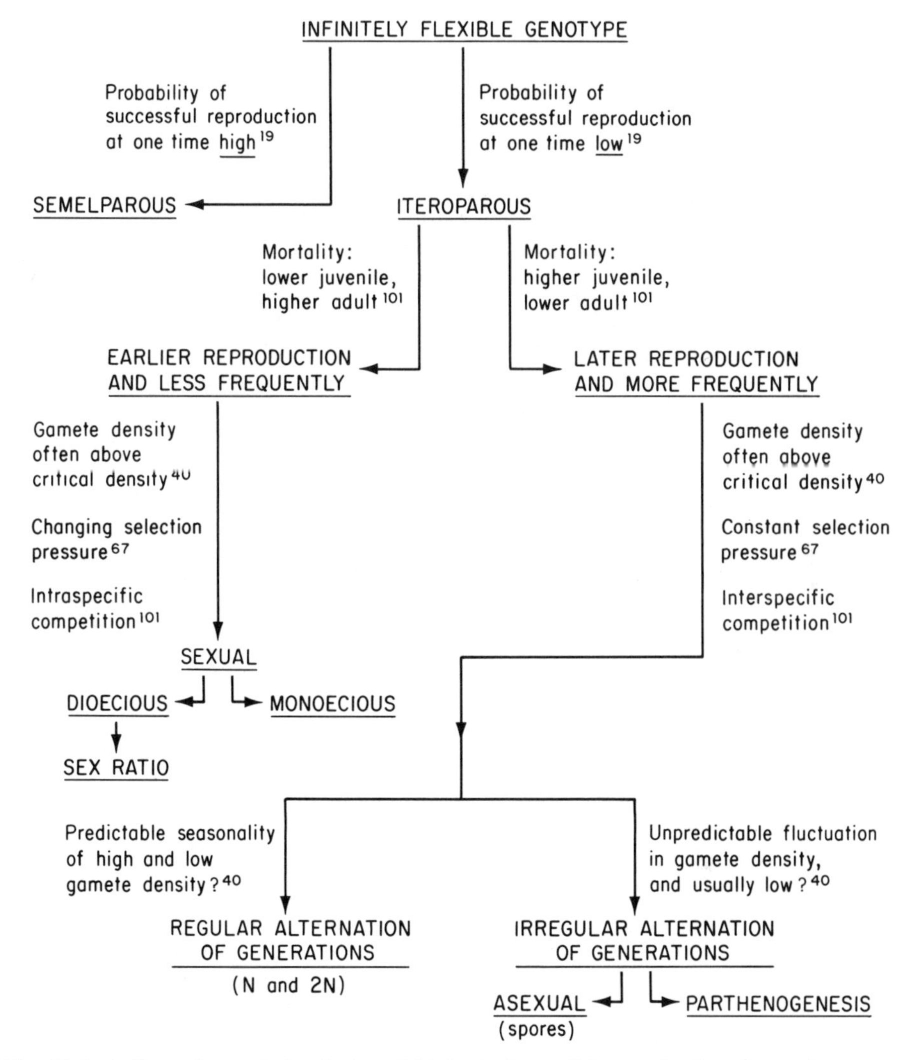

Fig. 13.1. A flow chart of physical and biological conditions selecting for various reproductive strategies in algae.

the kelps *Pelagophycus porra*[20] and *Macrocystis pyrifera*[12,81] can be described by such a type III curve. Type III curves have also been reported for *Lithophyllum incrustans,* an iteroparous red alga.[30,31,35] A type II curve (approximately constant risk of mortality with age) was reported for *Laminaria digitata* and *Laminaria longicruris.*[11,88]

Type III survivorship curves have also been reported for *Cystoseira osmundaceae, Laminaria farlowii,* and *Macrocystis pyrifera* in southern California, and type I curves are described for *Pterygophora californica* and *Eisenia arborea* at the same site (all five species are iteroparous, and all but the first are kelps).[23] Data used to construct these curves were interpolated from results collected during a 10-year field study; estimates of age are based on annual increments. Survivorship of recruits at the same site, calculated on a smaller time scale (3-month intervals) for a shorter duration (~24 months) showed very high initial mortality in all five species across all experimental treatments.

These data reflect a phenomenon often observed among kelps and other algae that recruit as microscopic sporelings or germlings and quickly grow to very large sizes. Massive recruitment is followed by high mortality; mortality rate subsequently declines among the prereproductive recruits that do ultimately persist. Thus, the prereproductive survivorship curve may be characterized by two segments of very different slope. Postreproductive survivorship can in turn remain high or decline. Documentation of this phenomenon is poor, owing to the difficulty of simultaneously estimating survivorship among tiny (often microscopic) plants at very high density and large plants at low density. Consequently, survivorship estimates may underestimate early mortality. The time scale on which the data are collected is crucial to accurate representation of the survivorship curves.

Few time-dependent survivorship curves have been reported for semelparous algae. A type II curve is reported for *Colpomenia peregrina*[95] and a type I for *Leathesia difformis.*[15] The calcified green alga *Halimeda* is long lived and apparently semelparous. According to Hillis-Colinvaux,[47] *Halimeda* produces a massive but unquantified number of gametes, and the adult dies following gamete liberation. In this case, semelparity is followed by high initial mortality of germlings, low prereproductive mortality, and high postreproductive mortality. In contrast, the short-lived semelparous kelp *Laminaria ephemera* exhibits high initial mortality of sporelings, very early reproduction, and high postreproductive mortality.[53]

Initiation of Reproduction and Age-Specific Reproduction

The onset of reproduction, relative to the life span of the organism, varies widely among algae. The published data are summarized in Table 13.1. *Pterygophora californica*[26,27] and *Laminaria setchellii*[53] are both iteroparous species, while *Laminaria ephemera*[53] and *Cymathere triplicata*[28] are semelparous. These species all belong to the order Laminariales and co-occur in Barkley Sound, British Columbia. Timing of the initiation of reproduction varies widely between them. It appears that such variation is not in this case attributable to site-specific differences.

The red alga *Iridaea cordata* produces reproductive structures on the annual blade within 3 months of blade initiation,[43] whereas the related species *Gigartina stellata* was found to be without reproductive structures (on the perennial blades) after 1 year.[10]

The genus *Sargassum* is common on many tropical and subtropical reefs. *Sargassum siliquosum* initiates reproduction in the second year of its estimated 3 year (minimum) life span,[3] while *Sargassum sinclairii* is generally reproductive in its first as well as its second year.[84] For other genera within the same family, 60% of *Landsburgia quercifolia* thalli first reproduced in their second year (none reproduced in the first year); reproduction in *Carpophyllum maschalocarpum* occurred only in the largest individuals in the second year (none in first year); and individuals of *Carophyllum angustifolium* never became reproductive in the first 3 years.[84]

The data summarized in Table 13.1 demonstrate that semelparous species may reproduce earlier than iteroparous species, but caution must be exercised in drawing

Table 13.1. A Comparison of Age of First Sport Production, Thallus Longevity, and Semelparity (S) or Iteroparity (I)[a]

Species	Age of first reproduction	Longevity	Semelparous/ iteroparous
Colpomenia peregrina[95]	2	42	S
Cymathere triplicata[28]	240	360	S
Iridaea cordata[44]	90	1460	S (blade)
Laminaria ephemera[53]	54	186	S
Leathesia difformis[15]	<78?	78	S
Carpophyllum maschalocarpum[84]	540	730 (min.)	I
Carpophyllum angustifolium[84]	>1080	>1080	I
Laminaria setchellii[53]	912	6570	I
Landsburgia quercifolia[84]	540	730 (min.)	I
Pterygophora californica[26]	1460	4380	I
Sargassum siliquosum[3]	600	1100 (min.)	I
Sargassum sinclairii[83]	270	730 (min.)	I

[a]All units are in days; (min.) denotes a minimum estimate.

comparisons between species, since we may be seeing canalized traits unrelated to age-specific reproduction.

There is evidence supporting both age-specific and age-independent reproduction in seaweeds. In *Pterygophora californica,* De Wreede[27] showed a highly significant positive correlation between age, sporophyll biomass, and number of spores produced. Over the 3-year study period, fecundity increased with age, even though the bulk of the population was distributed among the middle age classes. Younger plants were uncommon, and reproduction did not begin until the fourth year. In contrast, Klinger[53] reported no correlation between age and the magnitude of sorus production (and thus spore production) in *Laminaria setchellii,* another perennial kelp. In *Lithophyllum incrustans* there was a significant positive correlation between age of the individual and number of spores produced, but no such correlation was found between age class and number of spores produced.[35]

Size rather than age may determine the initiation of reproduction in some algae. In the annual kelp *Cymathere triplicata,* seasonal spore production began earlier in the season among the larger plants than it did among the smaller plants.[80] Large individuals of the short-lived kelp *Laminaria ephemera* initiated spore production earlier in the season than did small individuals; these large and precocious plants subsequently died earlier than did the small plants.[53] Schiel[84] concluded that *Landsburgia* in New Zealand becomes reproductive upon reaching some critical size, independent of age.

Reproductive Effort

Reproductive effort in *Laminaria* was estimated as the proportion (by surface area) of the vegetative blade transformed to sporangial sorus. The mean annual reproductive effort of the semelparous species *Laminaria ephemera* was found to vary from 13 to 32%; that of the co-occurring iteroparous species *Laminaria setchellii* varied from 1 to 32%.[53] Neushul[72] estimated a reproductive effort of 4% (by weight of fertile : sterile tissue) for the perennial *Macrocystis pyrifera* in southern California; Lobban[61] considered this value to be unusually high for kelps. Ford et al.[35] reported a mean reproductive effort of 14% for the perennial *Lithophyllum incrustans,* measured as the total volume of spore-producing conceptacles versus the total volume of growth during a 1-year period. Reproductive effort of the isomorphic sexual phase was estimated by these authors to be greater than that of the sporophyte phase. Reproductive effort increased with age for the first 12 years of a possibly 25 year life span, after which time reproductive effort remained relatively constant.[35]

Existing data indicate that some seaweeds experience high juvenile mortality and variable adult mortality. Large numbers of spores or gametes are typically liberated by algae such as the kelps (e.g., 1×10^7–1×10^{10}).[53,72] The evolution of semelparity versus iteroparity may be independent of the high initial mortality incumbent on the earliest juvenile stages, and may instead be related to mortality occurring in the later prereproductive period. Experiments appropriate to the evaluation of semelparity and iteroparity among the algae have yet to be performed. However, in general, first reproduction occurs earlier in semelparous than in iteroparous algae, and this is consistent with existing theory.

Reproduction may or may not be correlated with age in the algae, and is more closely correlated with size, according to the few age- and size-specific data that exist. Certainly more data are needed before age-specific processes in algae can be evaluated. Similarly, the data pertaining to reproductive effort in seaweeds are too few to allow identification of patterns that might exist.

ALTERNATION OF GENERATIONS AND SPOROPHYTE : GAMETOPHYTE RATIOS

The algae typically exhibit some alternation of free-living sporophyte and gametophyte phases, though the alternation may be facultative rather than obligate, and may proceed independently of alternation of ploidy levels.

The algae exhibit a wide range of life history variation that remains mostly unexplored in terms of evolutionary processes. Complex algal life histories may be appropriate systems for the evaluation of sexual versus asexual reproduction, gametic versus zygotic selection, and plastic versus canalized response to environmental variability, and may thus contribute to a better understanding of life history evolution.

Stebbins and Hill[90] have discussed the origin of the alternation of generations among plants, and cite an adaptive advantage accruing to the two-stage cycle in seasonal environments or in environments characterized by two alternate niches. These authors go on to invoke mechanisms similar to those of the Strawberry-Coral and Elm-Oyster models of Williams[101] in the evolution of the alternation of generations. An advantage of the Strawberry-Coral model that may be of importance among the marine algae is the capacity for both asexual dispersal of successful genotypes and generation of new genotypes via meiosis and syngamy.

The evolution of the triphasic life history was presented by Searles[86] in terms of maximization of potential recombination and genetic diversity resulting from a single fertilization event. Searles argues that fertilization may be a rare occurrence among the red algae, which have gametes that lack flagella.

According to Gerritson[40] the probability of successful sexual or asexual reproduction may be determined in part by some critical gamete density, defined as the minimum gamete density required for successful sexual reproduction. Since the probability of gamete encounter is reduced at low densities, below some critical density asexual reproduction or parthenogenesis may be favored. Gerritson's hypothesis has not been demonstrated to be operative among the algae, but is tenable for several reasons. First, the probability of gamete encounter may be reduced in marine systems experiencing strong local advection, especially if there is asynchrony in release of gametes. Second, the predominance of asexually reproducing populations at marginal sites has been observed repeatedly. Densities are expected to be reduced in marginal populations. Third, parthenogenesis and apogamy occur in many species (parthenogenesis is discussed later in this section).

There is evidence for both sporophyte and gametophyte domination of algal populations. Williams[102] reported that gametophytes of the brown algae *Taonia* and *Padina* were rare in Britain. Since then sporophyte predominance has been reported for brown and red algae worldwide. As shown in Table 13.2, reports of gametophyte dominance are much less common.

Gametophytes may be reduced in size as well as frequency. Mathieson[65] has reported the male gametophytes of *Taeonia lennebackeriae* to be present only as small

Table 13.2. Reports of Sporophyte and Gametophyte Dominance in Algal Populations

Genus and species	Location	Most common generation
Phaeophyta		
*Dictyopteris undulata*8	California (U.S.A.)	Sporophyte
Dictyota binghamiae[36]	California (U.S.A.)	Sporophyte
Elachista stellaris[98]	In laboratory culture	Sporophyte
Padina[102]	Britain	Sporophyte
Padina japonica[2]	Hawaii	Sporophyte
Padina pavonica[79]	Mediterranean	Sporophyte
Padina sanctae-crucis[60]	Carribean	Sporophyte
Taonia[102]	Britain	Sporophyte
Zonaria farlowii[4,46]	California (U.S.A.)	Sporophyte
Rhodophyta		
Chondrus crispus[66]	New Hampshire (U.S.A.)	Gametophyte
Dumontia incrassata[52]	Eastern U.S.A.	Sporophyte
Gigartina corymbifera[1]	California (U.S.A.)	Gametophyte
Iridaea cordata[44]	California (U.S.A.)	Sporophyte
Iridaea flaccida[1]	California (U.S.A.)	Gametophyte
Iridaea (2 species)[42]	Chile	Gametophyte
Lithophyllum incrustans[30,31,35]	Britain	Sporophyte
Lomentaria orcadensis[59]	European coast	Sporophyte
Plumaria elegans[55]	European coast	Sporophyte
Ptilota serrata[73]	Maine (U.S.A.)	Gametophyte
Rhodoglossum affine[1]	California (U.S.A.)	Gametophyte

antheridial sori on the larger sporophytic blades (which themselves produce sporangial sori). Female gametophytes have never been observed, in the field or in cultured isolates. Mathieson has proposed that the sporophyte populations are sustained strictly by production of adventitious thalli and by apomeiosis. The gametophyte phase has been eliminated from the life history of *Phaeostrophion irregulare,*[64] in which meiosis is apparently suppressed, and the liberated diploid spores give rise to new sporophytes directly.

There is also evidence for spatial variation in sporophyte and gametophyte abundance within populations. Powell[78] reported an increased dominance of tetrasporophytes with depth in *Constantinea rosea.* In *Bangia atropurpurea,* there was a higher frequency of gametophytes in the lower intertidal and a higher frequency of sporophytes in the high intertidal populations.[87] Tetrasporophytes of the commercially valuable seaweed *Chondrus crispus* occurred significantly more frequently in the deeper outer margins of the populations than in the center.[21] Three possible causes for the preponderance of the sporophytes were suggested: greater frequency of apomixis at the deeper sites, selection by the environment, or selection by commercial harvesting. Sporophytes of *Zonaria farlowii* were found to be more abundant than gametophytes in shallow water (upper distributional limits).[4] Sporophyte : gametophyte ratios in *Dictyopteris undulata* were greater in uncleared areas than in areas that had been experimentally cleared and subsequently recolonized.[8]

Such diploid dominance among algal populations has sometimes been attributed to the enhanced fitness of the diploid phase, presumably as a result of heterosis and of the masking of deleterious recessive alleles. However, numerical dominance of the sporophyte is not always the case among the algae. Reports of gametophyte-dominated populations are listed in Table 13.2.

There is some evidence of partitioning of seasonal environments by alternative phases of a two-stage life cycle, as predicted by Stebbins and Hill.[90] The upright (fleshy) morphs of some heteromorphic algae are not resistant to seasonally intense grazing, while the alternate crustose or boring phases may be grazer resistant.[63] It has been proposed[63] that such differential resistance to grazing can lead to the selective maintenance of heteromorphy under conditions of predictable seasonal variation in grazing pressure. However, the upright and crustose morphs of the *Ralfsia/Scytosiphon/Petalonia* group do not show such seasonal partitioning in reponse to grazing pressures,[25] and the generality of grazer resistance among crustose morphs is therefore uncertain.

Variability in sporophyte : gametophyte ratios is well documented. This variability may occur between and within populations, and between and within species. Both sporophyte-dominated and gametophyte-dominated populations have been reported; the former appears to be more frequent. Factors likely to result in sporophyte dominance are: (1) greater relative fitness (as fecundity and survivorship)of sporophytes over gametophytes; (2) production of adventitious thalli by sporophytes (common among some groups such as the Dictyotales); (3) apomeiosis, usually seen as the liberation of the entire (undivided) contents of the sporangium.

In theory, asexual reproduction should be more common in environments that are physically unpredictable or in which low gamete densities prevail.[40] The oberved spatial distributions of sporophytes and gametophytes are sometimes consistent with the second case: sporophytes are more common on the peripheries of some populations, and are assumed to be maintained asexually. Marginal or peripheral sites may be more likely to exhibit low population densities, though this has not been adequately demonstrated. In this context, it may be useful to examine an isomorphic species such as *Ulva* in order to determine the frequency of sexual versus asexual reproduction relative to the effective gamete densities.

There is no clear correlation between asexuality among the algae and unpredictable physical environments. Sporophytes may predominate at deeper sites (where conditions would seem to be more constant and predictable)[21,78] or at shallow sites (which may be less constant and predictable).[5,89]

Spatial and/or temporal partitioning of the environment by sporophytes and gametophytes may function in maintaining biphasic life histories. However, identification of such partitioning requires that measurements be made at the appropriate spatial or temporal scale, and failure to do so may result in some of the ambiguity we now see in the results of various investigations.

SEX RATIOS

The theory of sex ratios was first proposed by Fisher,[34] and has since been discussed by Maynard-Smith[67] and Charnov,[16] among others. Fisher's sex ratio refers to the ratio of parental investment in sons versus daughters; according to Fisher, this ratio should be 1 : 1. Fisher's construction assumes that sons and daughters are of equal cost. Charnov[16] has discussed some ramifications of unequal costs of sons versus daughters; unequal costs can cause sex ratios to deviate from 1 : 1.

Among the algae, meiosis may be gametic, zygotic, or sporic. Under conditions of gametic meiosis (e.g., as in *Fucus* and *Codium*), algal sex determination may be analogous to sex determination in diplontic animals. Gamete formation and subsequent fertilization in this case produce sexual diploid individuals.

Alternatively, among algae exhibiting sporic meiosis, sexual plants are haploid, and sex is presumably determined by the sex chromosome present in the haploid genome. There is some evidence of sex chromosomes in haploid gametophytes.[32,104] If the sex chromosomes of algae behave according to Mendelian inheritance,[32] then one would anticipate that the products of meiosis form a tetrad (or some mitotic derivation thereof), with males and females equally represented. Indeed, tetraspore formation is common throughout the phylum Rhodophyta, and among the order Dictyotales (phylum Phaeophyta).[9]

There exist few data pertaining to algal sex ratios. This may be due in part to the difficulties of distinguishing males from females, especially among spores and in prereproductive material. Among the data that do exist, there is some ambiguity in use of the term "sex ratio": the term has been used to describe the male-to-female ratio of spores[50] and also to describe the ratio of extant male-to-female gametophytes.[39,52,83,96] The latter usage is contrary to Fisher's[34] definition.

Members of the genus *Laminaria* produce 32 meiospores per sporangium. A 1 : 1 sex ratio among spores of *Laminaria saccharina* was demonstrated by Schreiber.[85] These findings appear to hold for all representatives of the genus.[50]

Additional reports of sex ratios and the site of meiosis are summarized in Table 13.3. In general, 1 : 1 ratios were most frequently observed but, when this was not the case, the female was more common than the male gametophyte. A temperature-dependent sex ratio was observed for *Laminaria religiosa*;[39] a 1 : 1 ratio of male and

Table 13.3. Species, Sex Ratio, and Location of Meiosis

Species	Sex ratio (female : male)	Site of meiosis
Phaeophyta		
Cystoseira osmundaceae[83]	1 : 1	Gametic
Dictyopteris undulata[8]	1 : 1	Sporic
Fucus spp.[96]	1 : 1 to >1 : 1	Gametic
Laminaria spp.[85,50]	1 : 1	Sporic
Laminaria religiosa[39]	1 : 1 (10–15°C)[a]	Sporic
Rhodophyta		
Bangia atropurpurea[87]	3 : 1 to 12 : 1	Sporic
Dumontia incrassata[52]	1 : 1 (site A)	Sporic
	2 : 1 (site B)	Sporic

[a]At temperatures above and below 10–15°C, there were more female thalli.

female gametophytes was found at the optimum growth temperature (10–15°C) but proportionately fewer males were present at both higher and lower temperatures.

Van der Meer[94] has reported occasional cytokinetic failure in *Gracilaria,* which may provide an explanation of varying sex ratios. In this species, incomplete cytokinesis may cause fewer than four spores to be produced, thereby increasing the number of nuclei in some spores. If male and female nuclei fail to separate at cytokinesis, monoecious individuals may be produced in a normally dioecious species. Alternatively, if male nuclei fail to separate from each other at a rate different from that at which female nuclei separate from each other, the resultant sex ratio may deviate from unity.

Mechanisms governing expression of sex ratio (as Fisher's sex ratio) may differ between species exhibiting gametic meiosis and those with sporic meiosis. Sex ratios among algal species with gametic meiosis may conceivably be determined by processes analogous to those determining sex ratios in diplontic animals, but there is, as yet, no evidence to support this.

Among species exhibiting sporic meiosis, Fisher's sex ratio should be 1 : 1 if meiosis and cytokinesis proceed normally. However, cytokinetic failure[96] within tetrasporangia of *Gracilaria* provides a possible mechanism for alternation of sex ratio. This has not been demonstrated in any other species.

Sex ratios among populations of mature gametophytes are variable, but in many cases are close to unity. Such ratios may, however, be confounded by differential survivorship among male and female gametophytes, and do not necessarily reflect sex ratios among the products of meiosis.

PARTHENOGENESIS AND ANDROGENESIS

Parthenogenesis describes the development of an individual from a female gamete without fertilization by a male gamete. This phenomenon may occur frequently among certain algal groups. Male parthenogenesis, or androgenesis, also occurs among the algae, and in the literature is often referred to merely as parthenogenesis. Where possible, we will distinguish between male and female parthenogenesis in the discussion that follows.

Parthenogenesis has been argued to be of advantage in low-density populations and in ecologically marginal environments, and has been proposed to be of advantage in obviating the cost of meiosis.[92] However, according to Williams,[102] a haploid parthenogenetic population derived from a sexual population suffers from expression of deleterious recessive alleles, and reversion to the sexual diploid condition is anticipated. Alternatively, evolution of a diploid parthenogen may require alteration of the meiotic products, by fusion or gamete duplication.[92] Templeton has identified at least two bottlenecks in the successful establishment of diploid parthenogenesis.[92]

Parthenogenesis has been demonstrated to occur in several groups of algae, at least in laboratory culture. There is little information regarding the ploidy of algal parthenogens and the nuclear processes that have produced them, and the frequency of parthenogenesis in naturally occurring populations has not been well documented. Most parthenogenesis in algae is apparently tychoparthenogenesis (accidental or rare parthenogenesis occurring in predominantly sexual species).[92]

Nakahara[71] found parthenogenesis to be common in the life histories of several brown algae. In *Analipus japonicus,* both male and female gametes developed parthenogenetically and erect fronds developed earlier from the females than from the males. The fronds developed from the female gametes produced predominantly plurilocular sporangia, whereas most of the male-derived fronds bore both unilocular and plurilocular structures. Unilocular structures are usually the site of meiosis in brown algae.[9] However, in this study, Nakahara[71] found that all plants bearing unilocs were haploid, and that meiosis failed to occur (meiosis did occur in diploid fronds produced by fertilized oogonia). Chromosome counts verified that plants derived from single gametes were haploid, and those derived from fused gametes were diploid. In subsequent generations, plants derived from female gametes again produced female gametes from plurilocs, and plants derived from male gametes similarly produced male gametes. Fusion of these male and female gametes produced normal diploid plants exhibiting meiosis in unilocs. Parthenogenetically derived plants can thus produce viable offspring that can in turn produce functional gametes.

Some plants derived from single gametes produced both unilocular and plurilocular reproductive structures.[71] The unilocs were more commonly formed at low temperatures. This result agrees with those for *Ectcarpus*[71] and for *Pogotrichum,*[82] though in these latter genera formation of sporangia occurs on diploid rather than haploid thalli.

Gametophytes of *Laminaria religiosa* were cultured under conditions of temperature and day length similar to, and contrasting with, those occurring in the field.[39] Parthenogenesis was found to be more frequent under culture conditions that were different from field conditions. In cultured *Laminaria* spp., 25% of isolated female gametophytes produced sporophytes parthenogenetically.[13] Parthenogenetic development was reported in two other kelps, *Agarum cribrosum* and *Alaria crassifolia.*[71]

Nakahara[71] demonstrated that the oogonia of *Desmarestia viridis* could develop parthenogenetically to produce haploid sporophytes bearing only unilocs; the swarmers from these unilocs subsequently produced gametophytes. Under identical culture conditions, Nakahara observed no parthenogenetic development among oogonia of the congeneric species *Desmarestia ligulata* or *Desmarestia tabacoides* (now synonymized).[14]

About 1% of the gametes of *Ulva mutabilis* were found to develop parthenogenetically into new gametophytes.[37,48] These in turn were likely to form gametes that subsequently produced gametophytes. In culture, 8% of the (+) mating-type gametes and 2% of the (−) mating-type gametes of this species developed parthenogenetically into either haploid or diploid sporophytes or single thalli with both haploid and diploid tissue.[62] Abortive mating reportedly increased and synchrony of gamete release declined with decreasing temperature, thus allowing the prediction that northern (marginal) populations should have a tendency to cease sexual reproduction and rely instead on parthenogenetic development.[62] Parthenogenesis has also been reported in *Ulva mutabilis*[37] and in *Percursaria percursa,*[54] while *Codium fragile* exists on some coasts as exclusively female populations.[55] This latter observation is suggestive of parthenogenetically maintained populations.

To our knowledge, parthenogenesis has not been unequivocally demonstrated in the red algae. Earlier suggestions of parthenogenesis in some species[73] were often based on the apparent absence of the tetrasporophyte phase in natural populations. In most cases, careful laboratory culture has demonstrated the existence of a heteromorphic, often cryptic, tetrasporophyte in the life history.[24] Patterns of sexuality in the red algae have been thoroughly reviewed by West and Hommersand.[100]

In the algae, parthenogenesis may occur in taxa exhibiting sporic meiosis; thus, meiospores germinate to produce gametophytes that in turn produce new sporophytes or gametophytes parthenogentically. This process clearly does not obviate the cost of meiosis, since meiosis has previously occurred in the sporangium. Alternative explanations are therefore necessary. It is attractive to postulate in this case that retention of the capacity for parthenogenesis is advantageous in low-density populations suffering from reduced probability of fertilization, though there is as yet no convincing evidence of this. Further, many diplohaplontic algae retain the option of apomeiosis through recycling of the sporophyte phase (see Alternation of Generations and Sporophyte : Gametophyte Ratios); such diploid recycling precludes the problem of expression of deleterious recessive alleles that would be manifest in haploid populations. Diploid recycling thus may be of greater value than parthenogenesis in low-density or ecologically marginal populations.

Finally, the establishment of exclusively parthenogenetic populations has not been reported in the algae. The predominance of the biphasic life history may facilitate the proximate reversion of parthenogens to sexuality.

SUMMARY AND CONCLUSIONS

Information pertaining to life history strategies in algae is piecemeal and scattered, and explicit tests of life history theory have not been performed. Gathering together the available data allows us to make a few generalizations and, more importantly, points the way toward problems that may merit investigation.

1. Reproduction has not yet been proved to impose a cost (as reduced probability of survival and future reproduction) on the organism. This may be attributable in part to the photoautotrophic habit, and to the manner in which reproductive structures are borne by the algae.

2. Age-specific processes of fecundity and survivorship are not yet apparent, perhaps because the scant data available do not allow the identification of existing patterns. Reproduction may or may not be age dependent; maturity may or may not be delayed. Fecundity (in terms of number of spores or gametes produced) generally seems to be very high, and the most appropriate models to describe the algae are prob-

ably the Elm-Oyster and Strawberry-Coral models.[100] Size-specific or stage-specific analysis may be appropriate for the macroalgae, and may prove to be successful in characterizing some life history patterns.

3. The complexity inherent in the bi- and triphasic life histories of the seaweeds is potentially suitable for testing sporophytic versus gametophytic selection, and selection for sexuality and asexuality. Many algal populations exhibit numerical dominance of the sporophyte, but gametophyte-dominated populations also exist. Apomeiosis and parthenogenesis have been reported in laboratory culture across a broad range of taxa; these processes tend to be facultative and inducible. Plasticity in expression of life history traits has not even been approached experimentally.

Finally, we would argue that rigorous experimental analysis of the complete life histories of algae will enhance our understanding of life history evolution in general. We also suggest that life history processes operative in the algae may be substantively different from those described for higher plants and animals.

ACKNOWLEDGMENTS

T. K. was supported in part by NSF grant #DPP8300189 to P. K. Dayton during the preparation of this manuscript.

REFERENCES

1. Abbott, I. A., Seasonal population biology of some carrageenophytes and agarophytes, in *Pacific Seaweed Aquaculture* (I. A. Abbott, M. S. Foster, and L. F. Ecklund, eds.) pp. 45–53. California Sea Grant College Program, 1980.
2. Allender, B., Ecological experimentation with the generations of *Padina japonica* (Yamada) (Dictyotales, Phaeophyta), *J. Exp. Mar. Biol. Ecol.* **26,** 225–234 (1977).
3. Ang, P., Phenology of *Sargassum siliquosum* J. Ag. and *S. paniculatum* J. Ag. (Sargassaceae, Phaeophyta) in the reef flat of Balibago, Calatagan, Philippines, presented at *5th Int. Coral Reef Symp., Tahiti,* 1985.
4. Barilotti, D. C., An ecological study of populations of a benthic alga: Genetic differences that affect life history strategies in diverse habitats. Ph.D. Dissertation, University of California Santa Barbara, 1973.
5. Barilotti, D. C., Ecological implications of haploidy and diploidy for the isomorphic brown alga *Zonaria farlowii* Setch. et Gardn. (Abstract), *J. Phycol. (Suppl.)* **7,** 4 (1971).
6. Bell, G., Measuring the cost of reproduction. I. The correlation structure of the life table of a planktonic rotifer, *Evolution* **38,** 300–313 (1984a).
7. Bell, G., Measuring the cost of reproduction. II. The correlation structure of the life table of five freshwater invertebrates. *Evolution* **38,** 314–326 (1984b).
8. Benson, M. R., The ecology of two sympatric species of *Dictyopteris* at Santa Catalina Island. Ph.D. Dissertation, University of Southern California, 1983.
9. Bold, H. C., and Wynne, M. J., *Introduction to the Algae,* pp. 1–305. Prentice-Hall, Englewood Cliffs, New Jersey, 1985.
10. Burns, R. L., and Mathieson, A. C., Ecological studies on economic red algae III. Growth and reproduction of natural and harvested populations of *Gagartina stellata* (Stackhouse) Batters in New Hampshire, *J. Exp. Mar. Biol. Ecol.* **9,** 77–95 (1972).
11. Chapman, A. R. O., Reproduction, recruitment and mortality in two species of *Laminaria* in southwest Nova Scotia, *J. Exp. Mar. Biol. Ecol.* **78,** 99–109 (1984).
12. Chapman, A. R. O., *Biology of Seaweeds; Levels of Organization,* pp. 1–82. Baltimore Park Press, Baltimore, 1979.
13. Chapman, A. R. O., The genetics of morphological differentiation in some *Laminaria* populations, *Mar. Biol.* **24,** 65–91 (1974).
14. Chapman, A. R. O., Morphological variation and its taxonomic implications in the ligulate members of the genus *Desmarestia* occurring on the West Coast of North America, *Syesis* **5,** 1–20 (1972).

15. Chapman, A. R. O., and Goudy, C. L., Demographic study of the macrothallus of *Leathesia difformis* (Phaeophyta) in Nova Scotia, *Can. J. Bot.* **61,** 319–323 (1983).
16. Charnov, E. L., *The Theory of Sex Allocation,* pp. 1–289. Monographs in Population Biology, 18, Princeton, New Jersey, 1982.
17. Clayton, M., Life history studies in the Ectocarpales (Phaeophyta): Contributions towards the understanding of evolutionary processes, *Bot. Mar.* **25,** 111–116 (1982).
18. Cody, M. L., A general theory of clutch size, *Evolution* **20,** 174–184 (1966).
19. Cole, L. C., The population consequence of life-history phenomena, *Q. Rev. Biol.* **29,** 103–137 (1956).
20. Coyer, J. A., and Zaugg-Haglund, C., A demographic study of the elk kelp, *Pelagophycus porra* (Laminariales, Lessoniaceae), with notes on *Pelagophycus* x *Macrocystis* hybrids, *Phycologia* **21,** 399–407 (1982).
21. Craigie, J. S., and Pringle, J. D., Spatial distribution of tetrasporophytes and gametophytes in four maritime populations of *Chondrus crispus, Can J. Bot.* **56,** 2910–2914 (1978).
22. Dawkins, R., *The Selfish Gene,* pp. 1–224. Oxford Univ. Press, New York, 1976.
23. Dayton, P. K., Currie, V., Gerrodette, T., Keller, B. D., Rosenthal, R., and ven Tresca, D., Patch dynamics and stability of some California kelp communities, *Ecol. Monogr.* **54,** 253–289 (1984).
24. De Cew, T. C., and West, J. A., Life histories in the Phyllophoraceae (Rhodophyceae, Gigartinales) from the Pacific Coast of North America. I. *Gymnogongrus linearis* and *G. leptophyllus, J. Phycol.* **17,** 240–250 (1981).
25. Dethier, M. N., Heteromorphic algal life histories: The seasonal pattern and response to herbivory of the brown crust *Ralfsia californica, Oecologia* **49,** 333–339 (1981).
26. De Wreede, R. E., Demographic characteristics of *Pterygophora californica* (Laminariales, Phaeophyta), *Phycologia* **25,** 11–17 (1986).
27. De Wreede, R. E., Growth and age class distribution of *Pterygophora californica* (Phaeophyta), *Mar. Ecol. Prog. Ser.* **19,** 93–100 (1984).
28. Druehl, L. D., The development of an edible kelp culture technology for British Columbia. II. Second Annual Report, Fisheries Development Report #26, pp. 1–43. Marine Resources Branch, Ministry of Environment, B.C., Canada, 1980.
29. Dunn, G. A., Development of *Dumontia filiformes* II. Development of sexual plants and general discussion of results, *Bot. Gaz.* **63,** 425–467 (1917).
30. Edyvean, R. G. J., and Ford, H., Population biology of the crustose red alga *Lithophyllum incrustans* Phil. 2. A comparison of populations from three areas of Britain, *Biol. J. Linn. Soc.* **23,** 353–363 (1984).
31. Edyvean, R. G. J., and Ford, H., Population biology of the crustose red alga *Lithophyllum incrustans* Phil. 3. The effects of local environmental variables, *Biol. J. Linn, Soc.* **23,** 365–374 (1984).
32. Evans, L. V., The Phaeophyceae: Part I, in *The Chromosomes of the Algae* (M. B. E. Godward, ed.), Chapter 5. St. Martin's Press, New York, 1966.
33. (Reference deleted in proof.)
34. Fisher, R. A., *The Genetical Theory of Natural Selection,* Chapter 6. Clarendon Press, Oxford, 1930.
35. Ford, H., Hardy, F. G., and Edyvean, R. G. J., Population biology of the crustose red alga *Lithophyllum incrustans* Phil. Three populations on the east coast of Britain, *Biol. J. Linn. Soc.* **19,** 211–220 (1983).
36. Foster, M., Neushul, M., and Chi, E. Y., Growth and reproduction of *Dictyota binghamiae* J. G. Agardh, *Bot. Mar.* **15,** 96–107 (1972).
37. Foyn, B. Geschlechtskontrollierte Verebung bei der marinen grunalga *Ulvan mutabilis, Arch. Protistenkd.* **104,** 236–253 (1959).
38. Fralick, R. A., and Mathieson, A. C., Ecological studies of *Codium fragile* in New England, U.S.A., *Mar. Biol.* **19,** 127–132 (1973).
39. Funano, T., The ecology of *Laminaria religiosa* Miyabe. I. The life history and the alternation of nuclear phases of *Laminaria religiosa,* and the physiological ecology of the gametophytes and the embryonal sporophytes, *Hokusui-Shiho* **25,** 61–109 (1983).
40. Gerritson, J., Sex and parthenogenesis in sparse populations, *Am. Nat.* **115,** 718–742 (1980).
41. Giesel, J. T., Reproductive strategies as adaptive to life in temporally heterogeneous environments, *Annu. Rev. Ecol. Syst.* **7,** 57–79 (1976).
42. Hanach, G., and Santilices, B. Ecological differences between the isomorphic reproductive phases of two species of *Iridaea* (Rhodophyta, Gigartinales), *Mar. Ecol. Prog. Ser.* **22,** 291–303 (1985).
43. Hansen, J. E., Ecology and natural history of *Iridaea cordata* (Gigartinales, Phodophyta) growth, *J. Phycol* **13,** 395–402 (1977).

44. Hansen, J. E., and Doyle, W. T., Ecology and natural history of *Iridaea cordata* (Rhodophyta, Gigartinaceae): Population structure, *J. Phycol.* **12,** 273–278 (1976).
45. Harper, J. L., *Population Biology of Plants,* Chapter 21. Academic Press, London, 1977.
46. Haupt, A. W., Structure and development of *Zonaria farlowii, Am J. Bot.* **19,** 239–259 (1932).
47. Hillis-Colinvaux, L., Ecology and taxonomy of *Halimeda:* primary producer of coral reefs, *Adv. Mar. Biol.* **17,** 1–127 (1980).
48. Hoxmark, R. C., Experimental analysis of the life cycle of *Ulva mutabilis, Bot. Mar.* **18,** 123–129 (1975).
49. Jones, W. E., The growth and fruiting of *Gracilaria verrucosa* (Hudson) Papenfuss, *J. Mar. Biol. Assoc. U.K.* **38,** 47–56 (1959).
50. Kain, J. M., A view of the genus *Laminaria, Oceanogr. Mar. Biol. Ann. Rev.* **17,** 101–161 (1979).
51. Kain, J. M., Aspects of the biology of *Laminaria hyperborea.* II. Age, weight, and length, *J. Mar. Biol. Assoc. U.K.* **43,** 129–151 (1963).
52. Kilar, J. A., and Mathieson, A. C., Ecological studies of the annual red alga *Dumontia incrassata* (O. F. Muller) Lamouroux, *Bot. Mar.* **21,** 423–437 (1978).
53. Klinger, T., Allocation of blade surface area to meiospore production in annual and perennial representatives of the genus *Laminaria.* M.Sc. thesis, University of British Columbia, Vancouver, 1985.
54. Kornmann, P., Zur morphologie und Entwicklung von *Percursaria percursa, Helgol. Wiss. Meeresunters.* **27,** 259 (1956).
55. Kornman, P., and Sahling, P.-H., *Meeresalgen von Helgoland,* pp. 1–289. Biologische Anstalt Helgoland, Hamburg, 1978.
56. Krebs, C. J., *Ecology,* pp. 202. Harper & Row, New York, 1985.
57. Law, R., Ecological determinants in the evolution of life histories, in *Population Dynamics* (R. M. Anderson, B. D. Turner, and L. R. Taylor, eds.), Chapter 4. Blackwell Scientific Publ., Oxford, 1979.
58. Levring, T., Some remarks on the structure of the gamete and the reproduction of *Ulva lactuca, Bot. Not.* **108,** 40–45 (1955).
59. Levring, T., Some modern aspects of growth and reproduction in marine algae in different regions, *Ann. Biol.* **33,** 57–65 (1957).
60. Liddle, L. B., Development of gametophytic and sporophytic populations of *Padina sanctae-crucis* in the field and laboratory, *Proc. Int. Seaweed Symp.* **7,** 80–82 (1972).
61. Lobban, C. S., The growth and death of the *Macrocystis* sporophyte (Phaeophyceae, Laminariales), *Phycologia* **17,** 196–212 (1978).
62. Lovlie, A., and Bryhni, E., On the relation between sexual and parthenogenetic reproduction in haplodiplontic algae, *Bot. Mar.* **21,** 155–163 (1978).
63. Lubchenco, J., and Cubit, J., Heteromorphic life histories of certain marine algae as adaptations to variations in herbivory, *Ecology* **61,** 676–687 (1980).
64. Mathieson, A. C., Morphology and life history of *Phaeostrophion irregulare* Setchell and Gardner, *Nova Hedwigia,* 13, 293–318, 1967.
65. Mathieson, A. C., Morphological studies of the marine brown alga *Taeonia lennebackerae* Farlow ex J. Agardh. I. Sporophytes, abnormal gametophytes, and vegetative reproduction, *Nova Hedwigia* **12,** 65–79 (1966).
66. Mathieson, A. C., and Burns, R. L., Ecological studies of economic red algae. V. Growth and reproduction of natural and harvested populations of *Chondrus crispus* Stackhouse in New Hampshire, *J. Exp. Mar. Biol. Ecol.* **17,** 137–156 (1975).
67. Maynard-Smith, J. M., The ecology of sex, in *Behavioural Ecology* (J. R. Krebs, and N. B. Davies, eds.), Chapter 6. Blackwell Scientific Publ., Oxford, 1978.
68. McCourt, R. M., Reproductive biomass allocation in three *Sargassum* species, *Oecologia* **67,** 113–117 (1985).
69. McLachlan, J., and Bidwell, R. G. S., Photosynthesis of eggs, sperm, zygotes and embryos of *Fucus serratus, Can. J. Bott.* **56,** 371–373 (1978).
70. Muller, D. G., Uber jahre and lunar periodishe Erscheinungen bei einige Braunalgen, *Bot. Mar.* **4,** 140–155 (1962).
71. Nakahara, H., Alternation of generations of some brown algae in unialgal and axenic cultures, *Sci. Pap. Inst. Algol. Res. (Hokkaido Univ.)* **7,** 77–94 (1984).
72. Neushul, M., Studies on the giant kelp, *Macrocystis.* II. Reproduction, *Am. J. Bot.* **50,** 349–353 (1963).
73. Newroth, P. R., and Markham, J., Observations on the distribution, morphology, and life history of some Phyllopharaceae, *Proc. Int. Seaweed Symp.* **7,** 120–126 (1972).

74. Norall, T. T., Mathieson, A. C., and Kilar, J. A., Reproductive ecology of four subtidal red algae, *J. Exp. Mar. Biol. Ecol.* **54,** 119–136 (1981).
75. Pearl, R., *The Rate of Living,* Chapter 3. Knopf, New York, 1928.
76. Peters, A. F., Observations of the life history of *Papenfussiella callitricha* (Phaeophyceae, Chordariales) in culture, *J. Phycol.* **20,** 409–414 (1984).
77. Pianka, E. R., *Evolutionary Ecology,* 2nd Ed., pp. 127. Harper & Row, New York, 1978.
78. Powell, J., The life history of the red alga, *Constantinea.* Ph.D. thesis, University of Washington, Seattle, 1964.
79. Ramon, E., and Friedman, I., The gametophyte of *Padina* in the Mediterranean, *Proc. Int. Seaweed Symp.* **5,** 183–196 (1966).
80. Roland, W. G., Resource management biology for the edible kelp *Cymathere triplicata, Can. J. Fish. Aquat. Sci.* **41,** 271–277 (1984).
81. Rosenthal, R. J., Clake, W. D., and Dayton, P. K., Ecology and natural history of a stand of giant kelp, *Macrocystis pyrifera,* off DelMar, California, *Fish. Bull.* **72,** 670–684 (1974).
82. Sakai, Y., and Saga, N., The life cycle of *Pogotrichum vezoense* (Dictysiphonales, Phaeophyceae), *Sci. Pap. Inst. Algol. Res. (Hokkaido Univ.)* **7,** 1–15 (1981).
83. Schiel, D. R., A short-term demographic study of *Cystoseira osmundaceae* (Fucales: Cystoseiraceae) in Central California, *J. Phycol.* **21,** 99–106 (1985).
84. Schiel, D. R., A demographic and experimental evaluation of plant and herbivore interactions in subtidal algal stands. Ph.D. dissertation, University of Auckland, 1981.
85. Schreiber, E., Untersuchungen uber parthenogenesis, geschlechtsbestimmung und bastiardierungsvermogen bei Laminarian, *Planta* **12,** 331–353 (1930).
86. Searles, R. B., The strategy of the red algal life history, *Am. Nat.* **115,** 113–120 (1980).
87. Sheath, R. G., Van Alstyne, K. L., and Cole, K. M., Distribution, seasonality, and reproductive phenology of *Bangia atropurpurea* (Rhodophyta) in Rhode Island, U.S.A., *J. Phycol.* **21,** 297–303 (1985).
88. Smith, B. D., Recovery following experimental harvesting of *Laminaria longicruris* and *L. digitata* in southwestern Nova Scotia, *Helgol. Wiss. Meeresunters.* **39,** 83–101 (1985).
89. Stearns, S. C., Life history tactics: A review of the ideas, *Q. Rev. Biol.* **51,** 3–47 (1976).
90. Stebbins, G. L., and Hill, G. J., Did multicellular plants invade the land?, *Am. Nat.* **115,** 342–353 (1980).
91. Tatewaki, M., and Iima, M., Life histories of *Blidingia minima* (Chlorophyceae), especially sexual reproduction. *J. Phycol.* **20,** 368–376 (1984).
92. Templeton, A. R., The prophecies of parthenogenesis, in *Evolution and Genetics of Life Histories* (H. Dingle, and J. P. Hegemann, eds.) pp. 75–102. Springer-Verlag, Berlin, 1982.
93. Thompson, K., and Stewart, A. J. A., The measurement and meaning of reproductive effort in plants, *Am. Nat.* **117,** 205–211 (1981).
94. Van der Meer, J. P., Genetics of *Gracilaria* sp. (Rhodophyceae, Gigartinales). II. The life history and genetic implications of cytokinetic failure during tetraspore formation, *Phycologia* **16,** 367–371 (1977).
95. Vandermeulen, H., The taxonomy and autecology of *Colpomenia peregrina* (Sauv.) Hamel (Phaeophyceae). Ph.D. Thesis, The University of British Columbia, Vancouver, 1984.
96. Vernet, P., and Harper, J. L., The cost of sex in seaweeds, *Biol. J. Linn. Soc.* **13.** 129–138 (1980).
97. Waaland, J. R. Experimental studies on the marine alga *Iridaea* and *Gigartina, J. Exp. Mar. Biol. Ecol.* **11,** 71–80 (1973).
98. Wanders, J. B. W., van den Hoek, C., and Schillern-van Nes, E. H., Observations on the life history of *Elachista stellaris* (Phaeophyceae) in culture, *Neth. J. Sea Res.* **5,** 458–491 (1982).
99. Watson, M. A., and Casper, B. B., Morphogenetic constraints on patterns of carbon distribution in plants, *Annu. Rev. Ecol. Syst.* **15,** 233–258 (1984).
100. West, J. A., and Hommersand, M. H., Rhodophyta: Life histories, in *The Biology of Seaweeds* (C. S. Lobban and M. J. Wynne, eds.), pp. 133–193. Blackwell Scientific Publ., Oxford, 1981.
101. Williams, G. C., *Sex and Evolution,* pp. 1–200. Princeton Univ. Press, Princeton, New Jersey, 1975.
102. Williams, J. L., Studies on the Dictyotaceae. II. The periodicity of the sexual cells in *Dictyota dichotoma, Ann. Bot. (London)* **19,** 531–560 (1905).
103. Wynne, M. J., and Loiseaux, S., Phycological reviews. 5. Recent advances in life history studies of the Phaeophyta, *Phycologia* **15,** 435–452 (1976).
104. Yabu, H., and Sanbonsuga, Y., A sex chromosome in *Cymathere japonica* Miyabe at Nagai, *Jpn. J. Phycol.* **29,** 79–80 (1981).

14

Reproductive Ecology of Bryophytes

BRENT D. MISHLER

The bryophytes, with some 25,000 species, are the second most diverse group of land plants (behind the angiosperms) in terms of number of species, yet many aspects of their biology remain relatively under-studied, and reproductive ecology is no exception.

This chapter has several goals, and since some of these differ from other chapters in this book, a few words of explanation are in order. The first goal is to present a picture of the evolutionary biology of bryophytes to general evolutionary ecologists who may not be fully aware of the unique or special features of this group. Thus the approach here is overtly taxonomic in both this empirical, advocative sense as well as in a further theoretical sense. I want to affirm the close and necessary connections between evolutionary ecology and systematics, in particular between the study of adaptation and the reconstruction of phylogenies. The inclusion of "taxonomic" contributions in the present volume is of considerable importance; in my opinion the future of evolutionary ecology depends upon much closer interaction with systematics than has been the case in the past.

In conjunction with the above arguments, a second goal is to critically examine the prevailing "strategic," "optimality," and "adaptationist" approach to plant evolutionary ecology. It will be argued that structural and physiological features observed in organisms can be due to a number of different causes, "adaptations" in the strict and precise sense may be rather rare, and therefore the widespread assumption of universal adaptation is counterproductive.

The third goal is to review what is known about the reproductive ecology of bryophytes. Since this subject has been reviewed so often (in fact, the number of recent reviews probably exceeds the number of recent empirical studies of the subject!), I will not attempt to be complete in terms of subject matter or literature surveyed. For further discussion see Anderson,[2] Anderson and Snider,[3] Longton,[62,64] Longton and Schuster, [68] Smith,[106] Wyatt,[139] and Wyatt and Anderson.[141] I will emphasize asexual reproduction, because this mode of reproduction is apparently important in bryophytes and because sexual reproduction in bryophytes has been emphasized by all recent reviews.

The fourth and final goal is to discuss the evolutionary significance of asexual reproduction in bryophytes. The widespread opinion that reliance on asexual reproduction has made the bryophytes genetically depauperate and/or evolutionarily stagnant will be challenged. A general discussion of asexual versus sexual reproduction and of the significance of somatic mutation will be presented that may perhaps be of wider interest.

REPRODUCTIVE "STRATEGIES" AND THE ADAPTATIONIST PROGRAM

The use of the work "strategy" in evolutionary biology has often been debated from a semantic viewpoint.[50,146] It has been charged that use of the term implies a conscious plan of action on the part of the species; the standard defense is that professional biologists may imply conscious teleology as a short-cut in discussion, but at least professional biologists understand what is really meant. My opinion is that this debate over semantics misses the main point.

Biologists, like any other group of scientists, can and should use language in whatever way they find useful. It does make one wonder why a word loaded down with as much baggage as "strategy" is useful to biologists, given the great pains which must be taken (but often are not) to explain to students and the general public what is *not* meant by the term. However, careful effort in biological education could in principle cure misunderstanding.

The truly damaging objection to the concept of evolutionary "strategy" on the professional level is to the mind-set that so often goes along with the term. Despite Zimmerman and Hicks' disclaimer that "it would *probably* be incorrect, for example, to assume that every phenotypic trait was important enough for natural selection to operate on it *directly*"[146] (italics added), evolutionary biology today is pervaded by a faith in the optimizing ability of natural selection. The first and foremost question asked by most evolutionary biologists when faced by some characteristic of an organims is, "For what purpose did this characteristic evolve?" Strong faith in natural selection leads in practice to the construction of an adaptive explanation for the existence of any characteristic and the search for another adaptive explanation if the first fails. A nonadaptive explanation is contemplated only if no adaptive explanation can be found (which is seldom, given the creativity of the human mind). A contemporary manifesto for plant population biology was recently given by Solbrig et al.[110] They listed two main objectives of population biology (p. xi): (1) "to understand in precise detail how natural selection operates," and (2) "to understand rigorously the mechanisms of adaptation." No concern for (or even awareness of) the fact that natural selection or mechanisms of adaptation may not universally be operating was evident. In the same volume, Horn[41] presented a defense of an optimality approach to adaptation and tried to "show that evolutionary history plays a surprisingly small role in the origin and maintenance of contemporary adaptive patterns" (p. 48).

An important paper by Gould and Lewontin[32] criticized this overriding "optimality" mind-set of contemporary evolutionary biology, one that they term the "adaptationist programme." Gould and Lewontin presented a list of alternative explanations for a given form or function, for example, a feature may occur because of purely random factors, it may occur because it is neutral (or even selectively deleterious) but developmentally correlated to some other feature under strong natural selection, or it may occur because of past natural selection for a different function than it currently has. Their general theme was not to deny that natural selection occurs but rather to point out the importance of other causal factors that act as constraints on natural selection. The theme of constraints in evolution has had a good deal of discussion recently.[31,88,89,109]

One primary cause of conflict over the importance of immediate adaptation in evolution is whether one takes an *equilibrium* or an *historical* approach.[55,60] The usual application of the adaptationist program is to assume that organisms have been adjusting to their environment and to each other for a long enough time for natural selection to have optimized the functions of all features of the organisms. In an equilibrium situation, there is no historical information present as to how the equilibrium was reached.[60] Horn,[41] quoted above, provides an explicit example of this approach. The alternative, historical approach recognizes that few if any ecological systems are anywhere near equilibrium, and that developmental or physical constraints (or simply lack of time for appropriate genetic variation to arise) may well have limited the optimizing efficacy of natural selection. In short, this second approach recognizes the importance of history in biological systems, both in terms of the need to understand initial conditions and to appreciate that, in a dynamic system, the current phenotypic position of a species depends on causes acting in the past as well as the present.

Another primary cause of conflict over adaptation is the lack of a precise and widely accepted definition. Gould and Vrba[33] presented an analysis of the situation and proposed a refined set of definitions that clarify the relationships of a number of important concepts subsumed under the heading "fit of organism to environment." They use "aptation" to refer to a structure that has some current utility to an organism. Aptations can be either "adaptations" (in a strict sense), i.e., structures originally shaped by natural selection for the current function, or *"exaptations,"* i.e., structures that were either originally shaped by natural selection for some other function or that were never the result of direct natural selection (hence originally "nonaptations"), but are co-opted for a current function. Their definition needs further refinement because it is clear that when a novelty first arises in a population it is a result of underlying genetic and developmental perturbations, not natural selection. Thus every adaptation could be viewed as an exaptation if examined early enough. It is necessary to add the distinction between whether or not the feature was originally shaped by natural selection *in becoming fixed in the species in which it arose.* Thus the original species' boundary is the dividing line between adaptation and exaptation (and the species problem raises its ugly head in yet another context[81,85]). An example of exaptation can be found in the biochemical evolution of land plants. Flavonoids, which occur in the land plants and their sister group Charales,[83,84] are thought to have originally been adaptations to protect aquatic ancestors of the land plants from ultraviolet light; flavonoids apparently were later co-opted for herbivore protection in terrestrial plants.[117]

A historical approach to evolutionary ecology, coupled with the restricted and precise definition of adaptation advocated by Gould and Vrba, requires a close connection between the study of evolution and systematics. In particular, a close connection is needed with the explicit analytical methods of phylogenetic systematics or cladistics.[136] Cladists (although in some cases themselves neglecting evolutionary implications of their systematic theories) are in the business of trying to determine just the parameters that are needed to take a historical approach to evolutionary ecology: discovery of homologies and their transformation series (i.e., evolutionary connections between homologies), determining evolutionary polarity of these transformations (i.e., establishing which homologies of a given transformation series are *plesiomorphic* or primitive at a particular hierarchical level and which are *apomorphic* or derived), and postulating genealogical relationships of taxa.

The utility of cladistic hypotheses for carrying out sound behavioral and ecological studies was recently discussed by Dobson.[22] My version of the argument boils down to this: (1) studies in evolutionary ecology necessarily make assumptions about homologies between features and their historical sequence of appearance; (2) cladistic methods, while not guaranteed to produce truth, are the most rigorous methods known for postulating homologies and their polarities; (3) these methods (unlike traditional methods of producing phylogenies) are also independent of particular evolutionary mechanisms, thus allowing noncircular inference between pattern and process.

Having a cladogram available for a group of organisms being studied ecologically is also very useful because it provides a minimum estimate of the number of independent (homoplasious) origins of a character of interest.[22] It is widely recognized that an hypothesis of adaptation is strengthened if a number of different groups have developed the same feature in response to similar environmental pressure (e.g., stinging hairs, succulence, dioecy). Arguments for parallel origin of a feature are usually made by evolutionary ecologists using the "comparative method" whereby phenotypes are compared across many taxa to find assocations with other phenotypes or ecological situations.[25] Felsenstein[25] points out a serious statistical problem with this method: taxa are arranged in a hierarchically structured phylogeny and cannot be considered independent samples from the same distribution. In order to discover parallel evolution, one needs the best available phylogeny.

To summarize, for a trait to be rigorously judged as an adaptation in the strict sense of a definition that is useful to biologists interested in evolutionary history, it should pass a series of tests. These tests, each of which can falsify a particular hypothesis of adaptation, obviously require both ecological and systematic studies.

1. The apomorphic state of some trait must be shown to function better (in an engineering sense) in solving some environmental problem than the plesiomorphic state does.
2. The trait must be heritable.
3. The trait must be shown to increase the fitness of organisms bearing it.
4. The apomorphic trait must have arisen when the improved function arose (this must be determined by mapping functions to cladograms); if the trait arose first and the function was added later, then the trait is an exaptation.

These tests are strict indeed and there are extremely few cases in plants where a putative adaptation has passed all four. This scarcity of suitably documented examples is enough in itself to cast serious doubt on the assumption that adaptations are ubiquitous in plants.

BRYOPHYTE MORPHOLOGY, ECOLOGY, AND PHYLOGENY

The bryophytes (while probably not monophyletic, as discussed below) share a number of features. Their life cycle has a more evenly balanced alternation of generations (in terms of size and duration of the two phases) than those of other land plants. The gametophyte is free living, usually persistent, and is the main assimilative phase. The form of the gametophyte ranges from thalloid in hornworts and some liverworts to "leafy" in most liverworts and all mosses (i.e., with a cylindrical stem and flat photosynthetic appendages one or a few cells thick—these are not thought to be homologous to sporophytic leaves in tracheophytes). In bryophytes, gametes are produced by mitosis and fertilization occurs in the archegonium. The sporophyte is ephemeral to perennial and is permanently epiphytic on the gametophyte, but is photosynthetic when immature (and remains so at maturity in the hornworts). There is always a single sporangium on each sporophyte, and meiosis occurs to produce spores. Further information about basic bryophyte biology can be found in the texts of Watson[129] and Schofield.[99]

The phylogenetic position of the bryophytes relative to the other land plants has been widely debated.[99] The bryophytes have been variously considered as completely unrelated to, ancestral to, or even derived from the tracheophytes. As discussed in the previous section, it is important to have the best available phylogeny for a group before discussing its evolutionary ecology, so the results of recent cladistic analyses of the problem will be summarized below.[83,84] The embryophytes (land plants), which include the bryophytes and the tracheophytes, are well supported as a monophyletic group by several synapomorphies (shared, derived characters). Closely related sister groups of the land plants are a number of lineages of green algae that have recently been lumped together in the paraphyletic (i.e., not containing all descendants of a common ancestor) group "charophytes."[74] The immediate sister group of the land plants appears to be the genus *Coleochaete.*[34] The classical group "bryophytes" also appears to be paraphyletic. The mosses alone are the sister group of the tracheophytes, as evidenced by a number of synapormophies, including xylem and phloem. Recognition of these characters as homologies at this phylogenetic level is controversial, but increasingly supported by ultrastructural studies.[39,83,96] The hornworts are the sister group of the moss–tracheophyte clade (which can, based on the above homologies, be called the vascular plants), but this placement is problematical because of a number of homoplasies involving the hornworts. Finally, the liverworts are the sister group of the hornworts plus the vascular plants.

The implications of this phylogeny for evolutionary transformations of land plant characters, particularly the origin of the embryophyte life cycle, have been discussed by Graham[34,35] and Mishler and Churchill.[83,84] The classic (but recently disfavored) "antithetic" theory[10] is overwhelmingly favored by recent phylogenetic analyses of the green algae and bryophytes. This theory holds that the ancestor to the land plants had a haplontic life cycle, and that the sporophyte generation arose by the retention of the zygote on the female gametophyte and interpolation of mitotic divisions between zygote formation and meiosis.

The prevailing view of bryophyte evolution is to see the group as evolutionary "failures" or "dead ends."[2,17,101,115] While fragmentary, the fossil record suggests that bryophyte species and higher taxa have remained virtually unchanged over long periods of time.[51,76,77,114] Furthermore, geographic ranges of extant bryophyte species tend to show broader ranges than for tracheophyte species.[17,40,97,99] Therefore, it is widely felt that bryophyte species evolve very slowly. Explanations that have been given for this include the widespread reliance on asexual reproduction, the haploid vegetative state in which mutations are presumably continuously exposed to selection, and the well-known ecological specificity of many bryophyte species that may involve strong stabilizing selection.[1,17,62,92] It is generally felt that bryophytes have occupied microhabitats that were more conservative over geologic time than the macroenvironments influencing tracheophytes. Tracking of relatively stable microhabitats is arguably enhanced by mechanisms thought to reduce genetic variability such as haploidy and asexual reproduction. Bryophytes appear to have high amounts of phenotypic plasticity,[63,66,100] and it is thought that this has provided an alternative to genetic adaptation in adjusting to environmental variations.

Testing such complex hypotheses is exceedingly difficult and will need to be attacked piecemeal. There clearly do appear to be some general differences between the ecology of bryophytes and tracheophytes. I also think the evident stasis of bryophyte species and higher taxa is a real phenomenon. However, I will argue that these phenomena can not be simply explained by a lack of genetic variation in bryophytes. An understanding of bryophyte reproductive biology will certainly be critical to finding a satisfactory explanation.

REPRODUCTION IN BRYOPHYTES

Sexual Reproduction

The bryophytes present a considerable diversity of breeding systems. Sets of standardized terms have recently been proposed by Anderson,[2] Wyatt and Anderson,[141] and Wyatt.[140] These recent sets of terms classifying sexual conditions to some degree differ from each other and from older sets of terms in the number of categories recognized and their hierarchical relationships. Another source of controversy is the appropriate ending for terms referring to sexuality. Wyatt[140] prefers "-oecious" endings, with analogy to terms referring to sexuality in seed plants. Zander[144] has expressed the opinion held by many American bryologists (and the present author) that "-oicous" should be used for terms referring to sexuality in gametophytes (as in bryophytes) and that "-oecious" should be reserved for terms referring to sexuality in sporophytes. Despite what may seem to be merely semantic arguments, these controversies over terminology are important since they reflect basic questions of homology. I will use the terminology of Anderson[2] in the discussion that follows, because I think his hierarchical grouping of terms in simpler and more logical than other recently proposed systems.

Gametophytes of a given bryophyte species are fundamentally either *dioicous, monoicous,* or *polyoicous.* In dioicous species, male and female inflorescences are produced on separate plants. ["Inflorescence" is another term commonly (but perhaps misappropriately) applied in bryology from analogy with higher plants, which simply refers to a cluster of gametangia borne at the end of a branch. This term is well entrenched in the literature; the present paper is not a suitable place to propose and defend an alternative.] Male and female plants usually are similar in size and morphology, but male plants may be smaller than females (but not attached to them, termed *heterodioicous* by Wyatt and Anderson[141]), or highly reduced and epiphytic either on the stems or rhizoids of female plants *(pseudautoicous)* or on the leaves of female plants *(phyllodioicous).*

In monoicous species, antheridia and archegonia are produced on the same plant: a *synoicous* condition is when the male and female gametangia occur mixed together in the same inflorescence; a *paroicous* condition is when male and female gametangia occur in the same inflorescence, but not mixed together (the antheridia occur peripherally); an *autoicous* condition is when male and female inflorescences occur separately on the plant; a *heteroicous* condition is when male and female gametangia occur in separate inflorescences and in mixed inflorescences on the same plant. The autoicous condition can be broken down into several interesting subcategories: *cladautoicous* if male and female inflorescences occur on separate branches, *gonioautoicous* if the male inflorescence is axillary on the same branch as the female inflorescence, and *rhizautoicous* if the sexes occur on separate stems that are joined by a rhizoid (hence belonging to the same genet sensu Harper[38]).

The polyoicous condition is when male and female gametangia occur sometimes on the same plant and sometimes on different plants. As pointed out by Wyatt,[140] a number of different situations are lumped together under this heading. In some species, for example *Tortula obtusissima,*[80] plants are either all female or gonioautoicous. This would be very different from a situation in which plants are either all male or autoicous, as pointed out by Wyatt,[140] who proposed a more refined terminology.

Sex determination in bryophytes has been reviewed by Ramsey and Berrie.[90] They discuss evidence pointing to genetic bases for sexual conditions in most bryophytes. Sex chromosomes are known from many bryophytes, although as pointed out by Smith,[106] the sex-determining status of such chromosomes has usually not been proven. There is a clear association between monoicy and polyploidy in bryophytes.[90,106] This may be simply due to the presence of both sex chromosomes in a single gametophyte.[90] Since polyploidy may serve to mask recessive deleterious alleles in the gametophyte,[62] an important implication of this association to the present discussion is that there may be a complicated interrelationship between the effects of breeding system and polyploidy on genetic diversity.

Genetic implications of different breeding systems in bryophytes have been widely discussed.[62,64,106,139] The fact that self-fertilization in gametophytes leads to a totally homozygous sporophyte is clearly significantly different from inbreeding in seed plants. Spores from a selfed gametophyte are effectively asexual propagules,[139] unless the gametophyte is diploid or polyploid and heterozygous, in which case some recombination could occur.[68] It has been argued that mechanisms to promote outcrossing should be especially favored in bryophytes.[2] For example, dioicy being one simple way to make outcrossing obligatory, it is interesting that slightly more than half of all moss and about two-thirds of all liverwort species are dioicous, as compared to about a 3–4% incidence of dioecy among angiosperms.[140]

Self-incompatibility would provide another mechanism for promoting outbreeding. Very little is known about breeding relationships within bryophyte species, but some cases of genetic incompatibility barriers within monoicous moss species are known.[5,16]

The evolutionary polarity of sexual conditions in bryophytes is clearly important to adaptive arguments meant to explain transformations in sexuality, but such polarities are rarely asserted with any rigor. It is usually assumed that dioicy is primitive in bryophytes, and prevailing evolutionary explanations are based on this assumption.[68] However, Wyatt[139] has presented arguments in favor of monoicy as the primitive condition, stating that "all other homosporous green plants produce bisexual gametophytes." His arguments are suggestive, but since some unisexual species are found in both the Charales and *Coleochaete,* the immediate outgroups of the embryophytes, the polarity decision for bryophytes as a whole remains equivocal. Polarities should be established on a case by case basis, because it is certain that evolution of breeding systems has proceeded differently in different lineages.

The potential advantages of dioicy in promoting outcrossing are likely to be balanced by reduction in the probability of sporophyte production. The bryophytes retain the need for free water to effect fertilization. Many observations have been made to demonstrate that sporophyte production is directly dependent on close proximity between male and female gametangia.[65,75,112,116] What little is known about gamete dispersal distances in bryophytes indicates that the distances are very short, of the order of a few centimeters.[2,138,139] A number of studies have been carried out, comparing frequency of sporophyte production in dioicous and autoicous species, in various geographical areas.[26,57,67] In all cases, dioicous species produce sporophytes significantly less often. It has been suggested that a trade-off is involved between the different advantages of dioicy and monoicy. Wyatt[139] suggested that monoicy may be especially advantageous in dry habitats where the ability to produce sporophytes is severely limited by lack of water, and indeed Stark[111] has demonstrated a significant correlation between monoicous species and desert habitats. According to data presented by Wyatt and Anderson,[141] no latitudinal trends are apparent in moss sexuality, but dioicous species of liverworts seem to be relatively more frequent in the tropics.

Population structure and gene flow in bryophytes is affected by spore dispersal in addition to gamete dispersal, but even less is known about the former than the latter. The durability of spores in relation to their potential ability to survive long-distance dispersal in the upper atmosphere has been studied by Van Zanten and co-workers.[123–125] There does seem to be a correlation between spore durability and the extent of a species' range in the southern hemisphere, and this perhaps can be taken as evidence for effective dispersal and establishment by spores. Studies of spore dispersal in nature in bryophytes are limited to those of McQueen[75] and some anecdotal data reported by Wyatt.[139] Based on these preliminary studies, even an estimate of neighborhood size for any species of bryophyte is premature; however, it is likely to be rather small.[75,139]

Until some understanding of recruitment in natural populations is gained, little can be said directly about the relative importance of asexual versus sexual reproduction or of various sexual breeding systems. For the present, all that can be done is to make inferences based on broad-scale comparisons and miscellaneous indirect observations.

Genetic Variation

Given the lack of direct evidence about the effectiveness of sexual reproduction in establishing new individuals in bryophytes, a number of attempts have been made to infer effective sexual reproduction, often based on the assumption that genetic variation and "evolutionary potential" are exclusively linked to sexual rather than asexual reproduction (a view that will be disputed below). Gemmell[26] presented an analysis of sexual reproduction and its correlates in the British moss flora as then known. As mentioned above, he showed that there is a marked reduction in "fruiting" (i.e., production of sporophytes) in dioicous species. He argued that selfing is the best explanation for more frequent fruiting in monoicous mosses, because if incompatability mechanisms were widespread, the frequency of fruiting should be similar to that for dioicous mosses. Gemmell took the number of taxonomically recognized varieties within a species and the species' frequency of occurrence to be indicative of evolutionarily significant variation. Using this measure, he pointed out that completely sterile species were significantly less variable and less widely distributed than either monoicous or dioicous species. Dioicous species, while not significantly different in variety production, were more widely distributed than monoicous species. Smith and Ramsay[107] updated the information presented by Gemmell, and showed that his conclusions still held even with further knowledge of the British flora. Longton and Schuster[68] pointed out cases in both mosses and liverworts in which totally asexual (non-fruiting) species tend to be less variable morphologically and more narrowly distributed than sexual species. Lefebvre[57] on the other hand provided evidence that, in the Plagiotheciaceae, monoicous species are more widely distributed and have a greater phenotypic variability than dioicous species. My own studies of the genus *Tortula* in Mexico,[80,81] indicate that while frequency of sporophyte production is strongly correlated with the monoicous condition, it is not at all correlated with the size of a species' geographic range, frequency of occurrence within the range, or morphological variability of the species.

It is of course problematic to equate the production of taxonomic variation or geographic frequency of occurrence with genetic variation. The genetic basis of characters used to distinguish bryophyte taxa has seldom been assessed experimentally.[80] In recent years, a number of electrophoretic studies have been made to attempt better estimates of genetic variation in natural populations of bryophytes.[18–20,52,142,143] In all cases, high levels of isozyme polymorphism were found within and between populations. There is a problem with interpreting these electrophoretic morphs as allelic variation, because genetic analysis of enzyme phenotypes of individuals has not been carried out due to the difficulty of making controlled crosses in bryophytes. Also, Taylor et al.[118] demonstrated significant environmental effects on numbers and mobility of enzyme bands detected. However, assuming that the enzyme polymorphisms detected in bryophyte populations represent genetic variation "fairly" (an assumption that may be warranted since the material tested is haploid[52]), levels of genetic variation are similar to those found in angiosperms and animals.

As yet, there are not enough electrophoretic data available to make general comparisons between genetic variation patterns of taxa with different breeding systems. It is of interest, however, that two of the aforementioned electrophoretic studies[19,52] dealt with bryophytes in which asexual reproduction is predominant (or, in the case of British *Sphagnum pulchrum,* apparently exclusive). In both studies, polymorphism within- and between-populations was as high as in species that presumably reproduce sexually much more frequently.

Asexual Reproduction

A number of very different processes occur to bypass part or all of the sexual cycle in reproduction. Harper[38] has cogently argued that these processes should be separated and considered individually. He rejects the use of the term "vegetative reproduction" and instead distinguishes between "growth" and "reproduction." In his usage reproduction refers to the production of a new individual from a single cell (a zygote or an unfertilized cell in an apomict), while growth refers to the development of an organized meristem. Thus a clone results from growth, not reproduction. The application of these ideas to bryophytes is clearly problematical. All growth of new branches results from a single cell (the apical cell), there are no structures closely analogous to apomictic seed of angiosperms, and there are a number of structures present (such as gemmae and elaborately ramified gametophytes) not found in angiosperms. I find it necessary to use a criterion of physiological independence to define reproduction.[14] The production of a new, physiologically independent plant is reproduction; reproduction is sexual if the new plant develops from a spore that itself resulted from cross-fertilization and meiosis, asexual if the new plant develops from a mitotically produced cell without cross-fertilization. I consider "growth" simply to be the increase in size of a single physiological individual. It must be kept in mind that a number of very different phenomena are lumped under "asexual reproduction" here, including branching of gametophytes (with later separation of the branches), apogamous spores, aposporous gametophytes, gametophyte fragmentation, and various specialized asexual propagules. These phenomena may well differ in their dispersability, effectiveness, and amount of resources required from the "parent" plant.

It is widely felt that a regression in functional sexuality has occurred during evolution of many bryophyte lineages.[1,68] However, this may represent an overgeneralization. Tiffney and Niklas[121] have presented paleobotanical evidence that a number of different types of clonal growth were common in early land plants, and argued that general evolutionary trends from clonal to nonclonal growth have occurred in some groups, such as in seed plants. Given this, and the fact that vegetative reproduction seems widespread in the green algal outgroups to the embryophytes, it appears that the occurrence of many types of asexual reproduction is plesiomorphic for the embryophytes. However, it appears that further lowering of the relative frequency of sexual reproduction, and origin of new modes of asexual reproduction, has occurred in a number of bryophytes. Here again, the evolutionary polarity of reproductive characters is critical to their explanation, and should be established rigorously on a case by case basis.

Based on admittedly anecdotal and incomplete evidence, most bryologists feel that asexual reproduction provides the bulk of new shoot establishment in natural bryophyte populations.[1,17,68] Indeed, Anderson[1] went so far as to state that "reproduction in perennial mosses (and perhaps in many annuals) is almost entirely other than by spores." There are a number of reasons for this conclusion: rarity of sporophyte production through large portions of many species' ranges, the wide variety of specialized asexual propagules found in bryophytes, and observations of the frequency of gametophyte regeneration versus spore germination.

Based on the geographical distribution of sporophyte production, a considerable or nearly exclusive reliance on asexual reproduction can often be inferred. It appears that only a rather small number of bryophyte species completely lack sporophytes (hence lack any opportunity for sexual reproduction). Longton and Miles[67] estimate that only 4% of the British moss flora totally lacks sporophytes; Lane[54] estimates that 9% of the eastern North American moss flora lacks sporophytes. Nonetheless, a substantial proportion of bryophyte species produce sporophytes rarely, or sometimes produce them only in a limited part of the species' range. For example, 43% of the British moss flora produces sporophytes rarely or never.[67] As discussed above, rarity of sporophyte production is strongly correlated with dioicy (a condition present in the majority of bryophytes). For example, based on the British data presented by Longton and Miles,[67] a significantly higher proportion of dioicous species produce sporophytes rarely or never than do monoicous species ($\chi^2 = 196.9$, $p < 10^{-6}$). Many examples exist of widespread and abundant species that must be interpreted as establishing new colonies via asexual propagules because of rarity of sporophyte production.[68] For example, the two most abundant species of *Tortula* in Mexico (which are also two of the most abundant bryophytes in this area) are *Tortula fragilis* and *Tortula amphidiacea.* Both species produce sporophytes only within a small part of their total range, yet are just as common where they do not produce sporophytes.[80] Another example is *Syrrhopodon texanus,* a North American endemic that produces sporophytes in only about 5% of the populations observed, and these only in a restricted geographical area.[91] Therefore rarity of sexual reproduction does not necessarily lead to a restricted range.

It is worth reiterating that production of sporophytes does not automatically imply effective mixis. The much higher frequency of sporophyte production in monoicous bryophytes certainly indicates a high level of selfing (but not complete selfing, of course, Zielinski[145] has demonstrated cross-fertilization in a monoicous liverwort). As discussed above, spores from a selfed sporophyte are essentially asexual propagules. In this context, it is interesting that in Mexican *Tortula,* specialized asexual propagules are only found in dioicous species (and, as noted above, these species produce sporophytes much less frequently than the monoicous species).[80] Longton and Schuster[68] similarly report a highly significant association of specialized asexual propagules with dioicous rather than monoicous species in the British moss flora. These correlations may well be explainable by the idea that spores of monoicous species serve the same ecological role as asexual propagules produced by gametophytes. Furthermore, sporophyte production by apogamy is known in bryophytes,[1,106] and has been observed to occur spontaneously in culture. The significance of apogamy in nature is unknown, but if frequent in a species, it would give an exaggerated impression of the frequency of sexual reproduction.

Further evidence for the importance of asexual reproduction is the variety of types of specialized propagules (interestingly, always borne on the gametophytes), that have evolved repeatedly in different bryophyte lineages. The tissues of bryophytes are totipotent; it has been demonstrated experimentally that virtually any part of the plants can regenerate new plants, either directly or via an initial protonema.[28] Besides this general capacity to regenerate, which makes any bryophyte a mass of potential asexual propagules, organs have evolved that appear to be specialized for especially efficient propagation. This has not in most cases been experimentally established, but is justifiably assumed because such organs (while very diverse in their structure and homologies) are smaller than vegetative organs, readily deciduous, and are frequently observed regenerating in nature.[68] The general importance of such organs is evident in that specialized asexual propagules occur in 17% of the British moss flora,[68] 12% of the North American flora,[54] and a somewhat larger proportion in several liverwort floras.[68]

As pointed out by Giles, regeneration is slower and more difficult in highly differentiated parts of the plant. I have suggested that neoteny (i.e., prolongation of juvenile stages into maturity in a descendant, relative to an ancestor) has been the predominant mechanism for producing asexual propagules in the genus *Tortula*.[82] To generalize this conclusion, many (or even all) of the types of asexual propagules outlined below result from an evolutionary process of neoteny, in the sense that organs are held in a juvenile state until suitable conditions for regeneration arise. Trade-offs within species between asexual reproduction in juvenile plants and sexual reproduction in mature plants have been widely reported.[24,98] It may be the case that such trade-offs are due to resource allocation by the plants (but there is no evidence for this as yet). Another interpretation is simply that the production of either gametangia or asexual propagules depends on the developmental stage of a particular shoot or branch.

Comprehensive treatments of asexual propagules in mosses and hepatics include those of Correns,[15] Buch,[11] Watson,[129] and Schuster.[100] Unfortunately, terminology has not been standardized, which can lead to confusion in interpreting literature. Such a standardization, while sorely needed, will not be attempted here because determination of homologies between different types of asexual propagules and normal vegetative organs is needed first.

Specialized asexual propagules in mosses include fragile or deciduous leaves, reduced deciduous branches, undifferentiated spherical or cylindrical bodies ("gemmae") borne on leaves or stem, and irregular masses of cells borns on rhizoids ("tubers"). Leafy liverworts may have deciduous or fragmenting leaves, reduced deciduous branches, or gemmae. Deciduous vegetative organs are found in some thalloid species of liverworts and in the hornworts. My experience has been that virtually never is more than one type of asexual propagule produced in a given species.

Additional support for the predominance of asexual reproduction in natural populations is given by comparative observations or regeneration from gametophyte fragments, in comparison to spore gemination. Most bryophytes will readily establish new gametophores from either gametophyte fragments or spores on agar or soil in a growth chamber or greenhouse. However, except for some miscellaneous observations of annual mosses,[68] no one has carefully documented establishment from spores in the field, despite some recent valiant attempts.[67] On the other hand, establishment from some form of asexual reproduction is frequently observed in nature.[7,66–68] Regenerating gametophyte fragments have been found in Pleistocene fossil deposits,[78] and are abundantly present in snow banks in the Arctic.[79] My own unpublished observations of *Tortula* in culture indicate that production of protonemata and new gametophores occurs much faster from regenerating gametophytic tissue than from spores. It might be expected that, especially in bryophytes of arid habitats, there may be a premium on rapid establishment during the short periods favorable for growth. If so, establishment from spores would be at a strong disadvantage.

One final observation has a potential bearing on the apparent regression in functional sexuality in many bryophytes. Duckett et al.,[23] who have been studying bryophyte sperm ultrastructure for purposes other than the study of reproductive ecology, report that their studies are hampered because apparently healthy antheridia are often necrotic. This appears to be a widespread phenomenon, especially in epiphytic bryophytes.[23] If confirmed by further observation, this many mean that sex organs are often produced, but degenerate before they are functional.

DISCUSSION

Sexual versus Asexual Reproduction

The question of the evolutionary significance of sex is a major theme in evolutionary theory. General literature on the subject has been reviewed in depth on a number of occasions.[8,13,37,108,122,137] I wish here to focus on just one aspect of this subject, one that is perhaps especially interesting from a systematic point of view. This is the question of the expected (or realized) effect of mode of reproduction on the "evolutionary potential" of a clade. Clearly, the matter of satisfactorily defining "evolutionary potential" is itself worth a good deal of discussion. But for my present purposes, I will take it to mean the ability of a clade to generate genotypic or phenotypic variation of evolutionary significance.

The supposed twofold advantage of asexual reproduction over sexual reproduction proposed by Maynard Smith[108] has served as the focus for a good deal of theoretical and empirical research addressing the question: Why does sex exist? As pointed out by Bell,[8] the real question, given the early origin of sex (perhaps a synapomorphy for all eukaryotes), is Why has sexuality not been lost? The importance of separating questions of the origin of sex from the maintenance of sex is underscored by a number of contributors to the volume edited by Halvorson and Monroy.[37] Whatever advantage sexual reproduction now confers upon organisms by generating genetic diversity may in fact be an exaptation; it has variously been suggested that the original adaptive function of sex was (1) the establishment of a haploid–diploid life cycle,[29] (2) as part of a DNA repair mechanism,[9] or (3) as the key to beginning an orderly morphogenetic process in complex multicellular organisms.[71] Bernstein et al.[9] and Margulis et al.[71] point out that the persistence of sex in most lineages may be due to the extreme difficulty of bypassing the original genetic or developmental functions of sex. Developmental problems causing lower fertility of parthenogenetic versus related bisexual insects are well known,[53,119] which suggests that the theoretical twofold advantage of asexuality is at least sometimes not realized. Thus, sex may be retained in some lineages because of the action of developmental contraints rather than current optimality in a selective sense.[72] Bell[8] discussed and justified the use of the customary equilibrium approach to the study of evolution of sex, but admitted that one is forced into invoking a strong historical component in the evolution of most aspects of sexuality.

In addition to the need to consider the phylogenetic history of organisms, there is a need to be concerned over proper units of comparison. This is another area where systematics can provide an essential viewpoint. For example, Hull[42] has argued that comparisons between sexual and asexual "species" must be made very carefully. In some senses a sexual species is comparable to one *clone* of an asexual "species." As stated by Hull[42]: "If like is compared to like, asexual lineages should be compared to sexual lineages, and in such a comparison, sexual reproduction becomes as rare as it should be" (p. 330). If Hull is right, the supposed ubiquity of sexual reproduction is really an illusion. On the other hand, one might argue that an asexual lineage is equivalent to a single genet in a sexual species. If this comparison is correct, then sexual reproduction is virtually universal! This issue is not a trivial play on words; we need to think as carefully about the basis for units of comparison as we do about the comparison itself. The important questions of what counts as an individual and what is the unit of selection are at their most difficult when dealing with clonal organisms.[43] The notorious species problem in plants is relevant here; many proposals have been made and consensus seems far away.[81,85]

What is the expected link between mode of reproduction and evolutionary potential? As discussed above, many bryologists view asexual reproduction as a direct cause of stenotypy, lack of an ability to evolve, and a dead end for a taxon in the long run. This view is shared by some angiosperm evolutionary ecologists.[21,113] However, a good deal of recent theoretical work suggests otherwise. Thompson,[120] in a detailed discussion of the subject, showed that sex would not cause any acceleration of evolution (under normal conditions in nature); in fact, sex may act most often as a damper to slow or moderate evolution. Marshall and Weir[73] showed that facultative apomixis has no effect on the maintenance of genetic variability in populations (although it does affect the rate at which equilibrium is reached). Lynch and Gabriel,[70] using computer simulations, showed that parthenogenesis with occasional sex actually was at a selective advantage to continuous bisexual reproduction, because of periodic release of hidden genetic variance. They made the intriguing suggestion that this could serve as a mechanism for quantum evolution. If geographically marginal populations of a species tend to show greater reliance on asexual reproduction (perhaps because of ecological conditions making completion of the sexual cycle difficult, as has been suggested for bryophytes by Longton and Schuster[68]), then such a mechanism may figure in relatively rapid allopatric speciation—a phenomenon of macroevolutionary interest.

It has been repeatedly shown that asexual species of animals show patterns of within- and between-population variability that are similar to those in sexual species.[6,87,95] Weimarck[133] found similar high amounts of variability in asexual species of angiosperms. Electrophoretic studies of facultatively asexual *Lycopodium*[59] and obligately asexual *Taraxacum*[86] have likewise shown the presence of considerable levels of variation. As mentioned in an earlier section, substantial isozyme variation was demonstrated in predominantly or exclusively asexual bryophytes.[19,52]

It appears likely that the question posed above has no universal answer. Knowing that a species relies strongly on asexual reproduction does not in and of itself allow one to predict its evolutionary potential. Such a prediction requires knowledge of other facets of the species' biology and habitat.

It is therefore important to consider geographical or ecological factors that favor sexual reproduction or asexual reproduction. This approach has been taken by a number of workers,[30,36,126] who agree that asexual reproduction is favored in transient, open, extreme, or unpredictable (in space, but not in time) environments where colonizing ability and rapid population growth are important (i.e., r-selecting environments) while sexual reproduction is favored in K-selecting environments (e.g., the tropics). In particular, there are theoretical and empirical reasons for linking sexual reproduction to environments in which biotic interactions are of major selective importance.[30,44,45,56,58,61] This is so in part because strong frequency- or density-dependent pressure by pathogens, predators, and competitors would seem to place a premium on an organism being different from nearby organisms of the same species.

Bryophytes in general appear to support the above distinctions. Few clear geographical trends in sexual versus asexual reproduction have been recognized in bryophytes.[68] However, it may well be the case that the habitats occupied by bryophytes throughout the world are such that biotic interactions are minimized compared to interactions with the physical environment. Preliminary studies have shown that compared to other groups of plants and animals, bryophytes have few pathogens or predators.[27,93] The extent to which bryophyte species competitively interact is as yet inadequately studied. Species in a community often have recognizable, if overlapping niches as defined by physical parameters such as pH, light, substrate, and water relationships.[104,127,132] However, competitive exclusion has not been satisfactorily demonstrated in any study.[103] On the contrary, Watson[130] found that plants growing in high densities (whether single species *or* in mixtures) had increased longevity. This unexpected result is no doubt dependent on the peculiar water relationships of many bryophytes that rely on growth in clumps or mats for external storage and conduction of the water film. Watson[131] suggested that competition may be less important in bryophytes because their growth rates are slow relative to the length of time that appropriate habitats are available, and because their dispersal abilities are poor enough that appropriate habitats are not saturated.

The evident regression of sexuality in many bryophytes thus may well be causally connected to occupation of *r*-selecting habitats where there is a minimal biotic environmental influence. Both Slack[103] and Watson[131] have suggested that many bryophytes are "fugitive" or "opportunistic" species that occur in relatively temporary habitats. During[24] provides a classification of life history traits in bryophytes, with many examples; of his six categories of "life strategies," five (fugitives, colonists, and three categories of shuttle species) have characteristics attributed to *r*-selected organisms. As discussed by Templeton,[119] it is difficult to determine the causal connection between *r*-selected phenotypes and asexual reproduction. In the case of bryophytes, any such correlation may be due to either the fact that primarily asexual species are at a selective advantage in *r*-selecting habitats, or simply to the fact that, at low population densities in colonizing situations, fertilization is less likely. Future studies should therefore focus on careful comparisons between the reproductive ecology of related species (or populations of one species) occupying different environments. Indeed, the bryophytes may provide ideal subjects for answering general questions about the function of sex in different physical and biotic environments because of their great diversity of reproductive modes and habitat specificities.

Somatic Mutations and Their Evolutionary Significance

The lack of any correlation between gross overall measures of genetic variation in species and their evolutionary "success" (i.e., abundance, variation, ecological dominance, and/or production of new species) has been widely recognized (and lamented). Phylogenetic "relicts" and members of rapidly radiating, diverse lineages have about the same amounts of electrophoretically detectable variation.[59] As discussed above, there seems to be no particular relationship between mode of reproduction and evolutionary success of a lineage. There are further reasons to think that even if raw genetic variation is evolutionarily important in some circumstances, asexual reproduction is not necessarily a drawback. This is because of the possibility of generation of variation through somatic mutation, a potential that is enhanced by apical cell growth as is present in the bryophytes.

As discussed by Buss,[12] the modern synthesis was predicated on Weismann's doctrine, a strict separation between the germ line and the soma. However, this doctrine can not be applied to many animals or to any protists, fungi, or plants![12] In these organisms, somatic cell lineages become reproductive. With the much greater number of mitotic events in vegetative growth than in a sequestered germ line (such as in the development of mammalian ova), even a small mutation rate can lead to many mutations, especially in long-lived or clonal organisms.

The special significance of somatic mutations to organisms that grow via an apical cell has been documented in an important series of papers on ferns by Klekowski.[46–49] Mutations in the apical cell will be present in all further growth of a branch, including gametangia. In contrast, mutations in a multicellular meristem will form a chimera that may not include gametangia. Klekowski[47] suggested that ferns (with a dominant sporophyte) should be much better than bryophytes at accumulating mutations since the latter are haploid. However, he has focused on lethal mutations; there is no reason to expect that all or even most mutations would act as cell lethals in a haploid organism. Presumably, many sporophytically expressed genes are inactive in the gametophyte generation, and it has been suggested that widespread polyploidy has led to duplication of genes in many haploid bryophytes.[62] Thus it would seem that many nonlethal mutations are to be expected in the bryophytes.

Since mutation rate is clearly a parameter that may be a target for selection,[69] it would be very interesting in the future to learn about mutation rates in bryophytes and determine whether there are differences between lineages with different reproductive biologies. In the modeling studies of Lynch and Gabriel,[70] if mutation rates were elevated and environmental sensitivity depressed, phenotypic evolution in obligate parthenogens was shown in simulations to exceed that of sexual reproducers.

White[134] reviewed important growth differences between animals and plants and developed the idea that many individual plants can best be regarded as a "metapopulation" of smaller, semi-independent parts such as leaves, branches, and flowers. This view is clearly appropriate for bryophytes, given their modular growth patterns. Considerable attention has recently been given to "metapopulation" genetic phenomena.[4,48,102,105,135] Competition between genetically different parts of the plant (or of a clone) results in somatic selection that could become heritable if the modified part becomes reproductive. Perhaps especially in plants, therefore, recent work suggests a number of alternatives to the old standby, sex, for generating genetic diversity.

SUMMARY AND CONCLUSIONS

Much empirical work is required before it will be possible to present a complete picture of bryophyte reproductive biology. For the present, I have discussed some theoretical issues that will need to be kept in mind as empirical work proceeds, and attempted to convey an outline of what the final picture may look like.

It is my opinion that future progress depends on avoiding certain ideological pitfalls. The "equilibrium" view of evolution, unless demonstrated to be applicable in a particular situation, should not be assumed because it leads to unjustified use of optimality criteria and a naive "adaptationist" mind-set. We must beware of unwarranted extrapolations from animal biology to plants, and from angiosperm biology to bryophytes, since these can lead to neglect of unique or special features. I have made some extrapolations in this chapter, due to lack of information from bryophytes in certain subjects, but these (as with all such extrapolations) should be taken with a grain of salt. Finally, systematics and evolutionary theory should be seen to be inseparable parts of an explanatory whole. A feature of a group of organisms (say reproductive biology) cannot be studied without regard to its historical context, a context which is provided by systematic analysis of all features of the group. In particular, a useful decision about the polarity of an evolutionary transformation series depends on the existence of a well-supported cladistic hypothesis.

Current knowledge of bryophyte reproductive ecology suggests that significant differences exist between it and the prevailing view of angiosperm reproductive ecology. In an evolutionary sense, the bryophytes appear to have responded less to direct competitive interactions with other species and more to the physical environment. Indeed, due to peculiarities of their water relationships (dependent on external reservoirs and conduction), bryophytes seem often to exhibit positive responses to crowding. In general, bryophytes may be unable to occupy many suitable microsites because of poor vagility; their reproductive abilities are far from "optimal" in any sense. Asexual reproduction of various types seems often to be more important than sexual reproduction in establishment in nature. Sex appears to be vestigial in many groups of bryophytes. Nonetheless, considerable genetic variation is present, perhaps due to facultative sexual reproduction or to somatic mutation. It has been argued that the latter may be especially important in a group, such as the bryophytes, that develops via an apical cell.

REFERENCES

1. Anderson, L. E., Modern species concepts: Mosses, *Bryologist* **66,** 107–119 (1963).
2. Anderson, L. E., Cytology and reproductive biology of mosses, in *The Mosses of North America* (R. J. Taylor, and A. E. Leviton, eds.), pp. 37–76. Pacific Division, American Association for the Advancement of Science, San Francisco, 1980.
3. Anderson, L. E., and Snider, J. A., Cytological and genetic barriers in mosses, *J. Hattori Bot. Lab.* **52,** 241–254 (1982).
4. Antolin, M. F., and Strobeck, C., The population genetics of somatic mutation in plants, *Am. Nat.* **126,** 52–62 (1985).
5. Ashton, N. W., and Cove, D. J., The isolation and preliminary characterization of auxotrophic and analogue resistant mutants of the moss, *Physcomitrella patens, Mol. Gen. Genet.* **154,** 87–95 (1977).
6. Atchley, W. R., Biological variability in the parthenogenetic grasshopper *Warramaba virgo* (Key) and its sexual relatives. I. The eastern Australian populations, *Evolution* **31,** 782–799 (1977).
7. Bedford, T. H. B., Sex distribution in colonies of *Climacium dendroides* W. & M. and its relation to fruit bearing, *North Western Naturalist* **13,** 213–221 (1938).
8. Bell, G., *The Masterpiece of Nature: The Evolution and Genetics of Sexuality.* Univ. of California Press, Berkeley, 1982.
9. Bernstein, H., Byers, G. S., and Michod, R. E., Evolution of sexual reproduction: Importance of DNA repair, complementation, and variation, *Am. Nat.* **117,** 537–549 (1981).
10. Bower, F. O., *The Origin of a Land Flora.* Macmillan & Co., London, 1908.
11. Buch, H., *Über die Brutorgane der Lebermoose.* Helsingfors, 1911.
12. Buss, L. W., Evolution, development and the units of selection, *Proc. Natl. Acad. Sci. U.S.A.* **80,** 1387–1391 (1983).
13. Case, T. J., and Taper, M. L., On the coexistence and coevolution of asexual and sexual competitors, *Evolution* **40,** 366–387 (1986).
14. Cook. R. E., Growth and development in clonal plant populations, in *Population Biology and Evolution of Clonal Organisms* (J. B. C. Jackson, L. W. Buss, and R. E. Cook, eds.), Chapter 8. Yale Univ. Press, New Haven, Connecticut, 1985.

15. Correns, C., *Untersuchugen über die Vermehrung der Laubmoose durch Brutorgane und Stecklinge.* Gustav Fischer, Jena, 1899.
16. Courtice, G. R. M., Ashton, N. W., and Cove, D. J., Evidence for the restricted passage of metabolites into the sporophyte of the moss *Physcomitrella patens* (Hedw.) Br. Eur., *J. Bryol.* **10,** 191–198 (1978).
17. Crum, H., The geographic origins of the mosses of North America's eastern deciduous forest, *J. Hattori Bot. Lab.* **35,** 269–298 (1972).
18. Cummins, H., and Wyatt, R., Genetic variability in natural populations of the moss *Atrichum angustatum, Bryologist* **84,** 30–38 (1981).
19. Daniels, R. E., Isozyme variation in British populations of *Sphagnum pulchrum* (Braithw.) Warnst, *J. Bryol.* **12,** 65–76 (1982).
20. Daniels, R. E., Isozyme variation in populations of *Sphagnum recurvum* var. *mucronatum* from Britain and Finland, *J. Bryol.* **13,** 563–570 (1985).
21. De Wet, J. M. J., and Stalker, H. T., Gametophytic apomixis and evolution in plants, *Taxon* **23,** 689–697 (1974).
22. Dobson, F. S., The use of phylogeny in behavior and ecology, *Evolution* **39,** 1384–1388 (1985).
23. Duckett, J. D., Carothers, Z. B., and Miller, C. C. J., Comparative spermatology and bryophyte phylogeny, *J. Hattori Bot. Lab.* **53,** 107–125 (1982).
24. During, H. J., Life strategies of bryophytes: A preliminary review, *Lindbergia* **5,** 2–18 (1979).
25. Felsenstein, J., Phylogenies and the comparative method, *Am. Nat.* **125,** 1–15 (1985).
26. Gemmell, A. R., Studies in the bryophyta I. The influence of sexual mechanism on varietal production and distribution of British Musci, *New Phytol.* **49,** 64–71 (1950).
27. Gerson, V., Bryophytes and invertebrates, in *Bryophyte Ecology* (A. J. E. Smith, ed.), Chapter 9. Chapman and Hall, London, 1982.
28. Giles, K. L., Dedifferentiation and regeneration in bryophytes: A selective review, *N.Z. J. Bot.* **9,** 689–694 (1971).
29. Glaser, V. H.-J., Zur Bedeutung der sexuellen Fortpfanzung in der Evolution, *Biol. Zentralbl.* **104,** 385–402 (1985).
30. Glesener, R. R., and Tilman, D. Sexuality and the components of environmental uncertainty: Clues from geographic parthenogenesis in terrestrial animals, *Am. Nat.* **112,** 659–673 (1978).
31. Gould, S. J., The evolutionary biology of constraint, *Daedalus, (Boston)* **109,** 39–52 (1980).
32. Gould, S. J., and Lewontin, R. C., The spandrels of San Marco and the Panglossian paradigm: a critique of the adaptationist programme, *Proc. R. Soc. London B* **205,** 581–598 (1979).
33. Gould, S. J., and Vrba, S. E., Exaptation—a missing term in the science of form, *Paleobiology* **8,** 4–15 (1982).
34. Graham, L. E., *Coleochaete* and the origin of land plants, *Am. J. Bot.* **71,** 603–608 (1984).
35. Graham, L. E., The origin of the life cycle of land plants, *Am. Sci.* **73,** 178–186 (1985).
36. Grassle, J. F., and Shick, J. M., Introduction to the symposium: Ecology of asexual reproduction in animals, *Am. Zool.* **19,** 667–668 (1979).
37. Halvorson, H. O., and Monroy, A., eds., *The Origin and Evolution of Sex.* Alan R. Liss, New York, 1985.
38. Harper, J. L., *Population Biology of Plants.* Academic Press, London, 1977.
39. Hébant, C., *The Conducting Tissues of Bryophytes.* J. Cramer, Vaduz, 1977.
40. Herzog, T., *Geographie der Moose.* Gustav Fischer, Jena, 1926.
41. Horn, H. S., Adaptation from the perspective of optimality, in *Topics in Plant Population Biology* (O. T. Solbrig, S. Jain, G. B. Johnson, and P. H. Raven, eds.), Chapter 2. Columbia Univ. Press, New York, 1979.
42. Hull, D. L., Individuality and selection, *Annu. Rev. Ecol. Syst.* **11,** 311–332 (1980).
43. Jackson, J. B. C., Buss, L. W., and Cook, R. E., eds., *Population Biology and Evolution of Clonal Organisms.* Yale Univ. Press, New Haven, 1985.
44. Jaenike, J., An hypothesis to account for the maintenance of sex within populations, *Evol. Theor.* **3,** 191–194 (1978).
45. Jaenike, J., and Selander, R. K., Evolution and ecology of parthenogenesis in earthworms, *Am. Zool.* **19,** 729–737 (1979).
46. Klekowski, E. J., Jr., Mutational load in a fern population growing in a polluted environment, *Am. J. Bot.* **63,** 1024–1030 (1976).

47. Klekowski, E. J., Jr., The genetics and reproductive biology of ferns, in *The Experimental Biology of Ferns* (A. F. Dyer, ed.), Academic Press, pp. 133–169. London, 1979.
48. Klekowski, E. J., Jr., Genetic load and soft selection in ferns, *Heredity* **49,** 191–197 (1982).
49. Klekowski, E. J., Jr., Mutational load in clonal plants: A study of two fern species, *Evolution* **38,** 417–426 (1984).
50. Kramer, P., Editorial, *BioScience* **34,** 405 (1984).
51. Krassilov, V. A., and Schuster, R. M., Paleozoic and mesozoic fossils, in *New Manual of Bryology* (R. M. Schuster, ed.), Vol. 2, Chapter 19. Hattori Botanical Laboratory, Nichinan, Japan, 1984.
52. Krzakowa, M., and Szweykowski, J., Isozyme polymorphism in natural populations of a liverwort, *Plagiochila asplenioides, Genetics* **93,** 711–719 (1979).
53. Lamb, R. Y., and Wiley, R. B., Are parthenogenetic and related bisexual insects equal in fertility? *Evolution* **33,** 774–775 (1979).
54. Lane, D. M., A quantitative study of *The Mosses of Eastern North America, Monogr. Syst. Bot. Missouri Bot. Gard.* **11,** 45–50 (1985).
55. Lauder, G. V., Form and function: Structural analysis in evolutionary morphology, *Paleobiology* **7,** 430–442 (1981).
56. Law, R., and Lewis, D. H., Biotic environments and the maintenance of sex—some evidence from mutualistic symbioses, *Biol. J. Linn. Soc.* **20,** 249–276 (1983).
57. Lefebvre, J., Fertilité et souplesse adaptative chez les Plagiotheciaceae de Belgique, *Rev. Bryol. Lichenol.* **36,** 162–166 (1969).
58. Levin, D. A., Pest pressure and recombination systems in plants, *Am. Nat.* **109,** 437–451 (1975).
59. Levin, D. A., and Crepet, W. L., Genetic variation in *Lycopodium lucidulum:* A phylogenetic relic, *Evolution* **27,** 622–632 (1973).
60. Lewontin, R. C., The bases of conflict in biological explanation, *J. Hist. Biol.* **2,** 35–45 (1969).
61. Lloyd, D. G., Benefits and handicaps of sexual reproduction, *Evol. Biol.* **13,** 69–111 (1980).
62. Longton, R. E., Reproductive biology and evolutionary potential in bryophytes, *J. Hattori Bot. Lab.* **41,** 205–223 (1976).
63. Longton, R. E., Climatic adaptation of bryophytes in relation to systematics, in *Bryophyte Systematics* (G. C. S. Clarke, and J. G. Duckett, eds.) pp. 511–531. Academic Press, London, 1979.
64. Longton, R. E., Reproductive biology and variation patterns in relation to bryophyte taxonomy, in *Bryophyte Taxonomy* (P. Geissler and S. W. Greene, eds.) *Beih. Nova Hedwigia* **71,** 31–37 (1982).
65. Longton, R. E., and Greene, S. W., The growth and reproductive cycle of *Pleurozium schreberi* (Brid.) Mitt., *Ann. Bot. (London)* **33,** 83–105 (1969).
66. Longton, R. E., and MacIver, M. A., Climatic relationships in antarctic and Northern Hemisphere populations of a cosmopolitan moss, *Bryum argenteum Hedw.,* in *Adaptations within Antarctic Ecosystems,* (G. A. Llano, ed.), pp. 899–919. Gulf Publ. Co., Houston, 1977.
67. Longton, R. E., and Miles, C. J., Studies on the reproductive biology of mosses, *J. Hattori Bot. Lab.* **52,** 219–240 (1982).
68. Longton, R. E., and Schuster, R. M., Reproductive biology, in *New Manual of Bryology* (R. M. Schuster, ed.), Vol. 1, Chapter 9. Hattori Botanical Laboratory, Nichinan, Japan, 1983.
69. Lynch, M., Spontaneous mutations for life-history characters in an obligate parthenogen, *Evolution* **39,** 804–818 (1985).
70. Lynch, M., and Gabriel, W., Phenotypic evolution and parthenogenesis, *Am. Nat.* **122,** 745–764 (1983).
71. Margulis, L., Sagan, D., and Olendzenski, L., What is sex? in *The Origin and Evolution of Sex* (H. O. Halvorson and A. Monroy, eds.), pp. 69–85. Alan R. Liss, New York, 1985.
72. Marshall, D. R. and Brown, A. H. D., The evolution of apomixis, *Heredity* **47,** 1–15 (1981).
73. Marshall, D. R., and Weir, B. S., Maintenance of genetic variation in apomictic plant populations. I. Single locus models, *Heredity* **42,** 159–172 (1979).
74. Mattox, K. R., and Stewart, K. D., Classification of the green algae: A concept based on comparative cytology, in *Systematics of the Green Algae* (D. E. G. Irvine, and D. M. John, eds.) pp. 29–72. Academic Press, London, 1984.
75. McQueen, C. B., Spatial pattern and gene flow distances in *Sphagnum subtile, Bryologist* **88,** 333–336 (1985).
76. Miller, N. G., Fossil mosses of North America and their significance, in *The Mosses of North America* (R. J. Taylor, and A. E. Leviton, eds.), pp. 9–36. Pacific Division, American Association for the Advancement of Science, San Francisco, 1980.

77. Miller, N. G., Tertiary and quaternary fossils, *New Manual of Bryology* (R. M. Schuster, ed.), Vol 2, Chapter 20. Hattori Botanical Laboratory, Nichinan, Japan, 1984.
78. Miller, N. G., Fossil evidence of the dispersal and establishment of mosses as gametophyte fragments, *Monogr. Syst. Bot. Missouri Bot. Gard.* **11,** 71–78 (1985).
79. Miller, N. G., and Ambrose, L. J. H., Growth in culture of wind-blown bryophyte gametophyte fragments from Arctic Canada, *Bryologist* **79,** 55–63 (1976).
80. Mishler, B. D., Systematic studies in the genus *Tortula* Hedw. (Musci:Pottiaceae). Ph.D. thesis, Harvard University, 1984.
81. Mishler, B. D., The morphological, developmental, and phylogenetic basis of species concepts in bryophytes, *Bryologist* **88,** 207–214 (1985).
82. Mishler, B. D., Ontogeny and phylogeny in *Tortula* (Musci:Pottiaceae), *Syst. Bot.* **11,** 189–208 (1986).
83. Mishler, B. D., and Churchill, S. P., A cladistic approach to the phylogeny of the "bryophytes" *Brittonia* **36,** 406–424 (1984).
84. Mishler, B. D., and Churchill, S. P., Transition to a land flora: Phylogenetic relationships of the green algae and bryophytes, *Cladistics* **1,** 305–328 (1985).
85. Mishler, B. D., and Donoghue, M. J., Species concepts: A case for pluralism, *Syst. Zool.* **31,** 491–503 (1982).
86. Mogie, M., Morphological, developmental, and electrophoretic variation within and between obligately apomictic *Taraxacum* species, *Biol. J. Linn. Soc.* **24,** 207–216 (1985).
87. Ochman, H., Stille, B., Niklasson, M., Selander, R. K., and Templeton, A. R., Evolution of clonal diversity of the parthenogenetic fly *Lonchoptera dubia, Evolution* **34,** 539–547 (1980).
88. Oster, G., and Alberch, P., Evolution and bifurcation of developmental programs, *Evolution* **36,** 444–459 (1982).
89. Rachootin, S. P., and Thomson, K. S., Epigenetics, paleontology, and evolution, in *Evolution Today* (G. G. E. Scudder, and J. L. Reveal, eds.), pp. 181–193. Carnegie-Mellon University, Pittsburgh, 1981.
90. Ramsay, H. P., and Berrie, G. K., Sex determination in bryophytes, *J. Hattori Bot. Lab.* **52,** 255–274 (1982).
91. Reese, W. D., Reproductivity, fertility, and range of *Syrrhopodon texanus* Sull. (Musci: Calymperaceae), a North American endemic, *Bryologist* **87,** 217–222 (1984).
92. Richards, P. W., The taxonomy of bryophytes, in *Essays in Plant Taxonomy* (H. S. Street, ed.), pp. 177–210. Academic Press, London, 1978.
93. Richardson, D. H. S., *The Biology of Mosses.* John Wiley & Sons, New York, 1981.
94. Rohrer, J. R., Sporophyte production and sexuality of mosses in two Northern Michigan habitats, *Bryologist* **85,** 394–400 (1982).
95. Saura, A., Lokki, J., Lankinen, P., and Suomalainen, E., Genetic polymorphism and evolution in parthenogentic animals, *Hereditas* **82,** 79–100 (1976).
96. Scheirer, D. C., Differentiation of bryophyte conducting tissues: Structure and histochemistry, *Bull. Torrey Bot. Club.* **107,** 298–307 (1980).
97. Schofield, W. B., Phytogeography of the mosses of North America (north of Mexico), in *The Mosses of North America* (R. J. Taylor, and A. E. Leviton, eds.), pp 131–170. Pacific Division, American Association of the Advancement of Science, San Francisco, 1980.
98. Schofield, W. B., Ecological significance of morphological characters in the moss gametophyte, *Bryologist* **84,** 149–165 (1981).
99. Schofield, W. B., *Introduction to Bryology.* Macmillan, New York, 1985.
100. Schuster, R. M., *The Hepaticae and Anthocerotae of North America,* Vol. 1. Columbia Univ. Press, New York, 1966.
101. Schuster, R. M., Evolution, phylogeny, and classification of the Hepaticae, in *New Manual of Bryology* (R. M. Schuster, ed.), Vol. 2, pp. 892–1070. Hattori Botanical Laboratory, Nichinan, Japan, 1984.
102. Silander, J. A., Microevolution in clonal plants, in *Population Biology and Evolution of Clonal Organisms* (J. B. C. Jackson, L. W. Buss, and R. E. Cook, eds.), Chapter 4. Yale Univ. Press, New Haven, 1985.
103. Slack, N. G., Bryophytes in relation to ecological niche theory, *J. Hattori Bot. Lab* **52,** 199–217 (1982).
104. Slack, N. G., and Glime, J. M., Niche relationships of mountain stream bryophytes, *Bryologist* **88,** 7–18 (1985).
105. Slatkin, M., Somatic mutations as an evolutionary force, in *Evolution: Essays in Honor of John Maynard Smith* (P. J. Greenwood, P. H. Harvey, and M. Slatkin, eds.), pp. 19–30. Cambridge Univ. Press, Cambridge, 1985.

106. Smith, A. J. E., Cytogenetics, biosystematics, and evolution in the Bryophyta, in *Advances in Botanical Research* (H. W. Woolhouse, ed.), pp. 195–277. Academic Press, London, 1978.
107. Smith, A. J. E., and Ramsay, H. P., Sex, cytology, and frequency of bryophytes in the British Isles, *J. Hattori Bot. Lab.* **52,** 275–281 (1982).
108. Smith, J. M., *The Evolution of Sex.* Cambridge Univ. Press, London, 1978.
109. Smith, J. M., Burian, R., Kauffman, S., Alberch, P., Campbell, J., Goodwin, B., Lande, R., Raup, D., and Wolpert, L., Developmental constraints and evolution, *Q. Rev. Biol.* **60,** 265–287 (1985).
110. Solbrig, O. T., Jain, S., Johnson, G. B., and Raven, P. H., eds., *Topics in Plant Population Biology.* Columbia Univ. Press, New York, 1979.
111. Stark, L. R., Bisexuality as an adaptation in desert mosses, *Am. Midl. Nat.* **110,** 445–448 (1983).
112. Stark, L. R., Reproductive biology of *Entodon cladorrhizans* (Bryopsida, Entodontaceae). I. Reproductive cycle and frequency of fertilization, *Syst. Bot.* **8,** 381–388 (1983).
113. Stebbins, G. L., *Variation and Evolution in Plants.* Columbia, Univ. Press, New York, 1950.
114. Steere, W. C., Cenozoic and mesozoic bryophytes of North America, *Am. Midl. Nat.* **36,** 298–324 (1946).
115. Steere, W. S., A new look at evolution and phylogeny in bryophytes, in *Current Topics in Plant Science,* pp. 134–143. Academic Press, New York, 1969.
116. Stoneburner, A., Fruiting in relation to sex ratios in colonies of *Pleurozium schreberi* in northern Michigan. *Mich. Bot.* **18,** 73–81 (1979).
117. Swain, T., and Cooper-Driver, G., Biochemical evolution in early land plants, in *Paleobotany, Paleoecology, and Evolution* (K. Niklas, ed.), Vol. 1, pp. 103–134. Praeger, New York, 1981.
118. Taylor, I. E. P., Schofield, W. B., and Elliott, A. M., Analysis of moss dehydrogenases by polyacrylamide disc electrophoresis, *Can. J. Bot.* **48,** 367–369 (1970).
119. Templeton, A. R., The prophecies of parthenogenesis, in *Evolution and Genetics of Life Histories* (H. Dingle, and J. P. Hegmann, eds.), pp. 75–101. Springer-Verlag, Berlin, 1982.
120. Thompson, V., Does sex accelerate evolution? *Evol. Theor.* **1,** 131–156 (1976).
121. Tiffney, B. H., and Niklas, K. J., Clonal growth in land plants: A paleobotanical perspective, in *Population Biology and Evolution of Clonal Organisms* (J. B. C. Jackson, L. W. Buss, and R. E. Cook, eds.), pp. 35–66. Yale Univ. Press, New Haven, 1985.
122. Uyenoyama, M. K., On the evolution of parthenogenesis: A genetic representative of the "Cost of Meiosis," *Evolution* **38,** 87–102 (1984).
123. Van Zanten, B. O., Preliminary report on germination experiments designed to estimate the survival chances of moss spores during long-range dispersal in the southern hemisphere, with particular reference to New Zealand, *J. Hattori Bot. Lab.* **41,** 133–140 (1976).
124. Van Zanten, B. O., Experimental studies of trans-oceanic long-range dispersal of moss spores in the southern hemisphere, *J. Hattori Bot. Lab.* **44,** 455–482 (1978).
125. Van Zanten, B. O., and Pócs, T., Distribution and dispersal of Bryophytes, in *Advances in Bryology* (W. Schultze-Motel, ed.), Vol. 1, pp. 479–562. Cramer, Vaduz, 1981.
126. Vepsäläinen, K., and Järvinen, O., Apomictic parthenogenesis and the pattern of the environment, *Am. Zool.* **19,** 739–751 (1979).
127. Vitt, D. H., and Slack, N. G., Niche diversification of *Sphagnum* relative to environmental factors in northern Minnesota peatlands, *Can. J. Bot.* **62,** 1409–1430 (1984).
128. Vries, A. de, Zanten, B. O. van, and Dijk, H. van, Genetic variability within and between populations of two species of *Racopilum* (Racopilaceae, Bryopsida), *Lindbergia* **9,** 73–80 (1983).
129. Watson, E. V., *The Structure and Life of Bryophytes,* 3rd Ed. Hutchinson & Co., London, 1971.
130. Watson, M. A., Age structure and mortality within a group of closely related mosses, *Ecology* **60,** 988–997 (1979).
131. Watson, M. A., Patterns of habitat occupation in mosses—relevance to considerations of the niche, *Bull. Torrey Bot. Club* **107,** 346–372 (1980).
132. Watson, M. A., Patterns of microhabitat occupation of six closely related species of mosses along a complex altitudinal gradient, *Ecology* **62,** 1067–1078 (1981).
133. Weimarck, G., Population structures in higher plants as revealed by thin-layer chromatographic patterns, *Bot. Not.* **127,** 224–244 (1974).
134. White, J., The plant as a metapopulation, *Annu. Rev. Ecol. Syst.* **10,** 109–145 (1979).
135. Whitham, T. G., and Slobodchikoff, C. N., Evolution by individuals, plant-herbivore interactions, and mosaics of genetic variability: The adaptive significance of somatic mutations in plants, *Oecologia* **49,** 287–292 (1981).

136. Wiley, E. O., *Phylogenetics: The Theory and Practice of Phylogenetic Systematics.* John Wiley & Sons, New York, 1981.
137. Williams, G. C., *Sex and Evolution.* Princeton University Press, Princeton, 1975.
138. Wyatt, R., Spatial pattern and gamete dispersal distances in *Atrichum angustatum,* a dioicous moss, *Bryologist* **80,** 284–291 (1977).
139. Wyatt, R., Population ecology of bryophytes, *J. Hattori Bot. Lab.* **52,** 179–198 (1982).
140. Wyatt, R., Terminology for bryophyte sexuality: Toward a unified system, *Taxon* **34,** 420–425 (1985).
141. Wyatt, R., and Anderson, L. E., Breeding systems in bryophytes, in *The Experimental Biology of Bryophytes* (A. F. Dyer, and J. G. Duckett, eds.) pp. 39–64. Academic Press, Orlando, Florida, 1984.
142. Yamazaki, T., Genic variabilities in natural population of haploid plant, *Conocephalum conicum.* I. The amount of heterozygosity, *Jpn. J. Genet.* **56,** 373–383 (1981).
143. Yamazaki, T., Genic variability in natural populations of the haploid plant, *Conocephalum conicum,* in *Molecular Evolution, Protein Polymorphism, and the Neutral Theory* (M. Kimura, ed.), pp. 123–134, 1982.
144. Zander, R. H., Bryophyte sexual systems: -oicous versus -oecious, *Bryol. Beitr.* **3,** 46–51 (1984).
145. Zielinski, R., Electrophoretic evidence of cross-fertilization in the monoecious *Pellia epiphylla,* N = 9, *J. Hattori Bot. Lab.* **56,** 255–262 (1984).
146. Zimmerman, M., and Hicks, D. J., Strategy: Misuse or insight? *BioScience* **35,** 66 (1985).

15

Reproductive Strategies of Pteridophytes

MICHAEL I. COUSENS

Pteridophytes include the Psilotales, (*Psilotum* and *Tmesipteris*), Sphenophyllales (*Equisetum*), Lycopodiales (*Lycopodium, Selaginella,* and *Isoetes*), and the fern orders Ophioglossales (*Botrychium* and *Ophioglossum*), Marattiales, Filicales, Salviniales, and Marsileales. The cardinal points of the pteridophyte life cycle are the meiotic production of spores by the sporophyte and the mitotic production of gametes by the gametophyte. Meiosis initiates the gametophyte generation and the fusion of mitotically formed gametes initiates the sporophyte. The homosporous life history pattern, with the production of a potentially bisexual gametophyte from a single spore is most common. However, heterospory, with production of separate archegoniate and antheridiate gametophytes, characterizes the Lycopsid genera *Selaginella* and *Isoetes* and the aquatic fern orders Marsileales and Salviniales.

The most numerous and ecologically diverse pteridophytes are the Filicales, with over 400 genera and perhaps 12,000 species.[29,30,70] Such species diversity, and the fact that many of the largest genera of Filicales are contemporary in origin with the angiosperms, belies the common perception that ferns as a whole are ancient and in decline.[1,70] Modern centers of pteridophyte diversity are mostly tropical.[27] Cenozoic and Pleistocene changes drastically reduced the diversity of temperate pteridophyte floras. This chapter reflects the fact that the origins, distribution, ecology, and reproductive biology of temperate species are much better studied than those of tropical species.[85] Such work may not properly represent the majority of pteridophytes since more recent species at the margin of distribution may express features of reproductive biology and physiology that reflect relative specialization when compared to generalist patterns that are perhaps more characteristic of the whole group.

The pteridophyte sporophyte is commonly termed the dominant generation. To systematists in seach of diagnostic characters, or to community ecologists seeking measures of productivity and community influence, the sporophyte is indeed dominant. To one interested in uncovering the special reproductive strategies of pteridophytes in nature, equal treatment of the generations is more useful.

Various facets of pteridophyte reproductive biology and ecology have been well reviewed. These include cytology,[70,90] phylogeny,[49] experimental biology,[27] and the perspectives shared by laboratory and field approaches.[28,87] The objectives of this chapter are (1) to gather from the broader literature those well-studied aspects of pteridophyte biology that relate to reproductive strategies, (2) to review work dealing with reproductive strategies in nature, and (3) to suggest lines of inquiry that may improve our understanding of pteridophyte reproduction in natural populations.

ALTERNATION OF GENERATIONS

Alternation of generations is the central feature of the pteridophyte life history. Typical alternation of gametophyte and sporophyte, and the atypical processes of apogamy and apospory, are well understood from a morphologic and morphogenetic perspective.[70] A traditional understanding of alternation places pteridophytes between bryophytes and seed plants with the implication that it is a life history pattern whose adaptive peak has passed. It is also traditional to label the sporophyte generation asexual and to label the gametophyte generation sexual. Since meiosis (a sporophyte function) and fertilization (a gametophyte function) each contribute to variation, this terminology may be misleading. Willson argues the need for alternation to be understood in terms of adaptive strategies in modern environments.[116] The present success of the modern families of the Filicales, measured by radiation of modern genera and by species numbers, supports the view that adaptation to present environments as well as phylogenetic constraints should be considered.[1,27]

Keddy suggested that strong selection for a horizontally disposed gametophyte with access to free water to effect fertilization, and for an erect sporophyte with increased capacity for spore production and distribution, explained both the origin and continued adaptiveness of distinctive generations.[50] Vitt suggested that, in the Bryophyta, gametophyte and sporophyte are subjected to selection in different ways but that ability of the bryophyte gametophyte and sporophyte to respond to divergent selection is strongly limited by permanent retention of the sporophyte.[105] Bryophytes are thus free from a second risky "germination event" of sporophyte establishment.[105] For pteridophytes, the sporophyte typically establishes independence and the gametophyte dies. In species where the gametophyte has a mycosymbiont, the gametophyte maintains connection between the generations for a longer period of time.[5,9,76] *Anogramma leptophylla,* a fern that was first identified as a bryophyte, bears a sporophyte that fails to become independent and dies after spore formation.[75] Its gametophyte appears to be perennial. Istock suggested that such ecologically distinct phases may evolve independently to a large degree and are limited in their independent evolution only by developmental and morphological features necessary for transition between phases.[48]

Additional features of the gametophyte generation include great numbers of spores with high mortality, lack of recessive allele storage, and relatively ephemeral life span. These features contrast with recessive storage and the very long life span of the sporophyte. These differences suggest the possibility of different evolutionary rates for the two generations. Klekowski observed 2,200 gametophytes of *Osmunda regalis* with normal appearance yet 83.5% of these carried recessive alleles deleterious or lethal in the sporophyte phase.[55,56] Deleterious mutations documented for the gametophyte generation function as gamete lethals and are thus immediately eliminated from the gene pool.[58] This argues strongly that deleterious genes whose expression is limited to gametophyte development are more rapidly eliminated from the genome than are comparable sporophyte genes. Ewing has proposed a genetic model in which selection at both the haploid and diploid levels occurs and maintains stability of the life cycle and variability over time.[30]

Variations in sporophyte life form, distribution[27] and ecology,[41] physiological features,[27] and microhabitat distribution have provided material for investigations of adaptiveness.[82] The gametophyte generation has often been labeled the "weak link" in the life cycle but this may not be the case. Evidence of gametophyte drought tolerance superior to that of the sporophyte,[34] gametophyte freedom from slug herbivory where young sporophytes are eaten,[71] winter survival of gametophytes,[20,21] allelopathic potential,[27] and persistence of gametophytes in habitats where sporophytes are not produced[28] give some credence to Willson's argument that complex plant life cycles are explained by selection to meet ecological problems. The gametophyte is commonly perceived to vary less than the sporophyte,[7] and the widely held view that the heart-shaped bisexual gametophyte illustrated in texts is typical, discouraged investigation of the relationship between gametophyte variability and adaptiveness. There are, however, several levels at which gametophyte variation occurs.[18,89] For homosporous pteridophytes, the great differences between buried mycosymbiotic and surficial photoautotrophic gametophytes, and the contrast between filamentous (Hymenophyllaceae, some Schizeaceae) and thalloid gametophytes (most Filicales) suggest quite different adaptations.[3,5,101,102] The great difficulties of working with less common mycosymbiotic gametophytes compared to the ease of establishing controlled cultures of photoautotropic gametophytes has resulted in an uneven distribution of research. A few workers of exceptional skill and patience have produced most of the investigations of mycosymbiotic gametophytes,[5,9,76,114] whereas an extensive number of workers with a great variety of emphases have studied the more easily cultured surficial gametophytes.[3,27]

Variation in gametophyte sex expression, its causes, and populational consequences of mating systems are areas that have received great attention in Filicales with surficial gametophytes.[2,17,18,45,66]

GAMETOPHYTE SEX EXPRESSION

Variations in patterns of sex expression were discovered in early investigations of the gametophytes of homosporous ferns.[3] Unisexual male gametophytes with high numbers of antheridia were routinely observed.[17] Almost all cultures contained at least some bisexual plants, and the proportions of sex expression types were not in simple Mendelian ratios. In addition, because sex expression types did not correlate with conspicuous differences in spore size (for exceptions, see Refs. 92 and 102), and were not reproducible from culture to culture, such patterns were typically not investigated further. Variations in sex expression were ascribed to environmental factors such as conditions of culture, crowding, or depauperate conditions for growth.[3] Following the work of Bower,[7] most workers sought features of gametophyte morphology that were useful for phylogenetic placement of families and genera. Open-grown fully developed gametophytes are more likely to yield consistent diagnostic characters. The work of Atkinson and Stokey stands out in comparative morphology of gametophytes.[3] Their studies included brief characterization of patterns of sex expression. Sex expression in the Osmundaceae[100] and Gleicheniaceae[101] is generally less variable than that of more recently derived families of Filicales. Gametophytes of the genus *Osmunda* are typically bisexual, although Stokey points out the presence of ameristic males in cultures,[100] and Klekowski documented unisexual females in cultures and in the field for *Osmunda regalis*.[56] In a recent study of five species of the more recent genus *Thelypteris*. Atkinson found that three species were characterized by distinctly different unisexual male and female gametophytes.[2] Other *Thelypteris* species bore both archegonia and antheridia on the same gametophyte.

Many terms have been used to describe the patterns of gametophyte sex expression. They may describe only the presence of gametangia, i.e., unisexual, male, bisexual; or include chronology, e.g., protandrous bisexual, protogynous bisexual, synchronous bisexual.[17] Although used frequently by early workers, and occasionally still, the terms monoecious and dioecious are best avoided as they have clear definitions regularly used for seed plants in which gametophytes are always unisexual.[3]

Pattern of sex expression at the family, species, and in fewer cases, population level, is now a well-investigated aspect of gametophyte variation.[17,18,47,66,81] Klekowski,[54] and Lloyd[66] reviewed the existing literature on variable sex expression and initiated research that assessed its possible adaptive value. Their cogent communication of genetic and life history differences between homosporous pteridophytes and seed plants greatly increased the tempo of research that sought to understand the population biology of pteridophytes. Klekowski's careful definitions of the levels of gamete exchange possible within populations of gametophytes are now widely accepted.[94] *Intragametophytic selfing* unites mitotically derived, and therefore genetically identical, sperm and egg. Barring somatic mutation, this produces a completely homozygous zygote. Inbreeding comparable to intragametophytic selfing is prohibited by the production of egg and sperm on morphologically distinct female and male gametophytes of heterosporous pteridophytes and seed plants. *Intergametophytic mating* includes *intergametophytic selfing* (gamete exchange between gametophytes meiotically derived from the same parent sporophyte) and *intergametophytic crossing* (exchange between gametophytes derived from different parent sporophytes). Intergametophytic selfing and crossing are comparable in genetic outcome to self- and cross-pollination for seed plants.

Klekowski initiated studies in which these patterns were more carefully described.[54] He proposed that separation of antheridia and archegonia on different gametophytes, chronology of their production on single gametophytes, position of antheridia relative to archegonia, disposition of the archegonium apex, and the populational mix that resulted from overlapping patterns were mechanisms that would reduce the frequency of intragametophytic selfing.

An important outcome of this work was the wide application of the simple experimental protocol of establishing isolated gametophytes and following the outcome of forced intragametophytic selfing. The degree to which isolated plants failed to produce viable sporophytes, especially when the selfing experiment was coupled with a search for swollen archegonia and aborted embryos, was interpreted as genetic load.[18,22,36,56,67,68,110] The degree of genetic load that characterized a population derived from a single sporophyte was then a measure of its heterozygosity, and indirectly, evidence of a history of degree of outbreeding. Such experiments tested predictions of breeding mechanisms based on morphologic sex expression. In one important series of experiments, *Osmunda regalis* gametophytes that were potentially bisexual indicated a history of outbreeding despite apparent morphological predisposition to inbreeding.[55,56] Clonal pteridophytes also provide efficient access to studies of mutational load accumulated during the life of a clone by screening gametophyte progeny of marked ramets.[59]

Studies that followed quickly established that patterns of sex expression and genetic load varied for individual sporophytes, and adjacent and distant populations, and that valid studies of fern reproductive biology necessitated a populational perspective.[18,68,110] The pattern of sex expression in populations of *Pteridium aquilinum* allowed the prediction of an initially outbreeding system, and the adaptive value of this system was supported by experiments in which lethality following intragametophytic selfing ranged up to 100%.[27] Holbrook-Walker and Lloyd[47] demonstrated species-specific patterns of sex expression in the genus *Sadleria,* and indicated that nearly obligate outbreeding occurred in one species, whereas the least widespread and ecologically most restricted species was predominantly inbreeding. They cautioned against too broad an application of laboratory determinations of patterns of gametophyte sex expression and recommended that parallel studies be made on field-collected specimens. Lloyd demonstrated that fern species that were dominant in pioneer habitats were more likely to have intragametophytic mating systems and to carry fewer lethal recessive alles.[67] In contrast, species from nonpioneer habitats were more likely to have intergametophytic breeding systems as indicated by pattern of sex expression and higher levels of genetic load.

The hypothesis that early gametophyte ontogeny and sex expression would vary in an adaptive pattern was investigated for 72 sporophytes representing five populations of *Blechnum spicant.*[17,18] Of interest were differences between a small disjunct population in Northern Idaho and extensive Pacific Coast populations. Different patterns of ontogeny and sex expression characterized each population. Variation in early ontogeny did not suggest any simple adaptive correlation with environment. The sex expression pattern that characterized the disjunct population predisposed it to intragametophytic selfing. Screening of isolated gametophytes revealed that genetic load was least in this population. This was consistent with either the biology of a population initiated by long distance dispersal, or a relict population that may have undergone depletion of numbers and genetic drift. Sex expression of gametophytes of one population was also determined for field collections. For the field observations most closely comparable to laboratory data (repeated observations of the same natural colony), the same sex expression types were found. However, the proportion of male gametophytes in the field was consistently greater than in culture. Several factors that differentiate analyses of sex expression for field populations from cultures are noted elsewhere.[20] Some variability in culture may be due to relaxation of short-term selection compared to natural populations. The possibility of continued spore shower and resultant range of germination times and the presence of gametophytes of several additional species are factors that will increase variability in nature. The sum of such complexities in nature may result in an increased proportion of unisexual males, even in taxa for which laboratory observations suggest an exclusively bisexual pattern.

In a critical review of sex expression in fern gametophytes, Willson suggested the need to relate gamete ratios to reproductive success, and to test whether or not sperm numbers limited reproductive success.[115] Since the term bisexual does not distinguish between a gametophyte that bears one antheridium and one archegonium and quite different ratios, the need to supply counts of gametangia is obvious. A few papers do present data on numbers of gametangia,[18,109] and others discriminate between dehisced and undehisced antheridia.[55,56,110] Warne and Lloyd derived a formula, the "Self Fertilization Index," in which proportions of males, bisexual plants, and dehisced antheridia are used to calculate a ratio that predicts degree of intragametophytic selfing.[110]

Some insight into effectiveness of gamete ratios may be obtained by a reexamination of data presented by Cousens for *Blechnum spicant.*[18,19] Reproductive success may be defined by the presence of an embryo. Sperm/egg ratios for multispore cultures, in which both intragametophytic selfing and intergametophytic selfing were possible, ranged from 10 : 1 to 113 : 1 over 35 to 90 days of culture. Genetic load experiments indicated that viable embryo formation was possible for 3 to 34% of gametophyte progeny of single sporophytes in these cultures. A few bisexual gametophytes with sperm/egg ratios as low as 14 : 1 bore embryos. Female gametophytes with as few as 16 archegonia bore embryos when sperm/egg ratios were at the maximum and embryo formation increased dramatically as sperm/egg ratios increased. The sperm/egg ratios for closely similar field collections ranged from 199 : 1 to 1015 : 1. The proportion of gametophytes bearing embryos was much greater, and these occurred on smaller plants. Reduction of embryo formation in the field collections due to genetic load would be much less than in cultures, as gametophyte colonies could have several sporophyte parents and intergametophytic crossing would be quite likely. Taken together, these data suggest that high sperm/egg ratios may be a factor in successful reproduction, but experiments with genetic lines that lack genetic load would be necessary to isolate the contribution of increased sperm/egg ratios to reproductive success.

Döpp was first to investigate the causes of variable sex expression that had been attributed to crowding in cultures.[23] He discovered that medium upon which gametophytes of *Pteridium aquilinum* had grown induced precocious antheridia formation on *Pteridium* gametophytes and hastened antheridia production by *Dryopteris filix-mas.* Subsequent research demonstrated at least five distinct chemicals that have been named antheridiogens.[81] The biology of the antheridium-inducing pheromone superimposes an additional level of variation upon patterns of gametophyte sex expression.[27,28] The structure of an antheridiogen from the genus *Anemia* has been characterized and its derivation from a gibberellin has been proposed.[80] Antheridiogen is secreted into the substrate by gametophytes shortly after they establish a notch meristem and become competent to form archegonia. Antheridiogen-secreting plants have themselves passed beyond the developmental stage susceptible to induction of antheridia by the pheromone. The degree to which other gametophytes are susceptible to induction depends on their stage of growth, with the youngest gametophytes being most susceptible and forming the greatest numbers of antheridia. Late-germinating spores may develop several antheridia on the first-formed somatic cells and cease further growth. Gemmae produced by gametophytes of the genus *Vittaria* are immediately susceptible to antheridia induction.[29]

Ceratopteris species have large spores, substantial range in spore size, and a more strongly dimorphic pattern of sex expression than other homosporous Filicales[43] (with the exception of *Platyzoma microphylla*[102]). Schedlbauer carefully investigated the range of gametophyte development prior to antheridiogen production in populations of *Ceratopteris thalictroides.*[92] He found that variation in spore size accurately correlated with the sequence of germination and further gametophyte growth. It also correlated with subsequent susceptibility of developing plants to antheridiogen. He suggested that size reflected spore contents and that spores with larger diameters germinated earlier and produced faster-growing large bisexual plants. It is important to note that variable sex expression has been documented for *Acrostichum danaeifolium,* which apparently lacks an antheridiogen.[68]

Antheridiogens have been shown to substitute for the light requirement for germination in some taxa, and Schneller has collected precociously male plants from nature that appear to document induction of germination in the dark and antheridium formation.[93] He suggested that the coupling of these actions of antheridiogen in nature would produce a populational system in which sperm for intergametophytic mating was nearly continuously available. Warne and Lloyd suggested that gametangia sequence on bisexual gametophytes of *Ceratopteris pteridoides* has an overriding influence on mating systems, and that the role of antheridiogens in effecting change in mating systems should be questioned.[110]

Many workers have demonstrated consistent differences in patterns of sex expression that characterize gametophyte progeny of single sporophytes maintained separately and grown under uniform culture conditions.[18,68] Scott has demonstrated genetically determined differences in response to antheridiogen by two strains of *Ceratopteris richardii.*[97] Comparable studies are needed to determine if genetic variation exists for rate of antheridiogen production, timing of germination, and for sex expression independent of antheridiogen.

Willson suggested that the presence of antheridiogen in a natural system might signal to smaller plants that another individual is ahead of them in development and that reproductive success as a male would be greater than that as a female.[115] She also suggested that antheridiogens may function as a form of allelopathy, by increasing resources available to a leading female, but not doing so to the extent that sperm availability becomes limiting.

The presence of bacteria, fungi that may produce chemicals closely similar to antheridiogens, several species of fern gamethophytes[19] with potential interactions, and leaching suggest that analysis of antheridiogen activity on soil in nature may be difficult. The pioneer work of Tryon and Vitale demonstrated the value of investigations of antheridiogen effects in nature and suggests that they do strongly condition sex expression.[104] A single gametophyte in soil may be surrounded by a hemisphere of high antheridiogen activity with a 10-cm radius.[96] The extreme chemical and biological stability of the molecule, and the power to induce antheridia by a few as 100 molecules,[96] argues that the biology of antheridiogen in nature merits further study.

Variations in pattern of gametophyte sex expression may ultimately be susceptible to adaptive explanations expressed in fitness values. This may be possible only after the contribution of other factors such as historical constraints (e.g., gametophytes of modern genera may have many archegonia because that is the ancestral condition), allometric growth (larger gametophytes may bear more gametangia as a consequence of continued growth), developmental sequence (e.g., archegonium initiation is linked to onset of three-dimensional growth), and traits that may be neutral in selective value (e.g., "sporophytic" trichomes on the gametophyte) are better understood.

The adaptiveness of the form of gametophytes in effecting fertilization has been little investigated. Cousens observed, on soil cultures and in the field, that the apical portion of early three-dimensional gametophytes of *Blechnum spicant* was raised above the substrate and that the wings of older female or bisexual gametophytes were elevated to form a cup-shaped depression in the center of the thallus.[18] Water pipetted dropwise upon the gametophyte formed a meniscus that allowed the cup to overfill. Ultimately an additional drop would cause the meniscus to collapse. Water cascaded through the notch region, wetted the archegonia, and then filled the space between the soil and plant. The water layer was thus continuous between archegonia and antheridia and fertilization occurred. A meniscus was also formed similarly for gametophytes of several species of *Dryopteris.*[17] In this genus it appeared possible that numerous trichomes with waxy caps reduced the wettability of the dorsal surface, enhancing formation of a large water drop. Atkinson observed that gametophytes of several species of *Thelypteris* were also supported above the soil surface by their rhizoids.[2] Large gametophytes of *Blechnum spicant* with wings flattened against the soil surface were often observed in nature. These always bore embryos and it appeared that wing flattening was caused by growth following fertilization.[18,19] More careful study of the form of the gametophytes (including those with lateral meristems, e.g., *Anemia* spp.), filamentous form,[56] and buried mycosymbionts,[28] in relation to water movement and fertilization is needed. Lloyd has utilized lines with phenotypically distinct first leaf pattern to demonstrate that bisexual gametophytes are likely to undergo intragametophytic-selfing when surrounded by fewer than five gametophytes regarded as sperm donors.[69] This suggests that gametophyte form may be adapted to favor selfing.

Chemotaxis was described in fern sperm more than a century ago.[95] Schneller has shown genus level specificity for chemotaxis toward the mucilage released by degeneration of neck canal cells following watering.[95] Sperm of *Athyrium filix-femina* were weakly if at all attracted by the mucilage of *Dryopteris filix-mas,* and *Dryopteris filix-mas* sperm were immobilized in the archegonial neck of *Athyrium filix-femina.* An understanding of the chemical that attracts sperm may contribute to our understanding of basic questions of how far fern sperm swim. It may be that dehiscence of a single archegonium is insufficient to attract sperm, or that dehiscence of a large number of archegonia serves to attract compatible sperm over greater distances than dehiscence of a single archegonium.

GENETIC AND POPULATIONAL ASPECTS

Analyses of gametophyte form, sex expression, antheridiogens, and genetic load allow predictions of population genetic structure that may then be tested. Models of population genetic structure of homosporous pteridophytes must take into account hypotheses for the adaptiveness of apparent polyploidy in the Filicales.[10,13,15,27,33,38,39,43,55,66,70,115] Following the description of reticulate evolution, by which reproductively isolated taxa form immediately following polyploidization of hybrid taxa, there has been a productive period in the study of pteridophyte evolution.[73,106,111] The diploid parents of many successful allopolyploid taxa are extant, and in some cases, recently extinct parental taxa have been proposed. A current and well-illustrated summary of the patterns of reticulate evolution in North American pteridophytes is provided by Lellinger.[62] Following the pioneering work of Wagner,[106] the genus *Asplenium* has been studied with a variety of techniques that have confirmed species origins following polyploidy of ancestral hybrids. Allozyme evidence has now convincingly shown that two allopolyploid species in this genus have originated more than once.[111] Diploid, tetraploid, and hexaploid cytotypes have been described for *Asplenium trichomanes.*[70] Additional complexities of reticulate evolution are suggested by evidence that two species in *Asplenium* may have arisen following hybridization between different populations of the same parental species.[8] Stein has shown that comparisons of DNA can help to identify parents of hybrid species and to elucidate the sequence of speciation within families.[99]

Polyploid systems increase gene dosage. Consequently, phenotypic expression of recessives is decreased, and for codominant genes a greater range of intermediate phenotypes results. A reduced rate of expression of recessives slows the rate of potential evolution. If pairing behavior is restricted by selection to homologues, heterozygosity between homoeologues will not segregate but will be fixed. Gene expression for such a system would be essentially similar to that of diploids and would not necessarily slow evolutionary rate. Species number in modern genera of the Filicales and reticulate relationship between them suggest that Cenozoic, and especially post-Pleistocene speciation of the Filicales has not been markedly slower than that for seed plants.[49]

The most significant and well-studied hypothesis for the adaptiveness of polyploidy is that of Klekowski and Baker.[52] They argue that polyploidy provides means of storage and release of heterozygosity that counters the homozygotizing influence of regular intragametophytic selfing. Several predictions follow: studies of effective mating in natural populations should uncover some intragametophytic-selfing; heterozygosity values for populations of pteridophytes, especially those that have recently invaded a stand, would be significantly lower than for heterosporous plants; heterozygosity will be buffered in its release by polyploid gene dosage, and some heterozygosity should be released by homoeologous pairing. Extensive studies of *Ceratopteris* species by Hickok resolved polyploid segregation ratios that demonstrated homoeologous pairing.[43] Cytological evidence for homoeologous pairing has been demonstrated in *Trichomanes.*[6] Evidence that homoeologous pairing occurs is also available from an allozyme study of *Pteridium aquilinum,*[13] although this interpretation has been contested.[33]

Recent advances in electrophoretic study of allozymes, well reviewed elsewhere,[28,33,38,39,40] provide means to test hypothesized mating systems, determine gene dosage, and map population genetic structure. A single gametophyte may provide sufficient material for the analysis of three or four enzyme systems, and the number of systems that can be studied using a single sporophyte frond is much greater.[28,33,39] As many as 20 enzyme systems per taxon have been surveyed and larger scale analyses for those systems that are heterozygous have been performed.[33] In an early demonstration of the insight afforded by allozyme analysis, Levin and Crepit studied genets of homosporous lycopod, *Lycopodium lucidulum.*[64] This species is thought to be the least derived in the genus. Thirteen of 18 loci analyzed were monomorphic. The mean number of alleles per locus was 1.39 and the mean proportion of polymorphic loci was 0.28. Many of the individuals were heterozygous for the same alleles and had an excess of heterozygosity over that expected from random mating. This excess of heterozygotes was interpreted as a reflection of fixed heterozygosity rather than outbreeding. Maintenance of heterozygosity was thought to be due to asexual reproduction. Such values are consistent with rare colonization by spores, followed by intragametophytic selfing, and clonal spread of populations.

McCauley et al. surveyed allozyme variation in three populations of the eusporangiate fern *Botrychium dissectum.*[74] Two enzymes were sufficiently polymorphic to allow a test of heterozygote frequencies expected for Hardy–Weinberg equilibrium. Allele frequencies were such that random mating would have resulted in up to 46% heterozygotes, yet fewer than 2% were found. An estimate of the inbreeding coefficient for each enzyme suggested that these populations showed some of the highest inbreeding coefficients yet measured for a natural population and that the rate of outcrossing was only 5%. Soltis and Soltis observed that 4 of 18 loci examined electrophoretically in *Botrychium virginianum* were polymorphic.[98] They demonstrated a very high inbreeding coefficient as well as lack of genetic evidence that the species is polyploid despite a $n = 90$ value. *Botrychium* and most *Lycopodium* species have subterranean mycosymbiotic gametophytes. It may be that this morphology is associated with low gametophyte population density or especially limited sperm travel. Wagner and Wagner, however, argue that the frequency with which hybrids form in these groups suggests that such gametophytes are not necessarily limited to intragametophytic selfing.[28]

Haulfer and Soltis reported heterozygosity values of 2.62 alleles per locus in *Bommeria hispida,* which has both an antheridiogen system resulting in strong separation of sex expression and high genetic load.[39] This value is close to averages for outcrossing seed plants. Large numbers of gametophytes derived from natural sporophyte populations were examined and predictions on the level of outcrossing were confirmed by allozyme inheritance.

Gastony and Gottleib examined genetic variability in nine sexual and three apogamous natural populations of the xeromorphic fern *Pellaea andromedifolia* and determined banding patterns for gametophytes by segregational analysis.[33] Of the sporophytes examined, 81.3% were heterozygous for one or more of five polymophic loci with a mean value of 2.60 alleles at the average heterozygous locus. This value was intermediate to mean values for selfed and outcrossed seed plants.[33] The heterozygosity was shown to be due to mating between gametophytes with different alleles and not to homoeologous pairing. Apogamous populations, in which sporophytes are produced by gametophytes without fertilization, were characterized by fixed heterozygosity and it appeared likely that each such population had an independent origin. Work with *Pellaea andromedifolia* demonstrated that the extensive literature and techniques available for population genetic analysis of seed plants may also be utilized for pteridophytes.

Low levels of homoeologous segregation, as documented by Hickok,[43] would be difficult to resolve by analysis of a limited number of enzyme systems. Broader surveys of enzymes, ideally a minimum of one coded by each chromosome, and much more extensive population samples may be needed to uncover homoeologous segregation at the 1–10% rate. Very low segregation ratios for homoeologous heterozygosity may be sufficient to influence evolutionary rate. Work by Buckley and Lloyd[10] and others,[15,115,117] suggests that additional hypotheses for the origin of polyploidy in homosporous pteridophytes merit investigation.

Apogamy is thought to characterize 5–10% of temperate ferns.[107] Some species are known to be obligately apogamous,[113] whereas in other species some populations of otherwise sexual species may be apogamous,[33,89] and apogamy may be induced in taxa in which it does not naturally occur.[112] Apogamous taxa normally fail to produce functional archegonia but may produce functional sperm. Three cytological mechanisms have been documented that result in shared ploidy level for both generations. These are well reviewed elsewhere.[57,70]

Apogamous species and races are more common in xerophytes.[113] Klekowski discussed the adaptiveness of apogamy in reducing the duration of the vulnerable gametophyte phase in xeric habitats.[57] Kornas and Jankun described an African species of *Selaginella* that is apogamous and completes its life cycle during the rainy season.[60] Apogamy may also be adaptive in that it preserves favorable gene combinations that can then proliferate by dispersal of viable spores and vegetative propagules. The demonstration of fixed heterozygosity in apogamous races of *Pellaea andromedifolia* supports this possibility.[33] Gastony suggested that the primary origin of apogamous taxa requires the correlation of archegonial dysfunction with modification of sporogenesis.[33] Since most apogamous taxa are triploid or of higher ploidy, allopolyploid origins have been assumed. Evidence from allozyme data for *P. andromedifolia* suggested that autoploidy may have initiated apogamy in nature. According to allozyme data, autopolyploidy preceded apogamy in *Bommeria*. In *Cystopteris* and *Woodsia*, unreduced sperm from triploid taxa could fertilize haploid eggs of the same species to yield reproductively competent tetraploids.[115]

Hickok derived a mutant strain of *Ceratopteris* following interspecific hybridization between two diploid species.[42] Gametophytes produced nonfunctional sperm and produced sporophytes apogamously. These sporophytes were initially haploid but portions of fronds doubled somatically. Klekowski suggested "subsexual" ways in which the meiotic forms of apogamy can release heterozygosity through meiosis of a restitution nucleus or by homoeologous pairing (in the Döpp–Manton scheme, the most common pteridophyte pattern), and crossing-over (Mehra–Singh scheme).[57] Assuming continued availability of mutations and potential aneuploidy, it appears that apogamous taxa may maintain some genetic diversity.

The genetic structure of a pteridophyte population may be strongly conditioned by the pattern of distribution of adults over the stand. Crist and Farrar noted than *Asplenium platyneuron*, growing on coal spoils in Iowa, was distributed as isolated individuals. This suggested that each individual was the result of sexual production from a single spore.[22] Genetic load studies strongly supported this. Similarly, some *Dryopteris* species are typically distributed singly in swamps.[77] Population densities of genera with and without horizontal rhizomes may be equally great, yet would have quite different population genetic structure.[19,21] Schneller suggested that distributions of *Dryopteris* spp. and *Athyrium filix-femina* in European forests are such that gametophytes produced by different individual sporophytes will regularly intermingle and result in increased population heterozygosity in subsequent sporophytes.[93]

REPRODUCTIVE ALLOCATION AND MORPHOLOGY

The morphologic diversity that characterizes pteridophytes suggests that varied reproductive strategies may be uncovered by analysis of allocation of biomass, minerals, and water to reproductive functions. Although such analyses have proved useful in the study of seed plants, few comparable studies exist for pteridophytes.[11,61,72] The hypothesis that reproductive allocation (RA) in pteridophytes differs strongly from patterns in seed plants merits investigation. Although pteridophyte spores have been well studied from several perspectives, neither dry weight nor caloric values for typical spores have been published, but it is still clear that allocation of resources to a single spore is less than to any seed.[44,79,90] Available weights for fern spores suggest a tremendous range. A gram of *Ceratopteris richardii* spores contains 1.25 million relatively large individuals,[44] and 17 million air-dried spores of *Onoclea sensibilis* weigh 1 g.[77] Most spores contain sufficient reserves to produce a several-celled gametophyte when triggered to germinate in the dark.[90] It has been estimated that there is a 30- to 50-fold increase of dry weight between germination of *Ceratopteris richardii* spores and 21-day-old sexual gametophytes grown under continuous light at 30°C.[44]

Although there is a substantial body of literature on gametangium and gamete structure[4,27,90] and of spore contents, studies of the precise energy and mineral costs of reproductive effort in pteridophytes are not available. Energy available to growth of gametophytes may be limited, as suggested by the decrease in somatic tissue concurrent with increased antheridia number under the influence of antheridiogen.[81] It is possible that allocation to a densely cytoplasmic antheridium initial exceeds that to a highly vacuolate somatic cell by an order of magnitude.

Willson suggested that antheridiogens may have been selected for their effect on reproductive allocation and as a potential means of competition between gametophytes.[115] Not only would susceptible gametophytes produce sperm quickly and at the appropriate time when influenced by antheridiogen-secreting females, but male size, and therefore competition for space and resources, would be reduced.

Large gametophytes of *Osmunda* and *Gleichenia* with relatively large gametangia are thought to be less derived than smaller gametophytes of more modern genera.[100] Decreased size of the archegonium in modern genera (e.g., *Dryopteris, Polypodium*) may indicate selection for more efficient RA since large and small archegonia both produce a single egg. Antheridia of modern genera are smaller than those of *Osmunda* and have fewer sterile cells. However, smaller antheridia produce fewer sperm. An hypothesis of selection for more efficient RA would require determination of allocation per gamete rather than gametangium to be tested. Gametophytes of *Osmunda* species found in nature are not only large, but appear to attain their large size very quickly. They then support relatively large juvenile sporophytes. Thus it may be that large size has been selected (or retained) for effective transfer of reserves from one generation to the next, and is not simply a relictual character. The transfer of nutrients from gametophyte to sporophyte, however, has been questioned.[4] Perhaps decay of a large gametophyte following sporophyte production provides increased local supplies of minerals to the sporophyte at an appropriate time and place.

Induction of an embryo by apogamy following addition of sucrose to culture media suggests that the carbohydrate budget of sporophytes is greater than that of gametophytes.[112] The addition of sucrose to cultured leaf primordia of *Osmunda cinnamomea* induces fertile rather than sterile frond formation[37] and fronds of *Pteridium aquilinum* typically produce spores only in full sun.[87] This suggests that fertile frond production may be energy limited. Dimorphism resulting in distinct fertile and sterile fronds, characterizes 20% of pteridophytes.[108] This topic has been well reviewed by Wagner and Wagner.[108] The range in amount of photosynthetic tissue borne by fertile leaves suggests that fertile leaves of some species may export energy (e.g., *Thelypteris* spp.) and others (e.g., *Onoclea sensibilis*) may import it. In *Dryopteris ludoviciana* and *Polystichum acrostichoides,* basal and medial pinnae are laminar and sterile, whereas apical pinnae are fertile, skeletonized, and die soon after spore release. Such fronds may be self-sufficient in energy production, and costs of transport of metabolites would be reduced relative to "metabolite importers" and "exporters."

No single pattern of RA characterizes pteridophytes.[11,61,72] Callaghan described a dynamic situation in terms of translocation and dry weight in the successful tundra plant *Lycopodium annotinum.*[11] Strobili constituted 43% of the dry weight for current-year ramets, but net RA was much less than this, as nutrients were imported to strobili from an extensive horizontal rhizome system during development and then exported back to rhizomes following spore release.

Lacey investigated RA in *Onoclea sensibilis* and *Lorinseria areolata.*[61] These are unrelated genera with closely similar morphologies. Dimorphic fronds allowed estimation of RA by dry weight of sporophylls. Reproductive allocation in *O. sensibilis* was 10.4–15.3% of total biomass. Relative RA for *O. sensibilis* was unchanged in m^2 plots where its total biomass was decreased dramatically due to co-occurrence with *L.*

areolata. Reproductive allocation of *L. areolata* in plots not shared with *O. sensibilis* was 19.2%, and decreased to 12.5% in plots in which *O. sensibilis* occurred. Under flooded conditions RA in *O. sensibilis* was unchanged, but *L. areolata* failed to produce sporophylls. It may be that RA in *O. sensibilis* is relatively fixed, whereas that of *L. areolata* is diminished under competitive or flooded conditions. Values for RA reported by Callaghan and Lacey do not differ greatly from those reported for seed plants. In contrast, RA for *Osmunda cinnamomea* measured by sporophyll dry weight may be as little as 1–2%.[72]

PHENOLOGY AND DEMOGRAPHY

Pteridophyte sporophytes are commonly thought to have a single phenologic pattern, that of the long-lived perennial. A closer examination of the variations in timing and duration of pteridophyte life cycle events suggests that there has been substantial divergence in phenologic features. Phenologic information is readily available from herbarium specimens that also document variation over geographic range and through time. The most substantial body of information yet available on phenology for temperate pteridophytes is by Page.[85] He presents diagrammatic "phenograms" that place periods of dormancy, frond expansion, spore shedding, and frond senescence on a calendar. More specific studies of phenology for one or a few species are presented elsewhere.[31,45,46,93,94] Klekowski demonstrated that, at least in culture, several genera of ferns complete their life cycle in 2 to 3 years.[53] Pteridophyte species within a single natural stand may have greatly different life spans related to the stability of the habitat.[31] Some genera, notably *Selaginella,*[60] *Ceratopteris,*[27] *Anogramma,*[75] and species of *Doodia*[86] complete their life cycles in a year or less. This suggests that the perennial pattern is not canalized, but rather has been generally adaptive.

Two contrasting phenologic and morphologic patterns for temperate pteridophytes bear special mention.[31,65,46] In the first of these, green spores are shed in early spring from sporophylls that were produced by stored photosynthate from the previous year. These sporophylls may either flush in early spring (as in several *Equisetum* species, and *Osmunda cinnamomea*) or may have persisted from the previous fall, with spores stored in contracted pinnae (as in *Equisetum hyemale, Onoclea sensibilis, Matteuccia struthiopteris*).[46] Green spores are capable of immediate germination and, in nature, have very brief periods of viability. In contrast, in a greater number of pteridophytes, fertile leaves are produced from the current year's photosynthate, and nongreen spores are shed from midsummer to fall. A intermediate strategy is seen in *Lorinseria areolata* in the Gulf Coastal Plain.[20,21] This fern sheds most of its nongreen spores in midwinter and stores a low proportion in dead erect sporophylls that release spores slowly through the following summer. Nongreen spores lack enforced dormancy and require a few days of hydration prior to germination. Such spores are known for their remarkable retention of viability over prolonged periods of laboratory and herbarium storage.[27,28,90] It is now clear that several mechanisms may operate to form a "spore bank" in nature. Spores stored in soil and on dead erect or prostrate fertile leaves have retained viability for at least a year.[21,24,32,93,94] Gametophytes that require a fungal associate in nature to grow beyond the first few divisions have recently received careful attention.[9,27,76] Quite different phenology and reproductive biology may characterize these mycosymbionts. Most taxa in which spores have an obligate fungal relationship

Table 15.1. Phenologic Patterns of Pteridophytes in Temperate Climates

Patterns	Representative Taxa
Characteristics of spores	
Green/nongreen	*Onoclea sensibilis, Lorinseria areolata*[61]
Brief/continuous period of release	*Dryopteris ludoviciana*[27], *Lorinseria areolata*[21]
Spring release/summer–fall release	*Equisetum telmateia, Equisetum palustre*[85]
Effective spore bank (yes/no)	*Dryopteris filix-mas*[81], *Osmunda cinnamomea*[65]
Dissemination of sporangia or clumped spores	*Cyathea arborea*[14], *Dryopteris filix-mas*[81]
Characteristics of gametophytes	
Spring/fall initiated	*Osmunda cinnamomea*[20], *Blechnum spicant*[19]
Frost or drought resistance	*Dryopteris villarii*[34], *Lorinseria areolata*[20]
Surficial autotroph/subterranean mycosymbiotic	*Dryopteris spp.*[17], *Lycopodium spp.*[9]
Characteristics of fertile fronds	
Dimorphic/hemidimorphic/monomorphic	*Onoclea sensibilis, Polystichum acrostichoides, Dennstaedtia punctilobula*[108]
Produced from past/current year's photosynthate	*Osmunda cinnamomea, Onoclea sensibilis*[62]
Single flush/spring and fall flush/continuous production	*Dryopteris ludoviciana*[51], *Osmunda cinnamomea, Dennstaedtia punctilobula*[62]
Rapid decay following shedding (yes/no)	*Polystichum acrostichoides*[108], *Polystichum munitum*[62]
Spore storage (yes/no)	*Lorinseria areolata, Onoclea sensibilis*[61]
General characteristics	
Strongly clonal/not	*Woodwardia virginica*[88], *Blechnum spicant*[19]
Evergreen/deciduous fronds	*Botrychium biternatum, Botrychium virginianum*[62]
Single flush/continuous frond production	*Osmunda spp., Gymnocarpium spp.*[62]
Regularly fertile/not	*Lorinseria areolata*[21], *Woodwardia virginica*[88]
Early/late successional	*Cystopteris fragilis, Dryopteris goldiana*[31]
Terrestrial/epiphytic/aquatic	*Dryopteris spp., Polypodium polypodioides, Azolla spp.*[62]

are shed in late summer.[46] On the Olympic Peninsula, in Washington, U.S.A., the epiphyte *Polypodium glycyrrhiza* grows throughout the winter and produces spores in early spring. Spores are apparently dormant until late summer when the majority of terrestrial ferns produce their spores.[19]

The period over which gametophytes survive in nature is not well known. For gametophytes produced by nongreen spores there are several reports of frost and drought hardiness. It appears that at least some of the population survives the winter.[20,31,32,93] A few studies of survival of newly initiated sporophytes suggests that these may be less frost- or drought-resistant than their gametophytes.[16,34]

Table 15.1, based on reports by several authors,[8,14,20,21,28,31,32,61,62,65,86,88,93] suggests phenological features by which pteridophytes may be distinguished. For example, in both eastern and western forests of North America, evergreen and deciduous pteridophytes, sometimes congeners, occur together.[62] When sympatric pteridophytes are categorized by several features of Table 15.1, they rarely share the same pattern.

Demographic analysis of pteridophyte populations by quantitative and genetic methods may ultimately allow direct access to changes in gene frequencies. Harper and White[36] point out that intrinsic rates of increase for plants depend on reproduction via spores, which allow for the dispersal, genetic variation and establishment of genets, and by subsequent establishment of daughter ramets that allow a "proven successful genotype to exploit a proven successful environment." The life history pattern of pteridophytes allows recognition and field-mapping of several stages antecedent to adult sexual plants. Determination of age and numbers of each stage allows a workable definition of cohort. Schneller was able to distinguish spores, gametophytes, and several sporophyte classes based on increasingly mature leaf morphology.[93] Cousens et al.[21] discriminated first- and second-season gametophytes, gametophytes with attached sporophytes, and four classes of increasingly mature sporophytes correlated with age classes up to 30 months. Marked individuals of the oldest life history stage, adult fertile plants, may revert to adult sterile plants in subsequent years.[51] Schneller[94] demonstrated that for several species taken together 10^8 spores and 10^3 gametophytes were found per m^2. Observations on density values of the life history stages available to date consistently demonstrate a logarithmic decrease from spore to spore-producing sporophyte, with the greatest mortality occurring between the spore and gametophyte stages, and the least occurring between subadult and adult sporophyte stages.[21,31,93,94] Once the adult fertile condition is reached, difficulties of determining age comparable to those for rhizomatous monocots occur.[36]

Annual and lifetime fecundity of pteridophytes has been estimated by spore counts and by field determination of spore distribution and density, utilizing horizontally placed slides with adhesive surfaces,[93] or by more complex and controlled means that estimate spore dispersal vertically and horizontally.[14,91] Eusporangiate pteridophytes bear relatively massive sporangia that may contain thousands of spores. These may be released passively (e.g., *Psilotum*), be aided in release by movement of the spores themselves (*Equisetum*) or actively ejected (megaspores of *Selaginella*).[25,26,86]

The homosporous Filicales have fewer spores per sporangium and active spore release. The number of spores per sporangium correlates well with antiquity of family and ranges from 512 to 16. Sixty-four is the most common number. Mickel has determined that annual spore output of a moderate size *Dryopteris intermedia* individual is 50 million spores and that lifetime output may range from 2.5–10 billion.[77] *Osmunda claytoniana*, with remarkably little tissue allocated to spore production, was estimated to produce 90 million spores annually.[77] Schneller estimated that *Dryopteris filix-mas* produced 100 million spores annually which would result in 2.5 million spores/m^2 in nature, based on densities of adults.[93] Similarly, *Athyrium filix-femina*[94] plants would produce densities of 11 million spores/m^2. Passive trapping experiments showed that the greatest numbers of spores fell within 5m of the parent sporophyte, a distribution pattern that is termed leptokurtic.[93]

Although it is usually assumed that single spores are dispersed, observations by Schneller[81] that spores were often dispersed in groups of two or more, and by Brownsey[8] that all spores of *Asplenium lepidum* were dispersed within sporangia are of great interest. Tryon documented the importance of long-distance dispersal of fern spores.[103] Raynor sonicated spores of *Osmunda* and *Dryopteris* to reduce their adhesion and ejected them by artificial means.[91] The fern spores were distributed in a strongly leptokurtic manner and did not travel as far as smaller ragweed pollen in the same experiments. Very few spores of *Osmunda* were found beyond 40m. Conant utilized radioisotope techniques to determine distribution of spores from an 8-m-tall tree fern.[14] He found two apparent strategies: a strongly leptokurtic distribution of spores contained in sporangia and more distant distribution of single spores. Carlquist suggested that spore size of island endemics increased due to selection for leptokurtic distribution.[12] The mid-frond position of fertile pinnae in *Osmunda claytoniana* suggests that a more precinctive spore dispersal pattern may have been selected relative to those of other *Osmunda* species that have entire skeletonized fertile fronds or bear fertile pinnae apically.[62] The prevalence of leptokurtic spore dispersal suggests that genetic neighborhood size for pteridophytes may be relatively small.

Spores are readily available to assess potential mortality and there are many reports on viability and longevity for laboratory-stored collections.[27] One study demonstrated greatly reduced viability for spores from a marginal population for a year in which optimal populations had nearly 100% viability.[19] Sporophytes in nature may be easily marked for studies of spore viability on a yearly basis. This would allow correlation of viability with environmental variables.

Successful germination of spores appears to be associated with safe sites that provide bare soil and may be distributed patchily.[20,93] Although gametophytes have been widely held to be both difficult to locate and to identify, recent studies and advances in technique should eliminate such reservations.[20,27,39] Soil slips, small soil clods turned up by herbivores,[19] tip-up mounds at the bases of fallen trees,[21] and "chimneys" excavated by crayfish[51] have been shown to provide safe sites for completion of the life cycle. Conspicuous populations of gametophytes on large disturbed areas are often entirely lost following erosion and thus may be less important in effective reproduction than populations on less conspicuous sites. The discovery of gametophyte densities of more than 75 plants/cm^2 indicate population densities that may be found soon after germination.[19,31,32,93] Morphology, especially the presence of diagnostic trichomes, of field-collected gametophytes provides a primary tool for identification to family and genus. Electrophoresis techniques are now sufficiently powerful to distinguish species for single field-collected gametophytes based on three enzyme systems.[51]

Substantial variation in the proportion of gametophytes that bear sporophytes has been reported. In some habitats sporophyte formation may occur for as few as 1/230 gametophytes, and for others as many as 1/16.[21] Even more dense populations of newly initiated sporophytes have been observed with as many as 1000 occupying area sufficient for only a single adult fertile plant.[21] Such colonies may be compared to dense patches of seedlings that occupy forest openings sufficiently large for the survival of a single tree. Decreases in the numbers of intermediate-age sporophytes are moderate and allow the inference that selection operates less strongly following the survival of the newly independent sporophyte.[19,20,93]

Sporophyte populations of pteridophytes may result exclusively from sexual reproduction (e.g., *Selaginella tenerrima,*[60] many *Isoetes* species[28]), but the majority of pteridophytes establish daughter ramets by vegetative means including horizontal rhizomes, stolons, buds produced by different parts of the fronds and roots, and by bulbils and gemmae.[29,64,78] *Pteridium aquilinum* owes its status as the most successful fern species to extensive clonal growth that may continue for centuries after genet establishment.[84,87] *Pteridium* clones have been documented to expand at a rate of 36 cm/year beginning 3 years after establishment by spore.[84] Sporophytes of *Lycopodium lucidulum,* which lacks a horizontal rhizome, establish ramets by bulbils.[64] These form fairy rings up to 2 m in diameter. Rhizomatous species of *Lycopodium* initiated clonal growth 15–22 years following colonization by spores.[84] Radial expansion of these clones ranged from 30–50 cm/year. Rare morphological markers for *Athyrium filix-*

femina, which lacks a horizontal rhizome system, allowed recognition of clones as large as 225 m^2 that were estimated to be 50 years old.[94] Where available, marker genes may be of great value in the recognition of clones.[58] Genets may need to be delimited by mapping of above-ground rhizomes, or excavation of subterranean rhizomes. The accuracy of such techniques is limited by eventual necrosis of connecting rhizomes. Repeated careful mapping of a genet, with a known point of origin, would eliminate these difficulties. This has not yet been utilized for shoot demographics of pteridophytes. Farrar had demonstrated that populations of a few pteridophyte species are maintained by vegetative proliferation of gametophytes only and their sporophytes are unknown.[28]

One approach to shoot demography when genotype is unknown, is to treat all adult units as ramets.[21] Genetic studies based on field collections usually avoid the problem of distinguishing genets by collecting spores from widely separated individuals.[64] Harper and White suggested that rhizome growth has both birth and death components that result in flux of the shoots through the habitat.[36] The rate of ramet formation by an advancing rhizome system may vary with habitat, thus genet size and age correlations should be determined on a stand by stand basis.[21] Mortality of adult sporophytes has been very little studied. Farrar and Gooch related adult death to succession and chronological availability of the habitat.[31] Critical long-term studies of adult mortality are under way by Willmot.[28]

Although studies that use repeated mapping are needed to document mortality of the several life history stages of pteridophytes, knowledge of mating systems and several environmental aspects suggests that quantitative studies alone will be insufficient for determination of selection at each life history stage.[21,35,39] Evidence of a spore bank suggests that annual estimates of spore mortality may be exaggerated.[24,94] Known aspects of mating systems indicate that documenting the decrease of gametophyte densities over time may overestimate their mortality. Although failure to persist to support an embryo defines mortality for a female gametophyte, male or bisexual gametophytes may contribute sperm and then be lost. Density determination would miss the fact that they had effectively contributed to the next generation. The high proportion of male gametophytes found in several studies strengthens this caution.[19]

The accessibility of discrete life cycle stages suggests that pteridophytes are especially favorable material for a direct genetic approach to selection in nature. As suggested by Hamrick, electrophoretic techniques, utilizing sufficiently large numbers of enzymes, may determine changes in gene frequencies for successively older surviving life history stages.[35] Although single fern spores have not yet been assayed for allelic variation, pooling of freshly collected green spores, or of hydrated nongreen spores, may allow assessment of population variation for comparison with surviving gametophytes. Separate analysis of single gametophytes and their attached young sporophytes may allow simultaneous determination of mating system, effective reproduction, and changes in allele frequencies between the generations.

Tropical Antillean species that occasionally reach the southeastern United States but fail to persist, and alien Old World temperate or subtropical species that do persist, provide opportunities for combined quantitative and electrophoretic analysis of demography. Novak and Cousens utilized allozyme studies and ramet mapping to test the hypothesis that a small newly discovered population of the Antillean fern *Dicranopteris flexuosa* was established by a single spore. Data available supported the hypothesis,[83] and progeny studies added further support. The Old World fern *Thely-*

pteris torresiana is extending its range in the southeastern United States.[63] Single individuals of this species are readily found at the edges of its present range. These would provide material for longitudinal study of integrated demographics. Broad evolutionary theory suggests that selection may be rapid in such species.

CONCLUSIONS

The pteridophytes have been studied from the perspectives of origins and phylogeny for nearly a century. A midpoint was marked by understanding speciation by way of reticulate evolution and by the perception that modern genera continue an adaptive radiation. This review demonstrates that several means of investigation are now available that may in the future provide an experimental understanding of selection at the population level. The relatively discrete life cycle stages of pteridophytes may allow investigation of selection for a short-lived and a long-lived phase. Both haploid and diploid phases may be analyzed in terms of their contribution to successful reproduction.

Homosporous pteridophytes are better studied than heterosporous groups. Breeding systems and consequences unique to these plants have been uncovered through pioneering work by Klekowski, Hickok, and Lloyd. The central position of polyploidy has been productive of hypotheses and opportunity for clear testing of alternatives. Genetic analysis utilizing enzymes strongly suggests that the reproductive biology of pteridophytes operates essentially as a diploid system despite high chromosome numbers. Such studies are not yet sufficiently extensive, however, to exclude homoeologous pairing and release of stored heterozygosity in natural populations.

Homosporous pteridophytes behave as diploids and thus allow investigators to exploit the extensive analytical methods of population genetics developed for seed plants. Studies of reproductive allocation, and the proportion of populations produced asexually, are in the earliest stages of investigation. Demographic analysis utilizing both numerical and electrophoretic techniques holds the promise of allowing simultaneous insight into mating systems, effective reproduction, selection at each life history stage, and short-term selection measured by changes in gene frequencies.

The subtle ways in which pteridophytes reproduce in nature are becoming more apparent. The validity of examining pteridophytes from perspectives of population biology developed for seed plants and animals is clear. The view that the biology of pteridophyte life history stages may be understood in terms of adaptations to present environments as well as historical constraints is likely to be productive, and will allow progress toward a unified theory of reproductive strategies in plants.

ACKNOWLEDGMENTS

The author wishes to thank L. Hickok, R. M. Lloyd, J. Miller, and V. Raghavan for discussion of specific points; E. M. Kelly, D. Lacey, S. Novak, and C. Page for extensive help in the field; and Jon and Lesley Lovett Doust and Edward Klekowski, Jr., for careful reviews of the manuscript.

REFERENCES

1. Arnold, C. A., Mesozoic and Tertiary fern evolution and distribution, *Mem. Torrey Bot. Club* **21,** 58–66 (1964).
2. Atkinson, L. R., The gametophyte of five old world Thelypteroid ferns, *Phytomorphology* **25,** 38–54 (1975).
3. Atkinson, L. R., and Stokey, A. G., Comparative morphology of the gametophyte of homosporous ferns. *Phytomorphology* **14,** 51–70 (1964).
4. Bell, P. R., Introduction: The essential role of the Pteridophyta in the study of land plants. *Proc. R. Soc. Edinburgh* **86B,** 1–4 (1985).
5. Bierhorst, D. W., Observations on *Schizaea* and *Actinostachys* spp., including *A. oligostachys* sp. nov. *Am. J. Bot.* **55,** 87–108 (1968).
6. Bierhorst, D. W., The apogamous life cycle of *Trichomanes pinnatum*—a confirmation of Klekowski's predictions on homoeologous pairing. *Am. J. Bot.* **62,** 448–456 (1975).
7. Bower, F. O., The Ferns, Vol. 1 Cambridge Univ. Press, Cambridge, 1923.
8. Brownsey, P. J., An example of sporangial indehiscence in the filicopsida, *Evolution* **31,** 294–301 (1977).
9. Bruce, J. G., and Beitel, J. M., A community of *Lycopodium* gametophytes in Michigan, *Am. Fern J.* **69,** 33–41 (1979).
10. Buckley, D. P., and Lloyd, R. M., A new homosporous genetic system hypothesis. *Am. J. Bot.* **72,** (Abstr.), 919 (1985).
11. Callaghan, T. V., Age-related patterns of nutrient allocation in *Lycopodium annotinum* from Swedish Lapland, *Oikos* **35,** 373–386 (1980).
12. Carlquist, S., The biota of long-distance dispersal. III. Loss of dispersibility in the Hawaiian flora, *Brittonia* **18,** 310–335 (1966).
13. Chapman, R. H., Klekowski, E. J., Jr., and Selander, R. K., Homoeologous heterozygosity and recombination in the fern *Pteridium aquilinum, Science* **204,** 1207–1209 (1979).
14. Conant, D. S., A radiosotope technique to measure spore dispersal of the tree fern *Cyathea arbores* Sm. *Pollen Spores* **20,** 583–593 (1978).
15. Conant, D. S., Allohomoploidy: A new hypothesis for speciation in homoploid homosporous ferns, *Am. J. Bot.* **72,** (Abstr.) 917 (1985).
16. Conway, E., Spore and sporeling survival in bracken (*Pteridium aquilinim* (L.) Kuhn.) *J. Ecol.* **41,** 289–294 (1953).
17. Cousens, M. I., Gametophyte sex expression in some species of *Dryopteris. Am. Fern J.* **65,** 39–42 (1975).
18. Cousens, M. I., Gametophyte ontogeny, sex expression, and genetic load as measures of population divergence in *Blechnum spicant. Am. J. Bot.* **66,** 116–132 (1979).
19. Cousens, M. I., *Blechnum spicant:* Habitat and vigor of optimal, marginal, and disjunct populations, and field observations of gametophytes, *Bot. Gaz.* **142,** 251–258 (1981).
20. Cousens, M. I., Lacey, D. G., and Kelly, E. M., Life history studies of ferns: A consideration of perspective. *Proc. R. Soc. Edinburgh* **86B,** 371–380 (1985).
21. Cousens, M. I., Lacey, D. G., and Scheller, J. M., Safe sites and the ecological life-history of *Lorinseria areolata. Am. J. Bot.,* in press.
22. Crist, K. C., and Farrar, D. R., Genetic load and long-distance dispersal in *Asplenium platyneuron, Can. J. Bot.* **61,** 1809–1814 (1983).
23. Döpp, W., Ein die Antheridienbildung bei Farnen fördernde Substanz in den Prothallien von *Pteridium aquilinum* (L.) Kuhn. *Ber. Dtsch. Bot. Ges.* **63,** 139–147 (1950).
24. During, H. J., and ter Horst, B., The diaspore bank of bryophytes and ferns in chalk grassland. *Lindbergia* **9,** 57–64 (1983).
25. Duthie, A. V. Studies in the morphology of *Selaginella pumila* Spring. *Trans R. Soc. S. Afr.* **II,** 131–144 (1923).
26. Duthie, A. V., The method of spore dispersal of three South African species of *Isoetes, Ann. Bot. (London)* **43,** 1–10 (1929).
27. Dyer, A. F., ed., The Experimental Biology of Pteridophytes. Academic Press, London, 1979.
28. Dyer, A. F., and Page, C. N., Biology of Pteridophytes. *Proc. R. Soc. Edinburgh* **86B,** vii-474 (1985).
29. Emigh, V. D., and Farrar, D. R., Gemmae: A role in sexual reproduction in the fern genus *Vittaria, Science* **198,** 297–298 (1977).
30. Ewing, E. P., Selection at the haploid and diploid phases: Cyclical variation. *Genetics* **87,** 195–208 (1977).

31. Farrar, D. R., and Gooch, R. D., Fern reproduction at Woodman Hollow, Central Iowa: Preliminary observations and a consideration of the feasibility of studying fern reproductive biology in nature, *Proc. Iowa Acad. Sci.* **82,** 119–122 (1975).
32. Farrar, D. R., Spore retention and release from overwintering fern fronds. *Am. Fern J.* **66,** 49–52 (1976).
33. Gastony, G. J., and Gottlieb, L. D., Genetic variation in the homosporous fern *Pellaea andromedifolia, Am. J. Bot.* **72,** 257–267 (1985).
34. Gilbert, O. L., Biological flora of the British Isles: *Dryopteris villarii* (Bellari) Woynar, *J. Ecol.* **58,** 301–313 (1970).
35. Hamrick, J. L., Plant population genetics and evolution, *Am. J. Bot.* **69,** 1685–1693 (1982).
36. Harper, J. L., and White, J., The demography of plants, *Annu. Rev. Ecol. Syst.* **5,** 419–463 (1974).
37. Harvey, W. H., and Caponetti, J. D., In vitro studies on the induction of sporogenous tissue on leaves of cinnamon fern: II. Some aspects of carbohydrate metabolism, *Can. J. Bot.* **51,** 341–349 (1973).
38. Haufler, C. H., and Gastony, G. J., Antheridiogen and the breeding system in the fern genus *Bommeria, Can. J. Bot.* **56,** 1594–1601 (1978).
39. Haufler, C. H., and Soltis, D. E., Obligate outcrossing in a homosporous fern: Field confirmation of a laboratory prediction, *Am. J. Bot.* **71,** 878–881 (1984).
40. Haufler, C. H., Electrophoresis is modifying our concepts of evolution in homosporous pteridophytes, *Am. J. Bot.* **74,** 953–966 (1987).
41. Hevly, R. H., Adaptation of cheilanthoid ferns to desert environments, *Ariz. Acad. Sci. J.* **2,** 164–175 (1963).
42. Hickok, L. G., Apogamy and somatic restitution in the fern *Ceratopteris, Am. J. Bot.* **66,** 1074–1078 (1979).
43. Hickok, L. G., Homoeologous chromosome pairing and restricted segregation in the fern *Ceratopteris, Am. J. Bot.* **65,** 516–521 (1978).
44. Hickok, L. G., and Schwarz, O. J., An in-vitro whole plant selection system: Paraquat tolerant mutants in the fern *Ceratopteris, Theor. Appl. Genet.* **72,** 302–306 (1986).
45. Hill, R. H., Comparative habitat requirements for spore germination and prothallial growth in three ferns from southern Michigan, *Am. Fern J.* **61,** 171–182 (1971).
46. Hill, R. H., and Wagner, W. H., Jr., Seasonality and spore type of the pteridophytes of Michigan, *Mich. Bot.* **13,** 40–44 (1974).
47. Holbrook-Walker, S., and Lloyd, R. M., Reproductive biology and gametophyte morphology of the Hawaiian fern genus *Sadleria* (Blechnaceae) relative to habitat diversity and propensity for colonization, *Bot. J. Linn. Soc.* **67,** 157–174 (1973).
48. Istock, C. A., The evolution of complex life cycles phenomena: An ecological perspective, *Evolution* **21,** 592–605 (1967).
49. Jermy, A. C., Crabbe, J. A., and Thomas, B. A., *The Phylogeny and Classification of the Ferns,* Academic Press, London, 1973.
50. Keddy, P. A., Why gametophytes and sporophytes are different: Form and function in a terrestrial environment, *Am. Nat.* **118,** 452–454 (1981).
51. Kelley, E. M., and Cousens, M. I., A population ecology study of *Dryopteris ludoviciana* (Kunze) Small, *Am. J. Bot.* **72,** (Abstr.) 923 (1985).
52. Klekowski, E. J., Jr., and Baker, H. G., Evolutionary significance of polyploidy in the pteridophyta, *Science* **153,** 305–307 (1966).
53. Klekowski, E. J., Jr., Observations on pteridophyte life cycles: Relative lengths under cultural conditions, *Am. Fern J.* **57,** 49–51 (1967).
54. Klekowski, E. J., Jr., Reproductive biology of the pteridophyta. II. Theoretical considerations, *Bot. J. Linn. Soc.* **62,** 347–359 (1969).
55. Klekowski, E. J., Jr., Populational and genetic studies of a homosporous fern—*Osmunda regalis, Am. J. Bot.* **57,** 1122–1138 (1970).
56. Klekowski, E. J., Jr., Genetic load in *Osmunda regalis* populations, *Am. J. Bot.* **60,** 146–154 (1973).
57. Klekowski, E. J., Jr., Sexual and subsexual systems in homosporous pteridophytes: A new hypothesis, *Am. J. Bot.* **60,** 536–544 (1973).
58. Klekowski, E. J., Jr., Mutational load in clonal plants: A study of two fern species, *Evolution* **38,** 417–426 (1984).
59. Klekowski, E. J., Jr., Mutations, apical cells, and vegetative reproduction, *Proc. R. Soc. Edinburgh* **86B,** 67–73 (1985).

60. Kornas, J., and Jankun, A., Annual habit and apomixis as drought adaptations in *Selaginella tenerrima, Bothallia* **14,** 647–651 (1983).
61. Lacey, D. G., Comparative life history of two sympatric fern species, *Onoclea sensibilis* and *Lorinseria areolata.* M.S. thesis. University of West Florida, Pensacola, 1983.
62. Lellinger, D. B., *A Field Manual of the Ferns and Fern-allies of the United States and Canada.* Smithsonian Institution Press, Washington, D.C., 1985.
63. Leonard, S. W., The distribution of *Thelypteris torresiana* in the southeastern United States, *Am. Fern J.* **62,** 97–99 (1972).
64. Levin, D. A., and Crepit, W. L., Genetic variation in *Lycopodium lucidulum:* A phylogenetic relic, *Evolution* **27,** 622–632 (1973).
65. Lloyd, R. M., and Klekowski, E. J., Jr., Spore germination and viability in pteridophyta: Evolutionary significance of chlorophyllous spores, *Biotropica* **2,** 129–137 (1970).
66. Lloyd, R. M., Reproductive biology and evolution in the pteridophyta, *Ann. Missouri Bot. Gard.* **61,** 318–331 (1974).
67. Lloyd, R. M., Mating systems and genetic load in pioneer and non-pioneer Hawaiian Pteridophyta, *Bot. J. Linn. Soc.* **69,** 23–35 (1974).
68. Lloyd, R. M., and Gregg, T. L., Reproductive biology and gametophyte morphology of *Acrostichum danaeifolium* from Mexico, *Am. Fern J.* **65,** 105–120 (1975).
69. Lloyd, R. M., Personal communication, 1986.
70. Lovis, J. D., Evolutionary patterns and processes in ferns, *Adv. Bot. Res.* **4,** 229–437 (1977).
71. Lovis, J. D., Fern hybridists and fern hybridising II. Fern hybridising at the University of Leeds, *Brit. Fern Gaz.* **10,** 13–20 (1968).
72. Maeda, O., On the dry matter productivity of two ferns, *Osmunda cinnamomea* and *Dryopteris crassirhizoma,* in relation to their geographical distribution in Japan. *Jpn. J. Bot.* **20,** 237–267 (1970).
73. Manton, I., Evolution in the pteridophyta. *Bot. Soc. Brit. Isles Conf. Rep.* **6,** 105–120 (1961).
74. McCauley, D. E., Whittier, D. P., and Reilly, L. M., Inbreeding and the rate of self-fertilization in a grape fern, *Botrychium dissectum, Am. J. Bot.* **72,** 1978–1981 (1985).
75. Mehra, P. N., and Sanhu, R. S., Morphology of the fern *Anogramma leptophylla, Phytomorphology* **26,** 60–76 (1976).
76. Mesler, M. R., The natural history of *Ophioglossium palmatum* in South Florida. *Am. Fern J.* **64,** 33039 (1974).
77. Mickel, J. T., Fern Spore? What for? *Fiddlehead Forum (Newsl. Am. Fern Soc.)* **9,**(5) (1982).
78. Mickel, J. T., The proliferous species of *Elaphoglossum* (Elaphoglossaceae) and their relatives, *Brittonia* **37,** 261–278 (1985).
79. Miller, J. H., personal communication, 1986.
80. Näf, U., Nakanishi K., and Endo, M., On the physiology and chemistry of fern antheridiogens, *Bot. Rev.* **41,** 315–359 (1975).
81. Näf, U., Antheridiogens and antheridial development, in *The Experimental Biology of Ferns* (A. F. Dyer, ed.), pp. 435–470. Academic Press, London, 1979.
82. Nobel, P. S., Microhabitat, water relations and photosynthesis in a desert fern, *Notholaena parryi, Oecologia* **31,** 293–309 (1978).
83. Novak, S. J., and Cousens, M. I., *Dicranopteris flexuosa* (Schrad.) Underwood in North Florida—Investigations of a disjunct population, *Am. J. Bot.* **72** (Abstr.) 925 (1985).
84. Oinonen, E., The size of *Lycopodium clavatum* L. and *L. annotinum* L. stands as compared to that of *L. complanatum* L. and *Pteridium aquilinum* (L.) Kuhn stands, the age of the tree stand and the dates of fire, on the site, *Acta. For. Fenn.* **87,** 5–53 (1968).
85. Page, C. N., *The Ferns of Britain and Ireland.* Cambridge Univ. Press, 1982.
86. Page, C. N., personal communication, 1986.
87. Perring, F. H., and Gardiner, B. G., eds. The biology of bracken, *Bot. J. Linn. Soc.* **73,** 1–239 (1976).
88. Pittillo, J. D., Wagner, W. H., Jr., Farrar, D. R., and Leonard, S. W., New pteridophyte records in the Highlands Biological Station area, Southern Appalachians, *Castanea* **49,** 263–272 (1975).
89. Pray, T. R., Interpopulational variation in the gametophytes of *Pellaea andromedaefolia. Am. J. Bot.* **55,** 951–960 (1968).
90. Raghavan, V. Cytology, physiology and biochemistry of germination of fern spores, *Int. Rev. Cytol.* **62,** 69–118 (1980).

91. Raynor, E. C., Ogden, G. S., and Hayes, J. V., Dispersion of fern spores into and within a forest, *Rhodora* **78,** 473–487 (1976).
92. Schedlbauer, M. D., Fern gametophyte development: Controls of dimorphism in *Ceratopteris thalictroides. Am. J. Bot.* **63,** 1080–1087 (1976).
93. Schneller, J. J., Untersuchungen an einheimischen Farnen, insbesondere der *Dryopteris filix-mas* Gruppe 3, *Teil. Ökologische Unters. Ber. Schweiz. Bot. Ges.* **85,** 110–159 (1975).
94. Schneller, J. J., Biosystematic investigations on the lady fern *(Athyrium filix-femina) Plant Syst. Evol.* **132,** 255–277 (1979).
95. Schneller, J. J., Evidence for intergeneric incompatibility in ferns, *Plant Syst. Evol.* **137,** 45–56 (1981).
96. Schraudolf, H., Action and phylogeny of antheridiogens, *Proc. R. Soc. Edinburgh* **86B,** 75–80 (1985).
97. Scott, R. J., and Hickok, L. G., Genetic analysis of antheridiogen sensitivity in *Ceratopteris richardii, Am. J. Bot.* In press.
98. Soltis, D. E., and Soltis, P. S., Electrophoretic evidence for inbreeding in the fern *Botrychium virginianum* (Ophioglossaceae) *Am. J. Bot.* **73,** 588–592 (1986).
99. Stein, D. B., Nucleic acid comparisons as a tool in understanding species relationships and phylogeny, *Proc. R. Soc. Edinburgh* **86B,** 283–288 (1985).
100. Stokey, A. G., and Atkinson, L. R., The gametophyte of the Osmundaceae, *Phytomorphology* **6,** 19–40 (1956).
101. Stokey, A. G., The gametophyte of the Gleicheniaceae, *Bull. Torrey Bot. Club* **77,** 323–339 (1950).
102. Tryon, A. F. *Platyzoma*—a Queensland fern with incipient heterospory, *Am. J. Bot.* **51,** 939–942 (1964).
103. Tryon, R., Development and evolution of fern flora of oceanic islands, *Biotropica* **2,** 76–84 (1970).
104. Tryon, R. M., and Vitale, G., Evidence for antheridogen production and its mediation of a mating system in natural populations of fern gametophytes, *Bot. J. Linn. Soc.* **74,** 243–249 (1977).
105. Vitt, D. H., Adaptive modes of the moss sporophyte, *Bryologist* **84,** 166–186 (1981).
106. Wagner, W. H., Jr., Reticulate evolution in the Appalachian Aspleniums, *Evolution* **8,** 103–118 (1954).
107. Wagner, W. H., Jr., Farrar, D. R., and Chen, K. L., A new sexual form of *Pellaea glabella* var. *glabella* from Missouri, *Am. Fern J.* **55,** 171–178 (1965).
108. Wagner, W. H., Jr., and Wagner, F. S., Fertile-sterile leaf dimorphy in ferns, *Gard. Bull. Singapore* **XXX,** 251–267 (1977).
109. Warne, T. R., and Lloyd, R. M., The role of spore germination and gametophyte development in habitat selection: Temperature responses in certain temperate and tropical ferns. *Bull. Torrey Bot. Club* **107,** 57–64 (1980).
110. Warne, T. R., and Lloyd, R. M., Inbreeding and homozygosity in the fern, *Ceratopteris pteridoides* (Hooker) Hieronymus, *Bot. J. Linn. Soc.* **83,** 1–13 (1981).
111. Werth, C. R., Guttman, S. I., Eshbaugh, W. H., Recurring origins of allopolyploid species in *Asplenium, Science* **228,** 731–733 (1985).
112. Whittier, D. P., The influence of cultural conditions on the induction of apogamy in *Pteridium* gametophytes, *Am. J. Bot.* **51,** 730–736 (1964).
113. Whittier, D. P., Obligate apogamy in *Cheilanthes tomentosa* and *C. alabamensis, Bot. Gaz.* **126,** 275–281 (1970).
114. Whittier, D. P., The organic nutrition of *Botrychium* gametophytes *Am. Fern J.* **74,** 77–86 (1984).
115. Willson, M. R., Sex expression in fern gametophytes: Some evolutionary possibilities, *J. Theor. Biol.* **93,** 403–409 (1981).
116. Willson, M. R., On the evolution of complex life cycles in plants: A review and an ecological perspective, *Ann. Missouri Bot. Gard.* **68,** 275–300 (1981).
117. Windham, M. D., and Haufler, C. H., Autopolyploid evolution among homosporous ferns, *Am. J. Bot.* **72** (Abstr.) 919 (1985).

Subject Index

Organism Index